Groundwater

Groundwater is invisible, but its impact is visible everywhere. Everything around us relies on groundwater, our drinking water and sanitation, our food supply and our natural environment. Yet because it is invisible, information, management and governance of groundwater is often poor and inadequate. This book contributes to UN Water Groundwater year (2022), and to the effort of "making the invisible, visible". Through worldwide case studies ranging from the Americas (California, Brazil), to Asia (India, Iran, Lao PDR, Nepal), Africa (Malawi, Tanzania, South Africa) and the MENA region (Lebanon, Morocco, Yemen), including cases of transboundary aquifers, the chapters in this edited volume reflect important recent advances in interdisciplinary knowledge on the governance, management, practice and science-policy interfaces of groundwater.

An insightful resource for researchers and planners in the field of environmental policies, water laws, climate change and groundwater governance, this book comes with a new Introduction. The other chapters were originally published in *Water International.*

Raya Marina Stephan, IWRA fellow and former Director, is an international consultant, expert in water law. She is the Deputy Editor-in-Chief of *Water International.*

James E. Nickum, IWRA Fellow, Global Reach Awardee, and former Vice-President, is the Editor in Chief of *Water International*, non-resident Professorial Research Associate at the School of Oriental and African Studies (SOAS), University of London and non-resident Adjunct Senior Research Fellow at the East-West Center, Honolulu, Hawaii.

Philippus Wester, IWRA Fellow and former Director, is Regional Programme Manager, Mountain Knowledge & Action Networks at ICIMOD and a former Deputy Editor-in-Chief of *Water International.*

Routledge Special Issues on Water Policy and Governance

Edited by: Cecilia Tortajada *(IJWRD) – Institute of Water Policy, Lee Kuan Yew School of Public Policy, NUS, Singapore,* and **James E. Nickum** *(WI) – International Water Resources Association, France*

Most of the world's water problems, and their solutions, are directly related to policies and governance, both specific to water and in general. Two of the world's leading journals in this area, the *International Journal of Water Resources Development* and *Water International* (the official journal of the International Water Resources Association), contribute to this special issues series, aimed at disseminating new knowledge on the policy and governance of water resources to a very broad and diverse readership all over the world. The series should be of direct interest to all policy makers, professionals and lay readers concerned with obtaining the latest perspectives on addressing the world's many water issues.

Global Water Resources
Festschrift in Honour of Asit K. Biswas
Edited by Cecilia Tortajada and Eduardo Araral

Source-to-Sea Management
Edited by Josh Weinberg, Qinhua Fang, Sarantuyaa Zandaryaa, Greg Leslie and James E. Nickum

Strengthening Cooperation over Transboundary Groundwater Resources
Edited by Gabriel Eckstein and Alice Aureli

Wicked Problems of Water Quality Governance
Edited by James E. Nickum, Raya Marina Stephan and Henning Bjornlund

Groundwater
Recent Advances in Interdisciplinary Knowledge
Edited by Raya Marina Stephan, James E. Nickum and Philippus Wester

Public Banks, Public Water
Exploring the Links in Europe
Edited by Thomas Marois and David A McDonald

For more information about this series, please visit: www.routledge.com/Routledge-Special-Issues-on-Water-Policy-and-Governance/book-series/WATER

Groundwater

Recent Advances in Interdisciplinary Knowledge

Edited by

Raya Marina Stephan, James E. Nickum and Philippus Wester

Routledge
Taylor & Francis Group
LONDON AND NEW YORK

IWRA

Designed cover image: © Shutterstock
First published 2023
by Routledge
4 Park Square, Milton Park, Abingdon, Oxon OX14 4RN

and by Routledge
605 Third Avenue, New York, NY 10158

Routledge is an imprint of the Taylor & Francis Group, an informa business

British Library Cataloguing in Publication Data
A catalogue record for this book is available from the British Library

ISBN13: 978-1-032-38607-2 (hbk)
ISBN13: 978-1-032-38609-6 (pbk)
ISBN13: 978-1-003-34586-2 (ebk)

DOI: 10.4324/9781003345862

Typeset in Minion Pro
by Newgen Publishing UK

Publisher's Note
The publisher accepts responsibility for any inconsistencies that may have arisen during the conversion of this book from journal articles to book chapters, namely the inclusion of journal terminology.

Contents

Citation Information

The following chapters were originally published in various volumes and issues of *Water International*. When citing this material, please use the original page numbering for each article, as follows:

Chapter 1

From an open-access to a state-controlled resource: the case of groundwater in the Kathmandu Valley, Nepal
Vishnu Prasad Pandey and Futaba Kazama
Water International, volume 39, issue 1 (2014), pp. 97–112

Chapter 2

Pathways for effective groundwater governance in the least-developed-country context of the Lao PDR
Paul Pavelic, Ounakone Xayviliya and Oualaphone Ongkeo
Water International, volume 39, issue 4 (2014), pp. 469–485

Chapter 3

Using backcasting to explore ways to improve the national water department's contribution to good groundwater governance in South Africa
P. Seward, Y. Xu and A. Turton
Water International, volume 40, issue 3 (2015), pp. 446–462

Chapter 4

Groundwater governance through institutional bricolage? Participation in Morocco's Chtouka aquifer contract
Annabelle Houdret and Rebecca Heinz
Water International, volume 47, issue 4 (2022), pp. 565–582

Chapter 5

Exploring the future impacts of urbanization and climate change on groundwater in Arusha, Tanzania
Tunde Olarinoye, Jan Willem Foppen, William Veerbeek, Tlhoriso Morienyane and Hans Komakech
Water International, volume 45, issue 5 (2020), pp. 497–511

Chapter 6

Impact of land use and occupation on potential groundwater recharge in a Brazilian savannah watershed
Arnaldo José Cambraia Neto and Lineu Neiva Rodrigues
Water International, volume 46, issue 3 (2021), pp. 348–364

Chapter 7

Problems and promise of managed recharge in karstified aquifers: the example of Lebanon
Wisam M. Khadra and Pieter J. Stuyfzand
Water International, volume 45, issue 1 (2020), pp. 23–38

Chapter 8

A multifaceted quantitative index for sustainability assessment of groundwater management: application for aquifers around Iran
Bahador Zarei, Esmaeel Parizi, Seiyed Mossa Hosseini and Behzad Ataie Ashtiani
Water International, volume 47, issue 3 (2022), pp. 338–360

Chapter 9

Whither collective action? Upscaling collective actions, politics and basin management in the process of 'legitimizing' an informal groundwater economy
Marta Rica, Aurélien Dumont, Fermín Villarroya and Elena López-Gunn
Water International, volume 39, issue 4 (2014), pp. 520–533

Chapter 10

Participatory rural appraisal to assess groundwater resources in Al-Mujaylis, Tihama Coastal Plain, Yemen
Wahib Al-Qubatee, Henk Ritzema, Adel Al-Weshali, Frank van Steenbergen and Petra J. G. J. Hellegers
Water International, volume 42, issue 7 (2017), pp. 810–830

Chapter 11

Federal reserved rights and California's Groundwater Management Act: resolving groundwater rights tensions in California and the western United States
Stefanie Viktoria Caroline Schulte
Water International, volume 44, issue 4 (2019), pp. 427–443

Chapter 12

Learning from the past to build the future governance of groundwater use in agriculture
Olivier Petit, Aurélien Dumont, Stéphanie Leyronas, Quentin Ballin, Sami Bouarfa, Nicolas Faysse, Marcel Kuper, François Molle, Charlotte Alcazar, Emmanuel Durand, Ridha Ghoudi, Aline Hubert, Selin Le Visage, Imane Messaoudi, Marielle Montginoul, Seyni Ndao, Audrey Richard Ferroudji, Jean-Daniel Rinaudo, Julie Trottier, Olivia Aubriot, Mohamed Elloumi, Marc Boisson, Rhoda Fofack-Garcia, Frédéric Maurel, Dominique Rojat, Bruno Romagny and Emmanuelle Salgues
Water International, volume 46, issue 7–8 (2021), pp. 1037–1059

Chapter 13

Drivers of groundwater utilization in water-limited rice production systems in Nepal
Anton Urfels, Andrew J. McDonald, Timothy J. Krupnik and Pieter R. van Oel
Water International, volume 45, issue 1 (2020), pp. 39–59

Chapter 14

Groundwater policies and irrigation development: a study of West Bengal, India, 1980–2016
Tapas Singh Modak
Water International, volume 46, issue 4 (2021), pp. 505–523

Chapter 15

Transboundary groundwater governance in the Guarani Aquifer System: reflections from a survey of global and regional experts
Zachary P. Sugg, Robert G. Varady, Andrea K. Gerlak, and Rafael de Grenade
Water International, volume 40, issue 3 (2015), pp. 377–400

Chapter 16

A methodology to identify vulnerable transboundary aquifer hotspots for multi-scale groundwater management
Christina M. Fraser, Robert M. Kalin, Modesta Kanjaye and Zione Uka
Water International, volume 45, issue 7–8 (2020), pp. 865–883

Chapter 17

Binational reflections on pathways to groundwater security in the Mexico–United States borderlands
Rosario Sanchez, José Agustin Breña-Naranjo, Alfonso Rivera, Randall T. Hanson, Antonio Hernández-Espriú, Rick J. Hogeboom, Anita Milman, Jude A. Benavides, Adrian Pedrozo-Acuña, Julio Cesar Soriano-Monzalvo, Sharon B. Megdal, Gabriel Eckstein and Laura Rodriguez
Water International, volume 46, issue 7–8 (2021), pp. 1017–1036

Chapter 18

A critical review of the transboundary aquifers in South-Eastern Europe and new insights from the EU's water framework directive implementation process
Charalampos Skoulikaris, Jacques Ganoulis and Alice Aureli
Water International, volume 46, issue 7–8 (2021), pp. 1060–1086

Notes on Contributors

Wahib Al-Qubatee, Water Resources Management Group, Wageningen University and Research, Wageningen, The Netherlands; Water and Environment Centre, Sana'a University, Sana'a, Yemen.

Adel Al-Weshali, Water and Environment Centre, Sana'a University, Sana'a, Yemen.

Charlotte Alcazar, Symcrau, Istres, France.

Behzad Ataie-Ashtiani, Department of Civil Engineering, Sharif University of Technology, Tehran, Iran; National Centre for Groundwater Research & Training and College of Science & Engineering, Flinders University, Adelaide, SA, Australia.

Olivia Aubriot, CNRS-CEH. Aubervilliers, France.

Alice Aureli, UNESCO Intergovernmental Hydrological Programme, Division of Water Science, Paris, France.

Quentin Ballin, Agence Française de Développement, Paris, France.

Jude A. Benavides, School of Earth, Environmental, and Marine Sciences, University of Texas Rio Grande Valley, Brownsville, TX, USA.

Marc Boisson, Artelia, Echirolles, France.

Sami Bouarfa, G-EAU, Cirad, INRAE, AgroParisTech, IRD, Montpellier Supagro, BRGM, University of Montpellier, Montpellier, France.

José Agustin Breña-Naranjo, Mexican Institute of Water Technology, Faculty of Engineering, Universidad Nacional Autonoma de Mexico Unam, Mexico City, Mexico.

Arnaldo José Cambraia Neto, Department of Agricultural Engineering (DEA), Federal University of Viçosa (UFV), Viçosa, MG, Brazil.

Aurélien Dumont, UNESCO, Intergovernmental Hydrological Programme, Paris, France.

Emmanuel Durand, GRET, Nogent-sur-Marne, France.

Gabriel Eckstein, Texas A&M School of Law, Texas A&M University, Fort Worth, TX, USA.

Mohamed Elloumi, Université de Carthage, INRAT, Laboratoire d'Economie Rurale, Ariana, Tunisia.

Nicolas Faysse, National Institute of Agronomy of Tunisia, Tunis, Tunisia.

Rhoda Fofack-Garcia, France Energies Marines, Plouzané, France.

Jan Willem Foppen, Water Science and Engineering Department, IHE-Delft Institute for Water Education, Delft, Netherlands.

Christina M. Fraser, International Groundwater Resources Assessment Centre (IGRAC), the Netherlands.

Jacques Ganoulis, UNESCO Chair Inweb, Department of Civil Engineering, Aristotle University of Thessaloniki, Thessaloniki, Greece.

Andrea K. Gerlak, Udall Center for Studies in Public Policy, University of Arizona, Tucson, USA.

Ridha Ghoudi, CRDA de Gabès, Ministère de l'Agriculture des Ressources Hydrauliques et de la Pêche, Gabès, Tunisia.

Rafael de Grenade, Udall Center for Studies in Public Policy, University of Arizona, Tucson, USA.

Randall T. Hanson, One-Water Hydrologic, San Diego, CA, USA.

Rebecca Heinz, Department of Geography, University of Bonn, Germany.

Petra J. G. J. Hellegers, Water Resources Management, Wageningen UR, Wageningen, the Netherlands.

Antonio Hernández-Espriú, Hydrogeology Group, Earth Science Division, Faculty of Engineering, Universidad Nacional Autonoma de Mexico Unam, Mexico City, Mexico.

Rick J. Hogeboom, Multidisciplinary Water Management, Faculty of Engineering Technology, University of Twente, Enschede, the Netherlands; Water Footprint Network, Enschede, the Netherlands.

Seiyed Mossa Hosseini, Physical Geography Department, University of Tehran, Tehran, Iran.

Annabelle Houdret, German Development Institute/Deutsches Institut für Entwicklungspolitik (DIE), Department of Environmental Governance and Transformation to Sustainability, Bonn, Germany.

Aline Hubert, GRET, Nogent-sur-Marne, France.

Robert M. Kalin, Department of Civil and Environmental Engineering, University of Strathclyde, Glasgow, UK.

Modesta Kanjaye, Ministry of Agriculture, Irrigation and Water Development, Government of Malawi, Lilongwe, Malawi.

Futaba Kazama, International Research Center for River Basin Environment, University of Yamanashi, Japan.

Wisam M. Khadra, Department of Geology, American University of Beirut, Lebanon.

Hans Komakech, Water Infrastructure and Sustainable Energy - Futures Centre, Nelson Mandela African Institution of Science and Technology, Arusha, Tanzania.

Timothy J. Krupnik, Sustainable Intensification Program, International Maize and Wheat Improvement Center, Dhaka, Bangladesh.

Marcel Kuper, G-EAU, Cirad, INRAE, AgroParisTech, IRD, Montpellier Supagro, BRGM, University of Montpellier, Montpellier, France.

Selin Le Visage, UMR Tree, CNRS, Université de Pau et des Pays de l'Adour, Pau, France.

Stéphanie Leyronas, Agence Française de Développement, Paris, France.

Elena López-Gunn, ICATALIST, Madrid, Spain.

Frédéric Maurel, Agence Française de Développement, Paris, France.

Andrew J. McDonald, School of Integrative Plant Sciences, Soil and Crop Sciences Section, Cornell University, Ithaca, NY, USA.

Sharon B. Megdal, Water Resources Research Center, University of Arizona, Tucson, AZ, USA.

Imane Messaoudi, Laboratoire Mosaïques Lavue, Université Paris Nanterre, Nanterre, France; Laboratoire Géosciences, Université de Fribourg, Fribourg, Switzerland.

Anita Milman, Department of Environmental Conservation, University of Massachusetts, Amherst, MA, USA.

Tapas Singh Modak, Foundation for Agrarian Studies, Bengaluru, India.

François Molle, G-EAU, Cirad, INRAE, AgroParisTech, IRD, Montpellier Supagro, BRGM, University of Montpellier, Montpellier, France.

Marielle Montginoul, G-EAU, Cirad, INRAE, AgroParisTech, IRD, Montpellier Supagro, BRGM, University of Montpellier, Montpellier, France.

Tlhoriso Morienyane, Water Science and Engineering Department, IHE-Delft Institute for Water Education, Delft, Netherlands.

James E. Nickum, Editor in Chief, *Water International*; IWRA Fellow; School of Oriental and African Studies (SOAS), University of London, UK; East-West Center, Honolulu, Hawaii.

Seyni Ndao, OFOR, Dakar, Senegal; mIndependent Sociologist, Montpellier, France.

Tunde Olarinoye, Chair of Hydrological Modeling and Water Resources, University of Freiburg, Germany.

Oualaphone Ongkeo, Natural Resources and Environment Institute, Prime Minister's Office, Vientiane, Lao PDR.

Vishnu Prasad Pandey, Department of Civil Engineering, Institute of engineering, Tribhuvan University, Pulchowk Campus, Nepal.

Esmaeel Parizi, Physical Geography Department, University of Tehran, Tehran, Iran.

Paul Pavelic, International Water Management Institute, Vientiane, Lao PDR.

Adrian Pedrozo-Acuña, Mexican Institute of Water Technology, Faculty of Engineering, Universidad Nacional Autonoma de Mexico Unam, Mexico City, Mexico.

Olivier Petit, Centre Lillois d'Études et de Recherches Sociologiques et Économiques (CNRS), University of Lille, France.

Marta Rica, Department of Social Sciences, University Carlos III de Madrid, Madrid, Spain.

Audrey Richard Ferroudji, Independent Sociologist, Montpellier, France.

Jean-Daniel Rinaudo, G-EAU, Cirad, INRAE, AgroParisTech, IRD, Montpellier Supagro, BRGM, University of Montpellier, Montpellier, France.

Henk Ritzema, Water Resources Management, Wageningen UR, Wageningen, the Netherlands.

Alfonso Rivera, Institut National de la Recherché Scientifique Inrs-ete Québec, Quebec, QC, Canada.

Laura Rodriguez, Texas Water Resources Institute, Texas A&M University, College Station, TX, USA.

Lineu Neiva Rodrigues, Department of Research and Development (CPD), Brazilian Agricultural Research Corporation (EMBRAPA – Cerrados), Planaltina, DF, Brazil.

Dominique Rojat, Maurel Agence Française de Développement, Paris, France.

Bruno Romagny, Aix Marseille University, IRD, LPED, Marseille, France.

Emmanuelle Salgues, CACG, Tarbes, France.

Rosario Sanchez, Texas Water Resources Institute, Texas A&M University, College Station, TX, USA.

Stefanie Viktoria Caroline Schulte, Independent researcher, Sydney, Australia.

P. Seward, Department of Earth Sciences, University of the Western Cape, Bellville, South Africa.

Charalampos Skoulikaris, UNESCO Chair Inweb, Department of Civil Engineering, Aristotle University of Thessaloniki, Thessaloniki, Greece.

Julio Cesar Soriano-Monzalvo, Hydrology Coordination, Mexican Institute of Water Technology, Jiutepec, Mexico.

Raya M. Stephan, Water law expert, Consultant, Deputy Editor-in-Chief, *Water International*; IWRA fellow.

Pieter J. Stuyfzand, Department of Geoscience and Engineering, Delft University of Technology, Netherlands; KWR Watercycle Research Institute, Nieuwegein, Netherlands.

Zachary P. Sugg, Babbitt Center for Land and Water Policy, Lincoln Institute of Land Policy, Arizona, USA.

Julie Trottier, CNRS, UMR PRODIG, Aubervilliers, France.

A. Turton, Centre for Environmental Management, University of the Free State, Bloemfontein, South Africa.

Zione Uka, Ministry of Agriculture, Irrigation and Water Development, Government of Malawi, Lilongwe, Malawi.

Anton Urfels, Sustainable Intensification Program, International Maize and Wheat Improvement Center, South Asia Regional Office, Kathmandu, Nepal.

Pieter R. van Oel, Water Resources Management Group, Wageningen University, Netherlands.

Frank van Steenbergen, MetaMeta Research, Hertogenbosch, the Netherlands.

Robert G. Varady, Udall Center for Studies in Public Policy, University of Arizona, Tucson, USA.

Fermín Villarroya, Faculty of Geology, Complutense University, Madrid, Spain.

William Veerbeek, Water Science and Engineering Department, IHE-Delft Institute for Water Education, Delft, Netherlands.

Philippus Wester, Mountain Knowledge & Action Networks, ICIMOD, Nepal; IWRA Fellow.

Ounakone Xayviliya, Department of Water Resources, Ministry of Natural Resources and Environment, Vientiane, Lao PDR.

Y. Xu, Department of Earth Sciences, University of the Western Cape, Bellville, South Africa.

Bahador Zarei, Political Geography Department, University of Tehran, Tehran, Iran.

Introduction

Raya M. Stephan, James E. Nickum, and Philippus Wester

Groundwater, comprising about 99% of the Earth's available freshwater, is hidden beneath our feet, invisible to our eyes, yet affects everything around us, including the one percent that is visible in the form of surface water. Groundwater provides almost half of all drinking water worldwide as well as nearly half of water consumed in irrigation (Siebert et al. 2010). It is essential for meeting the human right to water, especially in places where public or commercial water services are lacking or inadequate, yet where wells can obtain water easily (United Nations 2022). Groundwater maintains the baseflow of rivers and lakes, and it sustains wetlands, which provide important services for communities, and habitats for flora and fauna. As climate changes, groundwater can offer an option for adaptations. Yet as surface water bodies have been fully claimed, pressure has increased on groundwater, resulting in drying up of wetlands, land subsidence under cities and saltwater intrusion along coastal zones. It is no accident that even though groundwater is only mentioned explicitly in UN Sustainable Development Goal (SDG) target 6.6, about 53 SDG targets are linked to this unseen resource (United Nations 2022).

Because of its invisible and often de facto open access character, information, management and governance of groundwater remain often poor and inadequate. While much has been accomplished in the recent years, more still needs to be done to achieve satisfactory sustainable management of groundwater, in terms of either the related SDG targets or the human right to water.

This book presents a collection of chapters originally published as articles in *Water International* in recent years, that we have assembled here as a contribution to the UN Water Groundwater year (2022), and to contribute to the effort of "making the invisible, visible".[1] Through case studies ranging around the globe from the Americas (California, Brazil), to Asia (India, Iran, Lao PDR, Nepal), Africa (Malawi, South Africa, Tanzania) and the MENA region (Lebanon, Morocco, Yemen), as well as cases of transboundary aquifers, these chapters reflect important recent advances in interdisciplinary knowledge on the governance, management, practice and science-policy interfaces of groundwater. They are grouped into five sections: institutions, management, users, irrigation and transboundary aquifers.

In the first section, on institutions, we have included case studies from two Asian developing countries Nepal and Lao PDR, and from two middle income countries South Africa and Morocco. The first two deal with the institutional change that appeared necessary to introduce better management practices for groundwater, while the following ones are about improving groundwater institutions, either existing agencies, or attempting to go beyond the agency to create a multipartner contract.

Pandey and Kazama explore how the status of groundwater has changed from an open-access resource to a state controlled one in the rainfall abundant Kathmandu valley of Nepal where groundwater has been an important source of water supply since time immemorial, epitomized by the traditional stone spout systems (Molden, Griffin & Meehan, 2016). The open access status of groundwater led to its overexploitation resulting in depletion and quality degradation. This situation brought together governmental and non-governmental actors to try to reverse the trend, which impelled the move to making groundwater a state controlled resource. The authors discuss development dynamics, impacts and responses in the area, and suggest some "soft-path" approaches for groundwater management.

Laos is also a water rich Asian country with large major rivers and surface storages. Groundwater is used on a limited scale, but it contributes significantly towards the health and well-being of most of the population. Nonetheless, the levels of understanding, awareness and technical capacity remain extremely limited. In their chapter, Pavelic et al. discuss the current state of knowledge and management of groundwater and suggest some pathways forward. Whilst in Lao PDR the level of governance remains very low, there are promising signs of positive institutional change.

Seward et al. explore ways to improve the national government's contribution to good groundwater governance in South Africa. Using the South African Department of Water and Sanitation as a case study, they find that a lack of strategic thinking and implementation are the main impediments to service delivery. Although isolated cases of groundwater "overuse" and pollution can be found in South Africa, the main governance issues are currently competition between users for groundwater use, and inordinate delays in granting licensed authorization for that use.

Houdret and Heinz study an aquifer contract in a semi-arid part of Morocco which has witnessed a swift change from a rural subsistence economy to intensive agriculture. This development relied mainly on groundwater, causing a rapid and massive depletion of the aquifer. In view of the economic and hydrological urgency of the situation, and following a participatory process, an aquifer contract was elaborated, with a new mix-and-match "bricolage" of existing institutions. The aquifer contract resulted from the common awareness of water users and formal water institutions about the urgency to regulate groundwater use in the face of increased scarcity. Nonetheless, the authors find that existing socio-economic power relations divert these new institutions, even though they are based on users' participation, from their original purpose. These two studies bring us down to earth and remind us that nice ideas for an "all stakeholders" approach have to deal with resistance from existing power relations, social institutions, and cultural affiliations.

Our second section, on management, includes four chapters focusing on the use of scientific approaches to inform policy with cases studies from a developing country (Tanzania), middle income countries (Brazil and Iran), and a failed economy (Lebanon).

Olarinoyea et al. combine satellite imagery, urban growth modelling, groundwater modelling, and hydrogeological field expeditions to estimate potential impacts in 2050 in Arusha, Tanzania. Like many cities in Sub-Saharan Africa, Arusha relies on groundwater, has a rapidly growing population, and faces great uncertainty regarding the impacts of climate change. Groundwater reserves, if adequate, make cities or areas more resilient to climate change. In Arusha, however, as in other African cities such as Lagos, Dar es-Salaam, or Bamako, groundwater is already overdrafted, reducing resilience and leading to a number of adverse effects. Groundwater levels are predicted to continue to fall, worsening the general economic, social and environmental situation. In the absence of successful interventions, groundwater overuse is likely to be yet another factor hampering progress of urban Sub-Saharan Africa.

Cambraia Neto and Neiva Rodrigues use a water balance model (BALSEQ) to evaluate the impact of climate and land use and occupation changes in the estimation of potential groundwater recharge in the vast yet relatively still unknown reservoir of Cerrado Biome in Brazil. They have considered the current land use (baseline) and two scenarios: one with the replacement of the entire vegetation by soybean cropping (scenario 1), and one with the substitution of the agricultural areas by native savannah (scenario 2).

Moving to the MENA area, Khadra and Stuyfzand explore managed aquifer recharge, one of the more common management tools to cope with water scarcity, where geological conditions allow. It is seen as problematic in karstic and highly fractured aquifers due to their complex dynamics. In Lebanon, one of the most karstic areas in the Eastern Mediterranean, the national water strategy has proposed managed aquifer recharge as a means to augment water supplies. The authors seek to strengthen faith in managed recharge and raise awareness and acceptability of its application to karstic domains. They review the major types and associated obstacles and suggest several solutions towards successful applications in karst. They also provide examples of cases in Lebanon which are illustrative of viable managed recharge systems in a karstic dominated area. These examples could be generalised to similar hydrogeological settings elsewhere. Perhaps fractured karstic aquifers are not so problematic after all.

Zarei et al. close out this section with the development of a much-needed index to assess groundwater governance in Iran. As in all too much of the world, groundwater in Iran is abstracted at unsustainable rates, with about 90% of it going to agricultural use yet with a low irrigation efficiency. The declining level of groundwater seriously threatens socio-economic development, political stability, ecosystem integrity, the health and welfare of natural systems, and human communities. The index developed here evaluates the management of groundwater and uses quantitative indicators for four categories of sustainability: social, economic, environmental, and institutional. The authors apply the index to evaluate 443 of Iran's 536 aquifers which have appropriate quantity and quality monitoring networks. The results confirm that groundwater management in Iran is relatively poor. This is no surprise, but this index could help improve things by providing a means of evaluating progress of groundwater governance in Iran. This is a valuable contribution to the search for measures of the impact of water governance (see especially the special issue on measuring the impact of water governance in *Water International*, 47(2), 2022).

The chapters contained in the third section concentrate on groundwater users from three very different situations: war-torn Yemen, the developed economy of Spain, and California, a highly developed economy.

Rica et al. examine different forms and levels of collective action by aquifer users to secure access to over-allocated groundwater resources in La Loma, Úbeda (Jaén, Spain), one of the largest olive-growing areas in the world. They examine the role that water users have played and how they have evolved a development model that seeks to maintain existing socio-economic benefits. The authors question if the users act collectively to resolve a common pool resource problem or whether it is other motivations, not directly related to water conservation, that explain their coordination. They also reflect on the relationship with other key actors such as the Basin Water Authority and also the role political factors have played in either supporting or hindering user collectives. Finally they identify an opportunity to redesign the organisational and institutional configurations by both securing access to water and strengthening collaborative spaces at the basin level.

Today Yemen is devasted, but it has irrigation systems that have operated for more than two millennia using world-renowned water harvesting and diverting techniques. Al-Qubate et al. conduct a participatory rural appraisal in the Al-Mujaylis area of the Tihama Coastal Plain. They encourage stakeholders to participate in analysing and assessing measurements in land planning and to bring their own knowledge to the dialogue. The authors evaluate the use of a mixed approach (top down and bottom up) to participatory rural appraisal in areas with data scarcity and in a cultural context where Islamic religious values encourage people to consult among themselves, in conjunction with experts and specialists, and to be more cooperative and altruistic.

In California, Schulte considers the potential adverse impact of Federal reserved rights to groundwater on the implementation of California's Sustainable Groundwater Management Act 2014 (SGMA), through an analysis of the Agua Caliente Band of Cahuilla Indians v Coachella Valley Water District Case (decision by the Ninth Circuit court). California adopted the SGMA to address the overuse of groundwater resources and to establish a pathway for their sustainable use. By recognising that the Agua Caliente Indian tribe may have water rights that are outside state jurisdiction, the court decision could impact the implementation of this new legislation. Schulte explores these possible impacts and offers recommendations for reducing the tensions between the court's decision and the competing legal frameworks for groundwater management in California.

The fourth section deals with irrigation, which accounts for approximately 70% of global groundwater withdrawals, and even more in arid and semi-arid regions (Margat and Van der Gun, 2013), to produce food, fibres, livestock, and industrial crops (United Nations 2022).

Petit et al. provide a review on the consequences of the recent increase of groundwater use in agriculture. Their chapter, based on discussions between scientists and operational experts in two workshops held in 2018 and 2020, first takes stock of the dynamics and motivations of groundwater use in agriculture. It then examines the solutions most commonly proposed, in particular by public authorities, to regulate the access to and use of this resource and to limit its over-exploitation. These often optimistic solutions generally combine regulatory or economic instruments, or indirect measures linking water to other issues, with mechanisms based on the participation of all users. Yet they rarely question the intensive agricultural systems driving groundwater demand and fail to recognise the multifunctional nature of groundwater. The authors conclude that it is necessary to consider groundwater as a common patrimony in order to frame the discussion of problems and the finding of collective solutions that will raise awareness of the centrality of access to groundwater for vulnerable groups as a means of livelihood and poverty alleviation. Such an approach would also prove helpful in determining which territorial project "makes sense", particularly where the potential remains to increase groundwater exploitation, as in Sub-Saharan Africa.

As we have noted, in the Kathmandu Valley groundwater sources are under stress. In another area of Nepal, the rice-producing Terai, which has an extensive shallow groundwater irrigation infrastructure, water stress arising from an increasingly erratic monsoon has become one of the main factors limiting stable crop productivity. Urfels et al. delineate the drivers of shallow tube well irrigation use in order to identify development support pathways likely to increase irrigation use and climate resilience among rice farmers in this region of Nepal – so that they can transition from "life saving" to "productivity enhancing" irrigation use. Using a mixed-methods approach, the authors establish and rank the key factors influencing the use of shallow tube wells during that season in the Terai. They conclude a multi-scalar strategy is needed to elicit a "productivity enhancing" irrigation use among Nepali rice farmers at the farm, the community, and the regional level.

South of the Terai, in India's West Bengal, Modak explores the close relationship between the policies on groundwater use and electricity adopted there. He identifies two distinct phases of groundwater irrigation development in the state: (1) from 1975 to 1995, policies promoted a rapid expansion of tubewell irrigation; (2) since the mid-1990s, the government, inspired by policies in less water-abundant states, extended its direct control over the installation of groundwater structures while regulating electricity connections and pricing. The effect of these policies was to slow down the development of groundwater irrigation, and, consequently, the growth of agricultural production. Yet West Bengal still has abundant unexploited groundwater. One policy does not fit all places. Water is local, so its governance should be local.

Our final section is devoted to exploring the situation in some of the 468 transboundary aquifers that have been identified globally, not counting those within the European Union (IGRAC 2021). This topic is a major interest of the International Water Resources Association (IWRA). Specifically, we offer here four case studies on transboundary aquifers: the Guarani Aquifer System in South America between Argentina, Brazil, Paraguay and Uruguay (middle income and developing countries), in Malawi (developing country), along the US–Mexico border (developed and middle-income countries), and finally in South Eastern Europe (developed and middle-income countries).

In the Guarani Aquifer System, Sugg et al. survey global and regional experts about three key elements of governance: the region's groundwater quantity and quality, ownership and rights, and regulation and administration. Respondents perceive groundwater quality and withdrawal as under-regulated, and relevant information and data as inadequate; they suggest that contamination and overdrafting remain mostly incipient and localised along international borders; and they view groundwater as a shared resource administered by the state for the public, rather than as private property. Respondents observe that while there is progress towards implementing a formal transboundary aquifer agreement, local-to-national-scale governance is important. This study on the Guarani Aquifer System can help clarify the role of governance by highlighting specific challenges in local/regional contexts and check if those issues are addressed effectively through multilateral agreements. Studies such as this can prove helpful at the early stage of development of transnational groundwater agreements in general.

Fraser et al. take us to East Africa to identify vulnerable hotspots in transboundary aquifers, within the 38 aquifers shared between Malawi and its neighbouring countries. Using fuzzy logic and GIS overlays the authors generate "hotspot maps" of areas with the greatest risk of transboundary mismanagement, either in quantity or quality. This tool could be of direct policy use in prioritising the transboundary aquifers requiring management and international agreements, and determining where to direct the necessary funds. In Malawi, where water management is decentralised, the authors' tool could be of particular use in identifying areas where transboundary water management can take place at the local level.

Sanchez et al. focus on shared groundwater resources between Mexico and the United States which are facing unprecedented stresses. Their study begins with the state of groundwater knowledge, followed by an analysis of the challenges groundwater faces from a physical, societal, and institutional perspective. They conclude that the extent of ongoing cooperation frameworks, joint and remaining research efforts, from which alternative strategies can emerge, still need to be developed. The way forward necessitates a variety of cooperation models as the future presents rather complex, shared and multidisciplinary water challenges to the Mexico–US borderlands. The authors provide several relevant avenues to enhance the stewardship of transboundary water resources between Mexico and the United States.

As our final contribution, we move to South Eastern Europe where the collapse of former Yugoslavia has increased the number of states and thus of transboundary aquifers. Skoulikaris et al. provide a review article and an analysis of the main international programmes and initiatives over the last two decades directed towards the management of transboundary aquifers in South-Eastern Europe. The aquifers' delineation, identified pressures, and water quality and use are compared with those derived by the EU's Water Framework Directive implementation process. The final outputs reveal a solid conceptual follow-up of the programmes and initiatives with a gradual enrichment of information related to the status of transboundary aquifers over the years. Groundwater bodies' data analysis validates the initial outputs and provides enhanced information at finer scales. The conducted research demonstrates the continuity of collecting and sharing information related to transboundary aquifers in South East Europe, which has been achieved through a series of back-to-back international programmes and initiatives. This work is conceived as a comprehensive and up-to-date guide to the status of transboundary aquifers and groundwaters in South East Europe that can facilitate and foster shared groundwater management.

We hope that this collection of articles will contribute to the better understanding of the challenges groundwater faces around the world, and of the importance of tackling this resource in a manner resulting in its proper management. It is especially important to stress the need to ensure human rights, mainly to water and food, achieve the SDGs, adapt to climate change and face the current global crisis in groundwater governance. IWRA through its groundwater task force continues this effort and aims to contribute in building capacity and knowledge of the groundwater science-policy-practice interface.[2]

Notes

1 www.unwater.org/publications/un-world-water-development-report-2022/
2 For more information www.iwra.org/groundwater-task-force/

References

IGRAC (International Groundwater Resources Assessment Centre) (2021). Transboundary Aquifers of the World [map]. Edition 2021. Scale 1:50 000 000. Delft, Netherlands: IGRAC, 2021. Available at www.un-igrac.org/sites/default/files/resources/files/2021TBAMap_UNESCO%20Version.pdf

Margat J. and van der Gun J. (2013). Groundwater around the world, a geographic synopsis, IGRAC, UNESCO-IHP, CRC Press. Available at www.un-igrac.org/sites/default/files/resources/files/Groundwater_around_world.pdf

Molden, O., Griffin, N. and Meehan, K. (2016). The cultural dimensions of household water security: the case of Kathmandu's stone spout systems. *Water International*, 41(7), 982–997. https://doi.org/10.1080/02508060.2016.1251677

Siebert S., Burke, J., Faures, J.M., Frenken, K., Hoogeveen, J., Döll, P. and Portmann, F.T. (2010). Groundwater use for irrigation – a global inventory. Hydrology and Earth System Sciences, 14, 1863–1880. Available at www.hydrol-earth-sys-sci.et/14/1863/2010

United Nations (2022). The United Nations World Water Development Report 2022: Groundwater: Making the invisible visible. UNESCO, Paris. Available at https://unesdoc.unesco.org/ark:/48223/pf0000380721

Part I

Groundwater Institutions

From an open-access to a state-controlled resource: the case of groundwater in the Kathmandu Valley, Nepal

Vishnu Prasad Pandey and Futaba Kazama

Groundwater is no longer an open-access resource in Kathmandu. Perceptions of groundwater are changing from an "infinite" to a "finite" resource and the role of the government from supply developer to caretaker of the resource. In this context, this paper aims to unfold how perceptions of groundwater in Kathmandu Valley, Nepal, are changing from an open-access resource to an overexploited, depleted, degraded, vulnerable and state-controlled resource. In the process, it produces an aggregated picture of resource availability, development dynamics, impacts and responses in the area; suggests some "soft-path" approaches for groundwater management; and discusses implications of the experience for other areas.

Introduction

Groundwater, as a vast reservoir and strategic reserve of freshwater, has great significance for human development (Foster & Chilton, 2003). Compared to surface-water use, groundwater use often brings large economic benefits per unit volume because of ready local availability, high drought reliability and generally good quality requiring minimal treatment (Burke & Moench, 2000). This has resulted in groundwater being the world's most extracted raw material, with an estimated global withdrawal of 600–700 km^3/y (Zektser & Everett, 2004). The period 1950–1975 for many industrialized countries and 1970–1990 for most parts of the developing world witnessed a boom in groundwater abstractions (Foster & Chilton, 2003). Llamas and Martinez-Santos (2005) termed the boom a "silent revolution" because it led to dramatic changes in water use and food policies in regions including California, Italy, Mexico, Spain, India, China, and parts of Middle East and North Africa (Shah et al., 2007). During the revolution (i.e. intensive use of groundwater resources), focus was on rapid exploitation to ensure adequate supply rather than careful attention to sustainable use. The role of government agencies remained exclusively "supply developer" rather than "resource custodian" or "information provider" (Foster et al., 2000). Government agencies failed to fully engage groundwater users and relevant stakeholders in participatory management processes. As a result, groundwater escaped control, and therefore appropriate management, in many arid and semi-arid countries worldwide (Llamas & Martinez-Santos, 2005). This led to varying degrees of aquifer degradation, environmental impact, or both (Morris et al., 2003), making groundwater systems more vulnerable to (groundwater) resource depletion, quality degradation and land subsidence (Pandey et al., 2010).

Recently, the concept of groundwater – as a resource – is changing from development to management (Kemper, 2004; Shah et al., 2007). Groundwater is being changed from an open-access resource to a state-controlled resource. Evidence of excessive development and its consequences, together with increased advocacy from informal sectors (e.g. non-governmental organizations and civil societies), are helping shift the role of government agencies towards "resource custodian" and "information provider". The government agencies in many cases are realizing that social, institutional and political factors are the primary obstacles to sustainable groundwater management, as argued by Burke et al. (1999), and "changing communication dimensions" should be the starting-point for a new management design, as argued by Knegt and Vincent (2001). From experience it has been learnt that there is no simple and realistic blueprint for groundwater management action, mainly because the diversity of groundwater occurrence, coupled with changing patterns of human use of water and land, present a complex mosaic within which management and protection have to function (Foster & Chilton, 2003). Diversity in the hydrogeologic characteristics of groundwater aquifers is one of the challenges; however, the characteristics of human interaction with aquifers are equally important to consider. Although the scale of the problems and potential solutions may be location specific, understanding the nature and extent of the issues related to human interactions with aquifers in an area, the way they are being addressed, and the lessons learnt could provide important insights for other areas as well.

This study aims to explore the "communication dimension" (suggested by Knegt & Vincent, 2001) of groundwater management by unfolding how groundwater is being changed from an "open-access" resource to a "state-controlled" resource in Kathmandu Valley, Nepal (Figure 1). In the process, it produces an aggregated picture of groundwater resource availability and development dynamics, impacts and responses. It further discusses implications of the experience for other areas.

Groundwater as lifeline and open-access resource

In the Kathmandu Valley, in which nestles the capital of Nepal (Figure 1), groundwater has been an important source of water supply since time immemorial. The valley, founded on the sediments of the lake that once filled it, has two major groundwater aquifers, one shallow (unconfined) and one deep (confined), separated vertically by a clay aquitard layer of varying thickness. In ancient times, people relied on shallow groundwater abstracted through traditional means (e.g. stone spouts, springs, dug wells and infiltration galleries). From the early 1970s, technological developments led to mechanized abstraction of deep groundwater to cater to escalating water demands as the valley started to develop rapidly as an urban centre. Currently, up to 70% (during the dry season) of the water supplied by Kathmandu Upatyaka Khanepani Limited (KUKL, the water supply utility – called Nepal Water Supply Corporation [NWSC] until 2006) is produced from groundwater sources. KUKL's supply, however, is not enough to fulfil the valley's water demand of over 320 million litres per day (MLD). This gap in supply is met by groundwater privately abstracted by individual households, housing colonies, hotels and industries. Ownership of water resources is vested in the state by the Water Resources Act enacted in 1992; however, it was considered an open-access resource (Dixit & Upadhya, 2005) until the Kathmandu Valley Water Supply Management Board (KVWSMB) came into existence in 2006. This was because groundwater was perceived as an infinite

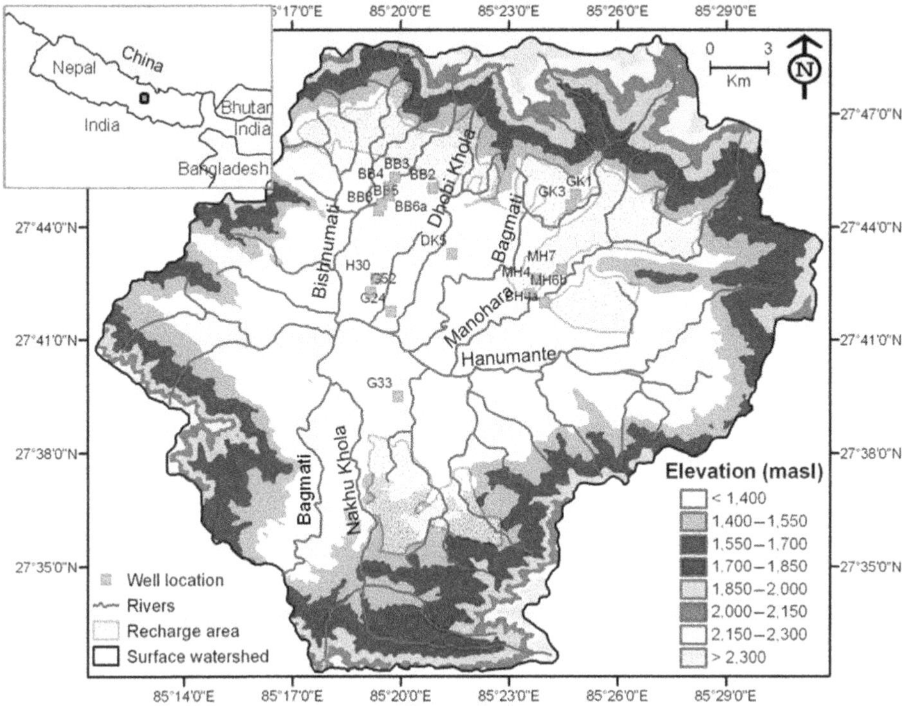

Figure 1. The Kathmandu Valley, in central Nepal. Green squares represent locations of the wells listed in Table 1.

Data sources: recharge areas from JICA (1990); river networks from NGIIP (2004); well locations from Acres International (2004) database; elevation from Tachikawa, Hato, Kaku and Iwasaki (2011).

resource and no effectively functional authority was present for its protection and management. The perception of groundwater as an open-access resource encouraged its uncontrolled abstraction over the years and subsequent overexploitation. The overexploitation, coupled with inadequate management of solid waste and wastewater from urban centres, has increased the vulnerability of the groundwater system to (groundwater) resource depletion, quality degradation and land subsidence (Pandey et al., 2010).

Groundwater as an overexploited resource

Overexploitation is defined as the situation when long-term average aquifer abstraction exceeds recharge (Custudio, 2002). In this study, "overexploitation" refers to a situation of stress on the aquifers. This section presents and discusses indicators suggesting groundwater overexploitation in the area.

Groundwater abstraction greater than natural replenishment

KUKL and the private sector (e.g. hotels, industries, embassies and tankers) are pumping groundwater with no limits on the rate, volume or timing of abstraction. As a result, groundwater discharge is more than twice the natural recharge, estimated at 4.61–14.6 MCM/y (Figure 2). Even considering the top end of this range, the current level of

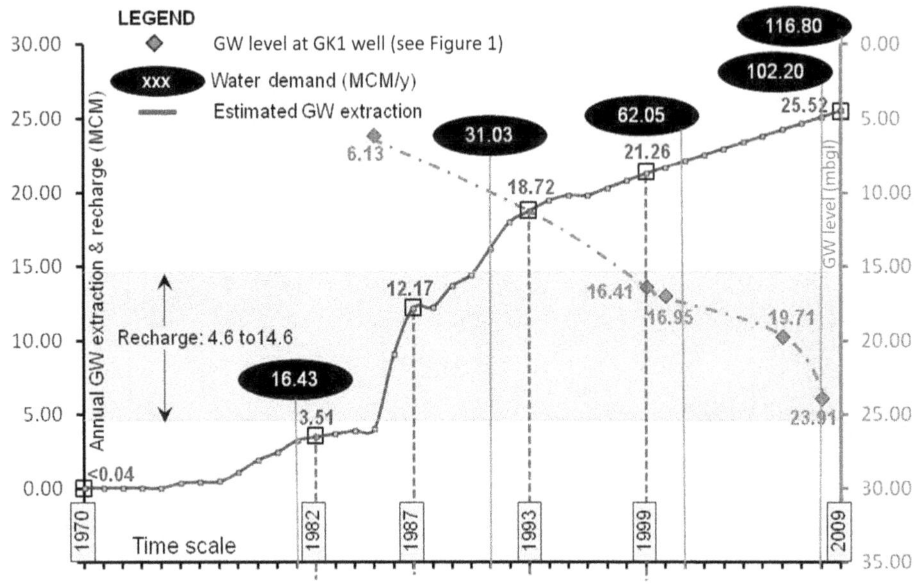

Figure 2. Groundwater development dynamics in the context of prevailing water demand and recharge (GW: groundwater; mbgl: metres below ground level).

Data sources: daily water demand for 1981/1991/2001 from ICIMOD (2007), for 2008 from KVWSMB (2008), for 2009 from Dhakal (2010); daily groundwater extraction for 1970–1999 from Metcalf and Eddy (2000), for 2009 from Dhakal (2010), between 1999 and 2009 calculated by linear interpolation; minimum and maximum recharge rates from Gautam and Rao (1991) and Binnie and Partners (1988), respectively; groundwater level before 1999 from Acres International (2004) and after 1999 from Groundwater Research and Development Project/Department of Irrigation, Kathmandu, Nepal. Annual GW extraction calculated assuming 360 days of well operation.

abstraction is far higher. During the last decade, not only did groundwater abstraction increase from 21.26 to 25.52 MCM/y (Figure 2) but also the greatest abstraction changed from KUKL (then NWSC), at 59.5% of abstractions in 1999, to the private sector, at 67.7% of abstractions in 2009 (Dhakal, 2010). This indicates added complexity in groundwater management, because controlling private-sector abstraction is more difficult than controlling the water supply agency.

Potential of the shallow aquifer

The shallow (unconfined) aquifer in the area has a huge groundwater storage potential compared to the deep (confined) one (Pandey & Kazama, 2011, 2012). This potential, estimated as a multiple of aquifer volume and storage coefficient (or specific yield for the shallow aquifer), shows that the shallow aquifer can store 1452 MCM of groundwater, while the deep one can store only 572 MCM (Figure 3).The northern part of the shallow aquifer has a relatively higher potential. Pandey and Kazama (2012) also estimated the storage potential above the current water level in the shallow aquifer as 226 MCM. If all the space could be filled, one cycle of filling could meet the valley's groundwater demand for 11 years. Though the estimate could be improved by considering the dynamics of groundwater flow and recharge, this preliminary estimate clearly indicates the potential of the shallow aquifer to combat groundwater overexploitation as well as to cover the gap in

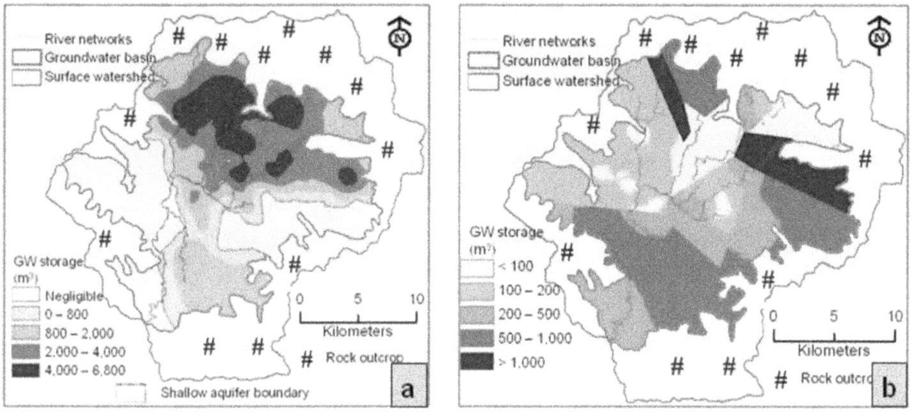

Figure 3. Groundwater storage volume per 20 m × 20 m grid cell in (a) the shallow aquifer and (b) the deep aquifer. Source: Pandey & Kazama (2012).

supply in the valley over the long run if its unused potentials are utilized wisely. However, there exist some addressable issues related to groundwater quality in the shallow aquifer (please refer to the section on groundwater as a degraded and vulnerable resource, below).

Imbalance between groundwater development and management

Producing more groundwater has been the prime focus ever since groundwater was first pumped in the early 1970s. Groundwater abstraction has continued to increase, with a sharp rise in the mid-1980s when groundwater was added to the supply system of the government-owned water supply agency (Metcalf & Eddy, 2000) to meet the escalating demand from growing population, urbanization and industrialization. Abstraction increased from 3.75 MCM/y in 1982 to 12.17 MCM/y in 1987, 21.26 MCM/y in 1999 and 25.52 MCM/y in 2009 (Figure 2). Several studies and projects conducted over the decades (see Shrestha, Pradhananga, & Pandey, 2012) have recommended groundwater management strategies. Unfortunately, the imbalance between groundwater development and management continued as measures to regulate abstraction and manage the aquifer systems remained low in priority over the years. If this imbalance continues and abstraction remains at the same rate as in 2001 (59.06 MLD), the valley's groundwater will be exhausted in less than 100 years (Pandey et al., 2010).

Informal water markets as a trigger for overexploitation

In the context of the gap in supply from the government-owned utility, informal groundwater markets – defined here as an arrangement through which entrepreneurs abstract and sell groundwater to customers in monetary terms (Mukherji, 2006) without defining (or practising) water rights and making transactions without legal sanction (Shah, 1993) – are evolving, expanding and functioning at different levels (individual, community, commercial and industrial) in the valley. The total transaction of groundwater for the market is estimated at 23 MLD and 13.8 MLD in the peak and off-peak seasons, respectively, resulting in average annual extractions of 6.35 MCM (Shrestha, 2011). The KVWSMB, the authority responsible for groundwater regulation, protection and management in the valley, took the initiative to bring them into the government's record by issuing licenses for groundwater development and distribution; however, many of them have yet to register with the KVWSMB. Though the markets cover the supply gap, they are triggering groundwater overexploitation. The extent of this effect has yet to be evaluated.

Groundwater as a depleted resource

Groundwater, long a lifeline for dwellers in the valley, is being depleted by uncontrolled abstraction and inadequate management. This section presents and discusses some facts revealing groundwater depletion in the area.

Decreasing groundwater levels

During 2000–2008, the groundwater level at KUKL wells fell by 0.37–7.50 m, with a gradual decrease over the years in all locations (Figure 4). Though depletion was low until 1982 (the earlier stage of abstraction), it became visible in 1982–1987 as the NWSC operated well-fields, increased in 1987–1993 as private wells increased, and became large after 1993 as unregulated pumping continued (Kharel et al., 1998).

Declining well yields

The yields of groundwater abstracting wells have also been declining since their installation. From the mid-1980s to 1999, the yields of KUKL wells have declined by 7.89, 4.97–36.17, 18.2–19.1, 31.31 and 15.17–27.81 L/s at the Dhobikhola, Bansbari, Gokarna, Bhaktapur/Bode and Manohara well-fields, respectively (Table 1). Greater declines are observed at the locations with greater decrease in groundwater levels.

Increasing drilling depth

An increasing trend in bore-well drilling depth over time has been observed. Depth has increased from about 20 feet (6.1 m) to about 40 feet (12.2 m) in a decade (Ramtel, 2010). According to a private bore-well contractor, drilling depth in the Gongabu area has

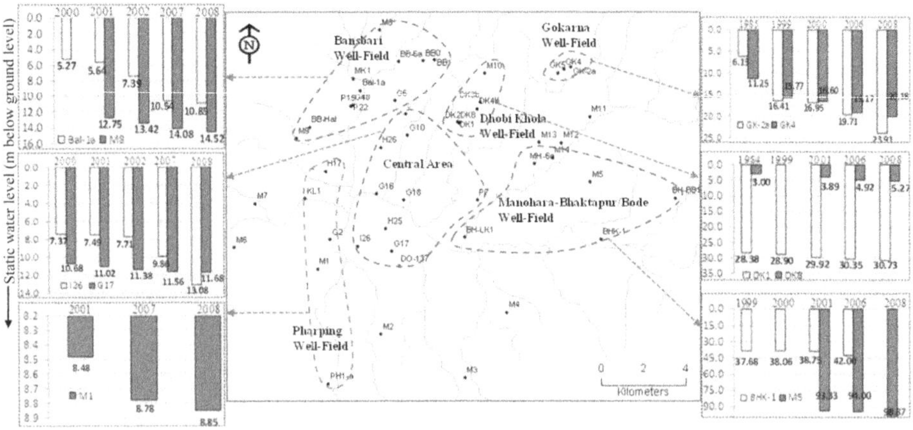

Figure 4. Aquifer depletion at selected locations during the dry season.

Source: Pandey et al. (2010).

Table 1. Decline in well yields in KUKL well-fields. Source: Pandey et al. (2010).

SN	WID	Well-field	Areas within GW basin	Previous discharge Year	Previous discharge Q (L/s)	Q in 1999 (L/s)	Decline in Q (L/s)
1	DK5	Dhobi Khola	Northern areas: KUKL well-fields	1983	27.89	20.00	7.89
2	BB2	Bansbari		1984	20.46	15.49	4.97
3	BB3	Bansbari		1984	43.24	21.67	21.57
4	BB4	Bansbari		1984	44.11	10.00	34.11
5	BB5	Bansbari		1985	41.00	15.00	26.00
6	BB6a	Bansbari		1984	43.67	7.50	36.17
7	BB8	Bansbari		1984	40.58	20.83	19.75
8	GK1	Gokarna		1985	35.77	16.67	19.10
9	GK3	Gokarna		1984	29.85	11.67	18.18
10	BH4a	Bhaktapur/Bode		1985	46.31	15.00	31.31
11	MH4	Manohara		1985	39.75	11.94	27.81
12	MH6b	Manohara		1983	35.00	16.67	18.33
13	MH7	Manohara		1985	38.50	23.33	15.17
14	H30	n/a	Central areas	1996	22.00	10.00	12.00
15	G24	n/a		1996	7.33	5.83	1.50
16	G33	n/a		1996	11.33	8.33	3.00
17	G52	n/a		1996	7.33	0.60	6.73

Note: WID: well identification number; GW: groundwater; *Q*: discharge. Well locations are shown in Figure 1.

increased from about 25 feet (7.5 m) to about 65 feet (19.8 m) in a decade, along with an increase in drilling-pipe size from 0.5 inches (13 mm) to 4 inches (102 mm), resulting in increased drilling costs.

Drying stone spouts and dug wells

There are over 400 stone spouts and over 10,000 dug wells in the valley. However, 110 spouts and several dug wells have already dried out or vanished. Several others are on the verge of drying; and only 45 stone-spouts are supplying water to KUKL pipes (NGOFUWS, 2006). This may be attributed to lowered groundwater levels and/or the construction of deep-foundation buildings that may have blocked groundwater flow to the spouts and dug wells.

Groundwater as a degraded and vulnerable resource

Groundwater in the area is not only depleted due to overexploitation but also degraded in quality. Groundwater is usually considered free of biological contaminants (e.g. pathogenic microbes) because of filtration of colloidal particles (Schwartz & Ibaraki, 2011). However, studies in the last two decades in the area have revealed vulnerability of groundwater to quality degradation. The shallow groundwater is contaminated by biological contaminants like *E. coli* and nutrients like nitrate; the deep groundwater, by ammonia, arsenic, iron and heavy metals (Chapagain et al., 2009, 2010). Bacterial contamination has also been observed in some deep groundwater samples (Shrestha et al., 2012). Drilling groundwater wells in or near agricultural fields increases the vulnerability of groundwater to contamination from pollutants with health implications (e.g. excess coliform, phosphorous, nitrogen and other organic compounds, spreading from manure, fertilizers and herbicides used in the fields) (Pradhanang et al., 2012).

Improper waste management in the valley is a primary reason for the increasing vulnerability of shallow groundwater to quality degradation. Pathak et al. (2009) found that even groundwater-rich areas like Gokarna (in the north of valley) are highly susceptible to pollution. Even deep groundwater has been found vulnerable to quality degradation from anthropogenic sources, mainly because of wells that are poorly designed or installed, improperly sealed (or not at all), or abandoned. Efforts are needed to seal off several abandoned wells. Beyond that, efforts to develop and implement locally suited, energy-efficient and cost-effective treatment technologies are necessary to address issues of water quality in promoting use of groundwater from the shallow aquifer.

Groundwater as an unsustainably developed resource

To move forward with groundwater management, apart from a good understanding of aquifer conditions, abstraction characteristics, and the state of groundwater quality, understanding the sustainability of groundwater development is equally essential. Sustainability, a time- and space-relevant perspective (De Carvalho et al., 2009), is a manifold concept approachable from many viewpoints (e.g. hydrological, ecological, economic, social, legal, institutional, inter- and intra-generational, and political) (Llamas et al., 2006). To evaluate progress in achieving groundwater sustainability in the area, the hydrological viewpoint is considered and the groundwater sustainability infrastructure index (GSII) (Pandey et al., 2011) is used as a tool. Here, "infrastructure" refers to the existing knowledge, practices and institutions whose adequate strength help achieve sustainability in groundwater management. The GSII consists of five components: groundwater monitoring (GwM), knowledge generation and dissemination (KgD), regulatory interventions (ReI), public participation (PuP) and institutional responsibility (InR). The components are disaggregated into 16 indicators. The results detailed by Pandey et al. (2011) showed that the overall situation of groundwater sustainability in the valley is relatively poor. The value (GSII = 0.22) falls at the low end of the scale (from zero to one, one representing the highest degree of sustainability). This reflects the need for significant attention to strengthening the sustainability infrastructure and subsequently achieving the goal of groundwater sustainability. All the components of the index are more or less poor and therefore need greater attention. They can be arranged in decreasing order of priority as follows: ReI, 0.08; PuP, 0.17; InR, 0.25; GwM, 0.25; KgD, 0.33. Within ReI, the situation of "economic instruments" is very poor and therefore it should get the highest priority.

Towards groundwater as a state-controlled resource

The depletion of groundwater and the degradation of its quality due to overexploitation coupled with improper waste-management practices have brought together formal (i.e. governmental) and informal (i.e. non-governmental) actors in the area to initiate activities aimed at the sustainable utilization and protection of groundwater. This eventually helped move towards state control of the resource. This section outlines those initiatives.

Informal-sector initiatives

Informal-sector initiatives can be broadly divided into two types: quiet advocacy (synthesizing available knowledge and disseminating it widely) and aggressive advocacy (organizing groundwater symposia, expert meetings and media campaigns, legal action, etc.). Examples of quiet advocacy include the publishing of at least 65 reports, papers and theses during 1966–2011 featuring groundwater in the area, in addition to a book titled *Kathmandu Valley Groundwater Outlook* (Shrestha et al., 2012). These publications suggest that a good amount of knowledge on the valley's groundwater system has already been accumulated, though more understanding is required on the complexities related to the dynamics of the groundwater flow, recharge and contamination processes. In addition to the publications, some aggressive advocacy has also been initiated to raise awareness and encourage actions to protect groundwater in the valley. Examples include a nationwide groundwater photography competition (December 2009); three consecutive national symposia called "Challenges and Opportunities for Sustainable Management of Groundwater Resources of Kathmandu Valley, Nepal" (December 2009, March 2011 and March 2012); and a groundwater expert meeting (July 2010). Those initiatives brought like-minded organizations (both national and international) and experts together to discuss potential solutions and strategies to protect and sustainably utilize groundwater in the valley. Another notable initiative is the filing of a lawsuit by Pro Public (http://propublic.org) in 2003 to ensure implementation of the law requiring permits for industrial and commercial groundwater abstraction and requiring that groundwater be used only for drinking and sanitation. This was in the context of the government's failure to enforce the Water Resources Act (1992), which created a situation of indiscriminate groundwater exploitation through illegal deep wells operated by factories, hotels, housing colonies and water markets. In response, the Supreme Court in 2010 issued a mandamus ordering the government agencies to follow the provisions of the Water Resources Act (1992), enforce the licensing system, and control the illegal exploitation of groundwater.

In addition, a number of newspaper articles and news clips on water shortage and the adverse impact of water quality on health appear every year, especially from March to the start of the monsoon season, and are helping raise the profile of groundwater in the policy agenda. Rainwater harvesting and recharge projects in the Kathmandu Valley through international NGOs like the United Nations Development Programme (e.g. Invaluable Drops - Recharge Kathmandu Project, 2011–2012, and the Community-Led Water Resource Management Project, 2012–2013) and WaterAid (a project with the Urban Environment Management Society) have also exerted more pressure on the authorities to take steps towards firm control of the valley's groundwater resources. A number of delegations and formal complaints to the KUKL by the community regarding uneven distribution of drinking water have put pressure on the authorities to act quickly. On top of that, the conflict between the community and water entrepreneurs in Gothatar, culminating in a strike by tanker and bottled-water associations, has convinced the authorities that the trouble may become unmanageable if not acted on quickly. The aforementioned initiatives of the informal sectors highlight the urgency of actions aimed at regulating and protecting groundwater through state control of the resource.

Institutional, legal and policy reforms

With the aim of wise development and conservation of groundwater, a number of reforms in institutional set-ups, laws and policies have been made since the late 1960s. Table 2 presents a chronology of institutional reforms in this regard. The very first institution, the Department of Groundwater Survey (established in 1968), was renamed the Groundwater Resources Development Board (GWRDB) in 1976 and mandated as the primary government agency responsible for groundwater investigation, basic data generation, monitoring and development. However, its effectiveness was limited to groundwater development for irrigation in the southern plains of the country through its implementation of the GWRD

Table 2. Summary of pre-existing institutional set-ups relating to groundwater in the Kathmandu Valley, Nepal (modified from Pandey et al., 2011).

Institutions	Established	Groundwater-related functions
1. Ministry of Irrigation (formerly Ministry of Water Resources)		
• Groundwater Resources Development Board	1976	• Oversees policies and plans related to groundwater development and implements them through its project offices located in different parts of Nepal
• Department of Irrigation	1987	• Develops groundwater resources for irrigation
2. Ministry of Physical Planning and Works		
• Department of Water Supply and Sewerage	1972	• Supplies drinking water from surface and underground sources in the areas outside the coverage of NWSC and KUKL
• Nepal Water Supply Corporation	1990	• Regulates, operates and manages drinking water supply (surface and groundwater) in urban centres outside the Kathmandu Valley
• Kathmandu Valley Water Supply Management Board	2006	• Regulates, manages and protects groundwater resources in the Kathmandu Valley; issues licenses to use groundwater for commercial purpose
• Kathmandu Upatyaka Khanepani Limited	2007	• Produces and supplies water from surface and underground sources in the Kathmandu Valley
3. Ministry of Industry		
• Department of Mines and Geology	1977	• Conducts geological surveys and maintains database
4. Ministry Environment	1995	• Protects groundwater quality
• Department of Hydrology and Meteorology	1988	• Collects, stores, manages and releases climate data
5. Ministry of Energy	1975 (1981)	• Serves as an apex body for water resources policy and planning
• Water and Energy Commission (Secretariat)		
6. District Water Resource Committee (most of the water-sector-related government agencies in the district are represented in this committee)	1993	• Issues licenses for the use of water resources (including groundwater)

Note: NWSC: Nepal Water Supply Corporation; KUKL: Kathmandu Upatyaka Khanepani Limited.

Project. Later, the Water and Energy Commission (established 1975) and its Secretariat (established 1981) were mandated to accelerate the integrated development of water and energy. Some of its mandates regarding groundwater development and management overlap with those of the GWRDB. In 1992, the Water Resources Act vested ownership of water resources in the state. The District Water Resource Committee was established in 1993 to regulate water resource use through permits and licenses. Despite this, no groundwater licenses were ever issued. With the lack of a clear mandate in a single authority for regulating and managing groundwater resources and also because of the overlapping groundwater-related responsibilities of several government agencies, groundwater regulation and management remained "nobody's responsibility". In 2004, the GWRDB developed a groundwater bill and sent it to Parliament. It included a provision to establish a Groundwater Authority to be in charge of groundwater data collection and processing, use planning, regulation, monitoring, research and management. Unfortunately, the Parliament dissolved and the bill remained pending. As the latest reform to transform the government's role towards resource custodian and information provider, the KVWSMB was established in 2006 as the institution with sole authority for groundwater regulation, licensing, management and protection in the Kathmandu Valley. The following section highlights the KVWSMB's initiatives.

KVWSMB initiatives towards groundwater monitoring and regulation

The KVWSMB first prepared a Strategic Plan (for 2008–2025). In 2010, it drafted a Groundwater Management and Regulation Policy, which came into effect in late 2012. The policy emphasizes recovering depleted groundwater resources by controlling haphazard abstractions, protecting groundwater quality and promoting rainwater harvesting and recharge programmes. From 2009, the KVWSMB concentrated on locating groundwater users, issuing licenses, and identifying wells for groundwater monitoring. In October 2009, it banned groundwater abstraction from below a depth of 100 m without a license. Out of an estimated 700-plus deep wells, the KVWSMB had issued 25 licenses by January 2011; this increased to 100 in March and 206 in December. Also, from August 2011, it started monitoring water level and quality at 41 wells. It then stepped up measures to book illegal groundwater users. As of December 2011, it has carried out inspections of various companies and five-star hotels and fined them up to NPR 50,000 each for illegal use. As a post-licensing step, the KVWSMB is preparing to restrict groundwater abstraction in critical zones and launching groundwater-recharge programmes.

It may take some time for full-fledged state control of groundwater, even though the policy is in effect. An enabling environment could be created by strengthening the KVWSMB with sufficient resources (for daily business and for research activities) and by providing adequate support to implement the groundwater management policy.

Discussion and summary

Evidence from across the world confirms that groundwater is no longer considered an infinite and hidden resource. Its occurrence, distribution, recharge, flow and contamination have become better understood in recent decades. Increasing scientific understanding of groundwater systems, together with advocacy from informal sectors, are supporting and encouraging government agencies in informed decision making. They are also helping transform the role of government agencies towards "resource custodian" and "information provider" rather than "supply developer". As an example, this study has described how science and advocacy are helping change perceptions of groundwater and its management in the Kathmandu Valley, Nepal.

In this area, groundwater, which used to be considered an infinite and open-access resource during the early stage of its abstraction some four decades ago, is now considered a state-controlled resource. The shallow aquifer has a huge potential to store and supply groundwater; however, significant efforts will have to be made to tap that potential and address supply gaps in the valley without overexploiting the resource. Several indicators (discussed under "Groundwater as a depleted resource") confirm that groundwater depletion has become common because of groundwater development without adequate consideration to management. Groundwater-based water markets, both legal and illegal, are causing further depletion. Though existing laws consider water a "state-owned" resource, efforts to bring the resource under the state's control had long been out of priority. This shows that not only the availability of laws but also an enabling environment for their implementation matter very much if we are to accomplish the goal of groundwater management. Efforts towards that end were stepped up only after the groundwater gradually changed its status from an open-access resource to an over-exploited, depleted, degraded and vulnerable resource. Since the KVWSMB came into existence in 2006, the role of this government agency has been changing towards "resource custodian" and "information provider". Substantial progress in this direction, however, has yet to be seen. As of now, groundwater is partially under state control, and the agency is struggling hard for full control. Low priority assigned to groundwater in national policies and programmes is the main cause. Keeping a groundwater bill, tabled in Parliament in 2004, pending until now (making it practically invisible); establishing a government agency with full authority for groundwater regulation and control only in 2006, after the influence of donor agencies; starting sustainable efforts to address the issue of overexploitation only in the late 2000s, even though it was noticed in the mid-1980s – these cases among others clearly speak of the low profile of groundwater in the political agenda. And though the Kathmandu Valley is just one location, there is no doubt that it is representative of many aquifers around the world.

This study also shows interesting examples where initiatives from informal sectors, through both quiet advocacy and aggressive advocacy, have played a vital role in raising the profile of groundwater in the political agenda. The success of Pro Public in urging the Supreme Court of Nepal to issue a mandamus ordering government agencies to follow the provisions of the Water Resources Act, enforce the licensing system and control illegal exploitation of groundwater can be considered an excellent example of how advocacy may help enforce existing legal provisions and safeguard the public interest. The initiatives from informal sectors are playing influential roles in raising awareness among concerned stakeholders of the urgency to regulate and protect the groundwater in the valley.

Responses to reduce the stress on groundwater resources could be "hard-path", "soft-path", or a combination of the two. As an example of the hard-path approach, the Melamchi Water Supply Project was initiated in the early 2000s with the aim of augmenting water supply in the valley from sources outside the valley to reduce dependency on groundwater. However, this costly project has yet to yield fruit and is expected to take several more years. Currently, social issues at and around the water source are delaying the project, and in future the cost associated with its use may encourage the valley people to continue groundwater use. Therefore, designing and implementing some "soft-path" approaches, a transition already occurring in many parts of the world (Gleick, 2003), might be one of the wise solutions to reduce stress on the valley's groundwater resources. Without a doubt, one of the goals of management interventions is ensuring the sustainable use of groundwater resources. Because the current status of groundwater sustainability in the area is at the lower end of the sustainability scale, several "soft-path" measures are suggested. Examples include: improvements in spatial and temporal coverage of groundwater monitoring; investment in research to enhance understanding of the dynamics of groundwater and its recharge system; the management of existing data/information/knowledge and their dissemination; groundwater regulation using markets and pricing; and policies and strategies for public participation in groundwater management and conservation activities. Another measure could be the promotion of rainwater harvesting. A couple of studies (e.g. Shrestha, 2009) have demonstrated and advocated for rainwater harvesting and aquifer recharge enhancement as an attractive measure to quench the thirst of people in the area and alleviate stress on the valley's groundwater resources. However, they require adequate incentives to local landowners and groundwater users, together with good planning, design and operation and appropriate monitoring to ensure that the selected method is effective and suitable (Foster & Chilton, 2003). Whichever soft-path measure or set of measures is selected, their overall goal should be to encourage efficient use, equitable distribution of the resource and sustainable system operation over time (Gleick, 2003), all of which must support meeting this basic human need.

However, we should learn from the failures of management intervention in places like Spain, where the legal situation of groundwater is uncertain almost 20 years after the 1985 Water Law vested ownership in the state (Fornes et al., 2007). A profusion of illegal wells has become a newer constraint which the existing legal and technical instruments seem unsuited to solving. The lesson is that political willingness is essential to success. Another example, the Mancha Occidental Aquifer, in the Upper Guadiana Basin in Spain, suggests that command-and-control approaches alone may not be adequate; active stakeholder involvement and adequate frameworks for social learning are also key elements in devising workable groundwater management plans (Martinez-Santos et al., 2008; Pahl-Wostl, 2006). Attention to these factors has led to more successful management approaches in neighbouring areas, such as with the Mancha Oriental Aquifer. Science will never know all there is to know. Therefore, we must devise our strategies applying the best of what is known today, and at the same time keep flexible enough to allow for what we have yet to learn. Instead of determining a fixed sustainable yield, managers should recognize that yield varies over time as environmental conditions vary (Sophocleous, 1997). A strong public-education programme may help improve understanding of the nature and complexity of groundwater resources and emphasize how this understanding must form the basis for operating conditions and constraints. In the context of the Kathmandu Valley, a public–private partnership modality may also work better, as evident in the community forestry sector. Communities having resources (spring sources or deep groundwater wells) in their geographical boundary have already started advocating and practising for sharing the resource with private entrepreneurs and/or KUKL. The Godavari community water supply, Matatirtha community water supply and Siddhipur community water supply are some of the successful examples in this regard. Similarly, many village development committees and district development committees have already started to work in that model with tanker operators, mineral-water plants and bottled-water plants to share the resource.

Acknowledgements

The authors would like to acknowledge the Global COE Program of the University of Yamanashi and the Takeda Foundation in Japan for financial support, and several organizations in Nepal (the Center of Research for Environment Energy and Water, the Kathmandu Valley Water Supply Management Board and the Groundwater Resources Development Project) for supporting this study with data, information and other logistics.

References

Acres International. (2004). Optimizing water use in Kathmandu Valley (ADB-TA) project, Final Report. Submitted to Government of Nepal, Ministry of Physical Planning and Works, Acres International in association with Arcadis Euroconsult Land and Water Product Management Group, East Consult (P) Ltd. & Water Asia (P) Ltd.

Binnie and Partners. (1988). Water supply for Kathmandu and Lalitpur from outside the valley. Final report on feasibility study, Appendix-L (Groundwater resources within the valley).

Burke, J. J., & Moench, M. H. (2000). *Groundwater and society: Resources, tensions and opportunities*. New York, NY: United Nations Publication ST/ESA/205.

Burke, J. J., Sauveplane, C., & Moench, M. (1999). Groundwater management and socio-economic responses. *Natural Resources Forum,* 23(4), 303–313.

Chapagain, S. K., Pandey, V. P., Shrestha, S., Nakamura, T., & Kazama, F. (2010). Evaluation of deep groundwater quality in Kathmandu Valley using multivariate statistical techniques. *Water Air and Soil Pollution,* 210(1–4), 277–288.

Chapagain, S. K., Shrestha, S., Nakamura, T., Pandey, V. P., & Kazama, F. (2009). Arsenic occurrence in groundwater of Kathmandu Valley, Nepal. *Desalination and Water Treatment,* 4(1–3), 248–254.

Custodio, E. (2002). Aquifer overexploitation: What does it mean?. *Hydrogeology Journal,* 10(2), 254–277.

De Carvalho, S. C. P., Carden, K. J., & Armitage, N. P. (2009). Application of a sustainability index for integrated urban water management in Southern African cities: Case study comparison - Maputo and Hermanus. *Water SA (Online),* 35(2), 44–151.

Dhakal, H. P. (2010). *Groundwater depletion in Kathmandu Valley: Need for management.* A paper presented at International Forum for Sustainable Asia and Pacific (ISAP 2010), Sharing groundwater issues in Asian countries, 12–13 July, 2010. Yokohama, Japan.

Dixit, A., & Upadhya, M. (2005). *Augmenting groundwater in Kathmandu Valley: Challenges and possibilities*. Kathmandu, Nepal: Nepal Water Conservation Foundation (NWCF).

Fornés, J., De la Hera, A., Llamas, M. R., & Martínez-Santos, P. (2007). Legal aspects of groundwater ownership in Spain. *Water International,* 32(4), 676–685.

Foster, S. S. D., & Chilton, P. J. (2003). Groundwater: The processes and global significance of aquifer degradation. *Philosophical Transactions of the Royal Society of London B,* 358, 1957–1972.

Foster, S. S. D., Chilton, P. J., Moench, M., Cardy, W. F., & Schiffler, M. (2000). Groundwater in rural development: facing the challenges of supply and resource sustainability. Technical Paper 463. Washington, DC: World Bank.

Gautam, R., & Rao, G. K. (1991). Groundwater resources evaluation of the Kathmandu Valley. *Journal of Nepal Geological Society,* 7, 39–48.

Gleick, P. H. (2003). Global freshwater resources: Soft-path solutions for the 21st Century. *Science,* 302, 1524–1528.

JICA. (1990). *Groundwater Management project in Kathmandu valley,* Final report Main report and supporting reports, November 1990. Japan International Cooperation Agency.

Kemper, K. E. (2004). Groundwater – From development to management. *Hydrogeology Journal,* 12(1), 3–5.

Kharel, B. D., Shrestha, N. R., Khadka, M. S., Singh, V. K., Piya, B., Bhandari, R., Shrestha, M. P., Jha, M. G., & Munstermann, D. (1998). *Hydrogeological conditions and potential barrier sediments in the Kathmandu Valley.* Final Report of the Technical cooperation project – Environment Geology, between HMG Nepal and Federal Republic of Germany.

Knegt, J. W. F., & Vincent, L. F. (2001). From open access to access by all: Restating challenges in designing groundwater management in Adhra Pradesh, India. *Natural Resources Forum,* 25, 321–331.

Llamas, M. R., & Martinez-Santos, P. (2005). Intensive groundwater use: Silent revolution and potential source of social conflicts. *ASCE Journal of Water Resources Planning and Management,* 131(5), 337–341.

Llamas, M. R., Martinez-Santos, P., & de la Hera, A. (2006). The manifold dimensions of groundwater sustainability: An overview. In S. Ragone, A. de la Hera, & N. Hernandez-Mora, (Eds.), *Proceedings of the international symposium on groundwater sustainability* (pp. 105–116). Westerville, OH: National Ground Water Association Press.

Martínez-Santos, P., De Stefano, L., Martínez-Alfaro, P. E., & Llamas, M. R. (2008). Wetland restoration in the Mancha Occidental aquifer, Spain: A critical perspective on water, agricultural and environmental policies. *Restoration Ecology,* 16(3), 511–521.

Metcalf and Eddy. (2000). Urban water supply reforms in the Kathmandu valley (ADB TA Number 2998-NEP), Completion report, Volume I & II – Executive summary, main report and Annex 1 through 7. Metcalf and Eddy, Inc. with CEMAT Consultants Ltd., 18 Feb 2000.

Morris, B. L., Lawrence, A. R., Chilton, P. J., Adams, B., Caylow, R. C., & Klinck, B. A. (2003). *Groundwater and its susceptibility to degradation: a global assessment of the problems and options for management.* UNEP Early Warning & Assessment Report Series RS, 03–3, Nairobi, Kenya

Mukherji, A. (2006). *Non-market interventions in private water market transactions: Evidence from West Bengal, India.* In Proceedings of International Symposium on Groundwater Sustainability (ISGWAS) (pp. 299–319). Madrid, Spain: Instituto Gelogico Y Minero de Espana.

NGIIP. (2004). *National geographic information infrastructure programme (NGIIP).* Nepal: Department of Survey.

NGOFUWS. (2006). *Traditional stone spouts enumeration, mapping and water quality.* Unpublished report by NGO Forum for Urban Water and Sanitation (NGOFUWS), Kathmandu, Nepal.

Pahl-Wostl, C. (2006). The importance of social learning in restoring the multifunctionality of rivers and floodplains. *Ecology and Society,* 11(1), 10.

Pandey, V. P., Chapagain, S. K., & Kazama, F. (2010). Evaluation of groundwater environment of Kathmandu Valley. *Environmental Earth Sciences,* 60(6), 1329–1342.

Pandey, V. P., & Kazama, F. (2011). Hydrogeologic characteristics of groundwater aquifers in Kathmandu Valley, Nepal. *Environmental Earth Sciences,* 62(8), 1723–1732.

Pandey, V. P., & Kazama, F. (2012). Groundwater storage potential in the Kathmandu Valley's shallow and deep aquifers. In S. Shrestha, D. Pradhananga, & V. P. Pandey (Eds.), *Kathmandu Valley Groundwater Outlook* (pp. 31–38). Kathmandu, Nepal: Asian Institute of Technology (AIT), The Small Earth Nepal (SEN), Center of Research for Environment Energy and Water (CREEW), and International Research Center for River Basin Environment-University of Yamanashi(ICRE-UY).

Pandey, V. P., Shrestha, S., Chapagain, S. K., & Kazama, F. (2011). A framework for measuring groundwater sustainability. *Environmental Science and Policy,* 14(4), 396–407.

Pathak, D., Hiratsuka, A., Awata, I., & Chen, L. (2009). Groundwater vulnerability assessment in shallow aquifer of Kathmandu Valley using GIS-based DRASTIC model. *Environmental Geology,* 57(7), 1569–1578.

Pradhanang, S. M., Shrestha, S. D., & Steenhuis, T. S. (2012). Comprehensive review of groundwater research in the Kathmandu Valley, Nepal. In S. Shrestha, D. Pradhananga, V. P. Pandey (Eds.), *Kathmandu Valley Groundwater Outlook* (pp. 6–18). Kathmandu, Nepal: Asian Institute of Technology (AIT), The Small Earth Nepal (SEN), Center of Research for Environment Energy and Water (CREEW), and International Research Center for River Basin Environment-University of Yamanashi(ICRE-UY).

Ramtel, A. (2010). *Groundwater resource dries up in the Kathmandu Valley.* A blog posted on guthi.net. Retrieved from http://guthi.net/cms/index.php?option=com_content&task=view&id=686&Itemid=59

Schwartz, F. W., & Ibaraki, M. (2011). Groundwater: A resource in decline. *Elements,* 7(3), 175–179.

Shah, T. (1993). *Groundwater markets and irrigation development: Political economy and practical policy.* Bombay, India: Oxford University Press.

Shah, T., Burke, J., & Villholth, K. (2007). Groundwater: A global assessment of scale and significance. In D. Molden (Ed.), *Water for food, water for life: A comprehensive assessment.* London: Earthscan, and Colombo: International Water Management Institute (IWMI).

Shrestha, R. R. (2009). Rainwater harvesting and groundwater recharge for water storage in the Kathmandu Valley. *Sustainable Mountain Development,* 56, 27–30. ICIMOD, Winter 2009.

Shrestha, D. 2011. *State and services of private water tanker operation in Kathmandu.* (Unpublished M.Sc. Thesis), Nepal Engineering College, Pokhara University, Nepal.

Shrestha, S. D. (2012a). Geology and hydrogeology of groundwater aquifers in the Kathmandu Valley. In S. Shrestha, D. Pradhananga, & V. P. Pandey (Eds.), *Kathmandu Valley Groundwater Outlook* (pp. 21–30). Kathmandu, Nepal: Asian Institute of Technology (AIT), The Small Earth Nepal (SEN), Center of Research for Environment Energy and Water (CREEW), and International Research Center for River Basin Environment-University of Yamanashi (ICRE-UY).

Shrestha S., Pradhananga D., & Pandey V. P. (Eds.). (2012). *Kathmandu Valley Groundwater Outlook.* Kathmandu, Nepal: Asian Institute of Technology (AIT), The Small Earth Nepal (SEN), Center of Research for Environment Energy and Water (CREEW), and International Research Center for River Basin Environment-University of Yamanashi(ICRE-UY).

Sophocleous, M. (1997). Managing water resources systems–Why "safe yield" is not sustainable. *Ground Water,* 35(4), 561.

Tachikawa, T., Hato, M., Kaku, M., & Iwasaki, A. (2011). *The characteristics of ASTER GDEM version 2.* IGARSS, July 2011.

Zektser, I., & Everett, J. (2004). *Groundwater resources of the world and their use.* UNESCO–IHP VI, Series on Groundwater No. 6. UNESCO, Paris. ISBN 92-9220-007–0.

Pathways for effective groundwater governance in the least-developed-country context of the Lao PDR

Paul Pavelic, Ounakone Xayviliya and Oualaphone Ongkeo

Instituting effective groundwater governance is highly challenging in a least developed country such as the Lao PDR where groundwater resources need to be developed for a variety of reasons but the levels of understanding, awareness and technical capacity are extremely limited. This paper discusses the current state of knowledge and management of groundwater and suggests some pathways forward. Whilst the level of governance remains very low, there is growing interest in tackling the entrenched technical and non-technical issues and constraints. Various initiatives have recently emerged, resulting in positive institutional change.

Introduction

Recognition of the importance of groundwater in achieving a broad range of social, economic and environmental benefits has increased at the global scale in recent years. Consequently, a good deal of attention has been directed towards groundwater governance as a paradigm by which to manage groundwater resources more holistically and inclusively (Foster & Garduño, 2013; Moench, Kulkarni, & Burke, 2012; Sharma, Villholth, & Mukherji, 2009; Villholth, 2006). Governance serves as a vehicle for improving the planning, development, operation and management of groundwater. It is largely about the process of decision making and the exercising of powers at all levels, from the individual user up, and ideally allows all stakeholders the space to exercise their political, economic and administrative authority and/or rights (Moench et al., 2012). Governance processes work well when they take into account the views of all stakeholders with an interest in resource development and sustainability issues and are translated into actions that seek to be effective, transparent and equitable (Foster & Garduño, 2013). The converse is of course equally true.

Although the overarching goals of the groundwater governance approach mentioned above are obviously worthwhile, there has been a clear emphasis on problems related to the consequences of overdevelopment, shaped by the thinking that has emerged from well-documented cases. However, for those actively engaged with groundwater issues in the world's most underdeveloped countries, somewhat removed from the possibility that extreme development will emerge as a major issue over the foreseeable future, many of the basic requirements for achieving an adequate level of governability are limited or absent, and a different set of perspectives and priorities may emerge.

Of the world's 193-odd countries, the 49 categorized by the United Nations as the least developed countries (LDCs) have a combined population of 851 million people (UNDESA-PD, 2011), or more than 10% of the global population. LDCs are characterized by their low incomes, poor state of human welfare and high economic vulnerability. The LDC that is the subject of this article is the Lao People's Democratic Republic (otherwise simply known as Laos). Groundwater has historically been highly under-represented in the water resources and related sectors in Laos, although this is gradually changing. This paper attempts to consolidate and document the current state of knowledge, along with current challenges, gaps and pathways forward for the country. It is not only intended to inform policy and decision making within the country but could also provide some level of guidance and perhaps a useful point of reference for other LDCs.

Country overview

Laos, on the Indochinese Peninsula in Southeast Asia, is home to around 6.5 million people. In this tropical location, rainfall is high, ranging from 1300 to 3700 mm per annum, but largely concentrated within the four-month wet season. The country is also mountainous, with about 50% of the native forest still retained and only 9% of the land under agriculture. It is these and other resources that provides the engine for socio-economic development and a means to lift many out of entrenched poverty and malnutrition, as Laos strives to rise above LDC status by the year 2020 (Government of Lao PDR, 2004). Most of the population earn their livelihoods as subsistence farmers, predominantly from the production of rice. One-third of the people live below the international poverty level of USD 1.25 per day, particularly those in upland villages that are remote from the major economic corridors. Rapid economic growth has been achieved in recent years, with support from a large number of donors, which has contributed positively to poverty alleviation and improvement in general health and well-being, as indicated by declines in child and infant mortality and in the prevalence of communicable diseases. The water resources sector has a vitally important role in the government's vision of Laos becoming the 'battery of South-East Asia' to feed local demand for electricity and that of neighbouring countries such as Thailand and Vietnam.

Developing water resources effectively to achieve short-term socio-economic development targets, whilst at the same time meeting longer-term sustainability goals, is a major challenge. An overall water policy has yet to be implemented, although incremental efforts are being made at the national level (see e.g. GHD, 2010) and also at the wider Mekong level (Mekong River Commission [MRC], 2012).

Is groundwater vital in a water-rich country?

In global as well as regional terms, Laos is a water-rich country. Per capita availability of renewable resources is more than 50,000 m^3 – at least 25 times that of India, for example, and the highest of the four Lower Mekong River Basin (LMRB) countries (Food and Agriculture Organization of the United Nations [FAO], 2013). Most of the available water resources support non-consumptive activities such as hydropower generation, navigation and fisheries, and thus possible future expansion of consumptive uses such as domestic water supply and irrigation is viable. Given Laos's position almost entirely within and at the core of the LMRB, much attention has naturally been given to surface-water management issues. Studies have tended to focus on issues associated with the numerous hydropower projects that are planned or underway (e.g. Lebel, Bastakoti, & Daniel, 2010; Molle, Foran, & Käkönen, 2009).

The relative abundance of water within major rivers and surface storages has meant that acute and widespread water shortages and competition have not necessarily been major concerns (Lacombe et al., 2011). However, owing to the marked seasonality and the extended dry season, parts of the country face regular shortfalls and occasionally droughts. Surface-water bodies, particularly smaller watercourses in mountainous areas, are prone to becoming polluted and to drying up for months at a time. Villages on upland terrain away from surface watercourses cannot always gain convenient access to water. Without locally available supplies from wells, they incur the burden of fetching domestic water, which is usually left to women and children (Chithtalath, 2007).

Traditionally, limited consideration has been given to groundwater resources. Some of the redeeming features of groundwater include a largely ubiquitous presence, large buffering capacity from climate shocks, relatively low establishment cost, on-demand water availability and high conveyance efficiency (Villholth, 2006). Over the past few decades, groundwater development has progressed quietly and steadily, with mixed success, under very rudimentary institutional arrangements. Groundwater is an important source of domestic supplies across most rural areas of Laos (Johnston et al., 2010) and is also an important source for small-town water supplies. Chanthavong (2011) reports that 10 of the 60 township areas rely on groundwater or spring flow which, combined, provide 20% of total water production. In the larger urban centres, where distributed surface-water supplies now cover most areas, bottled water from groundwater sources, sold by small and large private operators using their own infrastructure, is the preferred source for drinking from public health and aesthetic viewpoints. Soft drink and beer companies draw upon fresh groundwater reserves for production, whilst salt producers target the vast saline reserves stored within ancient sea beds for local consumption as well as for industrial uses. The use of irrigation water drawn from wells is limited but slowly emerging. Smallholder famers within the Mekong floodplain and in some favourable highland settings are using groundwater at the garden scale for local needs. Household-level survey data (personal communication, Dr J. Newby, University of Queensland, 2011) indicate that in Champassak Province about 70% of households have access to wells that are used for domestic uses, livestock and irrigation. In Savannakhet a similar picture emerges, with the corresponding figures ranging from 60% to 90%. In upland areas with good aquifers such as the Bolaven Plateau in Champassak Province, groundwater is beginning to be used to irrigate high-value cash crops for export. Groundwater also supports environmentally sensitive ecosystems by feeding springs and baseflow to rivers, streams and wetlands that are the natural resource base for the ecosystems that maintain the habitats of fish and aquatic vegetation harvested by communities. The intrinsic linkages between surface water and groundwater are not always clear but must be taken into account in water allocation planning processes. In the uplands where difficult terrain does not allow groundwater to be directly accessed, it is often the steady discharge of groundwater into streams and springs that can supply water for villages year round (Billving & Ågren, 2007).

Drivers of change

Overall, groundwater is widely used on a limited scale at present but still contributes significantly towards the health and well-being of most of the population of Laos. It is likely to be increasingly important for national development across most major socio-economic sectors. Groundwater may play a key role in addressing three major challenges facing the country.

Millennium Development Goals for improved water services

The government has clearly expressed its ongoing commitment to achieving its obligation towards the Millennium Development Goals (UN General Assembly, 2000). Targets include 80% provision of water supply nationally and 75% coverage in rural areas by 2015 (Nam Saat, 2012). Achieving this requires considerable upscaling of effort, particularly in remote areas with the highest levels of poverty. The success of the programme hinges on the provision of safe and reliable domestic supplies, which largely come from groundwater accessed via tube-wells or shallow wells that are sealed from polluted surface water to prevent contamination.

National and local development plans

Socio-economic development targets, detailed in the 7th National Socioeconomic Development Plan for 2011–2015 and the Strategy for Agricultural Development, 2011–2020 by the Ministry of Agriculture and Forestry, link development within the water resources and agriculture sectors with better harnessing of the country's abundant natural resources. Food and nutritional security, as well as longer-term development goals, are reliant on sustainable agricultural intensification that, in turn, hinges upon increasing the productivity of land and the total land area under irrigated agriculture. Currently the total irrigated area is just 224,000 ha – more than an order of magnitude less than Thailand or Vietnam, and on par with Cambodia (Hoanh et al., 2009). Even though Laos is a major rice producer, productivity increases are still needed in the poorest districts. Economic growth targeted towards poorer members of the community has a far larger impact on poverty alleviation than general economic growth. In that sense, it is well known that irrigation contributes to poverty alleviation and livelihood improvements through productivity gains, and that the use of groundwater, along with a range of other small-scale agricultural water interventions, provides a more cost-effective, democratic and pro-poor alternative to large-scale formal irrigation systems (Johnston et al., 2011; Moench, 2003). Research to establish the scope of these opportunities is currently underway in Laos (International Water Management Institute [IWMI], 2013).

Climate change adaption

Issues associated with intra- and inter-seasonal climate variability and climate change are universally recognized as necessitating greater and more diverse forms of storage to mitigate risks. Surface and subsurface storage, used in ways that are both conjunctive and strategic, is central to mitigating the risks (McCartney & Smakhtin, 2011). As a result, greater expectation is being placed on the capacity of groundwater resources to supplement and diversify the portfolio of water sources and thereby increase the resilience of communities in urban and rural areas with respect to current and future climate-related uncertainties. Groundwater resources are generally more resilient to climate impacts than surface waters, although not completely immune (Shresta & Kataoka, 2008).

Groundwater resources, availability and utilization

Laos is divided into two major geological regions: the Annamian region, in the north and east of the country; and the Indosinian region, mainly along the Mekong River. The Annamian region includes granitic and metamorphic rocks. The Indosinian region consists of sandstone, siltstone, shale, mudstone, limestone, conglomerate and basalt. Groundwater is contained within shallow and deep aquifers that are composed of fractured hard rock, consolidated marine or terrestrial sediments and alluvial/fluvial deposits along major watercourses. Groundwater is thought to be available in many areas, but aquifers are often very localized and variably productive (Landon, 2011). Well yields and water quality are highly dependent on the local geology and are highly variable spatially due to complex depositional and post-depositional processes. For example, in the uplands of Khamounan Province, pumping rates for individual wells can vary from 1–2 L/s for sandstones and siltstones to 10–20 L/s for limestones. In the fresh alluvial aquifers of the Vientiane Plains that overlie rock-salt layers, pumping rates are modest (1–3 L/s). Even wells spaced less than 100 metres apart can vary significantly in productivity and usability, as demonstrated in fractured rock (shale) terrain in Vientiane Capital Province (Nielsen, 2013).

The FAO's country-level AQUASTAT database estimates the annual groundwater replenishment for Laos at around 37,900 MCM/y, equivalent to 20% of the surface-water resources generated internally in the country (FAO, 2013). In areal-equivalent terms this amounts to 160 mm/y of groundwater recharge across the country, on average. The estimated groundwater use from rural and provincial water supplies (Figure 1), combined with irrigated agriculture (based on irrigated-area data from Siebert et al. (2010) assuming 2000 mm/y of irrigation), amounts to just 0.2% of the annual replenishment. Whilst the crudity of these estimates is acknowledged, the figures suggest that at the macro scale the resource is grossly underutilized, with significant scope for expansion in high-value economic sectors such as irrigation and industry, provided that continued reliability and expansion can be assured for the domestic-use sector.

Groundwater demand is dependent upon its productivity and accessibility and its attractiveness compared with alternative sources of supply. Areas where groundwater can have the greatest impact at the household or community levels for livelihoods and the economy appear to correlate with areas where the development potential is high and population density is highest (Figure 1). These areas are situated on the lowland floodplains of the Mekong and include large parts of the provinces of Vientiane Capital, Vientiane, Savannakhet, Saravan and Champassak. Some of these areas have limited surface-water supplies, and groundwater would offer a supplementary source. Development opportunities may also exist within the valleys of upland areas, which are sparsely populated and where groundwater may not be as abundant, yet there is a high need for good-quality, reliable water supplies in these communities.

Challenges in managing a 'hidden' resource

Groundwater is a largely invisible resource that is seen only by those who specifically look for it and is thus easily neglected, misunderstood, mismanaged or taken for granted. This, combined with the ease of access in some cases, can create issues through the perception that private, individual access equates to a sense of outright ownership of groundwater. In reality, groundwater is a common-pool resource that must be shared by multiple users. As a result, problems can emerge with falling water levels, increased pumping costs, reduced water quality, and competition or even conflict amongst competing water users.

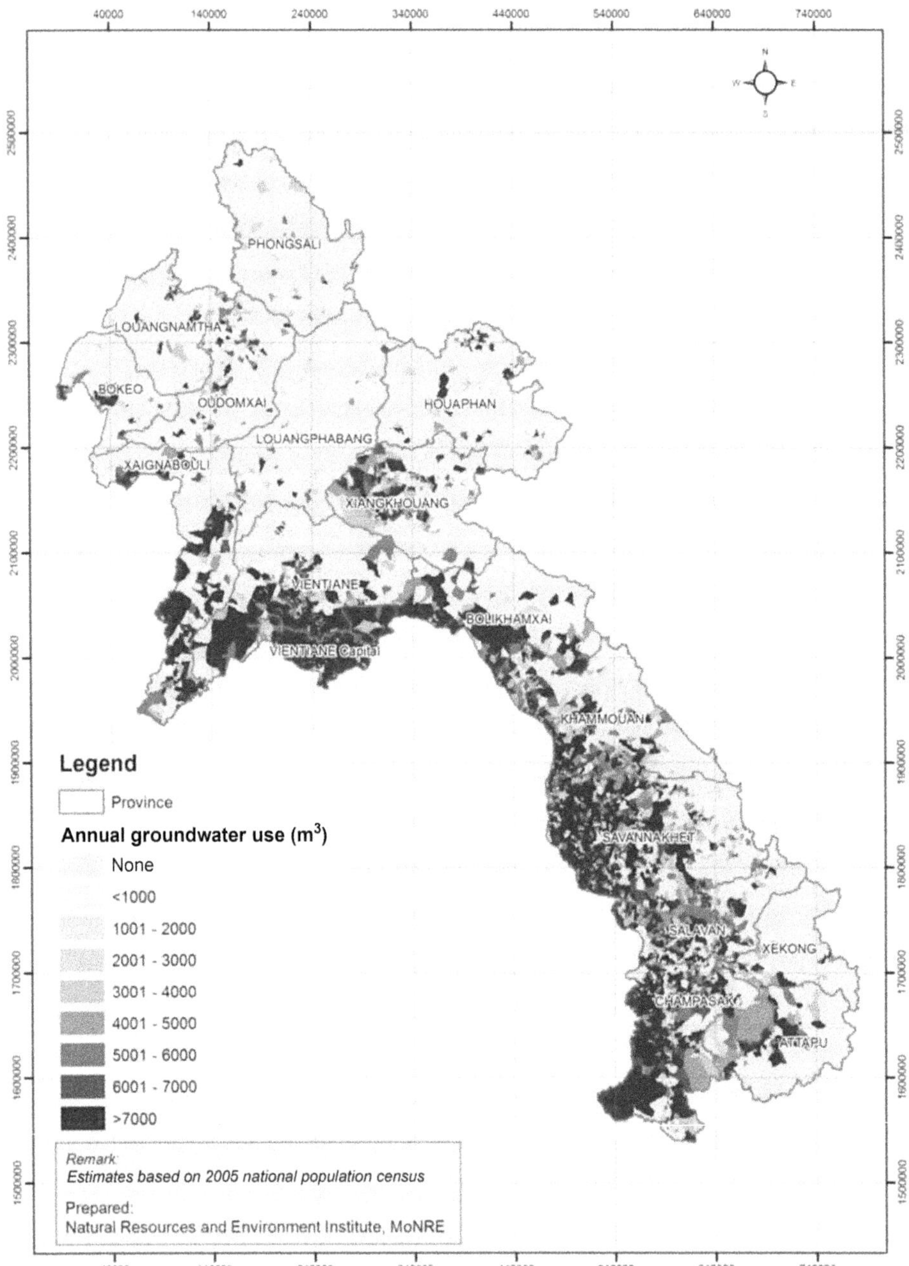

Figure 1. Distribution of groundwater use by householders in Laos, estimated from village-level census data in 2005 for the proportion of the population using groundwater and an assumed per capita use of 100 L/day.

Source: Natural Resources and Environment Institute.

Well-known and documented examples of such issues include those in the densely populated and agriculturally productive plains of India and North China (Giordano, 2009). Even within the LMRB region, localized and less well-known examples emerge. In parts of southern Cambodia recent rapid and unregulated growth of small-scale groundwater irrigation has raised concerns about competition with domestic supplies (International Development Enterprises [IDE], 2005). In the Mekong Delta region, over-pumping for irrigation, industrial and domestic uses is resulting in a lowering of the water table and saltwater intrusion (IUCN, 2011).

Reflection upon such cases evokes a number of questions. Some of the more fundamental and pertinent to Laos (and perhaps other LDCs more generally) include:

- In a scenario where greater utilization of groundwater is both necessary and likely over the longer term, how can use be encouraged, particularly for those groups most in need, without unduly compromising long-term sustainability and while maintaining environmental services?
- Can the level of risk in developing and operating water wells for water supply be reduced?
- Does significant and irreversible groundwater exploitation even pose a serious threat for a country characterized by high rainfall and relatively low population density?
- What is the level of governance needed to sustainably manage the resource, and what are the trade-offs across different community groups if it is not?

These questions also serve to provide useful points of reference for the remainder of this paper.

The present state of affairs

A number of methods have been developed to define the degree of groundwater governance. The 'checklist' type of approach described by Foster et al. (2010) is perhaps the most straightforward and appropriate for this case. These authors identify a set of 20 indicators of successful governance, grouped into four categories: technical; legal and institutional; cross-sector policy coordination; and operational success. Applying it to Laos reveals that only one of the indicators ("a dedicated groundwater agency") is addressed in a meaningful way, with two other technical indicators ("basic hydrogeological maps" and "aquifer body delineation") addressed to a limited degree. Thus, there is no shortage of issues that need to be addressed. Major issues include a lack of knowledge and understanding of the resource, making development risky; low capacity amongst field technicians and drillers; absence of monitoring network or reliable hydrogeological maps; and poor understanding of water quality (Table 1).

A subset of specific indicators that are easily enumerated and verifiable were selected to reveal the level of progress in knowledge generation and application within the groundwater subsector (Table 2). For example, there are currently no PhD-qualified personnel teaching groundwater-related topics in any academic institution; those teaching relevant graduate-level courses were trained at the MSc level in other disciplines. All five PhD or master's theses published to date were written by international students from developed countries enrolled in universities outside of Laos. The absence of take-up can be at least partly explained by the lack of accreditation for relevant postgraduate programmes at the National University of Laos or other, smaller universities. Students usually

Table 1. Major issues identified for the groundwater subsector in Laos.

Issue	Reference
General lack of knowledge of the resource amongst those tasked to manage it (except at the broadest scale and in more detail at a few localities) and a general absence of hydrogeological maps (again except in a few areas), making development a high-risk option	Knudsen et al. (2004), Chanthavong (2011)
Lack of capacity amongst field technicians/practitioners, resulting in poor drilling practices; lack of sufficient data collection and sound technical advice; high cost of groundwater development and high running and maintenance costs of bores	ADB and GHD (2012)
Lack of understanding of groundwater at the community level, leading to misconceptions and costly and/or failed projects	ADB and GHD (2012)
Inadequate or fragmented data records and limited information systems, leading to ineffective planning and evaluation	IWMI (2013)
Absence of a national or even regional monitoring-well network to support planning and development; hydrogeological maps to provide the basic tools for the initial planning of any investment are also largely absent	IWMI (2013)
Inadequate coordination between the concerned stakeholders, including within relevant government agencies, to pool efforts and to manage groundwater effectively	IWMI (2013)
Laws and regulations that are far removed from field realities (e.g. absence of a well permitting system)	GHD (2010)
Presence of potential water-quality problems and associated human health effects in specific regions from water constituents such as arsenic, iron, manganese, salinity, hardness, barium and suspended solids (turbidity)	Fengthong et al. (2002), Chanpiwat et al. (2011), Chanthavong (2011)
Rapid and poorly planned urbanization, combined with poor sewerage systems and population growth, resulting in contamination of shallow groundwater	Chanthavong (2011)
Scepticism and associated curtailment of investments due to lack of knowledge of locating water wells of sufficient quantity and quality for domestic supplies	Knudsen et al. (2004), Chanthavong (2011)
Absence of longer-term commitment after the life of a project, even where there would be clear benefits from doing so	Landon (2011)

Table 2. Indicators for the groundwater subsector in Laos.

Indicator	Number
PhD-qualified academics teaching groundwater-related curricula	0
Specialists practising permanently in the county	1
Papers published in peer-reviewed journals on the topic	3
Postgraduate theses produced at universities within Laos	0
Postgraduate theses produced at universities outside Laos	5
Conference papers published	6
Private drilling companies operating in the country	4

must attain those qualifications from regional universities or further afield. None of the three peer-reviewed journal papers produced to date was lead-authored by a Lao national, and only one of the six conference papers was. Whilst we have gone to considerable effort to identify and screen what limited information is available, the data used are not necessarily fully inclusive for every indicator. Still, the findings drawn from this analysis are believed to be robust and valid.

Current participating actors

An institutional map for the groundwater subsector has yet to be prepared; Table 3 is offered as a substitute. It lists actors within the government, international donors, academia/research and the private sector, segregated according to their major functions: development, research/investigation and management. The table only considers those actors known to be directly involved and who we may consider to be active 'participants' in the subsector.

Table 3. Indicative matrix of identified actors participating in different aspects of the groundwater subsector.

	Role		
Type	Development	Research/Investigation	Management
Government	DHUP, NS	DGM, DI, DWR, NAFRI, NREI	DWR, NREI, PONRE
International donors	ADB, AusAID, IFAD, JICA, SIDA, SNV, UNICEF, WB, WHO, WV	ACIAR, ADB, CF	ADB, IFC, WB
NGOs	ADRA, CDEA, DIC, SC, WE		
Academic, research		CSU, GWRC-KKU, ICE WaRM, IGES, IWMI, KTH, LUT, MRC, NUOL, UTS	
Private	BWSs, DCs, FMs, HPCs, LBC, MCs, RHHs	GHD, HG, NC	

Note. ACIAR: Australian Centre for International Agricultural Research; ADB: Asian Development Bank; ADRA: Adventist Development & Relief Agency; AusAID: Australian Agency for International Development; BWSs: bottled water suppliers; CDEA: Community Development and Environment Association; CF: Crawford Foundation; CSU: Charles Sturt University; DCs: drilling companies; DGM: Division of Groundwater Management of the DWR; DHUP: Department of Housing and Urban Planning; DI: Department of Irrigation; DIC: Deseret International Charities; DWR: Department of Water Resources; FMs: farmers; GHD: Gutteridge Haskins & Davey; GWRC-KKU: Groundwater Research Center at Khon Kaen University; HG: Hydrogeosci; HPCs: Hydropower companies; ICE WaRM: International Centre of Excellence in Water Resources Management; IFAD: International Fund for Agricultural Development; IFC: International Finance Corporation; IGES: Institute for Global Environmental Strategies; IWMI: International Water Management Institute; JICA: Japan International Cooperation Agency; KTH: KTH Royal Institute of Technology; LBC: Lao Brewing Company; LUT: Luleå University of Technology; MCs: Mining companies; MRC: Mekong River Commission; NAFRI: National Agriculture and Forestry Research Institute; NC: Norconsult; NREI: Natural Resources and Environment Institute; NS: National Centre for Environmental Health and Water Supply (Nam Saat); NUOL: National University of Laos; PONRE: Provincial Office of Natural Resources and Environment; RHHs: rural householders; SC: Save the Children; SIDA: Swedish International Development Cooperation Agency; SNV: SNV Netherlands Development Organisation; UNICEF: United Nations Children's Fund; UTS: University of Technology Sydney; WB: World Bank; WE: World Education; WHO: World Health Organization; WV: World Vision.

We see more than 40 participating institutions, even without specifically identifying many of the individual private users. The number and distribution of listings in the table is unlikely to be a good reflection of the magnitude of effort going into any particular domain space. However, it may provide a qualitative indication of the degree of interest. For example, we see a good distribution in each of the three functional roles.

There are of course a wider group of stakeholders that may directly or indirectly affect or be affected by the actions of the participants but do necessarily involve themselves in sector-centred management processes. The policies and practices of the energy sector, for example, were earlier identified as being central to groundwater development in the Mekong floodplains. Other stakeholders may be within conservation, tourism, agriculture, fisheries, mining or other sectors. The roles of participants or other stakeholders for some institutions may be dependent upon the particular circumstances.

The role of the government institutions

In Laos, various ministries, departments and divisions, operating independently or jointly with their line agencies at the provincial and district levels, are active in groundwater development and management under their respective mandates. The roles of the major government entities are summarized as follows.

- *Department of Water Resources (DWR).* A potentially significant development for management in the groundwater subsector is the recent establishment (in 2012) of the Division of Groundwater Management (DGM) within the DWR, a natural progression following the establishment of the Ministry of Natural Resources and Environment (MoNRE). It should be noted that at least two other divisions of the DWR are also tasked with groundwater-related duties. Whilst the method of implementation of the DGM (and related divisions) is still emerging, the general terms of reference are clearer: (1) to develop and manage groundwater resources (from the quantity and quality perspectives) at national, river-basin, provincial and district levels, in cooperation with line agencies; (2) to propose permits for resources access and use; and (3) to establish a database system and provide advice on technical matters, monitoring, evaluation and support activities to line agencies and other sectors.
- *Natural Resources and Environment Institute (NREI).* The mandate of the DGM is large and difficult to achieve in isolation; cooperation with national and international organizations is essential. Support is provided by the NREI, also under the MoNRE, which is tasked with providing technical and research-oriented support to the DGM.
- *Nam Saat.* The National Centre for Environmental Health and Water Supply, commonly known as Nam Saat, under the Ministry of Public Health, is the lead agency to facilitate and guide the rural water supply and sanitation sector in Laos. The provision of improved water supplies at the village and community levels must also involve a strong focus on planning and developing groundwater resources.
- *Department of Housing and Urban Planning (DHUP).* The DHUP, under the Ministry of Public Works and Transport, is tasked with planning and supply in towns and other urban centres. In some towns, this includes the provision of groundwater and spring flow. In urban areas groundwater is generally used when there is no alternative source (Chanthavong, 2011), which may be due to the energy needed to lift water, maintenance costs for maintaining effective performance in

boreholes, or risks to shallow groundwater due to natural or anthropogenic pollution.

- *Department of Geology and Mines.* The Department of Geology and Mines is a key information-creating agency through ongoing geological mapping efforts and the tools that are produced.
- *Department of Irrigation (DI).* The DI does not have a formal role in the groundwater subsector at present, but is currently exploring the possibility of utilizing groundwater resources to boost irrigation potential through two of the projects listed in Table 4 and referred to below.

This collective ensemble of government actors, with their overlapping sets of mandates and responsibilities, has the potential to create barriers for progress in terms of information and knowledge sharing amongst sectors and administrative areas and levels. However, if harnessed effectively, the converse could also be true.

Signs of progress

The low level of groundwater governance described earlier does not necessarily mean that the situation in Laos is bleak. Many promising signs have emerged in recent years that conscious and dedicated efforts are being made, and improvements are slowly starting to emerge.

There are seven ongoing or recently completed studies that represent investments of at least several million US dollars per year and have the potential to accelerate progress in groundwater governance. The characteristics of those projects are set out in Table 4. We see that one project focuses on technical resource assessment and monitoring needs at the regional level, another on national management and capacity building, three on agricultural water use, one on climate-smart agriculture and climate-resilient water supplies, and one on legal and regulatory aspects. Whilst significant in their own right, they are indicative of the higher-level goals of the government and donors to achieve sustainable development and management of land and water resources.

Past efforts by various organizations, although not explicitly recognized in Table 4, have helped shape the existing programmes. The drilling investigations and documentation of production wells by the Japan International Cooperation Agency in several provinces (e.g. Vientiane Province; Japan International Cooperation Agency [JICA], 1993) stand out. Another is the work of the Mekong Committee Secretariat (Johnson, 1983), and finally the Online Support and Training system described by Knudsen et al. (2004) and arising from the Asian Development Bank–funded Lao Water Supply and Sanitation Sector Project.

Conclusions and pathways forward

Laos is an LDC on a rapid development trajectory whose political economy is strongly focused on developing its water resources for hydropower and irrigation. The priority to date has been on surface-water development, which is appropriate; however, for a host of reasons that include the need to strengthen the resilience of water supply and agricultural production systems to climate variability and shocks, improved knowledge of groundwater resources is required.

Effective and sustainable management of groundwater systems means maintaining ongoing productivity of the resource for the suite of different water users where issues such as water-related conflicts, excessive use, aquifer degradation and groundwater-

Table 4. Overview of the current and recently completed projects.

	Donor	Partnering institutions	Period	Major aims	Further details
National integrated water resources management support program Component 3	AG, GS	ADB, DWR, GHD, NREI	2012–2013	Assess GW resources as a component of IWRM; propose improved monitoring arrangements and develop an initial national framework and action plan to manage GW	ADB and GHD (2012)
Agricultural groundwater use in Laos	ACIAR	DI, DWR, GWRC-KKU, IGES, IWMI, MDC, NREI, NUOL-FES, NUOL-WRED, THPC	2012–2015	Develop technologies and strategies to enhance use of shallow GW for irrigation by smallholder farmers to increase their food and nutritional security and livelihoods	http://gw-laos.iwmi.org
Developing improved farming and marketing systems in rainfed southern Laos (water and hydrology component)	ACIAR	CIAT, CSU, IRRI, NAFRI, UQ	2009–2014	Integrated water resources assessment; map flood- and drought-prone areas; trial water-saving measures; strengthen community management	Vote, Zeleke, and Eberbach (2013)
National water resources law revisions	IFC, WB	DWR	2012–2017	Revise the national laws for water resources for ratification by government and legislative drafting	http://www.monre.gov.la/dwr/index.php?option=com_content&view=category&layout=blog&id=108&Itemid=182&lang=en
Assessment of GW in the LMB		MRC, USGS	2012	Collate and review hydrogeological information within the Lower Mekong Basin	Landon (2011)
Capacity enhancement for coping with climate change	ADB	DDMCC, DWR	2010–2013	Relevant pilot demonstrations of CCA in two watersheds in Vientiane and Saravan Provinces, with associated capacity development	http://www.adb.org/projects/43443-012/main

(*Continued*)

Table 4. (Continued)

	Donor	Partnering institutions	Period	Major aims	Further details
Rapid appraisal of agricultural groundwater use in the four member countries	MRC	CWRPI, DGR, DI, MRD	2013–2014	Establishing protocols and building capacity to evaluate the potential and sustainability of agricultural groundwater use	http://www.mrcmekong.org/about-the-mrc/programmes/agriculture-and-irrigation-programme/

Note: ACIAR: Australian Centre for International Agricultural Research; ADB: Asian Development Bank; AG: Australian government; CCA: climate change adaptation; CIAT: International Center for Tropical Agriculture; CSU: Charles Sturt University; CWRPI: Center for Water Resources Planning and Investigation (Vietnam); DDMCC: Department of Disaster Management and Climate Change; DGR: Department of Groundwater Resources (Thailand); DI: Department of Irrigation; DWR: Department of Water Resources; GHD: Gutteridge Haskins & Davey; GS: Government of Spain; GW: groundwater; GWRC-KKU: Groundwater Research Center, Khon Kaen University; HG: Hydrogeosci; IFC: International Finance Corporation; IGES: Institute for Global Environmental Strategies; IRRI: International Rice Research Institute; IWMI: International Water Management Institute; IWRM: integrated water resources management; MDC: Mekong Development Center; MRC: Mekong River Commission; MRD: Ministry of Rural Development (Cambodia); NAFRI: National Agriculture and Forestry Research Institute; NREI: Natural Resources and Environment Institute; NUOL-FES: National University of Laos, Faculty of Environmental Sciences; NUOL-WRED: National University of Laos, Water Resources Engineering Department; THPC: Theun-Hinboun Power Company; UQ: University of Queensland; USGS: United States Geological Survey; WB: World Bank.

supported ecosystem decline do not become serious constraints. The implementation of groundwater management strategies or action plans is usually hindered by the complexity of aquifer systems and a lack of data.

This paper shows that the level of governance within the groundwater subsector in Laos is still in its infancy and somewhat commensurate with the brief history of modern development and generally low level of demand being placed on these resources. At the same time, this should not be taken to downplay their current importance and the even more critical importance in coming years.

Laos could now be seen as at a crossroads in groundwater management. The level of interest and activity is unprecedented, and institutional adjustments are taking place driven by government strategies and policies supported by donors and other development partners that create a sense of expectation that it can help deliver on the Millennium Development Goals and economic growth targets, and adapt to climate change. These present not only challenges but also distinct opportunities to improve the awareness and management of groundwater resources and to anticipate, prevent and mitigate potential problems. If technical and institutional improvements are not made they could threaten the resource and the array of benefits received by current and future groundwater users.

The most pressing problems and priority areas over the coming years relate to: (1) boosting technical capacity within government and non-governmental organizations; (2) regional groundwater mapping and monitoring, supported by a centralized database; (3) resource assessments to better understanding the nature and dynamics of the groundwater; and (4) awareness-raising of the stakeholders at all levels.

Governance itself is far from a stationary process and must continually adapt to changing information, priorities and circumstances. It should be noted that governments and other relevant authorities do not consciously choose to neglect the management of groundwater resources; it is more a case of existing priorities being directed elsewhere. With the increasing recognition of the value of groundwater arising from the national strategies and projects mentioned above comes greater willingness to make the investments needed to improve governance.

As the groundwater subsector in Laos matures and emerges with more clarity over the coming years, new sets of issues will emerge that may be more closely associated with optimizing development and better managing the externalities that may emerge with enhanced groundwater use. The general principles for effective governance would appear to be reasonably well founded and documented (see e.g. the GW-MATE series of publications at http://water.worldbank.org/gwmatepubs), and many approaches have been tried and tested. Formulating the 'correct' set of policies and strategies remains a challenge that has yet to be solved in many developed and developing countries around the world, but examples of 'workable' arrangements do exist, and policy makers, researchers, practitioners and donors in Laos should learn from the failures and take heart from the successes in other regions.

Acknowledgements

The work presented herein was made possible through the ACIAR-supported project: Enhancing the Resilience and Productivity of Rainfed Dominated Systems in Lao PDR through Sustainable Groundwater Use (Ref. No. LWR/2010/081) and contributes to the CGIAR research programs on Water, Land and Ecosystems and Climate Change, Agriculture and Food Security. The authors thank Mr Nicholas Lombardi (consultant for GHD) and Ms Anna (Snowy) Haiblen (IWMI) for many useful discussions. We are also appreciative of the invitation by Dr John Dore and Dr Lilao Bouapao to discuss the ideas for this paper in June 2013 as part of the 2nd Mekong Ganga Dialogue, hosted by the M-POWER network.

References

ADB and GHD. (2012, January 26). Working paper on training needs assessment for ADB CDTA 7780. Package 3 - groundwater assessment. Unpublished report.

Billving, A., & Ågren, A. (2007). Water resources in the cultivated uplands of Lao PDR: A case study of Huay Maha village. (Master's Thesis). Ref. LWR-EX-07-28, KTH Royal Institute of Technology, Department of Land and Water Resources Engineering, Stockholm, Sweden.

Chanpiwat, P., Sthiannopkao, S., Cho, K. H., Kim, K. W., San, V., Suvanthong, B., & Vongthavady, C. (2011). Contamination by arsenic and other trace elements of tube-well water along the Mekong River in Lao PDR. *Environmental Pollution, 159*(2), 567–576. doi:10.1016/j.envpol.2010.10.007

Chanthavong, P. (2011). Groundwater for water supply in Laos. On behalf of the Ministry of Public Works and Transport, Department of Housing Urban Planning. Presented at the launch meeting, Regional Water Knowledge Hub for Groundwater Management, 2–3 June 2011, Bangkok, Thailand. Retrieved from http://www.iges.or.jp/en/natural-resource/groundwater/index.html

Chithtalath, S. (2007). Community development projects and the status of ethnic minority women in the Moksuk-Tafa area, Bokeo province, Laos. *Community Development Journal, 42*(3), 299–316.

FAO. (2013). AQUASTAT country database – Lao PDR water resource sheet. Food and Agriculture Organization of the United Nations. Retrieved from http://www.fao.org/nr/water/aquastat/main/index.stm, last accessed 20 September 2013.

Fengthong, T., Dethoudom, S., & Keosavanh, O. (2002). Drinking water quality in the Lao People's Democratic Republic. Country report prepared for UN ESCAP – IWMI seminar on Environmental and public health risks due to contamination of soils, crops, surface and groundwater from urban, industrial and natural sources in South East Asia. Hanoi, Vietnam, 10–12 December.

Foster, S., & Garduño, H. (2013). Groundwater-resource governance: Are governments and stakeholders responding to the challenge? *Hydrogeology Journal, 21*(2), 317–320. doi:10.1007/s10040-012-0904-9

Foster, S., Garduño, H., Tuinhof, A., & Tovey, C. (2010). Groundwater governance: A conceptual framework for assessment of provisions and needs. *GW•MATE Strategic Overview Series Number 1*. Washington, DC: World Bank.

GHD. (2010). Updating the national water resources policy & strategy, ADB TA 7013-LAO. Mid-term report, volume 1.

Giordano, M. (2009). Global groundwater? Issues and solutions. *Annual Review Environmental Resources, 34*, 7.1–7.24.

Government of Lao PDR. (2004). Lao People's Democratic Republic: Poverty reduction strategy paper. Country report no. 04/393. Washington D.C: IMF. Retrieved from: http://planipolis.iiep.unesco.org/upload/Lao%20PDR/PRSP/Lao%20PDR%20PRSP.pdf

Hoanh, C. T., Facon, T., Thuon, T., Bastakoti, R. C., Molle, F., & Phengphaengsy, F. (2009). Irrigation in the Lower Mekong Basin countries: The beginning of a new era? In F. Molle, T. Foran, & M. Käkönen (Eds.), *Contested waterscapes in the Mekong Region: Hydropower, livelihoods and governance* (pp. 143–172). London, UK: Earthscan.

International Development Enterprises. (2005). Strategic study of groundwater resources in Prey Veng and Svay Rieng (Phase 1): Final report, Selia Task Force Secretariat (IFAD Loan No. 623-KH), Phnom Penh. Retrieved from: http://www.ide-cambodia.org/index.php?option=com_php&Itemid=37&lang=en

IUCN. (2011). Groundwater in the Mekong Delta. Discussion paper. International Union for the Conservation of Nature. Retrieved from: http://cmsdata.iucn.org/downloads/iucn___groundwater_in_mekong_delta_25_may_11_w.pdf

IWMI. (2013). Project website. International Water Management Institute. Retrieved from http://gw-laos.iwmi.org/ Last accessed:10 September 2013.

Japan International Cooperation Agency. (1993). Design study report on the Project for Groundwater Development in Vientiane Province in Lao People's Democratic Republic. Japan International Cooperation Agency.

Johnson, J. H. (1983). Preliminary appraisal of the hydrogeology of the Lower Mekong Basin. Interim Committee for Coordination of Investigations of the Lower Mekong Basin.
Johnston, R., Lacombe, G., Pavelic, P., Hoanh, C. T., Lefroy, R., & Fry, C. (2011). Managing water in rainfed agriculture in the GMS. Report prepared by International Water Management Institute (IWMI) for Swedish International Development Agency (SIDA), October 2011.
Johnston, R. M., Hoanh, C. T., Lacombe, G., Noble, A., Smakhtin, V., Suhardiman, D., ... Choo, P. S. (2010). Rethinking agriculture in the Greater Mekong Sub-region: how to sustainably meet food needs, enhance ecosystem services and cope with climate change. International Water Management Institute, 26p. doi:10.3910/2010.207
Knudsen, J. B. S., Ruden, F., & Smith, B. T. (2004). The online support and training project for the groundwater sector of Lao PDR. Proc. 30th WEDC International Conference "People-centred approaches to water and environmental sanitation", Vientiane, Lao PDR, 2004, pp. 434–437.
Lacombe, G., Douangsavanh, S., Thepphavong, B., Hoanh, C. T., Bounvilay, B., Noble, A., ... Phongpachith, C. (2011). Is there enough water in the Vientiane Plain? A water balance assessment of the Lower Nam Ngum Basin. Report prepared by International Water Management Institute (IWMI) for Commonwealth Scientific and Industrial Research Organisation (CSIRO) - Australian Agency for International Development (AusAID) Research for Development Alliance.
Landon, M. (2011). Preliminary compilation and review of current information on groundwater monitoring and resources in the Lower Mekong River Basin. U.S. Geological Survey Report to the Mekong River Commission, Version Dated July 2011.
Lebel, L. Bastakoti, R. C., & Daniel, R. (Eds.). (2010). Enhancing multi-scale Mekong water governance. CGIAR Challenge Program on Water and Food, CPWF Project report PN 50.
McCartney, M., & Smakhtin, V. (2011). Water storage in an era of climate change: Addressing the challenge of increasing rainfall variability. International Water Management Institute Blue Paper. Retrieved from: http://www.iwmi.cgiar.org/Publications/Blue_Papers/PDF/Blue_Paper_2010-final.pdf
Mekong River Commission. (2012, December). Lao National Indicative Plan (2011–2015) for Implementation of the IWRM-Based Basin Development Strategy (Final). Retrieved from: http://www.mrcmekong.org/assets/Publications/strategies-workprog/Lao-PDR-NIP-final-Dec042012.pdf
Moench, M. (2003). Groundwater and poverty: Exploring the connections. In R. Llamas & E. Custodio (Eds.), *Intensive use of groundwater: Challenges and opportunities*. Lisse, The Netherlands: Swets and Zeitlinger B.V.
Moench, M., Kulkarni, H., & Burke, J. (2012). Trends in local groundwater management institutions. Thematic paper 7 of Groundwater Governance: A Global Framework for Country Action, GEF Project ID 3726 [Food and Agriculture Organization of the United Nations (FAO), Global Environment Facility (GEF), International Association of Hydrologists (IAH), UNESCO-International Hydrological Programme (IHP), World Bank].
Molle, F., Foran, T., & Käkönen, M. (Eds.) (2009). *Contested waterscapes in the Mekong Region: Hydropower, livelihoods and governance* (448p). London: Earthscan.
Nam Saat (Centre for Environmental Health and Water Supply, Ministry of Health, Lao PDR). (2012). National Plan of Action for rural water supply, sanitation and hygiene. Rural Water Supply and Environmental Health Sector.
Nielsen, G. (2013). Groundwater evaluation, Somsanga Drug Rehabilitation Center, Vientiane, Lao PDR. Unpublished report by Deseret International Charities, Vientiane, Lao PDR, October 2013.
Sharma, B. R., Villholth, K. G., & Mukherji, A. (2009). The international training and research program on groundwater governance in Asia: Theory and practice. CPWF Project Report for Project Number: PN 42, March, 2009. International Water Management Institute, Colombo: Sri Lanka.
Shresta, S., & Kataoka, Y. (2008). Groundwater and climate change: No longer the hidden resource. In: *Climate change policies in the Asia-Pacific: re-uniting climate change and sustainable development*, series. (Ch. 7, pp. 159–184). Hayama: IGES.
Siebert, S., Burke, J., Faures, J. M., Frenken, K., Hoogeveen, J., Döll, P., & Portmann, F. T. (2010). Groundwater use for irrigation – a global inventory. *Hydrology and Earth System Science Discuss, 7*, 3977–4021. doi:10.5194/hessd-7-3977-2010
UN General Assembly. (2000, September 18). *United Nations millennium declaration.* Resolution A/RES/55/2. Retrieved from: http://www.un.org/millennium/declaration/ares552e.pdf

UNDESA_PD. (2011). *World population prospects: The 2010 revision*. New York: United Nations, Department of Economic and Social Affairs, Population Division. Retrieved from: http://esa.un.org/wpp/Documentation/pdf/WPP2010_Volume-I_Comprehensive-Tables.pdf

Villholth, K. G. (2006). Groundwater assessment and management: Implications and opportunities of globalization. *Hydrogeology Journal, 14*, 330–339. doi:10.1007/s10040-005-0476-z

Vote, C., Zeleke, K., & Eberbach, P. (2013). Assessment of groundwater resources in the Sukhuma District, Champasak Province, Southern Lao PDR. 2012/2013 Progress report for ACIAR Project No. CSE/2009/004, water and hydrology component. Charles Sturt University.

Using backcasting to explore ways to improve the national water department's contribution to good groundwater governance in South Africa

P. Seward, Y. Xu and A. Turton

Backcasting defines a vision of a desirable future and then analyzes the actions required to realize that vision. This article uses backcasting to explore ways to improve the national government's contribution to good groundwater governance in South Africa. Using the South African Department of Water and Sanitation as a case study, it is found that lack of strategic thinking and implementation are the main impediments to service delivery. Although isolated cases of groundwater 'overuse' and pollution can be found in South Africa, the main governance issues are currently competition between users for groundwater use, and inordinate delays in granting authorization for that use.

Introduction

Globally, there has been a dramatic increase in groundwater use in the past half-century, bringing significant social and economic improvements but also creating many consequences, such as ecosystem damage, drying-up of surface water, and seawater intrusions (Custodio & Llamas, 2003; Konikow & Kendy, 2005; Llamas & Martínez-Santos, 2005). Negotiating the best compromise between the benefits and consequences of groundwater use is primarily a governance challenge rather than a technical issue (Custodio & Llamas, 2003; Foster & Garduño, 2013; Llamas & Martínez-Santos, 2005).

Known examples of good groundwater governance are rare (López-Gunn & Cortina, 2006; Ross & Martínez-Santos, 2010; Wester, Minero, & Hoogesteger, 2011). Groundwater governance research is still in its infancy compared with the physical science aspects of hydrogeology (Llamas, Mukherji, & Shah, 2006; Mukherji & Shah, 2005), and there is a lack of systematic, non-anecdotal data on groundwater governance (Faysse, Errahj, Imache, Kemmoun, & Labbaci, 2014). Significant secondary research has, however, recently been carried out under the banner of Groundwater Governance: A Global Framework for Action, with numerous regional consultations held, and thematic reports issued (see e.g. Moench, Kulkarni, & Burke, 2012). This secondary research has proposed various general frameworks for improving groundwater governance, and has confirmed the prevailing wisdom that there is no blueprint for good governance that will work everywhere (Foster & Ait-Kadi, 2012; Llamas et al., 2006; Moench et al., 2012; Mukherji & Shah, 2005).

Although isolated cases of good local groundwater governance can be found where national-level bureaucracies play no role (Mukherji & Shah, 2005; Taher, Bruns, Bamaga, Al-Weshali, & van Steenbergen, 2012; van Steenbergen, 2006), there is a consensus that support by national public service departments can play a very important role in good groundwater governance (Aarnoudse, Bluemling, Wester, & Qu, 2012; López-Gunn & Cortina, 2006; López-Gunn, Willaarts, Rica, Corominas, & Llamas, 2013). (In South Africa, national, provincial and local government departments are referred to collectively as the Public Service.)

Where national (or regional) government's role in groundwater governance is found to be dysfunctional or semi-dysfunctional, one of two assumptions is usually made – either that the level of bureaucratic dysfunctionality is unlikely to change and one must therefore plan around it (López-Gunn & Cortina, 2006), or that by stating what the bureaucracy *should* be doing it will somehow be able to transform itself into a functional bureaucracy (Department of Water and Sanitation [DWS], 2010; Pietersen, Beekman, & Holland, 2011). In South Africa substantial research has been conducted on policies assigned to the Public Service and their success or failure, yet relatively little has been conducted on the inner workings of public service departments (von Holdt, 2010). Thus, little or no research has been done on the implementation practicalities of how a bureaucracy could be transformed and how that transformation could increase the contribution the bureaucracy could make to the improvement of groundwater governance. The aim of this article is to address that research gap.

To address this research gap the potential role that backcasting (Dreborg, 1996) could play in making the relevant bureaucracy more functional was investigated. Backcasting has two core elements: (1) defining a vision of a desirable future; and (2) analyzing the actions required to realize this vision (van Vliet & Kok, 2013; Wangel, 2011).

The article is structured as follows. The following section outlines the major groundwater governance issues in South Africa. This is followed by an explanation of and a motivation for the backcasting methodology and how it will be used in this article. The sources of data are explained. The analysis then uses backcasting to work backwards from an idealized definition of good groundwater governance to the current governance landscape, and identifies and discusses the interventions required. Then, a discussion is provided regarding whether the research objectives have been met.

The South African groundwater governance landscape

The South African groundwater governance landscape was chosen as a case study to investigate the relevance of backcasting. South Africa's water legislation has been described as "progressive", "advanced", "forward-looking" and even "revolutionary" (Burns et al., 2006). Yet the tangible implementation of groundwater governance has been characterized as weak to non-existent, according to both perception (Seward, 2010) and research (Knüppe, 2011; Pietersen et al., 2011). This incongruity between forward-thinking legislation and the retrograde implementation of that legislation is common in many other countries where groundwater is important (Aarnoudse et al., 2012; López-Gunn & Cortina, 2006; Mukherji & Shah, 2005).

As a whole, groundwater is not over-exploited in South Africa, and cases of severely declining water levels or increasing salinity are rare. Governance problems are more to do with competing uses rather than intensive use. A major issue is the continued dominance of (white) commercial farmers in the use of groundwater; little has been done to address the inequities that existed under apartheid, and very little groundwater is allocated to previously disadvantaged individuals.

Competition over groundwater use is not being effectively addressed because either the governance process takes an inordinately long time, or the local institutions for addressing groundwater governance are 'toothless', or there are no local institutions in place.

Methodology

Backcasting was used by the Texas Water Development Board (Gleeson et al., 2012) to reach sustainability goals, referred to as "desired future conditions", for groundwater levels, storage and spring flows. Maximum pumping rates for the desired sustainability goals were then determined using groundwater flows. This is a highly technical form of backcasting, but does illustrate its key components: identifying a desired future; and then identifying what needs to be done to realize that future. Thus, backcasting, unlike forecasting, does not attempt to predict what is likely to happen. Instead, it describes a desired future end-point, and then works backwards to determine the feasibility of that end-point and the policies required to reach it (Robinson, 1990).

Backcasting is more commonly used for broad societal aims (Robinson, 1990) and for the specific business and strategic plans of corporations (Dreborg, 1996). Backcasting can be regarded as a very specific type of scenario studies, since both are concerned with descriptions of what might happen rather than trying to forecast the future. Whereas scenario studies in general might typically look at a *range* of future conditions – desirable and undesirable – in order to facilitate preparedness and planning for any of those conditions, backcasting looks specifically at *desired* future conditions to identify what could be done to realize those conditions (Dreborg, 1996).

A limitation in the application of backcasting is that it does not always pay sufficient attention to the role of institutions responsible for implementing the backcasting plans. In this article 'institution' will be used in its broadest sense to mean any set of individuals with some common purpose and rules. Thus an institution could be a government department, a catchment management agency, a water user association (WUA), or an informal advisory/pressure group. Gleick et al.'s (1995) proposed backcasting solutions for California's water issues, while full of plans and programmes for water management, pay scant attention to the institutional changes needed to implement these plans and programmes. Such neglect seems a common failing (Nilsson, Nilsson, Hildingsson, Stripple, & Eikeland, 2011) in many environmental issues. For these reasons backcasting needs to be applied to institutions and not just to the plans and programmes for groundwater, environmental and/or socio-economic scenarios. It can be difficult to envisage what the desired sustainability landscape will be in the future. In these situations backcasting the *processes* for dealing with future negotiations over sustainability issues seems a much more credible, pragmatic and useful exercise than trying to envision actual sustainability issues.

The approach to backcasting used in this article was therefore to backcast desired institutional *processes* for determining desired groundwater scenarios rather than attempting to backcast the scenarios themselves. The specific processes addressed were those needed to improve groundwater governance. The first step in the backcasting exercise was a definition of good groundwater governance, since progression towards good groundwater governance is impossible unless it is known what 'good groundwater governance' means.

The second step in the backcasting process was to investigate what was needed to link the definition of good groundwater governance to a more detailed and recognizable good governance process. For the purposes of good governance, groundwater needs to be considered at the local level (Foster, Garduno, Tuinhof, & Tovey, 2010). Since a universal model of a good groundwater governance process that can be defended by empirical evidence has yet to be established, a generic common-pool model in the form of the Ostrom Design Principles (Table 2) was utilized for this investigation. Cox, Arnold, and Villamayor-Tomás (2010) conducted a meta-analysis of 91 case studies that had used the Ostrom Design Principles, and concluded that the principles were well supported empirically. The utilization of common-pool governance rules is defensible since groundwater is a common-pool resource (Foster et al., 2010; Giordano, Brugnach, & Vurro, 2012; Ostrom, 2005) characterized by high subtractability (one person's use subtracts from another person's use) and low excludability (it is difficult to exclude additional users and additional use).

The third step in the backcasting process considers the implementability of any interventions identified in Steps 1 and 2. Implementability in this case depends on the ability of the public service department responsible for water issues to implement these interventions. In South Africa this is the Department of Water and Sanitation (DWS). For the interventions to be implemented requires that two potential constraints be overcome: (1) whether the interventions identified are compatible with existing DWS strategies; and (2) whether DWS has the capability to implement these, or any other, strategies. The compatibility of the proposed interventions was assessed by comparing them with the National Groundwater Strategy (NGWS) (DWS, 2010).

The capability of DWS to implement these strategies was then assessed by considering (1) the current state of DWS in terms of its organizational culture (Ravasi & Schultz, 2006), and (2) the strength of DWS implementation strategies when change is required. The aspect of organizational culture relevant to this article is shared assumptions regarding appropriate responses to given situations. For example, responding to an email might be regarded as optional, not compulsory. The current state of the DWS organizational culture was assessed from (1) personal experience of DWS from 1986 to 2013 while employed; (2) the perspective of the NGWS on human and institutional capacity; and (3) a generic assessment of the South African Public Service based on documented research. The assumption was that the generic Public Service assessment could be extrapolated to DWS and its groundwater sections, since DWS is a Public Service department.

The strength of strategies for implementing change was assessed from the same three sources, namely personal experience, NGWS perspective and documented research on the Public Service in general. A summary of the methodology is provided in Table 1.

If the interventions identified in Steps 1–3 of the backcasting analysis were deemed to be significantly different from those currently in place, then the case for backcasting was regarded as proven. If the interventions identified by backcasting were to a large extent already planned for, then the case for backcasting was regarded as unproven.

Data

This investigation makes use of two sources of data, knowledge and judgement: (1) the existing literature; and (2) practitioner experience. The practitioner is the principal author, who worked as a hydrogeologist for DWS from 1978 to 1983 and from 1986 to 2013. The

Table 1. Summary of methodology.

Step	Overall objective	Explanation
1	Definition	Obtain a *definition* or *vision* of good groundwater governance.
2	Process	Determine what is needed to connect the definition to the *process* of good groundwater governance. (For this investigation, 'process' was defined by the Ostrom Design Principles.)
3	Implementation	Determine what is needed to connect the process of good groundwater governance to the current situation. This requires an assessment of *implementation strategies.*

bulk of the practitioner's work involved providing groundwater inputs to groundwater, surface water, ecosystem and land use governance processes. While stakeholder-based inputs to the backcasting process may be desirable (Wangel, 2011), obtaining these inputs was beyond the constraints of this investigation. Practitioner-based enquiries also have value in research because without them the stock of knowledge on any subject remains incomplete (McIntyre, 2006).

Analysis

Clear goal: definition of 'good groundwater governance'

Good groundwater governance can clearly not be realized if there is no definition of what good groundwater governance comprises. Groundwater governance can be regarded as the process of negotiating trade-offs between the benefits and consequences for competing groundwater users/stakeholders in a given situation (Llamas et al., 2006). Thus, good groundwater governance can be an elusive concept, open to a wide range of interpretations (Wijnen, Augeard, Hiller, Ward, & Huntjens, 2012). However, for the purposes of this exercise the following definition was adopted:

> A 'good' groundwater governance environment is one where governance processes equitably reflect the voices and interests of stakeholders (including regional and global stakeholders with interests in resource sustainability) and where broadly supported courses of action can be implemented in an effective and equitable manner. (Moench et al., 2012)

This definition was chosen because it makes it clear that groundwater governance must be both equitable and effective, and must include all stakeholders. It is accepted that the stakeholders and institutions involved might prefer a different definition. The point here, however, is not to advocate any specific definition but to indicate that the draft National Water Resource Strategy (DWS, 2012) and the NGWS do not contain such a definition, and this absence is a serious constraint in realizing good groundwater governance. An intervention is thus needed, namely inserting a definition of good groundwater governance in the relevant national strategies.

Clear understanding of the groundwater governance process

Now that we have defined what good groundwater governance is, the definition needs to be unpacked in more detail to outline a recognizable process, so that the constraints on this process can be evaluated and interventions to overcome these constraints can be

Table 2. The Ostrom design principles (Ostrom, 2009).

Principle	Comment/explanation
1a. *User boundaries.* Clear and locally understood boundaries between legitimate users and nonusers are present.	In the case of groundwater the user boundary is understood to mean, for example, membership or not in a water user association (WUA), rather than having a clear water right or permit.
1b. *Resource boundaries.* Clear boundaries that separate a specific common-pool resource from a larger social-ecological system are present.	In the case of groundwater this could be an aquifer unit.
2a. *Congruence with local conditions.* Appropriation and provision rules are congruent with local social and environmental conditions.	Rules about using groundwater are consistent with the capacity of the resource to provide that supply, and consistent with local socio-economic norms.
2b. *Appropriation and provision.* Appropriation rules are congruent with provision rules; the distribution of costs is proportional to the distribution of benefits.	In the case of groundwater this could mean that the benefits of belonging to a WUA must exceed the costs and disadvantages, and that the benefits are fairly distributed.
3. *Collective-choice arrangements.* Most individuals affected by a resource regime are authorized to participate in making and modifying its rules.	
4a. *Monitoring users.* Individuals who are accountable to or are the users monitor the appropriation and provision levels of the users.	
4b. *Monitoring the resource.* Individuals who are accountable to or are the users monitor the condition of the resource.	
5. *Graduated sanctions.* Sanctions for rule violations start very low but become stronger if a user repeatedly violates a rule.	
6. *Conflict-resolution mechanisms.* Rapid, low-cost, local arenas exist for resolving conflicts amongst users or with officials.	
7. *Minimal recognition of rights.* The rights of local users to make their own rules are recognized by the government.	'Rules' here would mean rules about the management of a groundwater resource, rather than (just) the internal institutional operating rules of a groundwater WUA. Higher levels of government must provide at least *some* recognition of these rights.
8. *Nested enterprises.* When a common-pool resource is closely connected to a larger social-ecological system, governance activities are organized in multiple nested layers.	In an earlier formulation (Ostrom, 1990) this rule was explained as "Appropriation, provision, monitoring, enforcement, conflict resolution, and governance activities are organized in multiple layers of nested enterprises." In some cases this rule is needed because a local collective is unable to meet its objectives without external support. In other cases it is needed because a local collective might not take broader societal objectives into account without external guidance.

identified. A generic common-pool governance model in the form of the Ostrom Design Principles (Table 2) was used for this evaluation.

Two main constraints to the implementation of the Ostrom Design Principles were identified and are discussed below.

Delegating genuine water management authority to water user associations

This is required by Ostrom's Principle 7, minimal recognition of rights. In South Africa, WUAs have no right to curtail or otherwise manage water use but exist only to optimize the entitlements allocated to them by higher institutions. Their role is essentially limited to that of a watchdog for the higher-level institution or of providing inputs to that institution. The National Water Act (NWA) (Republic of South Africa [RSA], 1998) explains the role of WUAs somewhat confusingly: "Although water user associations are water management institutions their primary purpose, unlike catchment management agencies, is not water management."

WUAs in South Africa have the right to create and modify their rules, but these rules, unless delegated to them by a catchment management agency or the minister of water and sanitation, would not include water management. In other words, they may not allocate or authorize water use. In theory, such authority could be delegated to WUAs. However, given the paternalistic nature of DWS, an institution that is reluctant to devolve even quite minor functional duties and responsibilities from the national head office to regional branch offices, delegating this authority in practice appears to be a major constraint. Possible workarounds for this constraint could be: (1) the allocation of water use licences to WUAs, which then become responsible for subdividing this allocation to their members; or (2) DWS's accepting that the WUA has the moral right to make its own decisions, and then rubber-stamping these decisions when the WUA has no legal authority to do so.

Giving effect to the broad societal aims of groundwater governance

This issue depends on Ostrom's Principle 8, nested enterprises. While a strong WUA might be able to allocate and control consumptive use by its members, there would be concerns as to whether broader societal aims are being met. There would be concerns that a WUA, created around a community of interest, would focus on its interest – for example optimal economic gain from a groundwater resource – and neglect broader societal aspects of sustainability.

Groundwater development creates a range of benefits and a range of consequences that depend on how intensively development occurs (Custodio, 2002; Pierce, Sharp, Guillaume, Mace, & Eaton, 2013). Sustainable groundwater development (hereafter abbreviated to 'sustainability') represents a subjective, value-driven decision on the trade-off between these benefits and consequences for a given situation (Llamas et al., 2006). While the list of the benefits and consequences that could be considered is essentially limitless, Llamas, Martínez-Santos, and de la Hera (2007) pin down nine aspects of sustainability: hydrological, ecological, economic, social, legal, institutional, inter-generational, intra-generational and political. Farmers, ecologists, social engineers and hydrogeologists might have interpretations of sustainability for every aquifer system. There is no one right or generic answer for these diverse requirements. Local negotiations to pursue unique, local solutions are necessary.

The broad array of competing requirements could be addressed if spokespeople for those competing requirements were an integral part of the WUA. These spokespeople would then be 'agents' for the specific water 'use' they represent – e.g. non-consumptive use for aquatic ecosystems – and would thus be allowed to participate in and vote on WUA matters. According to Thompson (2006) such a broad definition of a water user is permissible with South Africa's NWA: any interested and/or affected party could theoretically be allowed to be a member of a WUA, not just those that are direct, consumptive users.

Institutional capacity to implement strategies

Strategic compatibility of proposed interventions

The compatibility of the proposed interventions was assessed by comparing them with the NGWS (DWS, 2010). This strategy contains a list of what the NGWS regards as the most important strategic actions to overcome serious deficiencies in service delivery. The overall essence of these strategic actions has been distilled by grouping and compaction. The NGWS is contained in a 64-page document, so the distillation (presented in Table 3) is cryptic. While the interventions identified by the backcasting exercise are not explicitly mentioned in the NGWS, there appears to be no good reason why, in theory, they could not be included. "Implement regulations, policies and strategies" seems more than general enough to accommodate the proposed interventions. It would therefore appear that the proposed interventions are compatible with existing institutional strategies.

Strategic implementability of the proposed interventions

Personal experience (1986–2013). A distillation of the principal author's experience of DWS culture is provided in Table 4.

These characteristics are interconnected. For example: *staff shortages* might be used as an excuse for not implementing a plan. This *avoids responsibility* for failing to

Table 3. Distillation of the National Groundwater Strategy.

Strategic action theme	Example
Implement regulations, policies and strategies	Process and evaluate all groundwater licence applications within six months.
Provide groundwater availability and use data at a standard commensurate with surface water data	Update figures on groundwater availability and use as new data become available.
Implement a human capacity building strategy for the groundwater sector	Mobilize private-sector support where necessary to capacitate regional offices.
Establish a groundwater operations and management support section within DWS	Improve the maintenance and operation of groundwater infrastructure across sectors.
Ensure internal and external coordination and cooperation	Improve cooperation and coordination within the Department of Water and Sanitation and between government departments and the private sector to leverage available capacity and resources.
Provide an effective information system	Develop and implement an integrated groundwater information system to support water services provision at the municipal level.
Support groundwater research	Regularly assess the impact of research investment in groundwater.

Table 4. Culture of the Department of Water and Sanitation, as observed by the principal author.

Characteristic	Explanation/examples
Fear (of authority)	Avoiding reprimands and threats of disciplinary actionAvoiding responsibilityA culture of blame rather than of problem solving
Petty administrative issues override service delivery	Complying with petty administration rules is the primary concern – for example, completing monthly reports is prioritized over service delivery.
Employment equity overrides service delivery	For example, jobs requiring specialized skills tend to be filled by candidates who meet employment equity requirements but have lesser skills than candidates who do not meet employment equity requirements.
Service delivery is secondary or incidental	For example, in-house 'team building', meetings, sports days and social functions are regarded as more important than service delivery.
Using staff shortages to justify shortcomings	Shortages are used to justify poor service delivery.
Anarchistic individual actions	Objectives in the management-by-objectives system are so vague that virtually any individual action can be put under an official objective – petty administration does not prevent anarchy.
Resistance to change	Continuing programmes that provide minimal service delivery cause little anxiety. Modest reforms to improve service delivery provoke anxiety. For example, continuing to monitor 500 boreholes with no discernible benefit to society caused little anxiety, but trying to improve 'efficiency' by rationalizing the network and closing five monitoring points caused profound anxiety.
Minimal understanding of sustainability	There is inability to understand that sustainability is a complex issue and that each case requires diverse inputs that could be in conflict with each other and require negotiated solutions.Departmental groupings are organized in 'silos' – for instance there are groups to address 'economic development', and groups to protect 'aquatic ecosystems', and groups to protect 'the resource' that essentially work in isolation from and in conflict with each other.
Minimal appreciation of strategic thinking	Strategic thinking is poorly understood, believed to have no practical value, and seen as just a distraction from the task at hand.

implement the plan, or implementing it badly, and thus avoids *blame* and thus *reprimands and the threat of disciplinary action.* Not implementing a plan also helps strengthen *resistance to change.*

This interconnectedness helps explain the apparent anomaly of anarchistic individual actions in such a rule-based, authority-based culture. A system based on fear and petty administrative issues cannot align individuals to good service delivery because such a system cannot define what service delivery is when complex issues such as sustainability are involved. Thus, anarchistic individual action occurs not necessarily as a result of all the individuals' being anarchists – although some may well be – but because there is nothing substantial in place regarding good service delivery to conform to.

This anarchistic approach seems to have developed after the NWA of 1998 (RSA, 1998) was enacted. Before then the DWS had a relatively simple mandate and could provide service in an autocratic way. The complexities of the 1998 NWA, on the other hand, could not be reduced to simple rules, and DWS has yet to resolve this conundrum. The lack of appreciation for strategic thinking was not, however, a relatively recent development; it existed in pre-democratic South Africa. Then as now, the products generated by 'strategic planning' were wish lists of positive outcomes, without a realistic plan to achieve those outcomes. Thus, it is quite understandable that strategic planning was, and is, held in such low esteem.

National Groundwater Strategy perspective. According to the NGWS, the DWS in South Africa is desperately short of the hydrogeological skills and experience that are needed to give effect to the NWA. Reasons given for this include low salaries and poor working conditions that encourage experienced staff to leave and new recruits to move on as soon as possible. Anecdotal evidence cited by the NGWS to elaborate on what poor working conditions meant include "convoluted bureaucracy, institutional reorganization, uncertain career progression, disjointed policy and frequent changes of leadership", as well as the lack of "traditional" benefits of working for a Public Service department such as "job security, interesting and varied work, and the chance to contribute to the public good".

Overall public service perspective. While substantial research has been conducted on policies assigned to the public service and their success or failure, relatively little has been conducted on the inner workings of public service departments (von Holdt, 2010). What research has been done creates an impression of institutional failure and dysfunctionality (Sloth-Nielsen, 2007; Southall, 2007; von Holdt, 2010) for most departments, with notable exceptions such as the South African Revenue Service and the National Treasury. Sangweni and Mxakato-Diseko (2008) describe the South African Public Service as "groping in the dark in an infant state to fulfil its functions; implying mediocrity, weakness, ineptitude and a lack of cognizance of the urgency to 'raise the bar' on service delivery". Cameron (2009) suggests that the Public Service in South Africa still has an organisational culture that regards providing a service as a side issue, rather than its *raison d'être*, and mentions that a perception survey revealed a low level of commitment amongst public service workers, suggesting that much of the cause of this problem is managerial incompetence.

Von Holdt (2010) describes how in public hospitals health care services provided by doctors and nurses seem to be of secondary or even incidental importance compared with departmental bureaucracy, and that departmental bureaucracy has little interest in or patience for the practical, health care service provision problems that doctors or nurses experience. This appears to be a generic issue in South Africa's Public Service according to the National Development Plan (Manuel, 2011), which states that "There is a serious ambivalence towards skill in the public service. The skills that staff possess are not always valued, and status or connections are often prized more than expertise."

Von Holdt (2010, p. 255) describes some key dysfunctional features that tend to reinforce each other:

> High vacancy rates, affirmative action targets and mobility mean that at times personnel who lack the requisite skills and experience are employed in key jobs; this further undermines skill as a criterion, and reinforces the sense of fragility and the importance of deference and "face" to mask this; skills gaps and deference tend to elevate the importance of rules and procedures; assertiveness about black class formation and sovereignty further displaces a focus on skills and experience and may legitimate corrupt practices; the processes of meaning-formation through which these goals are legitimated discourage a focus on effective organizational performance; and the breakdown of discipline and the elevated significance of budgetary rituals further displace the clinical process to the margin of bureaucratic concerns.

Von Holdt's (2010) description of the relation between regional/provincial and national Public Service departments is also pertinent: "The structural relationship between province and institution is a disincentive for managerial innovation and responsibility, and rewards subservience, over-sensitivity to rules and a lack of focus on problem-solving."

Von Holdt suggests that many of the current Public Service failings may be traced back to pre-democratic days when public service departments were the implementers and/or consolidators of apartheid. Thus – at least according to von Holdt – even the professionalism and skills needed to provide services are currently seen in a negative, racist light, since these qualities were used to cement apartheid in place. As long as a black middle class is being created via the policy of affirmative action in the Public Services, this overrides the need to develop skills. Provision of a service is of negligible importance. Cameron (2009) has a simpler explanation, namely that a strong, rule-based culture of public service already existed under apartheid, and – because of the resilience of such institutional cultures – this culture persisted when more innovative approaches were needed to redress apartheid's ills.

This article suggests there may be a much less sophisticated explanation than von Holdt's for poor service delivery, namely that most employees currently in the Public Service are not aware that the Public Service, before 1994, used to provide a professional service, albeit to a racially defined minority. As a result of this lack of awareness, there is a general acceptance that the Public Service is primarily there to provide sheltered employment. A cruder possible explanation is that the ongoing purge of skills via affirmative action has created a Public Service that is unable to manage at anything more than a petty, bureaucratic level, whether it wants to or not. For example, it is relatively easy for a manager to check whether an employee is late. However, a manager requires depth of knowledge of a job and experience to check whether an employee has provided a professional service.

Implementation strategies. The preceding sections have attempted to characterize the public service culture that deals with groundwater in DWS from (1) 27 years of personal experience and (2) the NGWS assessment of human capacity, assuming that generic results for the South Africa Public Service will also apply to the groundwater responsibilities within DWS. While there are serious shortcomings with each of these approaches, taken together they yield results that are in reasonable agreement, suggesting that DWS is at worst dysfunctional, and at best severely incapacitated. Whatever position on the spectrum from dysfunctional to severely incapacitated is taken, it is clear that there is a wide gulf between the status quo and a future 'strong national institution' as informed by the NGWS. The second constraint is therefore the availability of strategies to get from the status quo to the desired end point.

It is suggested that the biggest implementation constraint is the lack of any systematic strategic implementation plan, or more generally the lack of attention given to implementation strategies. Evidence for this includes the following.

- Of the references in the NGWS, only 1 deals with institutional/implementation matters while the remaining 62 deal with (technical) water issues, suggesting a very strong bias towards defining the end state rather than planning for how to get there.
- The NGWS does not have an implementation plan as such and defers responsibility for the implementation of groundwater strategies to the overall National Water Resources Strategy 2 (NWRS2) (DWS, 2012). Yet the NWRS2 defers responsibility

for the implementation of groundwater strategy back to the NGWS. This suggests a strong desire to avoid responsibility for the implementation of any groundwater strategies.

- While the 293-page document containing the NWRS2 contains good ideas for the improvement of water management and water supply services, it devotes less than half a page to the implementation of those ideas. That half-page on implementation contains only vague exhortations to think and act differently, rather than any concrete, definable steps. It is thus clear that the plan is very biased towards how the country's water resources *should* be managed, and actually has no practical guidance on how to get to that end state.

Discussion

The analysis conducted in this article has shown that (1) specific groundwater governance strategies and (2) implementation strategies are needed before DWS can make significant improvements to groundwater governance. Neither of these is currently in place. In addition, DWS is plagued by staff shortages, lack of institutional capacity, and an obsession with petty administrative issues rather than service delivery. The difference between 'business as usual' and the strategic interventions identified by backcasting as needed to effect groundwater governance are sufficiently large that the case for backcasting can be regarded as proven according to the terms of reference of this article.

One of the limitations of this investigation is the use of one person's experience to characterize the institutional culture at DWS, and the use of the same person's perspective to identify constraints and interventions in the backcasting process. This clearly could introduce problems of subjectivity and bias. On the other hand, the personal perspective is that of a practitioner with 32 years' experience in providing hydrogeological inputs to the groundwater governance process. This "insider knowledge", while subjective, is very useful in increasing thought, debate and wisdom (McIntyre, 2006).

It is acknowledged that DWS employs some 10,000 people, all of whom could presumably have different perceptions regarding DWS's effectiveness. However, general assessments of the South African Public Service, and DWS's own assessment of its groundwater service, all concur with the principal author's characterization of dysfunctionality. Differences of opinion might exist regarding the degree of dysfunctionality, or its causes, but that it exists appears beyond dispute. The dysfunctionality can also be corroborated by its effect on groundwater governance, which has been assessed as weak to non-existent (Knüppe, 2011; Pietersen et al., 2011).

Likewise, a more consensus-based backcasting exercise might have delivered different strategies for giving effect to good groundwater governance. However, there appears to be no doubt that these strategies, in one form or another, are needed. There also appears to be no doubt that whatever groundwater governance strategies are chosen, there is also a strong need for implementation strategies to be developed, so that the governance strategies can be given effect to, rather than remaining as a planners' wish list in a strategic report. The key reason groundwater governance is weak in South Africa is not the legislative setting in the water area but the implementation gap.

While this investigation had the initial aim of exploring the role of backcasting in improving groundwater governance in South Africa, it would be more reasonable to accept that what has really been explored are ways to improve groundwater governance in South Africa using backcasting, together with practitioner experience and the Ostrom Design Principles. Thus it has been impossible to 'prove' or 'disprove' the value of backcasting. While accepting this limitation, the investigation has yielded many useful insights for further work.

With this in mind, the following areas are suggested for further research.

- *Backcasting using diverse stakeholders rather than just one practitioner.* The subjectivity of relying on one person's viewpoint has already been discussed. There is clearly a need for a stakeholder-driven backcasting process.
- *Increasing institutional capacity without increasing the number of staff employed.* The strategies identified by the backcasting exercise do not necessarily require any reduction in institutional staff shortages. Interventions might need to be limited to a few, high-priority areas. In these areas the backcasting strategies only require that the WUAs be allowed to do their work, rather than the state's contributing massive resources. Groundwater governance would be driven at the local level, and existing staff from DWS would merely be required to provide their inputs to that process. Each DWS staff member could conceivably be acting as de facto water user, or water use representative, representing their particular aspect of 'use', such as intergenerational or ecosystems 'use'.
- *Engendering a culture of effective strategic planning and implementation.* For groundwater governance to be improved in South Africa will require not just envisioning the desired water scenario, nor even envisioning the desired capacity and programmes of water institutions, but envisioning the desired organizational *culture* of those institutions, and a strategic plan to engender that culture.

It is suggested that research into strategic thinking and strategy implementation may also have value in giving effect to public service delivery in general, and not just to groundwater governance within DWS. Creative and innovative approaches to strategy implementation might make it possible to accommodate employment equity, and lack of capacity, *and* still provide an effective service, instead of the 'business as usual' approach of using these factors to justify lack of service delivery.

Support for this argument is that strategic thinking does not require high levels of technical expertise but good generalists who can think creatively. A strategic use of employment equity would be to appoint new people with good strategic-thinking skills rather than new people with no experience. Even if all the staff vacancies are filled, without better strategic thinking there will just be bigger sections and bigger operational 'silos', working disconnectedly or at cross-purposes with each other, and service delivery will not necessarily be improved.

- *Exploring the scenario of no groundwater management, on the grounds that groundwater management might not be economically justified according to the Gisser-Sanchez principle* (Gisser & Sanchez, 1980; Katic & Grafton, 2012). Hydraulic parameters are so low in many parts of South Africa that aquifers essentially "manage themselves" (Seward, 2010) since these parameters allow only very low pumping rates and/or limited radii of influence, which in turn prevent aquifers from being intensively exploited. Socio-economic development is the most pressing requirement for water use in Sub-Saharan Africa (Koppen, van der Zaag, Manzungu, & Tapela, 2014). Since groundwater is under-utilized in South Africa, possibly at only 6% of what is available (Woodford, Rosewarne, & Girman, 2005), attempted over-regulation using cumbersome permit systems does not seem an obvious way to support socio-economic development.

- *Further research on the Ostrom Design Principles and how they might be used to improve groundwater governance.* One of the key assumptions of this investigation is that because the Ostrom Design Principles apply to common-pool resources, and because groundwater is – or can be regarded as – a common-pool resource, the Ostrom Design Principles must apply to groundwater. This needs to be investigated by comparing well-documented areas of groundwater governance with the Ostrom Design Principles and evaluating their correlation with good groundwater governance.

One of the main benefits of backcasting is that it forces attention on a desired future and how to get there, rather than on current problems and the socio-technical constraints preventing change (Dreborg, 1996; Wangel, 2011). Backcasting therefore seems a very useful and possibly necessary tool for researching how to improve delivery in the Public Service. The observation that it is very difficult to get water experts and government water managers to think imaginatively and creatively, or to think beyond a current crisis (Claassen, Funke, & Nienaber, 2011), could well be applied to most public service departments in South Africa. Extrapolating from the results of other researchers in the groundwater governance field (Aarnoudse et al., 2012; López-Gunn & Cortina, 2006; Mukherji & Shah, 2005), it is suggested that the need for research using backcasting to improve public service delivery is by no means restricted to South Africa.

Conclusion

It has been demonstrated that backcasting has the potential to play a useful role in improving the state's contribution to good groundwater governance in South Africa. The backcasting method has revealed issues that need to be addressed that might not have been identified in the current organizational culture. These issues include the need for better strategic thinking and the need for better implementation of strategies. Whether backcasting is the *best* approach for improving groundwater governance in South Africa, or elsewhere, cannot be proven. However, the backcasting approach, in combination with practitioner experience and the Ostrom Design Principles, has helped identify many useful areas for further research. One of the recommended research areas is to investigate the degree of correlation between the Ostrom Design Principles and good groundwater governance using case studies.

References

Aarnoudse, E., Bluemling, B., Wester, P., & Qu, W. (2012). The role of collective groundwater institutions in the implementation of direct groundwater regulation measures in Minquin County, China. *Hydrogeology Journal*, *20*, 1213–1221. doi:10.1007/s10040-012-0873-z

Burns, M., Audouin, M., & Weaver, A. (2006). Advancing sustainability science in South Africa. *South African Journal of Science*, *102*, 379–374.

Cameron, R. (2009). New public management reforms in the South African public service: 1999–2009. *Journal of Public Administration*, *44*, 910–942. Retrieved from http://reference.sabinet.co.za/sa_epublication_article/jpad_v44_n4_si1_a2

Claassen, M., Funke, N., & Nienaber, S. (2011). *The water sector institutional landscape by 2025* (Report No. 1841/1/11). Pretoria: Water Research Commission

Cox, M., Arnold, G., & Villamayor-Tomás, S. (2010). A review and reassessment of design principles for community-based natural resource management. *Ecology and Society*, Retrieved from http://www.ecologyandsociety.org/vol15/iss4/art38/

Custodio, E. (2002). Aquifer overexploitation: What does it mean? *Hydrogeology Journal, 10*, 254–277. doi:10.1007/s10040-002-0188-6

Custodio, E., & Llamas, M. R. (2003). Intensive use of groundwater: Introductory considerations. In M. R. Llamas & E. R. Custodio (Eds.), *Intensive use of groundwater: Challenges and opportunities* (pp. 3–8). Lisse, Holland: A.A. Balkema Publishers.

Department of Water and Sanitation (DWS). (2010). *National groundwater strategy*. Retrieved from http://www.dwa.gov.za/Groundwater/Documents/GSDocument%20FINAL%202010_MedRes.pdf

Department of Water and Sanitation (DWS). (2012) *Draft national water resource strategy2*. Retrieved from http://www.dwaf.gov.za/nwrs/LinkClick.aspx?fileticket=M8NprZjscYw%3d&tabid=72&mid=435

Dreborg, K. H. (1996). Essence of backcasting. *Futures*, *28*, 813–828. doi:10.1016/S0016-3287(96)00044-4

Faysse, N., Errahj, M., Imache, A., Kemmoun, H., & Labbaci, T. (2014). Paving the way for social learning when governance is weak: Supporting dialogue between stakeholders to face a groundwater crisis in Morocco. *Society & Natural Resources*, *27*, 249–264. doi:10.1080/08941920.2013.847998

Foster, S., & Ait-Kadi, M. (2012). Integrated Water Resources Management (IWRM): How does groundwater fit in? *Hydrogeology Journal*, *20*, 415–418. doi:10.1007/s10040-012-0831-9

Foster, S., & Garduño, H. (2013). Groundwater-resource governance: Are governments and stakeholders responding to the challenge? *Hydrogeology Journal*, *21*, 317–320. doi:10.1007/s10040-012-0904-9

Foster, S., Garduno, H., Tuinhof, A., & Tovey, C. (2010). Groundwater governance – conceptual framework for assessment of provisions and needs. GW-Mate Strategic Overview Series No. 1, World Bank

Giordano, R., Brugnach, M., & Vurro, M. (2012) *System dynamic modelling for conflicts analysis in groundwater management*. International congress on environmental modelling and software, Sixth Biennial Meeting, Leipzig, Germany

Gisser, M., & Sanchez, D. A. (1980). Competition versus optimal control in groundwater pumping. *Water Resources Research*, *16*, 638–642. doi:10.1029/WR016i004p00638

Gleeson, T., Alley, W. M., Allen, D. M., Sophocleous, M. A., Zhou, Y., Taniguchi, M., & VanderSteen, J. (2012). Towards sustainable groundwater use: Setting long-term goals, backcasting, and managing adaptively. *Ground Water*, *50*, 19–26. doi:10.1111/j.1745-6584.2011.00825.x

Gleick, P. H., Loh, P., Gomez, S. V., & Morrison, J. (1995). *California water 2020: A sustainable vision*. Retrieved from http://pacinst.org/publication/california-water-2020-a-sustainable-vision/

Katic, P. G., & Grafton, R. Q. (2012). Economic and spatial modelling of groundwater extraction. *Hydrogeology Journal*, *20*, 831–834. doi:10.1007/s10040-011-0817-z

Knüppe, K. (2011). The challenges facing sustainable and adaptive groundwater management in South Africa. *Water SA*, *37*, 71–79. Retrieved from http://www.wrc.org.za/Knowledge%20Hub%20Documents/Water%20SA%20Journals/Manuscripts/2011/01%20January%202011/2566.pdf

Konikow, L. F., & Kendy, E. (2005). Groundwater depletion: A global problem. *Hydrogeology Journal*, *13*, 317–320. doi:10.1007/s10040-004-0411-8

Koppen, B., van der Zaag, P., Manzungu, E., & Tapela, B. (2014). Roman water law in rural Africa: The unfinished business of colonial dispossession. *Water International*, *39*, 49–62. doi:10.1080/02508060.2013.863636

Llamas, M. R., & Martínez-Santos, P. (2005). Intensive groundwater use: A silent revolution that cannot be ignored. *Water Science and Technology Series*, *51*, 67–174.

Llamas, M. R., Martínez-Santos, P., & de la Hera, A. (2007). The manifold dimensions of groundwater sustainability: An overview. In S. Ragone, A. de la Hera, & N. Hernandez-Mora (Eds.), *The global importance of groundwater in the 21st century: Proceedings of the international symposium on groundwater sustainability*. Westerville, OH: National Ground Water Association Press.

Llamas, M. R., Mukherji, A., & Shah, T. (2006). Guest editor's preface. Theme issue: Social and economic aspects of groundwater governance. *Hydrogeology Journal*, *14*, 269–274. doi:10.1007/s10040-006-0025-4

López-Gunn, E., & Cortina, L. M. (2006). Is self-regulation a myth? Case study on Spanish groundwater user associations and the role of higher-level authorities. *Hydrogeology Journal*, *14*, 361–379. doi:10.1007/s10040-005-0014-z

López-Gunn, E., Willaarts, B., Rica, M., Corominas, J., & Llamas, R. (2013). The Spanish water 'pressure cooker': Threading the interplay between resource resilient water governance outcomes by strengthening the robustness of water governance processes. *International Journal of Water Governance*, *1*, 13–40. doi:10.7564/12-IJWG8

Manuel, T. A. (2011). *National development plan vision for 2030*. The Presidency: National Planning Commission Department. Retrieved from http://www.npconline.co.za/medialib/downloads/home/NPC%20National%20Development%20Plan%20Vision%202030%20-lo-res.pdf

McIntyre, P. (2006). Creative practice as research: 'Testing Out' the systems model of creativity through practitioner based enquiry. In Queensland University of Technology (Ed.), *Speculation and innovation: Applying practice led research in the creative industries*. Brisbane, Australia: Queensland University of Technology.

Moench, M., Kulkarni, H., & Burke, J. (2012). *Trends in local groundwater management institutions*. Thematic Paper 7. Groundwater Governance: A Global Framework for Action. GEF ID 3726. Retrieved from http://www.groundwatergovernance.org/fileadmin/user_upload/groundwatergovernance/docs/Thematic_papers/GWG_Thematic_Paper_7.pdf

Mukherji, A., & Shah, T. (2005). Groundwater socio-ecology and governance: A review of institutions and policies in selected countries. *Hydrogeology Journal*, *13*, 328–345. doi:10.1007/s10040-005-0434-9

Nilsson, M., Nilsson, L. J., Hildingsson, R., Stripple, J., & Eikeland, P. O. (2011). The missing link: Bringing institutions and politics into energy future studies. *Futures*, *43*, 1117–1128. doi:10.1016/j.futures.2011.07.010

Ostrom, E. (1990). *Governing the commons: The evolution of institutions for collective action*. Cambridge: Cambridge University Press.

Ostrom, E. (2005). *Understanding institutional diversity*. Princeton, NJ: Princeton University Press.

Ostrom, E. (2009). *Beyond markets and states: Polycentric governance of complex economic systems*. Nobel Prize Lecture. Retrieved from http://www.nobelprize.org/nobel_prizes/economic-sciences/laureates/2009/ostrom_lecture.pdf

Pierce, S. A., Sharp, J. M., Guillaume, J. H. A., Mace, R. E., & Eaton, D. J. (2013). Aquifer-yield continuum as a guide and typology for science-based groundwater management. *Hydrogeology Journal*, *21*, 331–340. doi:10.1007/s10040-012-0910-y

Pietersen, K., Beekman, H. E., & Holland, M. (2011). *South African groundwater governance case study* (WRC Report No. KV 273/11). Report prepared for the WorldBank. Retrieved from http://www.wrc.org.za/Knowledge%20Hub%20Documents/Research%20Reports/KV%20273-11.pdf

Ravasi, D., & Schultz, M. (2006). Responding to organizational identity threats: Exploring the role of organizational culture. *The Academy of Management Journal*, *49*, 433–458. doi:10.5465/AMJ.2006.21794663

Republic Of South Africa (RSA). (1998). *National water act*. Pretoria: Government Printer.

Robinson, J. B. (1990). Futures under glass: A recipe for people who hate to predict. *Futures*, *22*, 820–842. doi:10.1016/0016-3287(90)90018-D

Ross, A., & Martínez-Santos, P. (2010). The challenge of groundwater governance: Case studies from Spain and Australia. *Regional Environmental Change*, *10*, 299–310. doi:10.1007/s10113-009-0086-8

Sangweni, S. S., & Mxakato-Diseko, N. (2008). It does not matter what slant or take you have on the developmental state: At the end of the day, a strong, coherent and astute public service is critical. In B. Turok (Ed.), *Wealth does not trickle down: The case for a developmental state in South Africa*. Cape Town: New Agenda.

Seward, P. (2010). Challenges facing environmentally sustainable ground water use in South Africa. *Ground Water*, *48*, 239–245. doi:10.1111/j.1745-6584.2008.00518.x

Sloth-Nielsen, J. (2007). The state of South Africa's prisons. In S. Buhlungu, J. Daniel, R. Southall, & J. Lutchman (Eds.), *State of the nation: South Africa 2007*. Cape Town: HSRC Press.

Southall, R. (2007). Introduction: The ANC state, more dysfunctional than developmental? In S. Buhlungu, J. Daniel, R. Southall, & J. Lutchman (Eds.), *State of the nation: South Africa 2007*. Cape Town: HSRC Press.

Taher, T., Bruns, B., Bamaga, O., Al-Weshali, A., & van Steenbergen, F. (2012). Local groundwater governance in Yemen: Building on traditions and enabling communities to craft new rules. *Hydrogeology Journal*, *20*, 1177–1188. doi:10.1007/s10040-012-0863-1

Thompson, H. (2006). *Water law – A practical approach to resource management and the provision of services*. Cape Town: Juta.

van Steenbergen, F. (2006). Promoting local management in groundwater. *Hydrogeology Journal*, *14*, 380–391. doi:10.1007/s10040-005-0015-y

van Vliet, M., & Kok, K. (2013). Combining backcasting and exploratory scenarios to develop robust water strategies in face of uncertain futures. *Mitigation and Adaptation Strategies for Global Change*, Published online. doi:10.1007/s11027-013-9479-6

von Holdt, K. (2010). The South African post-apartheid bureaucracy: Inner workings, contradictory rationales and the developmental state. In O. Edigheji (Ed.), *Constructing a democratic developmental state in South Africa: Potentials and challenges*. Cape Town: HSRC Press.

Wangel, J. (2011). Change by whom? Four ways of adding actors and governance in backcasting studies. *Futures*, *43*, 880–889. doi:10.1016/j.futures.2011.06.012

Wester, P., Minero, R. S., & Hoogesteger, J. (2011). Assessment of the development of aquifer management councils (COTAS) for sustainable groundwater management in Guanajuato, Mexico. *Hydrogeology Journal*, *19*, 889–899. doi:10.1007/s10040-011-0733-2

Wijnen, M., Augeard, B., Hiller, B., Ward, C., & Huntjens, P. (2012). *Managing the invisible: Understanding and improving groundwater governance*. Retrieved from http://water.worldbank.org/sites/water.worldbank.org/files/publication/ESW_Managing-the-invisible.pdf

Woodford, A., Rosewarne, P., & Girman, J. (2005, September 5–7). *How much groundwater does South Africa have?* Paper presented at the SANCIAHS Conference, Midrand. Retrieved from http://www.srk.co.za/files/File/newsletters/groundwater/PDFs/1_A_Woodford.pdf

OPEN ACCESS

Groundwater governance through institutional bricolage? Participation in Morocco's Chtouka aquifer contract

Annabelle Houdret and Rebecca Heinz

ABSTRACT

Controlling groundwater use has become one of the most urgent and difficult issues in common-pool resource governance. This article analyses an aquifer contract in Morocco as an innovative, participatory approach to the problem. Cleaver's concept of institutional bricolage allows us to understand how its implementation affected and interacted with existing social institutional arrangements for water use. The findings confirm that an awareness of insufficient rules is a precondition for new governance arrangements. They reveal how power relations and social institutions influenced the implementation of the aquifer contract. Finally, they show how social and cultural affiliations shape the implementation.

Introduction

Since the introduction of motor pumps in the 1950s and their widespread use since the 1980s, groundwater-dependent economies have emerged all over the world. In the Maghreb countries and Southeast Asia, agricultural intensification has accelerated, and groundwater resources have been heavily depleted (Shah, 2014). In most cases, irrigated agriculture is the main user of groundwater resources; nearly a quarter of freshwater consumed by global agriculture derives from aquifers (43% globally, 40% in Morocco) (Siebert et al., 2010;).

Groundwater-irrigated areas pose distinct management challenges. The numerous and diffuse abstraction points lead to high public transaction costs for the monitoring, control and application of sanctions (Del Vecchio & Barone, 2018). Authors have, therefore, claimed that groundwater is 'one of the most mismanaged resources globally' (Reddy et al., 2014, p. 276) as these characteristics make it practically impossible to control water exploitation through regulatory instruments. Given the urgent need to regulate groundwater use, and the inefficiency of existing top-down approaches, scholars have advocated self-governance systems through participatory decision-making (Lopez-Gunn & Martinez Cortina, 2006; De Stefano & Lopez-Gunn, 2012). As they argue, the evolution of institutions resulting from participatory interventions is expected to strengthen users' compliance with the rules and encourage local self-governance, while reducing public transaction costs for resource monitoring (Lopez-Gunn & Martinez Cortina, 2006; De Stefano & Lopez-Gunn, 2012).

Research on the design and evolution of institutions for common-pool resource management has been marked by the institutionalist school of thought around Elinor Ostrom (Ostrom, 2005). Ostrom's 'operational rules in use' directly affect day-to-day decisions, including those of groundwater users (Ostrom et al., 2014). However, critical institutionalists raise concerns about the underlying assumptions of the institutionalists' theories that rely heavily on rational-choice theories and, according to the critical institutionalists, neglect the impact of (unconscious) social and cultural factors on actors' behaviour. Critical institutionalists argue that beyond participation in formal policies and rules, natural-resource governance needs to assess institutional transformation via processes of social interaction and power relationships (Cleaver, 2012; Hall et al., 2014).

The present paper adopts a critical institutionalist perspective to scrutinize the potential of the participatory approach for a better understanding – and ultimately implementation – of it in groundwater management. More specifically, we analyse how social institutions at the local level (such as irrigation associations or informal water user groups sharing joint pumping stations, but also family networks or village communities) affect the implementation of an aquifer contract via institutional bricolage, a concept developed by Frances Cleaver (Cleaver, 2002, 2012). With some exceptions (such as De Bont et al., 2016), this concept has mostly been applied to study surface water or other natural resources (Cleaver, 2002, 2005).

The aquifer under study is located in southern Morocco's water-scarce Souss-Massa Valley, where policymakers and local stakeholders adopted a contract to impede further depletion of groundwater resources. Evaluations of the effectiveness of this Souss-Massa aquifer contract, which expired in 2017 after 10 years of implementation, differ. While representatives of water-related administrations highlight the success of increasing farmers' willingness to participate in formal groundwater institutions and comply with the official water law (authors' interviews; Faysse et al., 2012), other observers state that the implementation has 'been undermined by a general laisser-faire attitude' (Closas & Molle, 2016, p. 23). Against these contradictory assessments, we analyse the role of social aspects of the aquifer contract to understand the dynamics of implementation and the varying perceptions on its functioning. Adopting the concept of institutional bricolage to study these dynamics, the analysis reveals the impactful interference of power, social and cultural factors, and informal rules in the functioning of formal rule implementation.

This study is based on a thorough literature review and extensive field research that included 45 qualitative interviews conducted in Morocco in 2017. The paper is structured as follows. The next section introduces the concept of institutional bricolage as the theoretical approach for analysis of the aquifer contract's functioning and summarizes the methodological approach. The third section introduces the case study region and presents the findings. The discussion and conclusions section reviews these results in light of existing research, further research needs and policy implications.

Theoretical and methodological approach

Critical institutionalism and the institutional bricolage approach to analysing an aquifer contract

The analysis of groundwater governance often primarily relies on formal regimes and their problems of implementation but does not often fully consider social rules and their functioning. However, these factors frequently heavily influence the success or failure of groundwater governance. This lack of consideration for social aspects may be due to the assumption that the invisibility of the resource impedes the existence of social institutions for its regulation (Shah, 2009). The concept of institutional bricolage (Cleaver, 2002, 2012) bridges this gap because it allows for a comprehensive analysis of informal institutions and their interplay with formal ones. The following explanation will show the importance of this difference to the understanding of institutions for groundwater governance and related institutional change, such as an aquifer contract.

The critical institutionalist approach to bricolage emerged as a response to mainstream institutionalism. Both streams of thought, mainstream institutionalism and critical institutionalism, see the evolution of institutions at the centre of social systems that successfully self-govern open-pool resources. Institutions are defined as norms, rules for use and arrangements that structure the access, use and distribution of natural resources among their users (Crawford & Ostrom, 1995). Mainstream institutionalism is based on the assumption that actors behave according to rational and purposeful choices (Forsyth & Johnson, 2014). As such, institutional rules are understood as consciously crafted conventions that evolve from human interactions and past problem-solving strategies that become renegotiated and adapted to changing, for example, environmental conditions. Ostrom assumed that humans can discursively refer to applied working rules in social interactions (Ostrom, 2005). Although Ostrom recognizes that the repetitive application of rules can turn into a 'social habit', she argues that institutional evolution is based on 'fallible, norm-adopting individuals who pursue contingent strategies in complex and uncertain environments' (Ostrom, 1990, p. 185).

Critical institutionalists, in contrast, criticize the mainstream institutionalists' assumptions of human interaction as socially uninformed and blind towards cultural factors and power relationships that unconsciously shape human actions (Cleaver, 2002, 2012; Cleaver & de Koning, 2015; Hall et al., 2014). In opposition to Ostrom's assumption that rules are consciously practiced, rejected or negotiated based on human rationality, critical institutionalism understands institutions to be shaped by social and power relationships, psychological preferences and 'culturally and historically shaped ideas about the "right way of doing things"' (Cleaver, 2012, p. 15). People are thus able to plan actions purposefully and collectively, but non-conscious factors shape decisions through 'taken-for-granted everyday practices and embedded ideologies' (p. 15).

Cleaver (2012), following Douglas, 1986), defines institutions, therefore, as relatively stable social arrangements that rely on worldviews (e.g., religious beliefs, cultural convictions, etc.). As such, they cannot simply be replaced by new, rationally negotiated institutional impositions, such as the aquifer contract. Rather, institutional change works through analogies, which connect existing arrangements with emerging institutional innovations through 'institutional leakage'. 'Leaking' in this context implies that emerging institutions adapt logic from existing or previous institutional arrangements.

Following Cleaver, this leaking of meaning and drawing of analogies in the adaptation of institutions implies that existing and evolving institutional principles become blended. Subsequently, public regulative measures, such as the aquifer contract, are not simply adopted but transform via mechanisms of institutional blurring and blending to fit the social lifeworld of their 'bricoleurs'.

CI's broader understanding of the emergence and change of institutions sheds a different light on a bricoleur's motivation to engage with and transform institutions. While MI assumes that institutional design implies the crafting of particular mechanisms for particular functions, the bricolage approach leads to institutions that are multifunctional. Here, institutions evolve through improvised adaptation in which 'decision-making arrangements and relations of cooperation may be co-opted for new purposes' (Cleaver, 2012, p. 87). Cleaver labels this as institutional plurality induced by specific identities of institutional participants that affect preferences towards certain arrangements (Cleaver, 2012).

These assumptions have far-reaching implications for the understanding of water governance arrangements, including aquifer contracts, and may explain why public regulative measures are not adopted or are adopted in a different way than expected. The institutional plurality of an aquifer contract, for instance, is linked to the fact that water users have different individual identities beyond their water use (e.g., male or female, young or old, etc.) that involve different personal motivations, interests and concerns. This also affects the water users' perception of groundwater governance arrangements; diverging identities entail different interpretations of changing 'elements' of their specific lifeworlds, for example, changes in formal rules that affect institutional arrangements. As institutional arrangements are shaped by social interaction and tied to elements of people's identity, this implies that they are bound to social groups, their related identity features and their specific interests (Cleaver, 2012). Subsequently, it depends on the particular norms, beliefs and world views of these social spaces whether water users perceive an institution as legitimate or not and whether and to what extent they cooperate to support the arrangement.

In our case study, we understand the aquifer contract as an institutional arrangement formed by different water users and the administrations involved (such as the water basin agency, several agricultural producer associations, the local authority for agricultural development, the regional council). According to the literature, the motivation for new groundwater governance rules emerges out of a general awareness of the insufficiency and/or non-application of existing rules and subsequent scarcity; Shah (2009) and Petit (2004) describe how resource users tend to craft institutions for groundwater regulation in the face of increasing groundwater scarcity. The engagement of different actors in the transformation of institutions depends on the actors' identity and their specific lifeworlds and on their social position as influenced by their relative power in the local setting, which, in turn, influences their access to governance arrangements. For the sake of our study, we define power as the capacity of bricoleurs to engage in the transformation process of institutions (Cleaver, 2012). Social resources such as social networks and trustful relationships are of particular importance here as they may allow or deny access to specific social alliances in which bricoleurs negotiate particular rules and roles of institutional arrangements (Cleaver, 2005). We understand social alliances as exclusively organized networks of social relationships with a strategic understanding of these networks in which actors realize mutual favours through cooperation (cf. Long, 2001). In these alliances, bricoleurs have different capacities to engage in rule setting and benefit unequally from institutional outcomes. A group of water users that is a formal member of the aquifer contract, for instance, may be such a social alliance, but other alliances such as groups of villagers, other groups with shared identities and joint interests that exist independently from the contract may also impact the governance structure. Bricoleurs work within and through these relationships of power, using different social alliances to reshape institutional arrangements according to their specific priorities, and the drawing of analogies between and the blending of existing structural conditions.

The mechanisms deployed to reshape institutional arrangements may include practices of corruption. Rather than a form of illicit behaviour (cf. Teorell, 2007), we interpret corruption as an institutional mechanism that is part of the broader governance setting. Robbins (2000) defines corruption as 'a system of normalised rules, transformed from legal authority, patterned around existing inequalities, and cemented through cooperation and trust' (p. 424). In the process of institutional bricolage, corruptive practices evolve from social processes linked to the implementation of formal rules, such as the aquifer contract. As Robbins (2000, p. 428) states, policy interventions can favour the evolution of corruptive practices particularly when governance systems are marked by strong patterns of authority, as is the case in Morocco (Daadaoui, 2011). This raises the question of whether the implementation of the aquifer contract may have induced the evolution of corruptive practices as part of the reshaping of institutional arrangements.

Our analysis of the Chtouka aquifer contract as an institutional arrangement adopts the approach of institutional bricolage (Cleaver, 2002, 2012) to understand how institutional bricoleurs adopt and transform the formal rules of the contract through the blending and blurring of formal rules with existing institutional arrangements at the local level. This study focuses on the rules of the aquifer contract that relate to the regulation of groundwater access (monitoring of borehole drilling) but explicitly considers power issues and social/cultural factors as typical characteristics of critical institutionalist approaches. Based on these theoretical assumptions, the following three hypotheses guide the analysis:

- The aquifer contract as a new institutional arrangement emerged out of a general awareness about the insufficiency and/or non-application of existing rules and subsequent resource scarcity.
- A bricolage perspective entailing the analysis of power relations allows for better understanding of the adoption and reshaping of formal rule implementation related to the aquifer contract.
- Social and cultural factors such as group identity based on geographical or ethnic origin, on common language shape the implementation of the aquifer contract leading to multifunctional institutions.

Methodological approach

This study is based on the evaluation of qualitative data gathered during fieldwork in Chtouka and Agadir in October 2017. Qualitative research techniques and approaches stimulated wide-ranging conversations, while focusing on the previously defined fields of interests. Problem-centred interviews with local farmers, villagers,

Table 1. Overview of interviewees.

Local farmers	17
Villagers (representatives of the local drinking water association)	9
Representatives from local authorities in the agricultural and water sectors	13
Other experts (drilling companies, agricultural organizations)	6

representatives of farming organizations and officers from the different administrations related to the aquifer contract were at the heart of the research design. Problem-centred interviews were conducted with more than 45 respondents, 32 of which were recorded. The farmers and villagers interviewed were neither directly involved in the aquifer contract negotiations nor affiliated with participating associations. This selection assured an assessment of the extent to which institutional bricoleurs adopted and adapted the rules and mechanisms of the aquifer contract, including beyond the formal roundtable of the aquifer contract. The selection of interviewed farmers reflects their varied geographical origin, age and agricultural production schemes. This wide range of interviewees was chosen to represent a diverse and nuanced picture of how different institutional bricoleurs use different institutional mechanisms in the context of groundwater governance (Table 1).

The researchers adapted the methodological approach of problem-centred interviews to the highly varying social, professional and educational backgrounds of their interviewees. Thus, questions and their formulations were adapted according to both the fields of interests previously defined and the individual requirements of the interviewees in terms of social, educational and professional/thematic backgrounds to maintain empathy, interest and respect for the views of the interlocutors (Lamnek & Krell, 2010).

A Moroccan field assistant translated the interview sequences from Amazigh or Arabic into French during the interview recordings, and the researchers later translated them into English during the transcriptions. Lopez et al. (2008) argue that cross-cultural qualitative research is challenging because it is difficult to assess the reliability and the implicit meaning of information when conducting research outside one's own native language. To address these challenges, all scripts underwent a thorough joint-reflection process by the researchers and the translator.

The scripts were converted into research texts (Clandinin & Connelly, 2000) using MAXQDA software for qualitative data analysis. In doing so, qualitative data were interpreted along a hermeneutic process that included the alternate screening of field texts and review of literature. In this context, codes were developed that assigned meaning and attributes to selected passages of the field texts (Saldaña, 2016). The code creation also followed an iterative and hermeneutic approach whereby codes were inductively deduced from the qualitative material and derived deductively from the applied theory. The code system was developed during the transcription process and was readapted and modified during the theoretical elaboration and data analysis (Kelle, 2007). Along the data analysis, the research question and the generated hypothesis were revised, adapted and tested against the findings from the analysis.

The Chtouka aquifer contract as an institutional bricolage

Description of the case study region: from peasants to investors

The semi-arid Chtouka area is part of Southern Morocco's Souss-Massa Region and has witnessed a swift change towards agricultural intensification over the past 70 years. This has brought about two key trends that are crucial for understanding the emergence and evolution of the aquifer contract: important socio-economic transformations and a rapid depletion of groundwater resources (Bencherifa, 1980).

Until the 1950s, agriculture consisted of a complementary system of minor irrigated plots, extensive pastureland and rainfed cereal production (Bencherifa, 1980). In fewer than 50 years, Chtouka transformed from a rural subsistence economy to an agro-industrial production site with highly (ground)water-intensive greenhouse products cultivated for export. Between the 1950s and 1980s, Chtouka witnessed a massive emigration of rural inhabitants; every third man of working age migrated from Chtouka to France or Morocco's urban centres (Bencherifa, 1980). In the 1970s, the 80,000 ha Chtouka irrigation perimeter was inaugurated with specific mandatory cultivation plans and rotation cycles for the farmers. According to Pérennès (1993), this scheme presented a sharp rupture of the previous agricultural system and was a state-based attempt to transform collective farming systems into profit-oriented export enterprises. With the inability of many farmers to cope with the new requirements and large parts of land belonging to people who no longer occupied it and preferred to withdraw from the sector, a new dynamic of land selling and renting brought national and later international investors to the area (Popp, 1983). As with the evolution occurring in other parts of the country, this entailed a process of socio-economic marginalization of smallholders (Houdret et al., 2017). Two developments then further contributed to groundwater depletion: the deployment of large private farms, often beyond the limits of the public perimeter, that exclusively relied on groundwater (Pérennès, 1993) and unreliable water provision in the public irrigation scheme, which forced farmers to invest in groundwater pumping (Popp, 1983).

A culturally and socially diverse stakeholder landscape with large disparities in access to water and agricultural production resources emerged. Today, farmers range from international corporations to international and national export producers and smaller local investors/farmers (Sippel, 2016). According to a survey from 2015, 70% of the farmers in Chtouka operate on less than 10 ha and in total account for only 20% of the irrigated surface (Anzar Conseil, 2016a). As other authors have shown for the Souss-Massa Region, unequal access to groundwater reflects both income disparities and unequal access to different production resources, such as technical innovation and capital, as well as socio-political networks (Houdret, 2012).

The rapid extension and intensification of agriculture in Chtouka has provoked a massive depletion of groundwater resources: groundwater tables have fallen by 1.0 m per year on average since the 1990s and by 4.5 m per year since 2005 in the northern area (Anzar Conseil, 2016b). After the creation of the river basin authority (*Agence de Bassin Hydraulique* – ABH) in 2002, several hydrological studies revealed the severity of groundwater depletion and degradation in the whole region. In the following years, the ABH, in cooperation with provincial and local state authorities, attempted to enforce existing formal laws that have required an official licence for groundwater abstraction since 1995. Between 2002 and 2003, they controlled water abstraction more strongly, closed wells and seized drilling equipment, which resulted in heavy protests by farmer groups and drilling companies (Faysse et al., 2012).

In order to prevent conflict between water users and enable sustainable groundwater use, the ABH launched a participatory process, which resulted in the elaboration of an aquifer contract. According to the 1995 Moroccan Water Law, aquifer contracts are instruments to improve regulation mechanisms of groundwater management through stakeholder participation at the aquifer level (Closas & Villholth, 2016). An aquifer committee set up in 2005 developed the so-called 'framework convention' entailing central objectives and legal proposals (including *inter alia* simplified procedures for obtaining water-related permits, formal authorizations of previously 'illegal' wells or increased volumetric charges for groundwater extraction). A series of 'specific conventions', partly overlapping with the framework convention, referred to specific activities such as the enforcement of controls and capacity-building of the 'water police', improvements in water efficiency, the development of additional surface water resources by the state as well as the commission of a feasibility study for a desalination plant. Two official decrees were adopted in 2007 to prohibit the extension of existing irrigated agriculture, be it by using existent or the drilling of new wells and boreholes. In the present article, the term 'aquifer contract' refers to all these three types of regulations. These constitute the basis of the formal interactions between the different stakeholders analysed here (see above, among others, administrations, farmers, irrigation water-user associations) and set the scene for some of the contested issues and related power struggles around groundwater regulation in the Chtouka.

These developments in the Chtouka area, and especially the evolution of groundwater regulating instruments, characterize the context of our analysis of the three research hypotheses. In the present setting, policymakers often accuse farmers of being unaware of the urgency of groundwater depletion and lacking action in this sense. However, in accordance with findings in the literature (Shah, 2009), our first hypothesis is that the aquifer contract emerged as a new institutional arrangement due to an increased awareness of water users and formal water institutions about the urgency to regulate groundwater use in the face of increased scarcity.

Second, we seek to comprehend how adopting a bricolage perspective entailing the analysis of power relations allows for a better understanding of the adoption and reshaping of formal rule implementation of the aquifer contract. The history of the region suggests that in addition to the implementation of the aquifer contract as a new formal rule, this study needs to consider pre-existing rules. As previous studies have shown (Faysse et al., 2012: Houdret et al., 2017), power relationships and practices of corruption influence the (non-)implementation of such rules and therefore also inform our analysis of the rule implementation.

Our third hypothesis is that social and cultural factors shape the implementation of the aquifer contract, leading to multifunctional institutions. In the Chtouka area, migration, settlement and investment dynamics are key for understanding social and cultural factors that influence the implementation of the aquifer contract and the related processes of bricolage. In addition, the contract has faced the criticism that mainly large-scale farmer organizations (such as the *Association Marocaine des Producteurs Exportateurs de Fruits et Legumes* – APEFEL) were included in its negotiation and that farmers not involved in export, village organizations or representation of municipalities were marginalized from the 'participatory' negotiation of the new agreement (Faysse et al., 2012; Matz & Hoffmann, 2013).

Results

Awareness of insufficient rules as a precondition for new governance arrangements

In order to better understand the water users' awareness of the insufficiency and/or non-application of existing rules, we enquired about their knowledge regarding the regulations for groundwater use and the existence of the new aquifer contract. These existing rules are partly formal (such as the need for licences for groundwater pumping) and partly informal (such as agreements between farmers sharing pumping stations). The results suggest that farmers and villagers in Chtouka are well informed about formal obligations related to groundwater abstraction: during the entire field study, we did not encounter a single person who was not informed about regulatory restrictions, the necessary authorization required for borehole drillings and the licensing of extracted water volumes. Interviewees were able to define whether they (or their competitors) acted in line with formal water law or not. As one interviewee explained:

> I don't think that the ABH issues authorisation to every borehole of which we hear the drilling. We always hear the drilling gear at night. Drilling at night already implies that they work in the dark [illegally . . .] imagine an investor, who drills a well at night – that means they have no authorisation.

'You hear them drilling at night' was the most memorable sentence mentioned during various field conversations – not only because it was said so often, but also because one could hear the noise of the night-time drilling. As the above quotation reveals, respondents link drilling noises at night to the illegal activities of water users. These explanations imply that accessing groundwater has become an observable and noticeable event for neighbours and other resource users.

Interviewees were able to describe temporal changes in groundwater governance, as a local tomato farmer illustrated: 'In 2003–2004 the authority started chasing farmers who did not have an authorisation for their wells.' Only two of the respondents[1] knew the term 'aquifer contract'; they were able to recall related developments, measures and events that took place before and during the implementation process, for example, the start of control activities and awareness campaigns. Interviewees referred to control activities that started and became enforced between 2002 and 2008, which is consistent with descriptions provided by representatives of the ABH, ORMVA (Office Régional de Mise en Valeur Agricole) and the province and the literature (Faysse et al., 2012). In the above quotation, the word 'chasing' indicates that interviewees perceived the start of control activities as a rupture to previous public practices, where groundwater was rarely monitored via direct site controls by public organizations. These observations confirm our first hypothesis that the aquifer contract emerged in a context of high and commonly shared awareness of different water users about existing rules for resource governance and their insufficient application leading to the depletion of groundwater.

Power relations, including practices of corruption, influence aquifer contract implementation

Our analysis of the formal rule implementation of the aquifer contract shows that these rules are transformed or reshaped through processes of bricolage, involving power relations expressed *inter alia* through practices of corruption.

In the case of Chtouka, hierarchically organized social relationships (among villagers, farmers and administrations) enabled the formation of social alliances that fostered complex rule-making and role allocation among stakeholders for informal institutional mechanisms, such as corruption. Our interviews suggest that the implementation of the aquifer contract has contributed to the enforcement of corruptive practices in groundwater regulation: more than 18 of our field conversations linked groundwater control and monitoring to corruptive practices and provided detailed descriptions on related mechanisms and social-working rules. The aquifer contract ignored existing power asymmetries; some actors then used their power to circumvent the new regulations, which likely resulted in a reinforcement of corruption. The following quotation from a local, municipal-level political leader mirrors the gradual change of adoption and then circumvention of the new rules through bricolage – the blending of new rules with established practices, in this case corruption:

> In the beginning of the law, the people were more motivated, they took the issue seriously. In the beginning people are motivated, and pushed to do controls. But later, people adapted to the project.
>
> Researcher: What means adapted?
>
> I don't have any proofs, but *you know how things work out here* [emphasis added]. If you know that there is a law and at the same time you find all these drillings, so you know how things work. There is a law, you have the authority which passes by, which observes, which knows that there is a drilling – but nothing changes! So, you can be sure how things work out.

This politician was an elected representative from one of the municipalities visited during the fieldwork (municipalities, unlike *Caidat*s, were not invited to participate in the aquifer contract negotiations). What is special about this quotation is that the interviewee makes a chronological evaluation of the implementation process and points out that over time legally grounded controls have changed. While, in the eyes of the interviewee, controls initially met the legal requirements and objectives, they later became 'adapted'. We interpret the word 'adaptation' used by the interviewee as a reference to temporal processes through which the interviewee suspects the formal rules of the aquifer contract were appropriated and redefined within social practices at the field level. We conclude the latter from the phrase 'you know how things work out here', which the interviewee repeated several times. With different gestures and descriptions, he referred to payment flows that take place under the table, thereby avoiding the word 'corruption'. The statement suggests, on the one hand, that the interviewee views corruptive practices to be something ordinary and regular (i.e., bound to informal rules regarding the price, timing, etc.), a practice affecting different areas of everyday life (a statement explicitly supported by four other respondents). On the other hand, the interviewee avoided the word 'corruption', which we interpret as a precautionary measure in case our recording would be made public.[2] We interpret this observation made in three other cases as a reference to the strong power relationships linked to corruption, and the related fear of respondents raising the issue publicly.

The following interview quotation illustrates the role of power relationships and related social alliances in the negotiation of groundwater access and bribes:

> The controls turned into an investment. There are relationships in the field. That's why you call to say that someone is drilling – that's why there are no controls, because there are relationships. It [reports of illegal drilling] has become an affair of investment, to earn money. [. . .] There are no controls; I talk about it all the time. I already called [organisation and name redacted to protect the source] many times to inform them about an illegal drilling which is taking place near me or near another person who informed me. [. . .] So, what can I do? I make the calls, they arrive, they even seize the equipment and the very next day, you find the same equipment back in the field! I did this many times.

We interpret the phrase 'there are relationships in the field' as a reference to exclusive social alliances in which negotiated arrangements result in non-transparent regulation mechanisms for groundwater resources. In this context, formal law works as a reference point to provide legitimization to these arrangements. For instance, if one refuses to pay a bribe, access to water is restricted with reference to the application of formal law, as the granting of groundwater licences depends on an assessment process in which several administrations have to approve the drilling site. However, the quotation suggests that the interviewee accuses representatives of the administrations of participating in social alliances where corruptive mechanisms become organized along complex rules. As the quotation also indicates, the interviewee links phone calls (to report illicit drillings) to exclusive forms of cooperation between stakeholders from the field and those in administration. This implies that the interviewee perceives this form of cooperation (1) as tied to social relationships of well-organized alliances and (2) to offer the opportunity to participate in a financial allocation scheme of an extra-legal tax system (bribes).

Interviewees have highlighted institutional mechanisms and the complex social organization of corruption by pointing to different rules, roles and responsibilities that are in place and shared among different stakeholders, including farmers, administrations and drilling companies. The following quotation reveals the complexity of the social arrangements and how related institutional mechanisms function in Chtouka. It points to collective conventions that reflect relationships of trust within corruptive systems:

> Researcher: Why did you drill at night?
>
> Because of the *Caid*, the authority.
>
> Researcher: But didn't they know that you drilled a well?
>
> Yes, they knew, but it's *obligatory* [emphasis added] to pay some money and so I did.
>
> Researcher: How much did you pay?
>
> 10,000 MAD[3] for one well.
>
> Researcher: Didn't you need to show an authorisation to the drilling company?

> They don't ask, because they *know that you don't have an authorisation* [emphasis added]. If you have an authorisation, you must inform the company because drilling with an authorisation is cheaper than without one [500 and 800 MAD/m, respectively].
>
> Researcher: Can you please explain how the deal between the authority and you was arranged?
>
> You have to visit them before and to pay them in advance, then you can drill.
>
> Researcher: Why did you not drill a well without this deal?
>
> Because the people talk, the information passes very fast to the *Caid*, so you can't take that risk.
>
> Researcher: Have you already observed farmers who were stopped by the authority while drilling?]
>
> Yes, I have seen. They were several people who started drilling without an authorisation and the *Caid* came and took the equipment because they didn't give money. He comes to see if you have an authorisation and if you don't have, he confiscates the drilling equipment.
>
> Researcher: How passes the information to the *Caid*?
>
> So, everyone knows that drilling a well is illegal, so people talk about it and the *Caid* will know.

This quotation provides several interesting references that underline the institutional nature of corruption in this context. First, by stressing that giving money is *obligatory* the interviewee refers to corruption as a perceived fixed rule within groundwater governance. The note that drilling companies automatically expect one to drill without a licence underlines this argument. Further, the description highlights the different roles and tasks the interviewee ascribes to the different stakeholders, including farmers, the administration and the drilling company. As such, farmers need to make an arrangement with the authority prior to drilling or to inform the drilling company when the drilling has been officially approved by the administration. The role of the drilling company was also explicitly stressed by other interviewees, like this greenhouse tomato farmer:

> If you give honey to the inspectors, they will close their eyes.
>
> Researcher: Was an inspector sent to you?
>
> No, he didn't come. It's the drilling company that is occupied with the management of this affair. You only give the money and they will manage the affair for you [...].
>
> Researcher: So, when investing into agriculture it means that I already have to consider the corruption fee in my prior calculation?
>
> If you don't give backsheesh, you stay at home.

Hence, interviewees described illegal drillings as a common practice that are a result of a high level of social organization and mutual agreements between stakeholders and are linked to well-established rules among the participants. These predictable patterns of social interaction point to the presence of institutional mechanisms that structure corruptive practices and reduce uncertainties associated with formal rule violation. Although formal rules enforced and strengthened by the aquifer contract grant equal rights to all users, interviewees perceived those rules to have been transformed by social alliances into inequitable mechanisms that are selectively applied to water users according to their social (relationships or position within social alliances) or financial capacities (to pay bribery). Other respondents used words such as 'lobby' to point to the level of social organization beyond corruption. In an interview, two representatives of an agricultural organization pointed to the perceived unequal power relationships that restrict access to these social alliances where corruptive rules are set.

Corruption thus emerged as an indicator of power inequalities in the implementation of the aquifer contract, not only because of the financial capacities required, but also because of social resources needed such as influential networks and relationships to the authorities responsible for rule enforcement, etc. At the same time, the aquifer contract itself created a new opportunity for corruptive practices by providing them new legitimacy under the veil of a formal contract.

Social and cultural factors shape the implementation of the aquifer contract

Today, 'water users' as potential stakeholders of the aquifer contract are a socially, culturally and economically heterogeneous group due to, among others, emigration and immigration dynamics. Our analysis reveals that these actors reinvent the formal rules of the aquifer contract in different ways according to their social and cultural affiliations. The rules set in the aquifer contract transformed into multifunctional institutions for social regulation.

The following quotation exemplifies a normative attitude portrayed in 12 of the 45 interviews. The quotation is from an interview with two official representatives of a large agricultural organization with more than 600 farmers:

> The problems are the people who came from outside, the foreigners, who are not from Souss-Massa, so the migrants[4] from Morocco or from other countries. They are responsible because they sell the land afterwards or just leave it behind. So, for example, when a local inhabitant drills a well, I don't have a problem with that, because he has his land here, I have my land here, so each of us is responsible for his land. But they [the migrants] are not conscious about the value of the land, they are not attached to the land.
>
> Researcher: The members of your organisation are all locals, native inhabitants?
>
> Yes.
>
> Researcher: No migrants?
>
> No.

In this case, the interviewee distinguishes between 'native' farmers, whom he considers morally entitled to the groundwater, and 'non-native' farmers who he deems are not. Formal rules become blended with normative concepts of what is right and what is wrong. In this case, these concepts are linked to specific social and cultural affiliations, and formal mechanisms get selectively applied if they correspond to the worldviews and related norms of their applicants. Also, responsibility for groundwater protection by restricting its use is transferred to the 'outsiders' while the natives are perceived to be more legitimate to use the resource. The aquifer contract as a new institution is not only used for regulating groundwater, but also to exercise social control. Thus, formal institutions turn into multi-functional arrangements as they are applied against different backgrounds and objectives.

In the case of the quotation above, the reporting of unauthorized drillings depended on the origin of the farmer. Thereby, the motivation to report illicit drillings is less centred on formal rule compliance and general accounts for groundwater conservation. Rather, the willingness to participate in formal institutions can be interpreted as a means of defending a moral convention that defines who is entitled to the use of water and who is not. The reporting mechanism is interpreted as a means of excluding those farmers whose claim to water use is perceived as illegitimate.

Similarly, the same representatives of the agricultural organization explained that they faced challenges when reporting illicit drillings at the local *Caidat*. However, they argued that official declarations made at the *Caidat* have never been processed, in particular when declarations have affected farmers suspected to share the same cultural affiliations as the person in charge at the processing administration. Consequently, as they described in an interview, the farmer organization started forging new alliances with a representative of an administration who shares the same social and cultural affiliations:

> Concerning the declaration, I have already talked to a [person] at the [administration and name is kept anonymous to protect source], [they are] *correct* [emphasis added] [they] will manage this affair, because when I talked with [them] about the issue, the situation in the field, [they were] cooperative, so we are currently waiting for a response. [. . .] But I talked to the person, whom I already mentioned, [they are] from the region, [they have] this *affiliation* [emphasis added] to the land, and [they] told me that as soon as I observe the person drilling again, I should just call [them] and [they] will send someone immediately. The *Caid* is not taking actions, because the *Caid* is from Tiznit and the person who is drilling is also from Tiznit, so they have a relation.

The trustworthiness of a contact person is stressed by referring to their 'affiliation to the land'. We interpret shared cultural and social affiliations to be an important precondition for cooperation in this context. Thus, the interview suggests that institutional bricoleurs forge new social alliances to regulate groundwater access based on shared cultural and social affiliations.

Discussion and conclusions

Insights from this research support and complement existing findings. Our study confirms the work of Shah (2014) and Petit (2004) in their argument that water users develop new institutions for groundwater governance in response to water scarcity, but our findings also show a more complex picture. While the Chtouka aquifer contract did emerge in this context, not all relevant stakeholders had been included in its development (such as the municipalities or the irrigation and drinking water-user associations). Facing a new institution based on other actors' logics and priorities, some of these stakeholders transformed the contract through institutional bricolage to adapt it to their own purposes and logics. In their perception, water scarcity was thus not the only – and maybe not even the main – trigger and purpose of 'their' aquifer contract.

The different stakeholders' motivations for engaging in an aquifer contract may be very diverse and not always linked to water management. As our study shows, bricoleurs blend the rules of the aquifer contract with normative concepts, leading to new social alliances based on reinvented shared identity characteristics and tied to a perceived common cultural background. In the case of Chtouka, some groups apply formal water rules selectively to water users who they perceive as illegitimate and thus transform institutions designed to manage groundwater into multifunctional institutions aimed at exercising social control. This supports claims for improved awareness of power relations in research and the need for local case studies (Cleaver & de Koning, 2015; Hall et al., 2014).

Our results not only confirm the existence of Cleaver's 'bricolage arrangements', but also provide concrete insights on how power materializes in new water governance arrangements and impacts non-water-related issues. Cleaver (2012) points to the importance of identity characteristics for the way institutional bricoleurs shape and transform institutional mechanisms. In addition, we argue that to better understanding institutional transformation, researchers and practitioners need to analyse the processes through which the reinterpretation of existing, shared affiliations transforms collective identity characteristics, and how this contributes to the transformation of institutional arrangements. The case of Chtouka shows how bricoleurs blend the logic of formal institutional processes with their own normative values, which are bound to specific identity, social and cultural characteristics. When taking ownership of an aquifer contract, the stakeholders interpret its rules and functioning against their personal background and interests and act accordingly. This confirms Cleaver's (2012) hypothesis that the new institutions' legitimacy strongly depends on its embedding in the stakeholders' identity. Further research should expand existing bricolage approaches to focus more precisely on the interplay of identity-generating processes and institutional bricolage. The specific socio-cultural setting, encompassing norms and values, but also the role of corruption in bricolage should be given more attention, as it affects the legitimacy and functioning of water governance institutions.

Our research also shows that the widespread hypothesis of policymakers that aquifer contracts – in contrary to top-down approaches – are participatory instruments that allow for improved user compliance (for an overview on the Moroccan case, see Molle, 2017) is only partially true. User compliance requires legitimate institutions – however, as illustrated above, this depends on adequately addressing the users' diverse logics of engagement in new institutions. These findings are also relevant for policymakers. Based on our insights, we can summarize that the design and implementation of aquifer contracts should not only be based on inclusive approaches, but also needs to consider local socio-economic power relations and the role of identity that may impact water governance. Policymakers should also give more attention to the mobilization of (new) collective identities or shared narratives that could support adherence to the new institution and thus its legitimacy and functioning. Finally, a smooth implementation requires conflict prevention that could be supported by legitimate and independent (eventually donor-supported) complain and dispute settlement mechanisms.

Notes

1. Other than those employed by ABH, the province, *Caidat* and the Office de Mise en Valeur Agricole (agricultural development board).

2. While the interview was being recorded, the participating office staff warned the interviewee several times to be careful. He responded that it is important to him to report the truth.
3. Moroccan dirham = US$0.10.
4. They referred to investors and large farming companies.

Acknowledgements

We thank all those who kindly agreed to be interviewed and allowed us to assist their different activities in Morocco. We also thank the anonymous reviewers for their helpful comments.

Disclosure statement

No potential conflict of interest was reported by the authors.

Funding

We acknowledge that no financial interest or benefit arose from the direct applications of this research. Research for this article was conducted for the project 'Stability and Development in the Middle East and North Africa', funded by the German Federal Ministry for Economic Co-operation and Development (BMZ) [Grant number: 2014.4120.3]. The contents of this document are the sole responsibility of the authors and do not reflect the position of the BMZ.

References

Anzar Conseil. (2016a). *Etude et elaboration de contrat de nappe de Chtouka (Province Chtouka Ait Baha). Mission 3: Elaboration du plan d'action et du dossier du contrat de nappe (Part 2. Dossier technique: Synthèse hydrogéologique, état de connaissance sur les eaux souterraines)*. Internal Consultancy Report.

Anzar Conseil. (2016b). *Etude et elaboration de contrat de nappe de Chtouka (Province Chtouka Ait Baha). Mission 3: Elaboration du plan d'action et du dossier du contrat de nappe (Part 4. Synthèse et plan d'action)*. Internal Consultancy Report.

Bencherifa, A. (1980). *Chtouka et Massa: Une région su Sous en cours the Transformation. Étude géographie agraire*. Publications de la Faculté des Lettres et Science Humaines 5.

Clandinin, D. J., & Connelly, F. M. (2000). *Narrative inquiry: Experience and story in qualitative research*. Jossey-Bass.

Cleaver, F. (2002). Reinventing Institutions: Bricolage and the social embeddedness of natural resource management. *The European Journal of Development Research, 14*(2), 11–30. https://doi.org/10.1080/714000425

Cleaver, F. (2005). The inequality of social capital and the reproduction of chronic poverty. *World Development, 33*(6), 893–906. https://doi.org/10.1016/j.worlddev.2004.09.015

Cleaver, F. (2012). *Development through Bricolage: Rethinking institutions for natural resource management*. Routledge.

Cleaver, F., & de Koning, J. (2015). Furthering critical institutionalism. *International Journal of the Commons, 9*(1), 1–18. https://doi.org/10.18352/ijc.605

Closas, A., & Molle, F. (2016). *Groundwater governance in the Middle East and North Africa Region: Groundwater governance in the Arab World* (IWMI Project Publication No. 1). International Water Management Institute. http://gw-mena.iwmi.org/wp-content/uploads/sites/3/2017/04/Rep.1-Groundwater-Governance-in-MENA.pdf

Closas, A., & Villholth, K. (2016). *Aquifer Contracts: A means to solving groundwater over-exploitation in Morocco? (Groundwater Solutions Initiative for Policy and Practice (GRIPP) Case Profile Series 01)*. International Water Management Institute. http://gripp.iwmi.org/gripp/publications/case-profile-series/issue-01.pdf
Crawford, S. E., & Ostrom, E. (1995). A grammar of institutions. *American Political Science Review*, *89*(3), 582–600. https://doi.org/10.2307/2082975
Daadaoui, M. (2011). *Moroccan Monarchy and the Islamist challenge: Maintaining Makhzen power*. Palgrave Macmillan.
de Bont, C., Veldwisch, G. J., Komakech, H. C., & Vos, J. (2016). The fluid nature of water grabbing: The on-going contestation of water distribution between peasants and agribusinesses in Nduruma, Tanzania. *Agriculture and Human Values*, *33*(3), 641–654. https://doi.org/10.1007/s10460-015-9644-5
De Stefano, L., & Lopez-Gunn, E. (2012). Unauthorized groundwater use: Institutional, social and ethical considerations. *Water Policy*, *14*(S1), 47–160. https://doi.org/10.2166/wp.2012.101
Del Vecchio, K., & Barone, S. (2018). Has Morocco's groundwater policy changed? Lessons from the institutional approach. *Water Alternatives*, *11*(3), 25. https://www.water-alternatives.org/index.php/alldoc/articles/vol11/v11issue3/458-a11-3-11/file
Douglas, M. (1986). *How institutions think: The Frank W. Abrams lectures*. Syracuse University Press.
Faysse, N., El Amrani, M., El Aydi, S., & Lahlou, A. (2012). Formulation and implementation of policies to deal with groundwater overuse in Morocco: Which supporting coalitions? *Irrigation and Drainage*, *61*(3), 126–134. https://doi.org/10.1002/ird.1652
Forsyth, T., & Johnson, C. (2014). Elinor Ostrom's legacy. Governing the commons and the rational choice controversy. *Development and Change*, *45*(5), 1093–1110. https://doi.org/10.1111/dech.12110
Hall, K., Cleaver, F., Franks, T., & Maganga, F. (2014). Capturing Critical Institutionalism. A synthesis of key themes and debates. *The European Journal of Development Research*, *26*(1), 71–86. https://doi.org/10.1057/ejdr.2013.48
Houdret, A. (2012). The water connection: Irrigation and politics in southern Morocco. *Water Alternatives*, *5*(2), 284–303. https://www.water-alternatives.org/index.php/alldoc/articles/vol5/v5issue2/170-a5-2-6/file
Houdret, A., Kadiri, Z., & Bossenbroek, L. (2017). A new rural social contract for the Maghreb? The political economy of access to water, land and rural development. *Middle East Law and Governance*, *9*(1), 20–42. https://doi.org/10.1163/18763375-00901003
Kelle, U. (2007). Theoretisches Vorwissen und Kategorienbildung in der 'Grounded Theory'. In U. Kuckartz, H. Grunenberg, & T. Dresing Eds., *Qualitative Datenanalyse: Computergestützt. Methodische Hintergründe und Beispiele aus der Forschungspraxis* (2nd ed., pp. 32–49). VS Verlag für Sozialwissenschaften. https://doi.org/10.1007/978-3-531-90665-2
Lamnek, S., & Krell, C. (2010). *Qualitative Sozialforschung* (5th ed.). Beltz.
Long, N. (2001). *Development sociology: Actor perspectives*. Routledge.
Lopez-Gunn, E., & Martinez Cortina, L. (2006). Is self-regulation a Myth? Case study on Spanish groundwater user associations and the role of higher-level authorities. *Hydrogeology Journal*, *14* (3), 361–379. https://doi.org/10.1007/s10040-005-0014-z
Lopez, G. I., Figueroa, M., Connor, S. E., & Maliski, S. L. (2008). Translation barriers in conducting qualitative research with Spanish speakers. *Qualitative Health Research*, *18*(12), 1729–1737. https://doi.org/10.1177/1049732308325857
Matz, M., & Hoffmann, A. (2013). Appui-conseil à la « Convention Cadre » pour la préservation et le développement des ressources en eau dans le bassin hydraulique du Souss-Massa: Etat des lieux, analyse des points de blocage, proposition d'un plan d'action et structure de suivi. Unpublished manuscript.
Molle, F. (2017). Conflicting policies: Agricultural intensification vs. water conservation in Morocco. Working Paper/Rapport de Recherche No. 1. Montpellier, France. G-Eau. http://www.g-eau.net

Ostrom, E. (1990). *Governing the commons. The evolution of institutions for collective action.* (Canto Classics ed.) Cambridge University Press.

Ostrom, E. (2005). *Understanding Institutional Diversity.* Princeton University Press. https://doi.org/10.2307/j.ctt7s7wm

Ostrom, E., Cox, M. & Schlager, E. (2014). An assessment of the institutional analysis and development framework and introduction of the socio-ecological systems framework. In Sabatier, P. & Weible, C. (Eds.), *Theories of the policy process* (3rd ed., pp. 267–306). Westview Press.

Pérennès, J. J. (1993). *L'eau et les hommes au Maghreb. Contribution à une politique de l'eau en Méditerranée.* Karthala.

Petit, O. (2004). La surexploitation des eaux souterraines: Enjeux et gouvernance. *Natures Sciences Sociétés, 12*(2), 146–156. https://doi.org/10.1051/nss:2004020

Popp, H. (1983). *Moderne Bewässerungslandwirtschaft in Marokko: Staatliche und individuelle Entscheidungen in sozialgeographischer Sicht (Erlanger geographische Arbeiten, Sonderband 15).* Fränkische Geographische Gesellschaft.

Reddy, V., Reddy, M., & Rout, S. (2014). Groundwater Governance: A tale of three participatory models in Andhra Pradesh, India. *Water Alternatives, 7*(2), 275–297. https://www.water-alternatives.org/index.php/alldoc/articles/vol7/v7issue2/247-a7-2-1/file

Robbins, P. (2000). The Rotten Institution. Corruption in natural resource management. *Political Geography, 19*(4), 423–443. https://doi.org/10.1016/S0962-6298(99)00087-6

Saldaña, J. (2016). *The coding manual for qualitative researchers* (3rd ed.). SAGE Publications.

Shah, T. (2009). *Taming the Anarchy? Groundwater governance in South Asia.* Resources for the Future Press.

Shah, T. (2014). *Groundwater Governance and Irrigated Agriculture* (Tec Background Papers No. 19). Global Water Partnership (GWP). gwp.org/globalassets/global/toolbox/publications/background-papers/gwp_tec_19_web.pdf

Siebert, S., Burke, J., Faures, J. M., Frenken, K., Hoogeveen, J., Döll, P., & Portmann, F. T. (2010). Groundwater use for irrigation – A global inventory. *Hydrology and Earth System Sciences, 14* (10), 1863–1880. https://doi.org/10.5194/hess-14-1863-2010

Sippel, S. (2016). Breaking Ground: Multi-family farm entrepreneurs in Moroccan export agriculture. *Journal of Rural Studies, 45*, 279–291. https://doi.org/10.1016/j.jrurstud.2016.03.013

Teorell, J. (2007). *Corruption as an institution: Rethinking the nature and origins of the grabbing hand* (QoG Working Paper Series 2007:5). The Quality of Government Institute. https://www.sahlgrenska.gu.se/digitalAssets/1350/1350653_2007_5_teorell.pdf

Part II

Groundwater management

Exploring the future impacts of urbanization and climate change on groundwater in Arusha, Tanzania

Tunde Olarinoye, Jan Willem Foppen, William Veerbeek, Tlhoriso Morienyane and Hans Komakech

ABSTRACT

We combine satellite imagery, urban growth modelling, groundwater modelling and hydrogeological field expeditions to estimate the potential impacts in 2050 of rapid urbanization and climate change on groundwater in Arusha, Tanzania, and by extension similar areas in Sub-Saharan Africa. Our analysis suggests that a reduction of groundwater recharge by 30–44% will cause groundwater levels to drop by up to 75 m, mainly due to increased evapotranspiration and to an expansion in paved surface. If this scenario becomes reality, we predict that wells will run dry, creating health, social and environmental risks.

Introduction

There is a direct link between population growth, urbanization and water resources (Thomas & Zuberi, 2012). The growth of cities as a result of population growth and rural–urban migration is a global phenomenon (Kessides, 2006). In 2050, an expected 2.4 billion people will live in Africa, compared to 1.1 billion now. Assuming the urban population increases from 40% now to 55% in 2050 (UN/DESA, 2015), the urban population will grow from 440 million to 1320 million, a three-fold multiplication in only 35 years. With the exception of Asia, Africa will have more urban residents than any other continent in the world (Clark, 1998; Parnell & Walawege, 2011). Along with rapid growth in urban populations, there is a proportional rise in demand for potable water that typically exceeds the capacity of existing facilities (Taylor et al., 2004). Groundwater plays a fundamental but often unappreciated role in the economic and social well-being of urban areas (Taylor et al., 2009). More than 50% of the African population relies on groundwater for domestic, agricultural and industrial use (Carter & Parker, 2009). In many situations, without groundwater there would be a deficit in water supply. Groundwater is more readily available at or close to the demand point and is often potable at the source (Taylor & Barrett, 1999).

Besides the increase in water demand, urbanization also modifies the natural land cover. The long-term sustainability of urban groundwater relies on regular input of water to the subsurface as groundwater recharge (Taylor & Barrett, 1999). The amount of water that infiltrates into the aquifer during a precipitation event depends to a large extent on the land cover types. Land cover is frequently subjected to spatial and temporal change due to urbanization, and this alters the hydrological characteristics of the land surface, modifying infiltration rates and runoff processes (Foster et al., 1999; Mahmoud et al., 2015).

Where the city lacks good sanitation facilities, urban expansion often also leads to contamination of fresh groundwater (Foster et al., 1999). For example, the use of low-cost on-site sanitation facilities like pit latrines and septic tanks is very common in many countries in Sub-Saharan Africa (SSA). The subsurface is being used as containment for domestic and industrial waste, and sewerage systems and wastewater treatment plants are often not included in essential urban infrastructure because they are considered expensive. Therefore, the susceptibility of groundwater sources to contamination is very high in SSA (Taylor & Barrett, 1999). Yet, despite being a contamination source, on-site sanitation systems contribute to additional recharge to the aquifer (Morris et al., 2003).

Countries in SSA are among the most vulnerable to the impacts of climate change, which also threatens water availability (Case, 2006). Yet, climate conditions vary across Africa, and sub-regional climate change predictions are usually complicated (Christensen et al., 2007). While there is an agreement on temperature projections, projections of rainfall remain uncertain (Christensen et al., 2007; FCFA, 2017). Combined with the expected rapid socio-economic changes, predicting future water availability in SSA becomes a difficult exercise (Calow & Macdonald, 2009). The sustainability of turning to groundwater to ease the water scarcity brought by increased demand or climate change is therefore unclear, even though in many regions, urbanization is expected to have a much more profound impact on groundwater than climate change (Calow & Macdonald, 2009; Carter & Parker, 2009; Taylor et al., 2009).

Land use and land cover (LULC) change models, which attempt to extrapolate historical LULC transformations, provide a powerful lens for examining future urban development trends. Urban growth patterns are often determined by local geographic interactions, where neighbouring conditions determine the actual formation of urban land cover, instead of 'top-down' drivers (Batty, 2007). This is especially the case in rapidly growing urban areas in the developing world, where growth is predominantly uncontrolled. For climate change, the Intergovernmental Panel on Climate Change also provides documented reports for global and regional climate projections (Niang et al., 2014; Shukla et al., 2019). From a hydrogeological point of view, the results of urban and climate projections would be more relevant if their impacts on groundwater could be quantified. Fortunately, it is possible to assess the hydrogeological impacts of urbanization and climate change by integrating the outcomes of LULC change models and climate projections with a groundwater modelling approach in a multidisciplinary manner.

Studies assessing the combined impacts of urbanization and climate change on groundwater resources in rapidly growing and climate vulnerable SSA cities are rare. In this study, we combined knowledge of LULC classification, urban growth modelling and groundwater modelling to explore the combined quantitative impacts of urbanization and climate change on groundwater in SSA, with a special focus on the fast-growing city of Arusha, Tanzania. The outcomes provide an outlook on the state of future groundwater and should play a role in supporting local authorities to channel resources in the right direction.

Case study area

We used Arusha as an example city because it has characteristics similar to many typical SSA cities: principal reliance on groundwater; rapid urban growth; and location in a region with a large uncertainty in climate change predictions (FCFA, 2017). Arusha is the third-largest city in Tanzania, in the foothills of the volcanic plain of Mount Meru in the north-eastern part of the country. In 2012 it had a population of 416,000 (NBS and Office of Chief Statistician Zanzibar, 2013). More recent statistics are not available. Many of its residents are low and average income earners, living in many informal settlements scattered across the city. Elevations range from 1100 m in the south to 4566 m on the peak of Mount Meru. It is in a semi-arid zone, and its environmental and hydrological stability strongly depends on the evergreen forests of Mount Meru (Oikos, 2011). Rainfall distribution varies with altitude. The average annual rainfall is 794 mm, and potential evapotranspiration is 924 mm/y (GITEC, 2011). Temperature is highest in February, with an average of 25 °C, and 13 °C in the coolest month, July.

The landform is associated with the East African rift-related intrusive and extrusive event, which is responsible for the formation of Mount Meru, a young volcano of Pleistocene origin (Kashaigili, 2010). Several volcanic eruptions have developed a chronological interlayering of volcanic ashes, sands, gravels, tuffs and basalt. This has resulted in complex heterogeneous hydrogeological settings with varying permeability (Ghiglieri et al., 2008). The main aquifer occurs singularly in the volcanic alluvium or at times as a perched system with high permeability and an average thickness of 200 m (Ghiglieri et al., 2008). Hydraulic conductivity (k) varies from 0.5 to 5.9 m/day in different parts of the aquifer (Ong'or & Long-Cang, 2007).

Groundwater accounts for over 80% of the daily water supply in Arusha (EWURA, 2015). Recharge into the unconfined aquifer is mainly from infiltration of excess rainfall, with a significant amount coming from the slopes of Mount Meru (GITEC, 2011; Kashaigili, 2010). Climate variability threatens water availability in north-east Tanzania, as evidenced by dry spells in 2003 and 2005, for example (FCFA, 2017). The prolonged drought hurt farmers, causing severe livestock loss and repeated food crises. It resulted in water supply shortage and also affected the natural regeneration capacity of the forest (Oikos, 2011).

Arusha has demonstrated a great 'urban pull' effect, attracting internal migrants (Wenban-smith, 2015). In 2002, it became one of the largest regional capitals, after Dar es Salaam and Mwanza (NBC, 2014). The draw of this city for migrants is mainly due to its economic potential, driven by touristic and international presence. Much of the growth was unplanned, and only recently was a long-term development plan proposed, with a horizon of 2035. Yet, this proposed plan has never been evaluated to assess it potential impacts on water resources and future sustainability. The development between 1995 and 2015 was very rapid, and this had consequences for water supply and availability (Agwanda & Amani, 2014). The Arusha Urban Water and Sanitation Agency (AUWSSA), the local public water supply agency, has been unable to meet the growing demand of different sectors (EGIS, 2017). The infrastructural, social and developmental changes needed to dampen the effects of rapid urbanization and climate change are either not in place or happening at a very slow rate.

Methodology

- Groundwater level mapping

From November 2015 to January 2016, we carried out fieldwork, including a well inventory, in the Osunyai, Unga, and Sombetini wards of Arusha. When possible, the depth from the surface to groundwater was manually measured in open shallow wells. We also mapped spring locations. Springs were used assuming that groundwater head equals surface elevation. In addition to the fieldwork, we obtained groundwater level information from AUWSSA wells. Using kriging, the data were used to construct a groundwater-level contour map, which was then used to arrive at an initial head distribution in the groundwater model (see below) and to calibrate the groundwater model.

- Land use and land cover change model and analysis

Our estimates of future LULC changes were based on extrapolations of historic urban growth patterns, which were obtained by application of a dedicated LULC change model (Veerbeek et al., 2015). Using historic LULC maps as a training set, the model attempts to derive location-specific transition rules which explain LULC transitions (e.g. grassland to built-up area) as functions of the existence of specific spatial features (e.g. proximity to major roads). These weight-based rules are optimized using a machine learning algorithm that attempts to estimate the correct weight distribution to effectively produce a set of transition rules that mimic the observed LULC changes in the area of interest. Once sufficiently good fit was achieved between the projected LULC map and the observed LULC map of the area, the derived transition rules were used to develop a series of LULC projections for a given temporal horizon. The model was based on adaptation of the weights-of-evidence method (Agterberg & Bonham-Carter, 1990; Goodacre et al., 1993) for weight-based transition rules, while Veerbeek et al. (2015) incorporated a memetic-algorithm-based auto-calibration method to enhance the model's accuracy. Since its initial development in the early 2000s, the model has been applied to a broad range of projects (Rodrigues & Soares-Filho, 2018). The model's accuracy and performance have been extensively compared to other LULC models (e.g. Mas et al., 2014; Yi et al., 2012).

For this study, LULC data were based on Landsat 5TM and Sentinel-2 multispectral data covering 1995 to 2015. LULC classification was performed by a multi-temporal maximum-likelihood classification (Congedo, 2017) at a resolution of 30 m. The classification scheme was based on an adapted version of the NLCD land cover class definitions (Homer et al., 2004), emphasizing classes for built-up areas. Projections of the greater Arusha area were developed for 2015–2050 at 5-year intervals. To obtain an expressive characterization of the trends in the growth projections, the maps were post-processed using a methodology developed by Angel et al. (2007) in which the projected built-up areas and urban development were classified into meaningful components (e.g. infill, urban extension and leapfrogging development). This ensured consistency when delineating the urban footprint and the associated descriptive statistics.

- Groundwater recharge and abstraction estimation

Spatially distributed groundwater recharge for 2015 and 2050 from direct rainfall infiltration was estimated using WetSpass-M, a distributed model which computes water balance components (evapotranspiration, surface runoff and groundwater recharge) at the raster cell level (Abdollahi et al., 2012). The model takes into account land use type, slope, soil, depth to groundwater, evapotranspiration and temperature to estimate surface–subsurface water balances (Batelaan & De Smedt, 2007). LULC fractions are used as criteria weighing factors for estimating the water balance (Abdollahi et al., 2017). This approach makes WetSpass-M very appropriate for this study to estimate recharge and recharge variations due to LULC changes, which were then later used as input for the groundwater model.

Soil information was obtained from the Harmonized World Soil Database. Rainfall and evapotranspiration reported by GITEC (2011) were used. Røhr and Killingtveit (2003) derived equations for correcting the altitudinal effect on precipitation for the slopes of Kilimanjaro, ca. 50 km from Arusha. These equations were applied to estimate rainfall on the slopes of Mount Meru, which is an important recharge zone. To accommodate the uncertainties in the region's climate projections, groundwater recharge was estimated for two climate change scenarios: (A) 10% *increase* in precipitation by 2050; and (B) 11% *decrease* in precipitation by 2050. These were the extreme cases based on the projections for Tanzania with RCP 8.5, the worst emission pathway (FCFA, 2017). For both scenarios, a clear warming trend is noticeable: mean annual temperature is projected to increase by between 0.8°C and 1.8°C by 2050 (FCFA, 2017). Therefore, an average rise of 1.3 °C in temperature was used in estimating recharge for 2050 with the WetSpass-M model.

Groundwater abstraction was mainly from water supply agency boreholes and individual private wells. The average daily groundwater abstraction and location of boreholes were obtained from AUWSSA. Abstractions from private wells and groundwater return flows were estimated at the household level. Average household size, average daily groundwater abstraction and return from unlined pit latrines and septic tanks were estimated from data obtained from a household survey carried out during a field campaign in May 2017. To avoid overestimation, groundwater abstraction and return flow were only estimated for households within the urbanized areas of the aquifer boundary. The net groundwater abstraction from private wells estimated for 2015 and 2050 (see 'Calculation of Characteristic values' in the online supplemental data) was used as input for the steady-state groundwater model by evenly distributing the values over urbanized cells. Historically, limited resources have been directed to water infrastructure in Arusha, especially for groundwater development. This has encouraged the widespread construction of deep and shallow private wells, which were also perceived as a more reliable source of water than AUWSSA. We therefore assumed a worst-case scenario whereby AUWSSA would not expand its groundwater abstraction capacity and so would be unable to satisfy future water demand. Therefore, groundwater abstraction by the water agency between 2015 and 2050 remained unchanged.

- Groundwater model setup and description

A steady-state MODFLOW (McDonald & Harbaugh, 1988) groundwater model was used to simulate the water balance for 2015 and 2050. The two years were then compared to assist in exploring spatial changes in groundwater levels as the combined effect of urbanization and climate change. Our steady-state approach is essentially incorrect, since a significant part of the abstracted groundwater in both years was probably derived from storage. However, to build a useful transient model, three main categories of data are required (Zhou & Li, 2011): static data to construct a detailed 3D hydrogeological framework, including aquifer storage parameters; dynamic data to assess groundwater recharges and discharges (e.g. river outflows, spring discharges); and temporal data or time series of groundwater levels in the aquifer system. In the Arusha region, most of these data were not available, so the construction of a transient model was not useful. In the apparent state of groundwater depletion the aquifer complex is in, we considered the results of the steady-state model to represent a worst-case scenario, whereby the transient effect of groundwater release from storage into the model, which reduces spatial groundwater level decline patterns compared to a no-storage steady-state model set-up, was absent.

The aquifer was considered a heterogeneous unconfined/semi-confined system (Ghiglieri et al., 2010), and, guided by the available geophysical survey reports, the model was divided into four layers (see Figures S1 and S2 in the online supplemental data for a conceptual aquifer cross section and map view of the model set-up) with different permeability fields. All AUWSSA boreholes were located in the lowest layer, and shallow private wells in the second layer. Several studies have established that inflow into the aquifer was mainly from rainfall infiltration and fluxes from Mount Meru (Ghiglieri et al., 2012; GITEC, 2011; Kashaigili, 2010; Ong'or & Long-Cang, 2007). Head-dependent flow boundaries (Cauchy conditions; general head boundary package in MODFLOW) were used to simulate a horizontal flux from the forest area of Mount Meru and outflow in the south-east part of the aquifer. Similarly, head-dependent flow boundaries were used to simulate the two main rivers: the Burka and Themi, both flowing throughout the year, were included in the model (river package in MODFLOW).

The surface elevation was derived from Aster GDEM (https://earthexplorer.usgs.gov) with a 30 m resolution, resampled to the groundwater model grid size of 250 m. Measured groundwater level data (see above) was used as initial groundwater head and to calibrate the 2015 model.

The Parameter ESTimation (PEST) package of MODFLOW was used to determine optimized aquifer parameters, whereby general head boundary conductance, riverbed conductance, and horizontal and vertical conductivity were optimized. Several different solutions were achieved during optimization of aquifer parameters. The combination of parameters that gave the best correlation between observed and calculated head was evaluated and used for scenario calculations.

Results

Groundwater level mapping

The groundwater head distribution pattern (Figure S2a in the online supplemental data) obtained from kriging measured groundwater levels from wells, springs and AUWSSA wells indicated that flow of groundwater was from north-east to south-west, whereby in the northern part of the area flow was directed more to the west, downhill from Mount Meru, and in the southern part, the flow was south to south-east. Measured groundwater level data only covered elevations between 1100 m and 1500 m, which mostly belong to the three wards, Osunyai, Unga and Sombetini. A spring at 1750 m elevation, which was captured by AUWSSA, was used to complement groundwater head observations.

LULC change model

Analyses of the derived LULC maps revealed that Arusha's total urban footprint grew by 287% from about 37.2 km^2 in 1995 to 144.2 km^2 in 2015 (Table 1; Figure 1(a)). This represented an annual growth rate of 7% during this period, which could be attributed to Arusha's rapid development into a major city. Urban development during this period connected the two urban cores and extended urban built-up areas primarily towards the south. An additional urbanized 'finger' developed on the eastern side of Arusha, creating a strip of rural open land along the north-south axis (Figure 1(a)).

Extrapolation of the derived growth trends resulted in a projected 180% increase of the urban footprint in 2050 (Figure 1(b)) compared to its 2015 size (Table 1), with an associated 23.2-year doubling period. This suggested that the urban development rate might be decreasing compared to the initial growth in the 1995–2015 interval, indicating a maturation of the initial rapid transition into the consolidation of Arusha's growth into

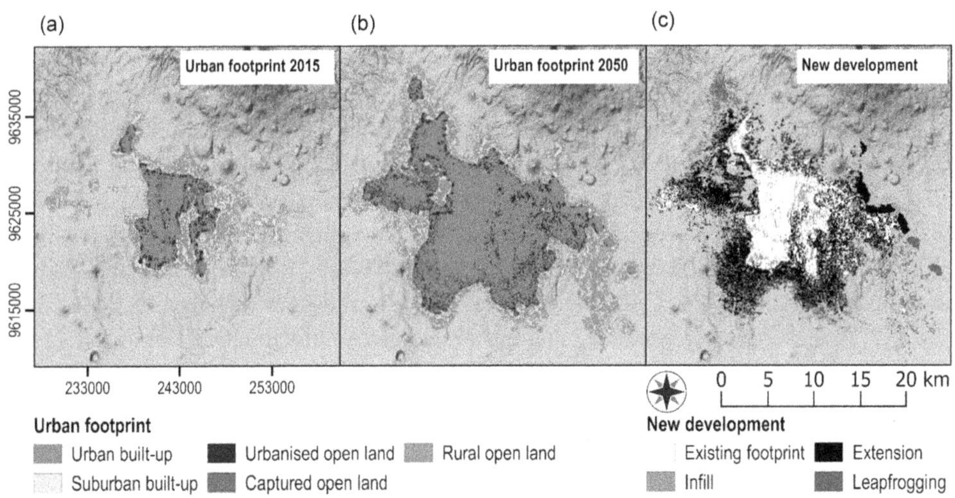

Figure 1. Urban area and new development between 2015 and 2050.

Table 1. Estimated urban footprint area in 1995 and 2015 and new development.

Land cover type	Area in 1995 (km^2)	Area in 2015 (km^2)	Area in 2050 (km^2)	Difference, 2015–2050 (%)	Annual change, 2015–2050 (%)
Urban built-up	7	48	208	332	4.3
Suburban built-up	9	24	42	76	1.6
Urbanized open land	2	12	42	240	3.6
Captured open land	0.5	3	11	275	3.9
Rural open land	19	57	101	77	1.6
Total footprint	37	144	404	180	3.0

a regional urban hub. This was also exemplified by the projected transition of low-density suburban areas into higher-density urban built-up areas, which was expected to increase by over 300%, resulting in a vast urbanized area of approximately 208 km^2 (Table 1). Pattern analysis (Figure 1(c)) of the new development projected about 8% of urban infill and 14% of leapfrogging (disconnected growth). However, most of the new development (78%) consisted of growth along the western and southern fringes of the existing city (urban extension). Extension into a contiguous urbanized area seemed to be the result of a two-step development cycle initiated by uncontrolled leapfrogging, after which, over time, these areas connected to the existing urban core. This expansive growth behaviour often increases land fragmentation, giving rise to depreciation of land resources and declining population density, and it is an important attribute of a 'sprawling' city.

In Arusha, groundwater is the main water source, and on-site sanitation is very common. Leapfrogging will spatially increase the number of drilled wells and potential groundwater contamination sources, mostly pit latrines and septic tanks. The hillslopes of Mount Meru were predominantly agricultural land, and urban development initiated by leapfrogging might force agricultural lands to move further uphill to the forest areas. Since considerable recharge comes from this area (Figure S3, in the online supplemental data), displacing and forcing farmland uphill will have an effect on recharge and therefore on the development of groundwater systems below Arusha.

Recharge estimation

Groundwater recharge was estimated to be 107 mm/y (Table 2), close to the groundwater recharge estimate of 114 mm/y for the Arusha area by GITEC (2011). The estimated recharge decreased by 23% in 2050 (Table 2) due to LULC changes resulting from the predicted urban growth (business as usual, or without climate change; Table 2). In climate change Scenario A (10% precipitation increase, 1.3 °C temperature increase) groundwater recharge further decreased by 30% of the 2015 value (Table 2), while in Scenario B (11% precipitation decrease, 1.3 °C temperature increase) groundwater recharge decreased 44% (Table 2). Irrespective of a predicted increase or decrease of precipitation (Figure S4, in the online supplemental data), increasing air temperature played a key role in reducing groundwater replenishment.

Groundwater model

Details on the groundwater model calibration and water balance results are given in the online supplemental data. The changes in groundwater levels between 2015 and 2050 were compared for scenarios including business as usual only, as well as combined scenarios of business as usual and climate change. The changes in groundwater levels between 2015 and 2050 in business as usual plus Scenarios A and B yielded a groundwater drawdown pattern, which increased towards the west part of the aquifer and elongated north–south (Figure 2). The north–south elongation of the depression is due to the high difference in head gradient in the slopes of Mount Meru. For Scenario A, in which future rainfall increased by 10%, groundwater levels decreased by 10–55 m in the southern part of the area (Figure 2). This decrease was larger than observed when only urbanization scenarios were considered (40–55 m; Figure S6 in the online supplemental data).

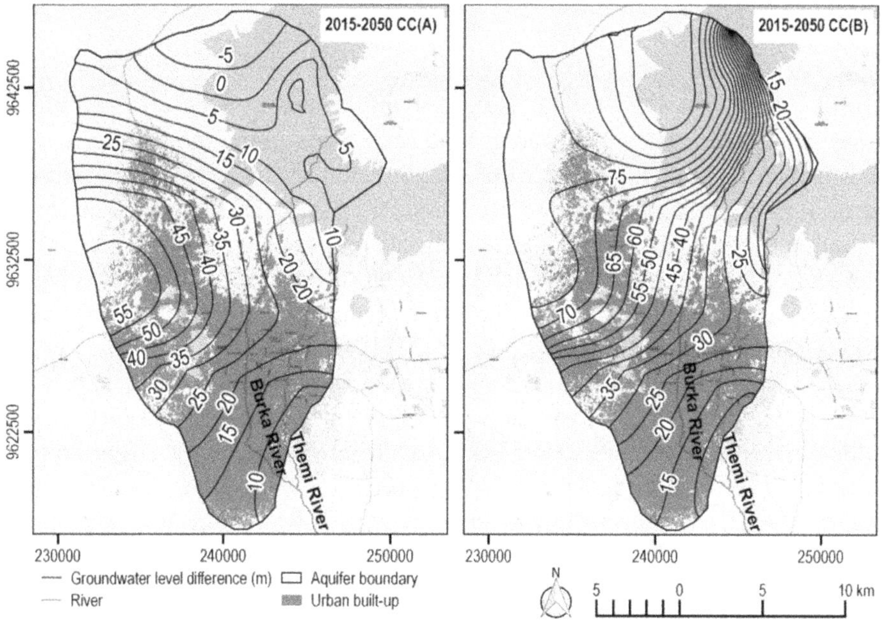

Figure 2. Combined impacts of predicted urban growth and projected climate change scenarios for 2050 on modelled groundwater level decline.

Table 2. Groundwater recharge from rainfall.

Year	Amount (mm/y)	Difference (%)
2015	107	–
2050 Business as usual	82	23
2050 Climate change scenario A	74	30
2050 Climate change scenario B	59	44

The southern part was dominated by dense urban clusters, and much of the predicted growth occurred there as well. Arusha largely relies on groundwater, and the predicted urban growth had a dual impact: more groundwater abstraction and less groundwater recharge, resulting from LULC changes. The increasing future temperature trend also limited the amount of rainfall infiltrating into groundwater, despite a 10% increase in rainfall. If the observed urban growth pattern persists and future rainfall shifts towards a positive side, this scenario might be the best case to expect for future groundwater in Arusha.

The opposing rainfall direction in Scenario B only further intensified the effects of urbanization on groundwater level drop. Due to the combined effects of urbanization and climate change Scenario B, groundwater levels dropped between 15 m and 75 m in the southern part of the aquifer (Figure 2). This implies that the effects of urbanization might be exacerbated by those of future climate change and that more impacts will be felt if the climate shifts towards Scenario B.

Discussion

We find that between 2015 and 2050, Arusha will rapidly grow to 179% the size of its 2015 urban footprint. Due to the increased impervious surface area, groundwater recharge will be reduced by 23%, and groundwater levels in the modelled groundwater basin, which encompasses a large part of western and south-western urban Arusha and the surrounding area, are expected to drop 55 m at most. Considering climate change as well, groundwater recharge may fall 30–44% by 2050, causing groundwater levels to drop at most 75 m. Interestingly, this greater reduction (30–44% vs. 23%) was mainly due to a stronger warming trend in the Arusha region, giving rise to more evapotranspiration and less groundwater recharge.

The scenarios we considered were the extreme ones based on the projections for Tanzania under RCP 8.5, which is the worst emission pathway (FCFA, 2017). A clear warming trend is apparent in both scenarios, with mean annual temperatures expected to increase by between 0.8 °C and 1.8 °C by 2050 (FCFA, 2017). Other general circulation models (Christensen et al., 2007; Serdeczny et al., 2017) also show a clear warming trend in SSA. Similar and higher temperature increases were calculated for West Africa (FCFA, 2016), South Africa (Daron, 2014) all of Africa (Engelbrecht et al., 2015; Niang et al., 2014). Urbanization rates in SSA have been reported by UN/DESA (2015) and Dos Santos et al. (2017), among others, whereby usually present and future population numbers are estimated. In our case, we arrived at a modelled annual change of the urban footprint of 3% in 2015–2050 (Table 1). The average annual population growth of urban agglomerations in SSA is projected to vary between 4% and 5% in 2016–2030 (UN/DESA, 2015; Dos Santos et al., 2017). Although there is no linear relationship between urban population and urban footprint, we are confident that the modelled increase in footprint would produce figures that are comparable to the reported population growth.

Our results imply that by 2050, climate change will reduce groundwater recharge by 7–21%. But most (23%) of this reduction is caused by urbanization, due the transformation of vegetated area into impervious areas (roads, roofs and other paved surfaces). As a result, surface runoff and evapotranspiration increased, while groundwater recharge decreased. In reality, the conversion from vegetated to impervious area and associated changes in water balance terms is more complex, and the outcomes are not straightforward. On the one hand, because parts of the catchment's surface will become impermeable to water, urban expansion will lower the total infiltration capacity of the soil surface in the catchment; but on the other hand the urbanized area also contains gardens and parks, which have a higher infiltration capacity (Poelmans et al., 2010). Furthermore, as was pointed out by Gwenzi and Nyamadzawo (2014), in the urbanized area, utility trenches, tunnels and buried structures form a complex network resembling a shallow urban karst system, which may provide preferential flow pathways for groundwater recharge. Please note that here we discuss only changes in recharge from precipitation. As pointed out in the Methods section and in the online supplemental data, recharge from wastewater infiltration was taken into account in the abstraction term of the water balance. Furthermore, leakage from the water supply mains is likely to be substantial, but, due to the limited network in the modelled area, leakage was assumed to be insignificant.

The estimated lowering of groundwater levels by on the order of 50–75 m by 2050 as a result of urbanization and future climate change will likely have major consequences. If such a scenario becomes reality, we predict that springs will dry up, wells will run dry, wetlands and urban green areas will disappear, the price of water will increase, agricultural produce will become more expensive, and the number of cash crops will shrink. As a result, tensions will increase, and power struggles for water will intensify. We predict that the general economic, social and environmental situation in Arusha will worsen. It should also be noted that most of the shallow groundwater resources are in use by the poor part of the population, mostly living in informal areas scattered throughout Arusha and having limited access to basic services like water, sanitation and electricity. Furthermore, this shallow groundwater is contaminated due to the on-site infiltration of wastewater. Also, based on fieldwork we carried out in the region (Nastar et al., 2018), at present, the poor communities already pay more for water than richer communities.

As in any forecast or projection, the limitations of our study and the robustness of the outcomes rely heavily on data, modelling conceptualization and associated simplifications and assumptions. For instance, the LULC cover projections are largely dependent on the quality of the LULC maps. In turn, the interval between the base maps used to 'train' the model (1995 and 2015) determine the robustness of the growth projections. Veerbeek (2017) provides a deeper discussion of these issues. Second, the outcomes of the entire groundwater modelling exercise depend heavily on the quality of the data inputs. In this regard, as mentioned in the Methodology section, the precipitation distribution we used was based on precipitation data for Mount Kilimanjaro, as these data were lacking for Mount Meru. In addition, as stated above, the conversion from vegetated to impervious area is a complex process (effect of increased paved surfaces, effect of increased infiltration capacity in gardens, parks, and small agricultural plots, effect of urban karst) for which there is no data. Furthermore, as mentioned in the Methods Section, the choice of a steady-state model in our work was conceptually incorrect, as some of the groundwater is abstracted from storage. However, as explained, due to data limitations, a transient model was not considered useful. Although the use of a steady-state model in this groundwater depletion situation could be considered a worst-case scenario, the results we obtained with the groundwater model merely indicate the order of magnitude of groundwater level decline.

As mentioned in the Introduction, groundwater reserves make cities or areas more robust to climate change. In Arusha, however, groundwater is over-consumed, with adverse effects, as described above. The question then is, how should Arusha adapt to and/or mitigate the changes? The ideal vehicle for both types of interventions seems to be the Arusha 2035 urban plan, but it makes no mention of how the future growth of the city should be supported in terms of satisfying future water demands, other than drilling more groundwater wells, and thus increasingly depleting the resource. In fact, the plan seems to be echoing the desire to leapfrog systematic transition to modernity (Lwasa et al., 2018), which will surely cause harm in terms of speeding up degradation of ecosystems, limiting economic opportunity, and accentuating social inequality. Such pathways will most certainly lead to a lock-in situation whereby increasing returns to adoption can lead to lock-in of incumbent technologies, which, on their turn, would eventually lead to a more sustainable, climate and change resilient future.

The situation observed in Arusha seems not to be unique. Many of the largest urban agglomerations in SSA (Dos Santos et al., 2017; UN/DESA, 2015) are in areas where groundwater is an important resource for a (large) part of the urban population, and more so by the urban poor. These agglomerations include Lagos, Nigeria (at present 13.6 million inhabitants), Dar es Salaam, Tanzania (5.4 million), Ouagadougou, Burkina Faso (2.8 million), Bamako, Mali (2.7 million), Abuja, Nigeria (2.5 million), Mogadishu, Somalia (2.3 million), Lusaka, Zambia (2.3 million), Kampala, Uganda (2.0 million), and N'Djamena, Chad (1.3 million), to name a few. All suffer from high to very high average annual rates of population growth, and in all of them rising temperatures are foreseen with climate change. We foresee that by 2050, the limited availability of groundwater will be another factor hampering the progress of urban Sub-Saharan Africa.

Acknowledgments

This work was carried out in the framework of the T-GroUP project. T-GroUP was funded by the UK Department for International Development, the Economic and Social Research Council and the Natural Environment Research Council (NERC) under the UPGro Programme, NERC grant no. NE/M008045/1. We thank the people of Arusha for their support.

Disclosure statement

No potential conflict of interest was reported by the authors.

Funding

This work was supported by the Natural Environment Research Council [NE/M008045/1].

ORCID

Jan Willem Foppen http://orcid.org/0000-0002-1112-2383

References

Abdollahi, K., Bashir, I., & Batelaan, O. (2012). WetSpaSS graphical user interface. Cartography of Higher Swiss Education. http://www.e-cartouche.ch/content_reg/cartouche/ui_access/en/html/UnitGUI_DefGUI.html

Abdollahi, K., Bashir, I., Verbeiren, B., Harouna, M. R., Van Griensven, A., Huysmans, M., & Batelaan, O. (2017). A distributed monthly water balance model: Formulation and application on Black Volta Basin. *Environmental Earth Sciences*, *76*(5). https://doi.org/10.1007/s12665-017-6512-1

Agterberg, F. P., & Bonham-Carter, G. F. (1990, September). Deriving weights of evidence from geoscience contour maps for the prediction of discrete events. *Proceedings of the 22nd APCOM Symposium, Berlin, Germany* (Vol. 2, pp. 381–395).

Agwanda, A., & Amani, H. (2014). Population growth, structure and momentum in Tanzania. Economic and Social Research Foundation. THDR 2014 ESRF discussion paper 61. Retrieved 26 July 2017 from http://www.thdr.or.tz/docs/THDR-BP-7.pdf

Angel, S., Parent, J., & Civco, D. (2007). Urban sprawling metrics: An analysis of global urban expansion using GIS. *Proceedings of ASPRS 2007 annual conference* (pp. 1–12). Tampa, Florida.

Batelaan, O., & De Smedt, F. (2007). GIS-based recharge estimation by coupling surface-subsurface water balances. *Journal of Hydrology*, *337*(3–4), 337–355. https://doi.org/10.1016/j.jhydrol.2007.02.001

Batty, M. (2007). Cities and complexity: Understanding cities with cellular automata, agent- based models, and fractals. MIT press.

Calow, R., & Macdonald, A. (2009). What will climate change mean for groundwater supply in Africa? Overseas Development Institute. . https://www.odi.org/sites/odi.org.uk/files/odi-assets /publications-opinion-files/4120.pdf

Carter, R. C., & Parker, A. (2009). Climate change, population trends and groundwater in Africa. *Hydrological Sciences Journal, 54*(4), 676–689. https://doi.org/10.1623/hysj.54.4.676

Case, M. (2006). Climate change impacts on East Africa. Report. WWF-World Wide Fund for Nature. https://doi.org/10.1007/978-3-642-14776-0_36

Christensen, J. H., Hewiston, B., Busuioc, A., Chen, A., Gao, X., Held, I., Jones, R., Kolli, R. K., Kwon, W. T., Laprise, R., Magana, R. V., Mearns, L., Menendez, C. G., Räisänen, J., Rinke, A., Sarr, A., & Whetton, P. (2007). Regional climate projections. In S. Solomon et al. (Eds.), *Climate change 2007: The physical science basis. Contribution of Working Group I to the fourth assessment report of the Intergovernmental Panel on Climate Change.* Cambridge University Press. https://doi.org/10.1080/07341510601092191

Clark, D. (1998). Interdependent urbanization in an urban world: An historical overview. *Geographical Journal, 164*(1), 85–95. https://doi.org/10.2307/3060547

Congedo, L. (2017). Semi-automatic classification plugin documentation. *Release, 4*(1), 29. https://doi.org/10.13140/RG.2.2.29474.02242/1.

Daron, J. D. (2014). *Regional climate messages for Southern Africa. Scientific report from the CARIAA Adaptation at Scale in Semi-Arid Regions (ASSAR) Project, 1–30.* International Development Research Centre. http://www.assar.uct.ac.za/sites/default/files/image_tool/images/138/RDS_reports/climate_messages/Southern%20Africa%20Climate%20Messages%20-%20Version%201%20-%20Regional%20Level.pdf

Dos Santos, S., Adams, E. A., Neville, G., Wada, Y., de Sherbinin, A., Mullin Bernhardt, E., & Adamo, S. B. (2017). Urban growth and water access in Sub-Saharan Africa: Progress, challenges, and emerging research directions. *Science of the Total Environment, 607–608*(2017), 497–508. https://doi.org/10.1016/j.scitotenv.2017.06.157

EGIS. (2017). *Consultancy service for additional water sources investigations (quantity and quality including treatment options) and study on lake Duluti.* Final Study Report. Arusha Urban Water Supply and Sanitation Authority (AUWSA).

Engelbrecht, F., Adegoke, J., Bopape, M. J., Naidoo, M., Garland, R., Thatcher, M., McGregor, J., Katzfey, J., Werner, M., Ichoku, C., & Gatebe, C. (2015). Projections of rapidly rising surface temperatures over Africa under low mitigation. *Environmental Research Letters, 10*(8), 085004. https://doi.org/10.1088/1748-9326/10/8/085004

EWURA. (2015). Water Utilities Performance Review Report 2014/2015. Energy and Water Utilities Regulatory Authority, Tanzania.

Foster, S. S. D., Morris, B. L., & Chilton, P. J. (1999). Groundwater in urban development: A review of linkages and concerns. In *Proceedings of impacts of urban growth on surface water and groundwater quality.* IUGG 99 Symposium, Birmingham. IAHS Publication no. 259. http://hydrologie.org/redbooks/a259/iahs_259_0003.pdf

FCFA. (2016). *Africa's climate: Helping decision makers make sense of climate.* Future Climate for Africa. www.futureclimateafrica.org/wp-content/uploads/2016/11/africas-climate-final-report-4nov16.pdf www.futureclimateforafrica.org

FCFA. (2017). *Future climate projections for Tanzania.* Future Climate for Africa. www.futureclimateafrica.org/resource/future-climate-projections-for-tanzania/ www.futureclimateforafrica.org

Ghiglieri, G., Balia, R., Oggiano, G., Ardau, F., & Pittalis, D. (2008). Hydrogeological and geophysical investigations for groundwater in the Arumeru District (Northern Tanzania). *Rendiconti Online Della Societa Geologica Italiana, 2,* 1–3.

Ghiglieri, G., Balia, R., Oggiano, G., & Pittalis, D. (2010). Prospecting for safe (low fluoride) groundwater in the Eastern African Rift: The Arumeru District (Northern Tanzania). *Hydrology and Earth System Sciences, 14*(6), 1081–1091. https://doi.org/10.5194/hess-14-1081-2010

Ghiglieri, G., Pittalis, D., Cerri, G., & Oggiano, G. (2012). Hydrogeology and hydrogeochemistry of an alkaline volcanic area: The NE Mt. Meru slope (East African Rift-Northern Tanzania). *Hydrology and Earth System Sciences, 16*(2), 529–541. https://doi.org/10.5194/hess-16-529-2012

GITEC. (2011, May). *Groundwater assessment of the Pangani Basin, Tanzania.* Pangani Basin Water Board and International Union for Conservation of Nature. Draft Final Report.

Goodacre, A. K., Bonham-Carter, G. F., Agterberg, F. P., & Wright, D. F. (1993). A statistical analysis of the spatial association of seismicity with drainage patterns and magnetic anomalies in western Quebec. *Tectonophysics, 217*(3–4), 285–305. https://doi.org/10.1016/0040-1951(93)90011-8

Gwenzi, W., & Nyamadzawo, G. (2014). Hydrological impacts of urbanization and urban roof water harvesting in water-limited catchments: A review. *Environmental Processes, 1*(4), 573–593. https://doi.org/10.1007/s40710-014-0037-3

Homer, C., Huang, C., Yang, L., Wylie, B., & Coan, M. (2004). Development of a 2001 national land-cover database for the United States. *Photogrammetric Engineering & Remote Sensing, 70* (7), 829–840. https://doi.org/10.14358/PERS.70.7.829

Kashaigili, J. J. (2010). *Assessment of groundwater availability and its current and potential use and impacts in Tanzania.* Report prepared for the International Water Management Institute (IWMI), Colombo, Sri Lanka. Retrieved 14 April 2020 from http://www.suaire.sua.ac.tz:8080/xmlui/handle/123456789/1484

Kessides, C. (2006). *The urban transition in Sub-Saharan Africa: Implications for economic growth and poverty reduction.* The Cities Alliance.

Lemaire, B. (2017, June). Arusha Sustainable Urban Water and Sanitation Delivery Project Consultancy Services for Additional Water Sources Investigations (Quantity And Quality Including Treatment Options) and Study on Lake Duluti. Final study report.

Lwasa, S., Buyana, K., Kasaija, P., & Mutyaba, J. (2018). Scenarios for adaptation and mitigation in urban Africa under 1.5 °C global warming. *Current Opinion in Environmental Sustainability, 30*, 52–58. https://doi.org/10.1016/j.cosust.2018.02.012

Mahmoud, S. H., Alazba, A. A., & Chapman, M. G. (2015). Hydrological response to land cover changes and human activities in arid regions using a geographic information system and remote sensing. *PLoS One, 10*(4), e0125805. https://doi.org/10.1371/journal.pone.0125805

Mas, J. F., Kolb, M., Paegelow, M., Olmedo, M. T. C., & Houet, T. (2014). Inductive pattern-based land use/cover change models: A comparison of four software packages. *Environmental Modelling & Software, 51*, 94–111. https://doi.org/10.1016/j.envsoft.2013.09.010

McDonald, M. G., & Harbaugh, A. W. (1988). *A modular three-dimensional finite-difference ground-water flow model* (Techniques of Water-Resources Investigations 06-A1). Retrieved 14 April 2020 from https://pubs.er.usgs.gov/publication/twri06A1

Morris, B. L., Lawrence, A. R. L., Chilton, P. J. C., Adams, B., Calow, R. C., & Klinck, B. A. (2003). *Groundwater and its susceptibility to degradation: A global assessment of the problem and options for management* (Early Earning and Assessment Report Series, RS. 03-3). United Nations Environment Programme. http://nora.nerc.ac.uk/id/eprint/19395/1/Groundwater_INC_cover.pdf

Nastar, M., Abbas, S., Aponte Rivero, C., Jenkins, S., & Kooy, M. (2018). The emancipatory promise of participatory water governance for the urban poor: Reflections on the transition management approach in the cities of Dodowa, Ghana and Arusha, Tanzania. *African Studies, 77*(4), 504–525. https://doi.org/10.1080/00020184.2018.1459287

NBS and Office of Chief Statistician Zanzibar. (2013). 2012 *Population and housing census: Population distribution by administrative areas.* United Republic of Tanzania, Dar es Salaam

NBC. (2014). *Basic demographic and socio-economic profile statistical tables, Tanzania mainland.* National Bureau of Statistics, Ministry of Finance, Dar es Salaam, The United Republic of Tanzania.

Niang, I., Ruppel, O. C., Abdrabo, M. A., Dube, P., Leary, N., Schulte-Uebbing, L., … White, L. (2014). Africa. In V. R. Barros, C. B. Field, D. J. Dokken, M. D. Mastrandrea, K. J. Mach, T. E. Bilir, M. Chatterjee, K. L. Ebi, & Y. O. Estra (Eds.), *Climate change 2014: Impacts, adaptation, and vulnerability. Part B: Regional aspects. Contribution of Working Group II to the fifth assessment report of the Intergovernmental Panel on Climate Change* (pp. 1199–1265).

Cambridge University Press. https://www.ipcc.ch/pdf/assessment-report/ar5/wg2/WGIIAR5-Chap22_FINAL.pdf

Oikos, I. (2011). *The Mount Meru challenge: Integrating conservation and development in Northern Tanzania*. Ancora Libri. www.istituto-oikos.org

Ong'or, B. T. I., & Long-Cang, S. (2007). Groundwater overdraft vulnerability and environmental impact assessment in Arusha. *Environmental Geology, 51*(7), 1171–1176. https://doi.org/10.1007/s00254-006-0408-9

Parnell, S., & Walawege, R. (2011). Sub-Saharan African urbanisation and global environmental change. *Global Environmental Change, 21*(SUPPL. 1), S12–S20. https://doi.org/10.1016/j.gloenvcha.2011.09.014

Poelmans, L., Van Rompaey, A., & Batelaan, O. (2010). Coupling urban expansion models and hydrological models: How important are spatial patterns? *Land Use Policy, 27*(3), 965–975. https://doi.org/10.1016/j.landusepol.2009.12.010

Rodrigues, H., & Soares-Filho, B. (2018). A short presentation of Dinamica EGO. In M. T. Camacho Olmedo, M. Paegelow, J.-F. Mas, & F. Escobar (Eds.), *Geomatic approaches for modeling land change scenarios* (pp. 493–498). Springer. https://doi.org/10.1007/978-3-319-60801-3_35

Røhr, P. C., & Killingtveit, Å. (2003). Rainfall distribution on the slopes of Mt Kilimanjaro. *Hydrological Sciences Journal, 48*(1), 65–77. https://doi.org/10.1623/hysj.48.1.65.43483

Serdeczny, O., Adams, S., Baarsch, F., Coumou, D., Robinson, A., Hare, W., Schaeffer, M., Perrette, M., & Reinhardt, J. (2017). Climate change impacts in Sub-Saharan Africa: From physical changes to their social repercussions. *Regional Environmental Change, 17*(6), 1585–1600. https://doi.org/10.1007/s10113-015-0910-2

Shukla, P. R., Skea, J., Slade, R., van Diemen, R., Haughey, E., Malley, J., Pathak, M., & Portugal Pereira, J. (eds.) (2019). *Technical Summary, 2019* (Climate Change and Land: an IPCC special report on climate change, desertification, land degradation, sustainable land management, food security, and greenhouse gas fluxes in terrestrial ecosystems). Retrieved 12 April 2020 from https://www.ipcc.ch/site/assets/uploads/sites/4/2019/11/03_Technical-Summary-TS.pdf

Taylor, R., & Barrett, M. (1999). Urban groundwater development in Sub-Saharan Africa. *Proceedings of 25th WEDC Conference, Integrated Development for Water Supply and Sanitation*, Addis Ababa, Ethiopia.

Taylor, R. G., Barrett, M. H., & Tindimugaya, C. (2004). Urban areas of Sub-Saharan Africa; weathered crystalline aquifer systems. In D. Lerner (Ed.), *Urban groundwater pollution* (pp. 155–179). Balkema.

Taylor, R. G., Koussis, A. D., & Tindimugaya, C. (2009). Groundwater and climate in Africa: A review. *Hydrological Sciences / Journal des sciences hydrologiques, 54*(4), 655–664. https://doi.org/10.1.1.693.6273&rep=rep1&type=pdf

Thomas, K. J. A., & Zuberi, T. (2012). *Demographic change, the IMPACT model, and food security in Sub-Saharan Africa*. Working paper. United Nations Development Program, Regional Bureau for Africa.

UN/DESA. (2015). *World urbanization prospects: The 2014 revision* (ST/ESA/SER.A/366). United Nations, Population Division. http://esa.un.org/Unpd/Wup/

Veerbeek, W. (2017). *Estimating the impacts of urban growth on future flood risk: A comparative study*. CRC Press. http://resolver.tudelft.nl/uuid:4f081b83-b03f-4358-abb3-efab1111aff4

Veerbeek, W., Pathirana, A., Ashley, R., & Zevenbergen, C. (2015). Enhancing the calibration of an urban growth model using a memetic algorithm. *Computers, Environment and Urban Systems, 50*, 53–65. https://doi.org/10.1016/j.compenvurbsys.2014.11.003

Wenban-smith, H. (2015). *Population growth, internal migration and urbanization in Tanzania, 1967-2012*. Working paper. International Growth Centre, Phase 2 Final Report.

Yi, W., Gao, Z., & Chen, M. (2012). Dynamic modelling of future land-use change: A comparison between CLUE-S and Dinamica EGO models. In *Remote sensing and modeling of ecosystems for sustainability IX* (Vol. 8513, pp. 85130H). International Society for Optics and Photonics.

Zhou, Y., & Li, W. (2011). A review of regional groundwater flow modeling. *Geoscience Frontiers, 2* (2), 205–214. https://doi.org/10.1016/j.gsf.2011.03.003

Impact of land use and occupation on potential groundwater recharge in a Brazilian savannah watershed

Arnaldo José Cambraia Neto and Lineu Neiva Rodrigues

ABSTRACT

This study evaluates the impact of land use and occupation changes on potential groundwater recharge in a Brazilian savannah watershed. The results show that potential groundwater recharge was directly influenced by changing scenarios in the watershed. The impact of land use and occupation changes was assessed by considering current land use and the replacement of all agricultural crops by soybeans (scenario 1) and natural savannah (scenario 2). We conclude that the processes of land occupation of the watershed should be understood in order to minimize reductions in the potential groundwater recharge in a watershed located in the Brazilian savannah biome.

Introduction

Representing approximately 24% of Brazilian territory, the Cerrado Biome (Brazilian savannah) is a vast water reservoir in South America that is responsible for maintaining the river springs of eight of the 12 national watersheds and the Guarani Aquifer (MMA, 2013). Groundwater in this region has been poorly studied, and there is still great uncertainty regarding its potential use, which has made it difficult to allocate water for users.

Long drought periods observed recently contributed to increasing water scarcity in watersheds in this biome, thus aggravating problems such as water supply problems in regions such as in the capital Brasilia and its surroundings (ANA, 2017). Conflicts regarding water tend to increase and worsen because irrigated agriculture, the main water user, is booming in the region (Maneta et al., 2009). Since the 1970s, the Brazilian savannah region has been considered Brazil's agricultural frontier, accounting for about 70% of national grain and beef production (Silva, Monteiro, et al., 2015; Klink, 2014).

Groundwater is an essential source of water and is fundamental for the maintenance of minimum flows in watercourses (Gaspar et al., 2007) and of native vegetation in the region (Oliveira et al., 2005), especially during periods of drought. As a result, reliable estimates of recharge are important in order to have adequate knowledge of groundwater potential use.

Groundwater exploitation is mainly based on recharge which, in many cases, is not commensurate with its exploitation and that may result in a risk of depletion of parts of permanent aquifer reserves (Albuquerque et al., 2015). The reduction in aquifer recharge can result in the drawdown of the water table, compromising river flows, fundamental for the maintenance of economic activity in the region during the drought period.

Groundwater recharge is directly impacted by watershed land use and occupation. Studies conferred that savannah areas in Brazil (Anache et al., 2019; Oliveira et al., 2015) have shown changes in recharge levels when the substitution of natural areas by agricultural areas took place. The same was demonstrated in the African (Ilstedt et al., 2016) and Australian savannas (Chen et al., 2014). Other sites, such as coniferous forests in the United States which in large part have been replaced by pasture (Adane et al., 2018) along with different land-use forms in the Chinese lowlands (Li et al., 2018), also show that recharge was impacted by the conditions of land use and occupation.

There are several ways to calculate groundwater recharge, but due to the complex interaction between factors involved in the process, such as climate and land-cover conditions, the application of computational models for the estimation of groundwater recharge has become increasingly common. These models are valuable tools for water resources planning and management because they enable, among other things, to simulate the impact of climate and land use on recharge and to establish management strategies. Several models can be used for this purpose (Kambhammettu et al., 2014; Sadaf et al., 2018; Tilahun & Merkel, 2009).

The BALSEQ model (Lobo-Ferreira, 1981) was developed to calculate potential recharge and has been applied in several regions of the world (Oliveira, 2004; Paralta et al., 2003). The model has the advantage of being simple to apply and has been applied successfully in several Brazilian biomes (Brito et al., 2009; Camargo et al., 2011; Cambraia Neto & Rodrigues, 2020; Lopes et al., 2012; Pontes et al., 2016; Souza et al., 2019).

Besides, this model allows us to assess the isolate impact of each land use on recharge, which is important to drive environmental policies. The fact that this model does not need much input data turn it a good model for data-scarce regions, such as the Brazilian savannahs. This study evaluates the impact of climate and land-use and occupation changes in the estimation of potential groundwater recharge in a Brazilian savannah watershed.

Materials and methods

Study area and data collected

The Buriti Vermelho River watershed has a drainage area of 10 km^2, an average altitude of 921 masl and is entirely within the Brazilian savannah region (Figure 1). It is part of the São Francisco River watershed. The region is characterized by tropical altitude climate, an average temperature of 21°C, an average annual rainfall ranging from 900 to 1500 mm, and two well-defined climatic seasons: a dry season from April to September, and a rainy season from October to March (Wendt et al., 2015).

The predominant soil class in the Buriti Vermelho River watershed is latossolo vermelho (or oxisols) with 94.9%, followed, respectively, by cambissolos (or inceptisols) and gleissolos (or entisols), with 3.0% and 2.1%, respectively. The Buriti Vermelho River

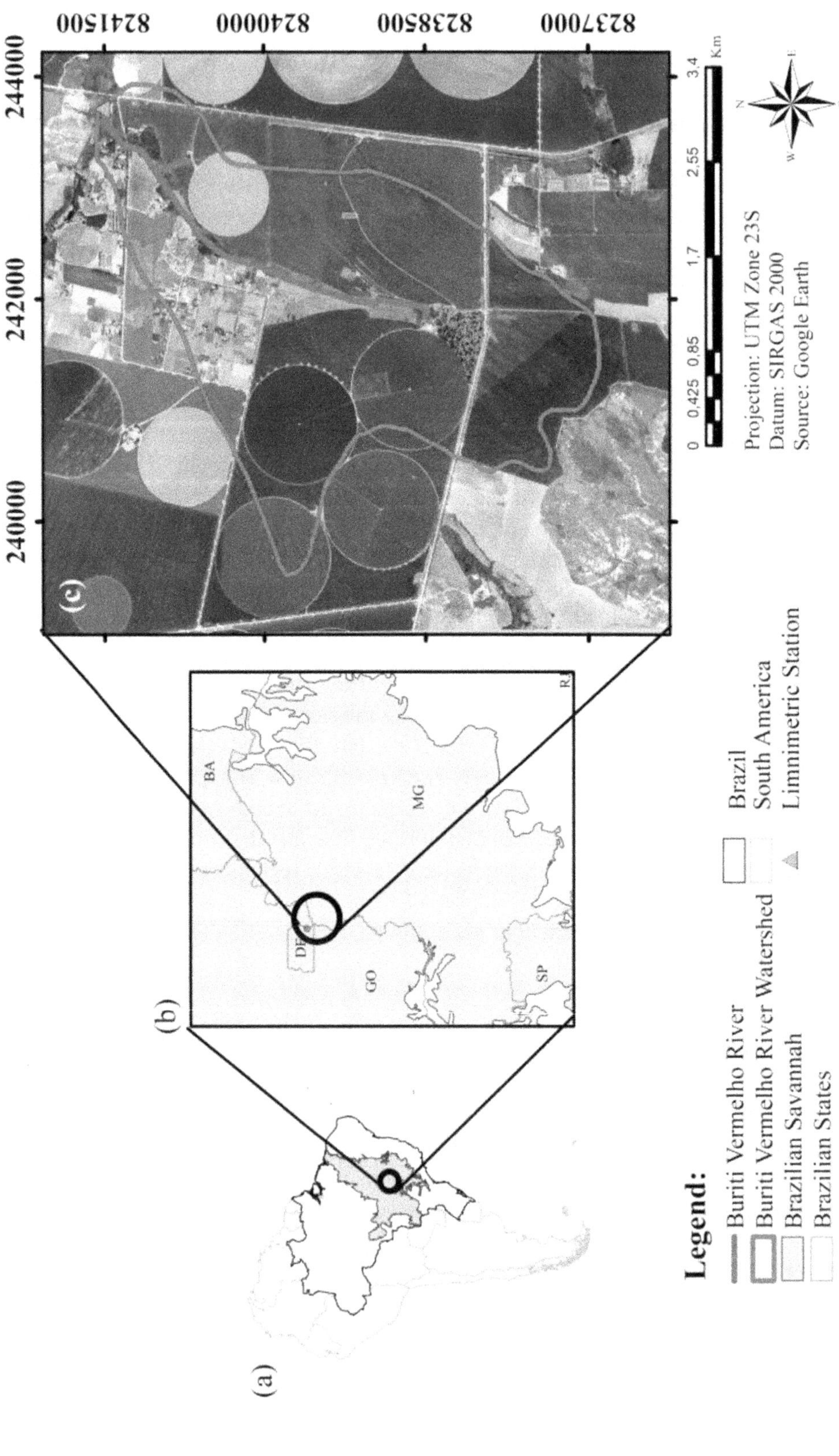

Figure 1. Land use and occupation in the Buriti Vermelho River watershed (c), its location in relation to the Federal District (b), and in relation to Brazil, the Brazilian savannah and South America (a).

Table 1. Land use and occupation of the Buriti Vermelho River watershed.

Land use and occupation	Area (km^2)	Area (%)
Irrigated crops	2.3	22.6%
Rainfed crops	5.3	53.7%
Brazilian savannah vegetation	0.3	2.6%
Farms	1.4	13.7%
Riparian	0.5	5.4%
Pasture	0.2	2.0%
	10.0	100%

is located in the centre–south of the Brasilia Folding Belt, where there is a predominance of rocks from the Bambuí group, dominated by meta-sedimentary units dating from the Neoproterozoic period. In smaller amounts there are limestone and dolomite layers (Campos et al., 2006; Castro et al., 2009).

Land use in the Buriti Vermelho River watershed is predominantly agricultural (rainfed and irrigated agriculture), with soybeans as the main crop, followed by corn, wheat and beans. Rainfed areas occupy the other regions of the watershed with pasture, riparian forest and native savannah vegetation. The watershed has a small housing core with a population of approximately 300 people and is composed mainly of small farms growing vegetables and fruits that occupy 13.7% of the area (Maneta et al., 2009; Moreira et al., 2010; Wendt et al., 2015). The land use and occupation identified in the Buriti Vermelho River watershed as well as their respective areas are shown in Table 1.

Land-use impact assessment on the potential groundwater recharge rate

Potential groundwater recharge was calculated using climate data collected at the Cerrado Agricultural Research Center (CPAC) weather station (UTM 209211.32; 8273041.18), installed around 40 km from the Buriti Vermelho River watershed from October 1987 to September 2017.

After confirming the consistency of the climatic data, potential groundwater recharge was calculated for the period October 1987–September 2017 using the BALSEQ model (Lobo-Ferreira, 1981). The calculation was performed considering current land use and occupation and two proposed scenarios.

In order to represent the dynamics of agriculture practiced in the watershed, we used the planting calendar shown in Table 2, according to CONAB (2017).

Table 2 shows the organization and sequence of the planting season usually followed in the region (current situation). The agricultural area was then divided into rainfed and irrigated crops. For example, from November to February, 80% of the rainfed area of the Buriti Vermelho River watershed is cultivated with soybean and 20% with corn. From March to June, 80% of this area is cultivated with corn and 20% with wheat. This same logic was used for irrigated areas. Between harvesting and planting of a new crop in the same area, a fallow period of 10 days was considered.

In addition to the current land-use situation, two new land-use scenarios were proposed:

Table 2. Planting calendar of rainfed and irrigated crops in the Buriti Vermelho River watershed.

	Month													Cycle duration (days)
	Cultures	Oct	Nov	Dec	Jan	Feb	Mar	Apr	May	Jun	Jul	Aug	Sep	
80% of rainfed area	Soybean													110
	Corn													115
	Fallow													140
20% of rainfed area	Corn													115
	Wheat													120
	Fallow													130
60% of irrigated area	Soybean													110
	Bean													105
	Wheat													120
40% of irrigated area	Corn													115
	Corn													115
	Bean													105

Table 3. Planting calendar of rainfed and irrigated crops in the Buriti Vermelho River watershed, considering scenario 1.

	Month													
Scenario 1	Cultures	Oct	Nov	Dec	Jan	Feb	Mar	Apr	May	Jun	Jul	Aug	Sep	Cycle duration (days)
Rainfed area	Soybean													110
	Soybean													110
	Fallow													145
Irrigated area	Soybean													110
	Soybean													110
	Corn													115

- Scenario 1: considers that all agricultural crops were replaced by soybeans (Table 3). It was used considering the fact that soybeans have been the fastest growing crop in the Brazilian savannah region in recent decades (Klink, 2014). Two crop cycles were simulated in the rainfed area and two in the irrigated areas, interspersed with a corn crop cycle, respecting the sanitary void (Juhász et al., 2013).
- Scenario 2: considers that all agricultural crops were replaced by natural savannah vegetation. It was designed to compare groundwater recharge that took place in an agricultural watershed with groundwater recharge in a preserved watershed.

During the simulations for scenarios 1 and 2, soil attributes such as curve number, water content in field capacity and permanent wilting point, and global density, as well as the relative dynamics of vegetation cover parameters such as root system depth and crop coefficient were corrected for land use.

The impact of land use on potential groundwater recharge was assessed by the difference between the percentage of total rainfall converted into potential groundwater recharge for current land use and occupation in the watershed and the other scenarios.

Once the scenarios were established, we considered for the time scale of simulations that there were no changes in land use and occupation. The result can help manage the rational occupation of the watershed, minimizing impacts on water resources and assisting in the formulation of environmental policies.

BALSEQ model

Potential groundwater recharge was estimated using the BALSEQ model (Lobo-Ferreira, 1981), which is a daily sequential water-balance model based on the premise that there are no less permeable layers in the unsaturated zone of the soil profile. The potential recharge of the following day is equal to the deep percolation calculated for each day (Camargo et al., 2011; Oliveira, 2004; Paralta et al., 2003).

Potential groundwater recharge was calculated for each day, and each land use using the following equations:

$$Rp_{i,j} = \max\left(0, DRp_{i,j}\right) \tag{1}$$

$$DRp_{i,j} = Rp_{i-1,j} - ETa_{i-1,j} + \Delta S_{i-1,j} - Rd_{i-1,j} \tag{2}$$

where Rp is the potential groundwater recharge (mm); DRp is the auxiliary variable (mm); R is the total daily rainfall (mm); ETa is the current daily evapotranspiration (mm); ΔS is the soil water storage (mm); Rs is the daily surface runoff (mm); i is the index for the day of the simulation; and j is the index for the type of land use and occupation.

The total groundwater recharge in the watershed ($Rp_{watershed}$, mm) over a given time interval was calculated using the following equation:

$$Rp_{watershed} = \sum_{i=1}^{nd} \sum_{j=1}^{k} Rp_{i,j} \tag{3}$$

where nd is the number of days; and k is the number of land uses and occupation in the watershed.

Crop potential evapotranspiration ($ETP_{i,j}$, mm d – 1), for a given land use and day, was calculated using the following equation:

$$ETP_{i,j} = ETo_i \; Kc_{i,j} \tag{4}$$

where ETo is the reference evapotranspiration (mm d – 1); and $Kc_{i,j}$ is the crop coefficient (non-dimensional).

The ETo was calculated using the Penman–Monteith method (PM-FAO56) (Allen et al., 1998). Crop coefficients (Kc) were estimated individually for each watershed area based on land use and occupation characteristics. In agricultural areas, Kc varied with the type and stage of crop development. The Kc values used for each land use are shown in Table 4.

Runoff was calculated using the US-SCS curve number method (US-SCS, 1972). This study considers that all the soil of the watershed was composed of oxisols to define the curve number. Surface runoff calculation was performed based on an estimated number as a function of land use and management, soil group, hydrological condition, and antecedent moisture condition. To define NC values for this study, we considered as a criterion for the differentiation of soil groups the establishment of limits of water infiltration rates (Pruski et al., 1997).

In accordance to the water infiltration rates for each use and occupation of the Buriti Vermelho River watershed obtained by Passos et al. (2009), the curve numbers of the watershed area were calculated according to antecedent moisture condition II. These

Table 4. Crop coefficients (Kc) used for crop evapotranspiration estimation in the Buriti Vermelho River watershed.

Land use and occupation	Kc
Soybean[a]	0.5–1.1
Corn[b]	0.03–1.63
Bean[b]	0.03–1.52
Wheat[b]	0.04–1.57
Brazilian savannah vegetation – rainy season (October–March)[c]	1.0 (October–March) and 0.30 (April–September)[d]
Pasture – rainy season (October–March)[d]	0.80 (October–March) and 0.30 (April–September)[d]
Riparian[c]	1.00
Farms[e]	1.00
Fallow[e]	0.10

Sources: [a]Silva, Sediyama, et al. (2015); [b]Guerra et al. (2003); [c]adapted from Allen et al. (1998); [d]adapted from Oliveira et al. (2005); and [e]adapted from Alencar et al. (2009).

values of conditions I and III were corrected according to rainfall during the previous five days in the watershed (Brito et al., 2009). Limits of 35.0–52.5 mm were considered for the definition of conditions I–III: condition I = rainfall in the last five days is < 35 mm; condition II = rainfall in the last five days is between 35.0 and 52.5 mm; and condition III = rainfall in the last five days is > 52.5 mm.

Surface runoff (SR, mm/day) was calculated using the following equation:

$$SR = \frac{25,4\left(\frac{R}{25,4} - \frac{200}{CN} + 2\right)^2}{\frac{P}{25,4} + \frac{800}{CN} - 8} \tag{5}$$

The maximum amount of available soil water for evapotranspiration (AGUT, mm) was calculated using the equation:

$$\mathrm{AGUT} = \frac{(\mathrm{FC} - \mathrm{PWP})}{10}\mathrm{SDZ} \tag{6}$$

where FC is the soil moisture at field capacity (g/g); PWP is the soil moisture at permanent withering point (g/g); SD is the global soil density (g/cm^3); and Z is the soil depth evapotranspiration (cm).

The attributes related to the soil's physical characteristics such as FC, PWP and SD were obtained from the Embrapa Cerrados database (Rodrigues, 2016). These attributes were determined at 99 sites and then interpolated to allow for the estimation of an average amount for each region and land use in the watershed (Table 5).

Table 5. Curve number (CN), field capacity (FC), permanent withering point (PWP) and global soil density (SD) values used for soil versus land-use relations for antecedent moisture conditions – AMC II in the Buriti Vermelho River watershed.

Land use and occupation	*Ks* (mm/h)[a]	Soil hydrological group[b]	CN (AMC II)[c]	FC (g/g)	PWP (g/g)	SD (g/cm)
Irrigated crops	165.3	B	77	0.42	0.26	0.99
Rainfed crops	187.0	B	77	0.42	0.26	0.97
Brazilian savannah vegetation	207.3	A	45	0.45	0.27	0.94
Farms	227.0	A	56	0.41	0.27	0.97
Riparian	344.0	A	36	0.40	0.26	0.98
Pasture	290.3	A	25	0.43	0.26	0.95

Sources: [a]Passos et al. (2009); [b]Pruski et al. (1997); and [c]Tucci (2012).

Table 6. Maximum soil depth that can be used for evapotranspiration for major land uses and land cover in the Buriti Vermelho River watershed.

Land use and occupation	Z (cm)	Land use and occupation	Z (cm)
Soybean[a]	50	Pasture[b]	280
Corn[b]	115	Riparian[b]	250
Bean[a]	40	Farms	100
Wheat[a]	40	Fallow	10
Brazilian savannah vegetation[b]	300 (October–March) and 500 (April–September)		

Sources: [a]Adapted from Rodrigues et al. (2009); and [b]Rodrigues et al. (2009).

In the case of soil vegetation cover, the soil layer subject to evapotranspiration is equal to the depth of the crop root system. Evapotranspiration soil depth amounts used in this study are shown in Table 6.

For agricultural crops, the value of Z varied as a function of the crop's root growth (Santos & Ferreira, 2017). A linear root growth model with an initial root value of 5 cm at emergence until a maximum was used, as shown in Table 6. A value of 15% was added those shown in Table 6 for rainfed crops since they have greater root system development in comparison with irrigated areas due to water restriction compensation (Asseng et al., 1998; Benjamin & Nielsen, 2006; Gheysari et al., 2017). In the native savannah areas, Z varied in the growing season, being higher in the dry season (Bucci et al., 2008; Christoffersen et al., 2014).

The sequential water balance was then applied for the period, considering that on the first day of the simulation, the soil contained moisture in its field capacity and was subsequently adjusted to the average amount of water stored in the last five days of the hydrological year. This occurred from the first to the last year of the simulation and on the first five days of the second to the last year of the simulation (Camargo et al., 2011).

Model calibration was performed based on surface runoff simulation. Therefore, the curve number of the curve number method was calibrated from October 2007 to September 2010 and validated from October 2010 to September 2013. The surface runoff values simulated by the BALSEQ model were calculated using the base runoff separation method, considered as a reference.

A computational system was used to aid the separation of the runoff and groundwater hydrographs (SepHidro), based on the local minimum method (Miranda et al., 2014).

To evaluate model performance, the Nash–Sutcliffe efficiency (NSE) coefficient was used (Nash & Sutcliffe, 1970), where NSE $\geq$ 0.75 = good performance, 0.36 $\geq$ NSE > 0.75 = satisfactory performance, and NSE> 0.36 = unsatisfactory performance (Baltokoski et al., 2010; de O. Silva et al., 2008; Viola et al., 2010).

With the curve number calibrated and validated, the water balance was carried out for the period from 1987 to 2017, considering current land occupation scenarios and scenarios 1 and 2.

Results and discussion

Data

The average monthly total rainfall and monthly reference evapotranspiration values for the period 1987–2017 are shown in Figure 2. The highest average rainfall was recorded between October and March, with the highest average registered in December (220.8 mm/month) Reference evapotranspiration had its highest daily average recorded in September (4.8 mm/day) and its lowest average in June (3.2 mm/day).

Average daily flow observed in the river watershed outlets varied widely throughout the year, ranging from 0.20 m^3/s in the rainy season to 0.02 m^3/s in the dry season, with maximum amounts in December and minimum amounts in September.

Calibration and validation of the BALSEQ model

The NSE coefficient in the calibration and validation steps of the BALSEQ model were 0.71 and 0.67, respectively. This indicates that the BALSEQ model had a satisfactory performance in the runoff simulation based on the classification proposed by Schaefli and Gupta (2007).

The curve numbers used at the beginning of the simulation and those obtained after model calibration are shown in Table 7. The curve number changed only in agricultural and farm areas. This can be explained by the fact that agricultural land use has proportionally much larger areas than other uses and, therefore, has a greater influence on the calibration process than do other uses.

The curve numbers calibrated in the areas were lower than those initially used. In the other areas (native savannah, riparian and pastures), the calibrated curve numbers did not change.

Even when the amounts are lower than those low reported by Tucci (2012), coefficients obtained are consistent, that is, curve numbers are larger for agricultural areas in comparison with those of the Brazilian savannah, riparian and pasture areas.

<B>*Impact of land use and occupation on the potential groundwater recharge*

Having estimated potential groundwater recharge for current land use and land-use conditions, the BALSEQ model was adjusted to simulate recharges considering land use

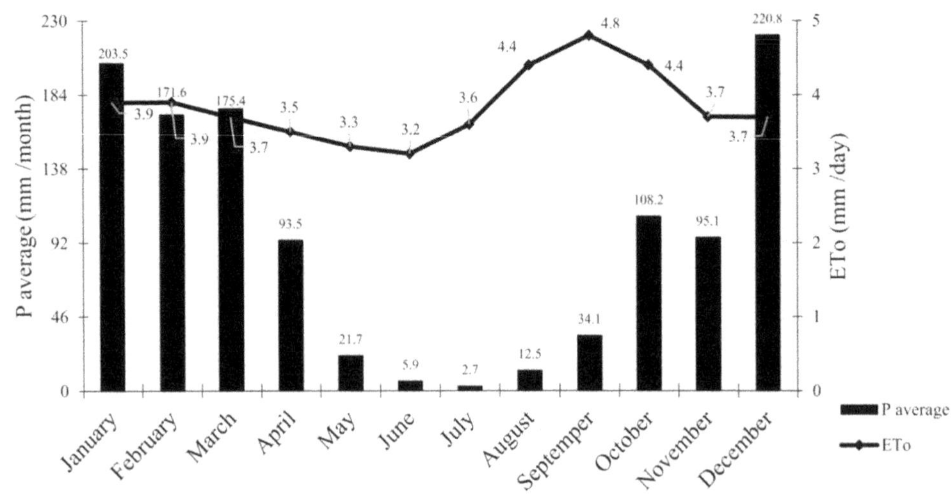

Figure 2. Average monthly total rainfall (mm/month) and average monthly reference evapotranspiration (mm/day) for the period 1987–2017 in the Buriti Vermelho River watershed.

Table 7. Curve number (CN) values for the antecedent humidity conditions – AMC II used at the beginning of the simulation and the values obtained in the BALSEQ calibration step for the Buriti Vermelho River watershed.

Land use and occupation	CN (AMC II), not calibrated[a]	CN (AMC II), calibrated
Irrigated crops	77	50
Rainfed crops	75	58
Brazilian savannah vegetation	45	45
Farms	56	52
Riparian	36	36
Pasture	25	25

Source: [a]Tucci (2012).

and land-use scenarios 1 and 2 over the period 1987–2017. The average amounts of potential groundwater recharge, runoff and actual evapotranspiration as a function of land use and scenario are shown in Figure 3.

Average annual total rainfall in the period was 1250.3 mm/year. For current use, 46.3% of average rainfall was converted to potential groundwater recharge, while ETa represented 50.6% and Rs was 1.7% (Figure 3). In scenario 1, 46.4% of average rainfall was converted to potential groundwater recharge, ETa represented 50.8% and Rs was 1.7%. The amounts of 25.4%, 73.1% and 1.0% of rainfall were converted into potential groundwater recharge, ETa, and Rs, respectively, when scenario 2 was considered.

In scenario 1, no significant changes were observed in relation to current land use. There was a 0.1% and 0.4% increase in potential recharge and current evapotranspiration, respectively, and a 2.9% reduction in runoff. This little variation in groundwater recharge was due to annual crops which accounted for 76% of land use in the current scenario. That is, changes in land use and coverage totalled 24%.

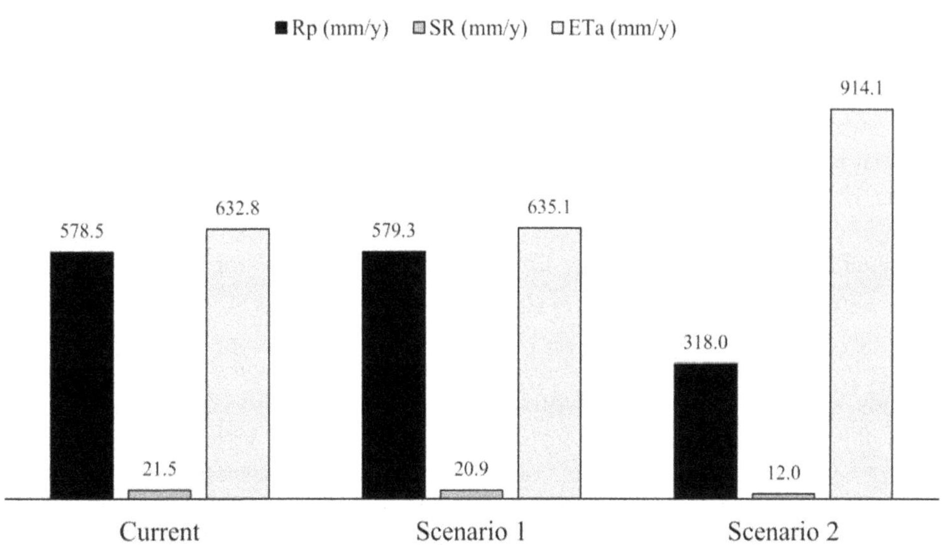

Figure 3. Mean total potential recharge (Rp), mean total surface runoff (SR) and mean total current evapotranspiration (ETa) (mm/year) for current and simulated land use and occupation scenarios of the Buriti Vermelho River watershed from 1987 to 2017.

When compared with changes in current land use for scenario 2, potential groundwater recharge was 45.0% lower. The increase in the soil layer that can be used for evapotranspiration and the smaller curve number of the savannah vegetation were also reflected in a 44.2% reduction in runoff and a 44.5% increase in current evapotranspiration. The greatest impact observed on recharge in scenario 2 can be attributed to the fact that in current use native savannah areas represent only 3% of the total watershed area. That is, changes in land use and land cover equivalent to 97% of land use in the Buriti Vermelho River watershed took place. This greater change in the area for use in scenario 2 was one of the factors responsible for the higher significant variation in groundwater recharge.

A similar situation was reported by Oliveira et al. (2015), who observed higher groundwater recharge amounts in pasture areas than in the native savannah in the State of São Paulo, Brazil. In the transition areas from Brazilian savannah to agricultural areas in São Paulo (Anache et al., 2019) and Baixo Xingu (Dias et al., 2015), reductions in the order of 40–50% were identified in evapotranspiration rates. In the short term, groundwater recharge did not change significantly during transitions of agricultural areas. Native vegetation cover density also influenced groundwater recharge processes, as reported in the African (Ilstedt et al., 2016) and Australian (Chen et al., 2014) savanna areas. During the dry season, areas with less dense vegetation cover had about 20% higher groundwater recharge than areas with more vegetation cover.

Average monthly amounts of potential groundwater recharge, runoff and current evapotranspiration as a function of current land use and proposed scenarios are shown in Figure 4.

Runoff was low, ranging from 0.0 to 0.2 mm/day, and was similar in the different scenarios, although slightly higher in current use and scenario 1. Factors related to the very structure of the BALSEQ model may explain the lower runoff amounts. The amount of runoff water varies as a function of the curve number. For example, there will be no runoff in irrigated agriculture areas when rainfall is < 15 mm; in pasture areas, the minimum amount of rainfall to produce runoff is 152 mm. The low runoff amounts can also be explained by the high-water infiltration rates observed in the study area (Ks > 150 mm/h).

Potential groundwater recharge and ETa for current land use and scenario 1 showed similar behaviour during the months of the rainy season. In scenario 2, potential groundwater recharge was lower than current use and in scenario 1. However, ETa presented higher amounts. Another factor that may explain this more significant difference observed in scenario 2 is the way the BALSEQ model was structured. This model states that for groundwater recharge, the amount of water stored in the soil throughout the day must be greater than available soil water for evapotranspiration (AGUT), which depends on soil water retention capacity and root system depth (*Z*-value) That is, the larger the soil layer to be used for evapotranspiration, the lower the groundwater recharge and the greater the current evapotranspiration. As in scenario 2, all agricultural crops were replaced by native savannah vegetation, the soil layer used for evapotranspiration increased considerably, and this consequently increased AGUT and reduced potential groundwater recharge.

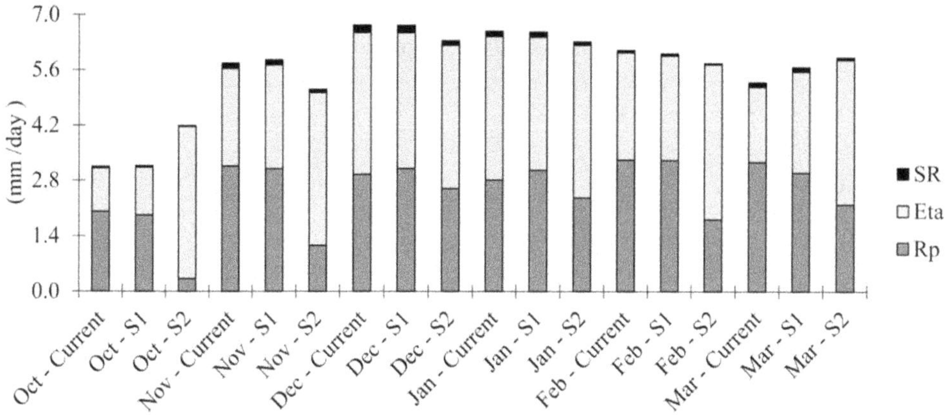

Figure 4. Monthly average potential groundwater recharge, current evapotranspiration and runoff (mm/day) for the rainy season from 1987 to 2017, estimated from the BALSEQ model with the current scenario and simulated land use and occupation (S1, scenario 1; S2, scenario 2) in the Buriti Vermelho River watershed.

Simulations show that, for example, with 10% increments of *Z*-value, potential groundwater recharge will be reduced by 2.2%, while current evapotranspiration may increase by 1.8%.

When considering the increase in depth of the crop root system in scenario 2, it was possible to observe that the potential groundwater recharges increased after a longer period of rain because the first rains were not enough to fill the existing water deficit and to constitute deep percolation.

Variables that directly interfere with groundwater recharge can be grouped into two groups. The first group, called non-controllable factors, is composed of variables that cannot be modified and include vegetation cover, especially the depth of the root system, among others. The higher the *Z*-value, the lower the water stress the crop may suffer from water restrictions. On the other hand, there will be an increase in evapotranspiration due to a greater availability of water and a reduction in potential groundwater recharge, with a direct impact on minimum flow rates. The second group, made up of controllable factors, is composed of variables that can be modified and include soil conservation practices, such as no-tillage, which is fundamental to increase groundwater recharge, thus contributing to reduce runoff and to increasing water infiltration into the soil.

The land use and occupation of the watershed assessed is pretty much the same for the last 15 years. Basically, this is an agricultural watershed very much anthropized that has only 2.6% of natural vegetation. Considering the three scenarios evaluated and the basin drainage area we found for both current scenario and scenario 1. a total recharge equal to 17,379.2 mm. Meanwhile, in scenario 2, the total recharge was equal to 9539.1 mm. Despite the 45% difference between the values, it is observed that the infiltrated water is stored in the region of the root zone of the vegetation, contributing to less runoff and greater availability for evapotranspiration in scenario 2. It is noteworthy that the phyto-physiognomy of the native savannah vegetation, with deciduous or semi-deciduous species, contributes to less evapotranspiration in the dry season (Cabral et al., 2015; Ribeiro & Walter, 2008), requiring less water in relation to agricultural crops (Veloso et al., 2017). Thus, according to Leão et al. (2020), the water stored in the soil profile is important for the maintenance of the native vegetation of the Biome during the dry season.

Conclusions

The water balance model presents itself as a very viable method for estimating the potential groundwater recharge, obtained above the saturated zone. The sequential water balance model (BALSEQ) was applied to evaluate the impact of land use and occupation changes on potential groundwater recharge in a watershed located in the Brazilian savannah biome. The potential groundwater recharge for the basin was calculated for a period October 1987–September 2017, considering the baseline scenario (current land use) and two other land-use scenarios (1 and 2). Potential groundwater recharge, considering current land use and occupation, represented 46.3% of total rainfall over the period.

In scenario 1, when the entire vegetation cover of the Buriti Vermelho River watershed was replaced by soybean cropping, potential groundwater recharge represented 46.4% of the total rainfall of the period. In scenario 2, when the entire agricultural area of the watershed was replaced by native savannah vegetation, potential groundwater recharge represented 26% of the total rainfall that took place during the period. Changes in land use and occupation in the Buriti Vermelho River watershed, with the substitution of current crops with soybeans (scenario 1) and the complete substitution of agricultural areas by areas of native savannah (scenario 2), accounted for reductions in 0.1% and 45.0% in potential groundwater recharge estimated by the BALSEQ model, respectively. In scenario 2, current evapotranspiration estimated by the BALSEQ model increased by 44.5%.

The development of the Cerrado region depends heavily on river low flows, which, in turn, depends directly on recharge. A first analysis of the scenarios suggests that scenario 2 would be the worst strategy to adopt if the interest is groundwater recharge, but one must take into account the importance of natural vegetation for the region's biodiversity and the maintenance of moisture in the soil profile.

The increase in evapotranspiration observed in scenario 2, which contributes to reducing recharge, seems to be overestimated, since natural vegetation reduces its evapotranspiration during the dry period.

Acknowledgements

The authors thank the data providers: the Brazilian Agricultural Research Corporation (EMBRAPA Cerrados) for providing the climatic data collected on Cerrado Agricultural Research Center (CPAC) weather station, and soil parameters collected on Buriti Vermelho River watershed.

Disclosure statement

The authors have no conflict of interest in publishing this research.

Funding

This study was supported by the Brazilian Agricultural Research Corporation (EMBRAPA Cerrados), the Federal University of Viçosa (UFV), and funded in part by the Federal District Research Support Foundation (FAP-DF) and the Coordination for the Improvement of Higher Education Personnel (CAPES - Finance Code 001).

ORCID

Arnaldo José Cambraia Neto http://orcid.org/0000-0001-8944-131X
Lineu Neiva Rodrigues http://orcid.org/0000-0001-5971-3441

References

Adane, Z. A., Nasta, P., Zlotnik, V., & Wedin, D. (2018). Impact of grassland conversion to forest on groundwater recharge in the Nebraska Sand Hills. *Journal of Hydrology: Regional Studies, 15*, 171–183. https://doi.org/10.1016/j.ejrh.2018.01.001

Albuquerque, C. G., Montenegro, S. M. G. L., Montenegro, A. A. D. A., & Fontes Júnior, R. V. D. P. (2015). Recarga de aquífero aluvial sob uso agrícola. *Águas Subterrâneas, 29*(1), 60–71. http://dx.doi.org/10.14295/ras.v29i1.27931

Alencar, C. A. B., Cunha, F. F., Da, Martins, C. E., Cóser, A. C., Rocha, W. S. D. D., & Araújo, R. A. S. (2009). Irrigação de pastagem: Atualidade e recomendações para uso e manejo. *Revista Brasileira De Zootecnia, 38*, 98–108. https://doi.org/10.1590/S1516-35982009001300012

Allen, R. G., Pereira, L. S., Raes, D., & Smith, M. (1998). *Crop evapotranspiration - Guidelines for computing crop water requirements - FAO Irrigation and drainage paper 56*. https://doi.org/92-5-104219-5

ANA. (2017) . *Conjuntura dos recursos hídricos no Brasil 2017: Relatório Pleno/Agência Nacional de Águas.*

Anache, J. A. A., Wendland, E., Rosalem, L. M. P., Youlton, C., & Oliveira, P. T. S. (2019). Hydrological trade-offs due to different land covers and land uses in the Brazilian Cerrado. *Hydrology and Earth System Sciences, 23*(3), 1263–1279. https://doi.org/10.5194/hess-23-1263-2019

Asseng, S., Ritchie, J. T., Smucker, A. J. M., & Robertson, M. J. (1998). Root growth and water uptake during water deficit and recovering in wheat. *Plant and Soil, 201*(2), 265–273. https://doi.org/10.1023/A:1004317523264

Baltokoski, V., Tavares, M. H. F., Machado, R. E., & De Oliveira, M. P. (2010). Calibração de modelo para a simulaçãoo de vazão e de fósforo total nas sub-bacias dos rios Conrado e Pinheiro - Pato Branco (PR). *Revista Brasileira De Ciencia Do Solo, 34*(1), 253–261. https://doi.org/10.1590/S0100-06832010000100026

Benjamin, J. G., & Nielsen, D. C. (2006). Water deficit effects on root distribution of soybean, field pea and chickpea. *Field Crops Research, 97*(2–3), 248–253. https://doi.org/10.1016/j.fcr.2005.10.005

Brito, G. L. M., Schuster, H. D. M., & Srinivasan, V. S. (2009). Sustainability of the aquifer system of the sedimentary basin of the Peixe River in the semi-arid region of Brazil. *Trends and sustainability of groundwater in highly stressed aquifers. Proceedings of Symposium JS.2 at the Joint IAHS & IAH Convention, Hyderabad, India, September 2009*. IAHS Publ. *329*, 28–35.

Bucci, S. J., Scholz, F. G., Goldstein, G., Meinzer, F. C., Franco, A. C., Zhang, Y., & Hao, G. Y. (2008). Water relations and hydraulic architecture in Cerrado trees: Adjustments to seasonal changes in water availability and evaporative demand. *Brazilian Journal of Plant Physiology, 20* (3), 233–245. http://dx.doi.org/10.1590/S1677-04202008000300007

Cabral, O. M. R., Da Rocha, H. R., Gash, J. H., Freitas, H. C., & Ligo, M. A. V. (2015). water and energy fluxes from a woodland savanna (cerrado) in southeast Brazil. *Journal of Hydrology: Regional Studies, 4*, 22–40. https://doi.org/10.1016/j.ejrh.2015.04.010

Camargo, R. A., Couto, E. A., Alves, L. G. S., Calijuri, M. L., & Silva, M. D. F. M. (2011). Aplicação de modelo de balanço hídrico BALSEQ para estimativa da infiltração profunda em região cárstica. *Ambiente & Água - Interdisciplinary Journal of Applied Science*, *6*(3), 303–317. http://dx.doi.org/10.4136/ambi-agua.456

Cambraia Neto, A. J., & Rodrigues, L. N. (2020). Evaluation of groundwater recharge estimation methods in a watershed in the Brazilian Savannah. *Environmental Earth Sciences*, *79*(6), 1–14. https://doi.org/10.1007/s12665-020-8884-x

Campos, J. E. G., Monteiro, C. F., & Rodrigues, L. N. (2006). *Geologia e zoneamento hidrogeológico da bacia do rio Preto, DF/GO/MG - Documentos 172*. Embrapa Cerrados.

Castro, K. B., Martins, E. S., Reatto, A., Lima, L. A. S., Rodrigues, L. N., Carvalho Júnior, O. A., Borges, M. E. S., Souza, V. V., & Gomes, M. P. (2009). *Compartimentação geomorfológica da bacia hidrográfica do rio Buriti Vermelho, Distrito Federal, DF - Boletim de Pesquisa e Desenvolvimento 244*. Embrapa Cerrados.

Chen, C., Eamus, D., Cleverly, J., Boulain, N., Cook, P., Zhang, L., Cheng, L., & Yu, Q. (2014). Modelling vegetation water-use and groundwater recharge as affected by climate variability in an arid-zone Acacia savanna woodland. *Journal of Hydrology*, *519*(PA), 1084–1096. https://doi.org/10.1016/j.jhydrol.2014.08.032

Christoffersen, B. O., Restrepo-Coupe, N., Arain, M. A., Baker, I. T., Cestaro, B. P., Ciais, P., Fisher, J. B., Galbraith, D., Guan, X., Gulden, L., Van Den Hurk, B., Ichii, K., Imbuzeiro, H., Jain, A., Levine, N., Miguez-Macho, G., Poulter, B., Roberti, D. R., Sakaguchi, K., Schaefer, K., ... Saleska, S. R. (2014). Mechanisms of water supply and vegetation demand govern the seasonality and magnitude of evapotranspiration in Amazonia and Cerrado. *Agricultural and Forest Meteorology*, *191*(February), 33–50. https://doi.org/10.1016/j.agrformet.2014.02.008

CONAB. (2017). *Calendário de Plantio e Colheita de Grãos no Brasil 2017*. https://doi.org/2318-6852

Dias, L. C. P., Macedo, M. N., Costa, M. H., Coe, M. T., & Neill, C. (2015). Effects of land cover change on evapotranspiration and streamflow of small catchments in the Upper Xingu River Basin, Central Brazil. *Journal of Hydrology: Regional Studies*, *4*(PB), 108–122. https://doi.org/10.1016/j.ejrh.2015.05.010

Gaspar, M. T. P., Campos, J. E. G., & Cadamuro, A. L. D. M. (2007). Condições de infiltração em solos na região de recarga do sistema aqüífero Urucuia no oeste da Bahia sob diferentes condições de usos. *Revista Brasileira De Geociências*, *37*(3), 542–550. https://doi.org/10.25249/0375-7536.2007373542550

Gheysari, M., Sadeghi, S., Loescher, H. W., Amiri, S., Javad, M., Majidi, M. M., Asgarinia, P., & Payero, J. O. (2017). Comparison of deficit irrigation management strategies on root, plant growth and biomass productivity of silage maize. *Agricultural Water Management*, *182*, 126–138. https://doi.org/10.1016/j.agwat.2016.12.014

Guerra, A. F., Rodrigues, G. C., Rocha, O. C., & Evangelista, W. (2003). *Necessidade hídrica no cultivo de feijão, trigo, milho e arroz sob irrigação no bioma cerrado - Boletim de Pesquisa e Desenvolvimento 100*. Embrapa Cerrados.

Ilstedt, U., Bargués Tobella, A., Bazié, H. R., Bayala, J., Verbeeten, E., Nyberg, G., Sanou, J., Benegas, L., Murdiyarso, D., Laudon, H., Sheil, D., & Malmer, A. (2016). Intermediate tree cover can maximize groundwater recharge in the seasonally dry tropics. *Scientific Reports*, *6* (February2015), 1–12. https://doi.org/10.1038/srep21930

Juhász, A. C. P., Pádua, G. P., Wruck, D. S. M., Favoreto, L., & Ribeiro, N. R. (2013). Desafios fitossanitários para a produção de soja. *Informe Agropecuário*, *34*(276), 66–75. http://www.alice.cnptia.embrapa.br/alice/handle/doc/978383

Kambhammettu, B. V. N. P., King, J. P., & Schmid, W. (2014). Grid-size dependency of evapotranspiration simulations in shallow aquifers: An optimal approach. *Journal of Hydrologic Engineering*, *19*(10), 4014018. https://doi.org/10.1061/(ASCE)HE.1943-5584.0000957

Klink, C. A. (2014). Policy Intervention in the Cerrado Savannas of Brazil Changes in Land Use and Effects on Conservation. In A. G. Consorte-mccrea & E. F. Santos (Eds.), *Ecology and conservation of the maned Wolf: Multidisciplinary perspectives* (pp. 293–308). CRC Press.

Leão, T. P., Freire, B., Bufon, V. B., Fidel, F., & Aragón, H. (2020). Using time domain reflectometry to estimate water content of three soil orders under savanna in Brazil. *Geoderma Regional, 21*, e00280. https://doi.org/10.1016/j.geodrs.2020.e00280

Li, H., Si, B., & Li, M. (2018). Rooting depth controls potential groundwater recharge on hillslopes. *Journal of Hydrology, 564*(February), 164–174. https://doi.org/10.1016/j.jhydrol.2018.07.002

Lobo-Ferreira, J. P. (1981). Mathematical model for the evaluation of the recharge of aquifers in semiarid regions with scarce (lack) hydrogeological data. In A. Verruijt & F. B. J. Barends (Eds.), *Proceedings of Euromech* (Vol. 143, pp. 2–4). Rotterdam: AA Balkema.

Lopes, J. W. B., Pinheiro, E. A. R., & Araújo Neto, J. R. D. (2012). Modelagem da recarga de águas subterrâneas em uma bacia semiárida: Influências da ocupação agrícola. *Revista Brasileira De Agricultura Irrigada, 6*(1), 33–40. https://doi.org/10.7127/rbai.v6n100078

Maneta, M. P., Torres, M., Wallender, W. W., Vosti, S., Kirby, M., Bassoi, L. H., & Rodrigues, L. N. (2009). Water demand and flows in the São Francisco River Basin (Brazil) with increased irrigation. *Agricultural Water Management, 96*(8), 1191–1200. https://doi.org/10.1016/j.agwat.2009.03.008

Miranda, A. C. R., Pruski, F. F., Martinez, M. A., & Cecon, P. R. (2014). Métodos de separação dos escoamentos superficial direto e subterrâneo: Estudo de caso para a bacia do Rio das Velhas. *Revista Brasileira De Recursos Hídricos, 19*(2), 169–181. https://doi.org/10.21168/rbrh.v19n2.p169-181

MMA. (2013) . *Programa nacional de conservação e uso sustentável do bioma cerrado. Programa cerrado sustentável.*

Moreira, J. M. M. A. P., Sousa, T. C. R., Souza, M. A., Aguiar, J. L. P., & Rodrigues, L. N. (2010). *Caracterização dos produtores do Núcleo Rural do Buriti Vermelho: Aspectos sociais, geográficos e de uso do solo e da água - Boletim de Pesquisa e Desenvolvimento 292.* Embrapa Cerrados.

Nash, J. E., & Sutcliffe, J. V. (1970). River flow forecasting through conceptual models part I – A discussion of principles. *Journal of Hydrology, 10*(3), 282–290. https://doi.org/10.1016/0022-1694(70)90255-6

Oliveira, M. M. (2004). Aplicação de dois métodos diferentes para a estimativa da recarga de águas subterrâneas numa área de Estarreja. *VII Congresso Da Água.*

Oliveira, P. T. S., Wendland, E., Scott, R., Rosolem, R., & Rocha, H. R. (2015). The water balance components of undisturbed tropical woodlands in the Brazilian Cerrado. *Hydrology and Earth System Sciences, 19*(6), 2899–2910. https://doi.org/10.5194/hess-19-2899-2015

Oliveira, R. S., Bezerra, L., Davidson, E. A., Pinto, F., Klink, C. A., Nepstad, D. C., & Moreira, A. (2005). Deep root function in soil water dynamics in cerrado savannas of central Brazil. *Functional Ecology, 19*(4), 574–581. https://doi.org/10.1111/j.1365-2435.2005.01003.x

Paralta, E., Oliveira, M. M., Lubczynsky, M., & Ribeiro, L. F. (2003). Avaliação da recarga do Sistema Aquífero dos Gabros de Beja segundo os critérios múltiplos: Disponibilidades hídricas e implicações agro-ambientais. *VI Simpósio de Hidráulica E Recursos Hídricos Dos Países de Língua Oficial Portuguesa - 6° SILUSBA*, Praia, Cabo Verde, 495–510.

Passos, D. P., Rodrigues, L. N., Cardoso, W. S., Martins, E. S., & Ratto, A. (2009). Relação do uso do solo com a taxa de infiltração na Bacia Hidrográfica do Rio Buriti Vermelho - DF. *XXXII Congresso Brasileiro de Ciência Do Solo*, Fortaleza, Brazil, 1–4.

Pontes, L. M., Coelho, G., Mello, C. R. D., Silva, A. M. D., Oliveira, G. C., & Viola, M. R. (2016). Avaliação de modelo de balanço hídrico com base na estimativa da recarga potencial. *Ambiente & Água - an Interdisciplinary Journal of Applied Science, 11*(4), 915–928. https://doi.org/10.4136/1980-993X

Pruski, F. F., Ferreira, P. A., Ramos, M. M., & Cecon, P. R. (1997). Model to Design Level Terraces. *Journal of Irrigation and Drainage Engineering, 123*(1), 8–12. https://doi.org/10.1061/(ASCE)0733-9437(1997)123:1(8)

Ribeiro, J. F., & Walter, B. M. T. (2008). As principais fitofisionomias do Bioma Cerrado. In S. M. Sano, S. P. Almeida, & J. R. Ribeiro (Eds.), *Cerrado: Ecologia e flora. Volume 1* (pp. 151–212). Embrapa Cerrados.

Rodrigues, L. N. (2016). Bacia Experimental do Rio Buriti Vermelho, na ecorregião do Planalto Central. In L. N. Rodrigues & A. E. Schuler (Eds.), *Água: Desafios para a sustentabilidade da agricultura* (Vol. 1a, pp. 233–255). Embrapa.

Rodrigues, L. N., Marioti, J., Vargas, L. M. P., Sano, E. E., & Martins, E. S. (2009). Bacia experimental do Rio Buriti Vermelho: Instrumentação, monitoramento e análise preliminar de dados. *XVIII Simpósio Brasileiro de Recursos Hídricos*, Campo Grande, Brazil, 1–19.

Sadaf, R., Mahar, G. A., & Younes, I. (2018). Appraisal of ground water potential through remote sensing in River Basin, Pakistan. *International Journal of Economic and Environmental Geology*, *9*(3), 25–32. https://doi.org/10.46660/ijeeg.Vol0.Iss0.0.135

Santos, P. S., & Ferreira, L. G. (2017). Análise descritiva dos aspectos biofísicos associados aos padrões de cobertura e uso da terra na bacia hidrográfica do Rio Vermelho, bioma Cerrado. *GeoFocus Revista Internacional de Ciencia Y Tecnología de La Información Geográfica*, *20*(2), 3–28. https://doi.org/10.21138/GF.395

Schaefli, B., & Gupta, H. V. (2007). Do Nash values have value. *Hydrological Processes: An International Journal*, *21*(15), 2075–2080. https://doi.org/10.1002/hyp

Silva, A. J., Monteiro, M. S. L., & Silva, M. V. (2015). Contrapontos da consolidação do agronegócio no cerrado brasileiro. *Sociedade E Território*, *27*(3), 95–114. https://periodicos.ufrn.br/sociedadeeterritorio/article/view/7302

Silva, F. C. S., Sediyama, T., Sediyama, G. C., Borem, A., & Silva, A. F. (2015). Manejo da Irrigação. In T. Sediyama, F. Silva, & A. Borem (Eds.), *Soja: Do plantio à colheita* (pp. 214–233). Editora UFV.

Silva, P. M. D. O., Mello, C. R. D., Silva, A. M. D., & Coelho, G. (2008). Modelagem da hidrógrafa de cheia em uma bacia hidrográfica da região Alto Rio Grande. *Revista Brasileira De Engenharia Agrícola E Ambiental*, *12*(3), 258–265. https://doi.org/10.1590/s1415-43662008000300006

Souza, E., Pontes, L. M., Fernandes Filho, E. I., Schaefer, C. E. G. R., & Santos, E. E. D. (2019). Spatial and temporal potential groundwater recharge: The case of the Doce River Basin, Brazil. *Revista Brasileira De Ciência Do Solo*, *43*, e0180010. https://doi.org/10.1590/18069657rbcs20180010

Tilahun, K., & Merkel, B. J. (2009). Estimation of groundwater recharge using a GIS-based distributed water balance model in Dire Dawa, Ethiopia. *Hydrogeology Journal*, *17*(6), 1443–1457. https://doi.org/10.1007/s10040-009-0455-x

Tucci, C. E. (2012). Escoamento superficial. In C. E. Tucci (Ed.), *Hidrologia: Ciência e aplicação* (Vol. 4*a.reimp*, pp. 391–441). Editora da UFRGS/ABRH.

US-SCS. (1972) . *National Engineering Handbook, section 4, Hydrology*. U.S. Dept. of Agriculture, available from U. S. Government Printing Office.

Veloso, G. A., Ferreira, M. E., & Silva, B. B. (2017). Determinação da evapotranspiração real diária em áreas irrigadas do projeto Jaíba (Minas Gerais, Brasil), mediante imagens Landsat 5 – TM. *Revista Cerrados*, *15*(1), 53–76. https://doi.org/10.22238/rc24482692v15n12017p53a76

Viola, M. R., Mello, C. R., De, Acerbi Júnior, F. W., & Silva, A. M. D. (2010). Modelagem hidrológica na bacia hidrográfica do Rio Aiuruoca, MG. *Revista Brasileira De Engenharia Agrícola E Ambiental*, *13*(5), 581–590. https://doi.org/10.1590/s1415-43662009000500011

Wendt, D. E., Rodrigues, L. N., Dijksma, R., & Dam, J. C. V. (2015). Assessing groundwater potencial use for expanding irrigation in the Buriti Vermelho watershed. *Irriga*, *1*(2), 81–94. https://doi.org/10.15809/irriga.2015v1n2p81

Problems and promise of managed recharge in karstified aquifers: the example of Lebanon

Wisam M. Khadra and Pieter J. Stuyfzand

ABSTRACT

Managed aquifer recharge can store surface water as safe and reliable groundwater for later recovery. However, most options are problematic in karstic aquifers due to complex hydrodynamics reducing their effectiveness and hence general applicability. River bank filtration and urban stormwater infiltration systems are among the main managed recharge approaches to cope with this complexity. Experiences in Lebanon demonstrate the viability of these and other options in karstic domains.

Introduction

The distribution of water resources varies at regional and global scales, with intensifying changes mainly due to population growth, rapid urbanization, increasing living standards, pollution of water resources, and changing climates (UNESCO-IHP, 2011). Water scarcity affects megacities worldwide, notably in the belt between 10°N and 40°N, e.g., in southern Florida, central India, western Mexico, and Lebanon (Figure 1). With a 1% annual increase in global water demand, it is anticipated that around 5 billion people will face water shortages in 2050 (i.e., nearly half the global population expected by then). This necessitates urgent actions to reduce stress on surface and ground water resources (UN-Water, 2018).

Managed aquifer recharge is one of the foremost management tools to cope with water scarcity. It has multiple objectives, for instance to augment water supplies and reduce saltwater intrusion (Missimer, Guo, Woolschlager, & Maliva, 2017). It is definitely on the rise after successful application at many sites in the world (David & Pyne, 2015; Dillon & Stuyfzand, 2019; Hartog & Stuyfzand, 2017; Maliva, 2020). Geologic units usually act as natural filters that transform water to better quality, so even polluted water can be used as a recharge source (Asano, 1985; Bouwer, 2002; Stuyfzand, 1989a, 1989b; Stuyfzand, 2002). Numerous studies have shown the effectiveness of managed recharge in the removal or degradation of microorganisms, turbidity, pesticides, dissolved organic carbon, and organic micropollutants (Dillon & Stuyfzand, 2019; Maliva, 2020).

However, managed aquifer recharge is problematic in karstic and highly fractured aquifers, due to complex dynamics (Bakalowicz, 2011; Khadra, 2017). Hence, it is rarely

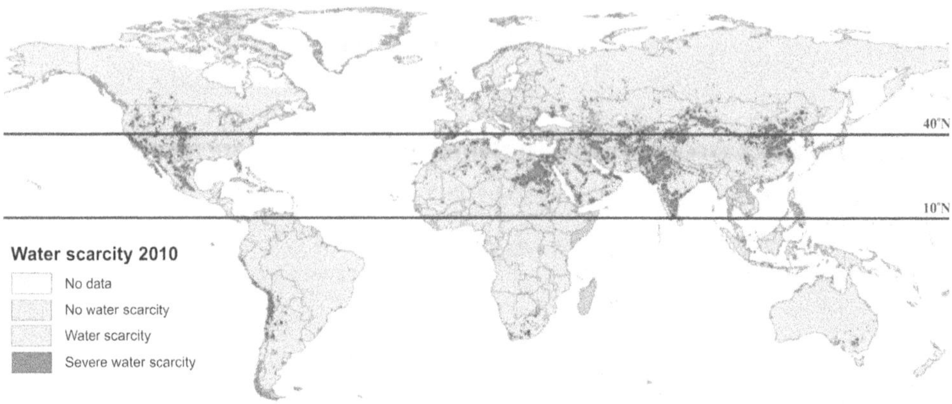

Figure 1. World map of water scarcity in 2010. An area is water-scarce or severely water-scarce when human demand is 20–40% or above 40% of available surface water, respectively (Burek et al., 2016).

applied in karst (Daher, Pistre, Kneppers, Bakalowicz, & Najem, 2011), and only few sites exist in Europe (Sprenger et al., 2017) and elsewhere in the world. The most famous pilot projects are the Wala reservoir in Jordan (Xanke et al., 2015; Xanke, Jourde, Liesch, & Goldscheider, 2016), the karstic Gambier limestone aquifer in South Australia (Vanderzalm et al., 2014), the Caldas Novas aquifer in Brazil (Tröger, 2010), the Floridan karst aquifer system in the US (Bacchus, Bernardes, Xu, & Madden, 2015a, 2015b), the Querença-Silves limestone aquifer in Algarve, Portugal (Leitão et al., 2017), and the Nardò aquifer in Italy (Masciopinto & Carrieri, 2002).

Lebanon is one of the most karstic areas in the Eastern Mediterranean (Figure 2). It is facing water shortage, especially since 2011, because the great number of Syrian refugees (around 1.4 million) has increased the pressure on the water resources (MoE/EU/UNDP, 2014). The national water strategy considers managed aquifer recharge one solution to augment water supplies (MoEW, 2010). However, in practice it has not yet gained confidence due to the different hurdles to its applicability in karst.

This article aims to strengthen faith in managed recharge by raising awareness and acceptability of its application in karstic domains. We review the major types and associated obstacles, and suggest several solutions for successful application in karst. Cases from Lebanon are selected and described as examples of viable managed recharge systems in a karstic-dominated area, which could be generalized to similar hydrogeological settings elsewhere.

Managed aquifer recharge types and their prerequisites

In recent decades, managed aquifer recharge has witnessed many developments to meet different conditions and needs (Dillon, 2005; Dillon & Stuyfzand, 2019; Maliva, 2020; Stefan & Ansems, 2018). The main systems (Figure 3) include aquifer storage recovery (water injected, stored and then extracted from the same well); aquifer storage transfer recovery (water injected and then extracted by another well down-gradient); river bank filtration (induced water recharge from a hydraulically connected river); basin recharge (artificial recharge via basin(s)); rainwater harvesting (rainfall collected at rooftops, then

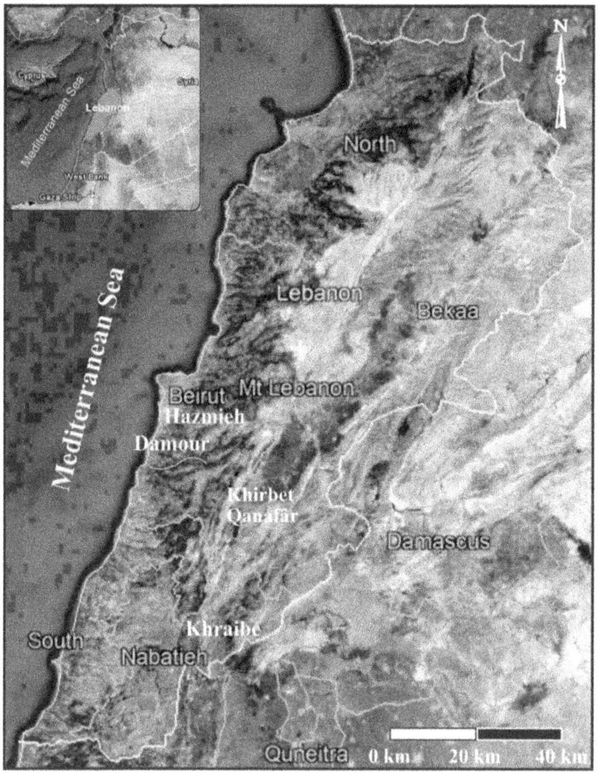

Figure 2. Location map of Lebanon.

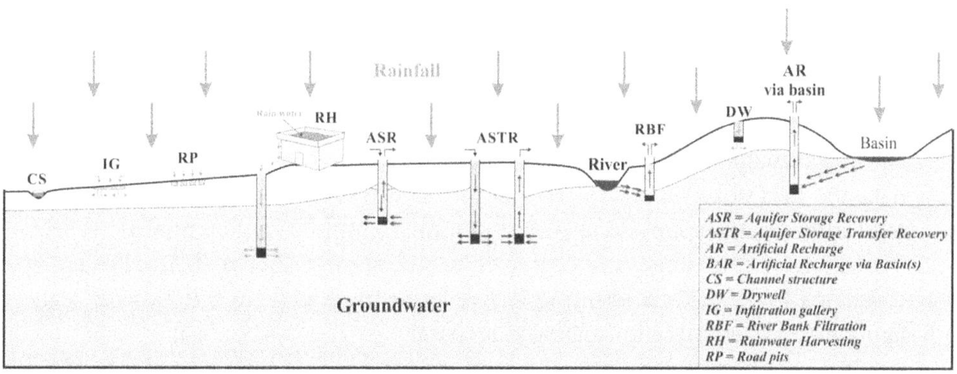

Figure 3. The common types of managed aquifer recharge (adapted from Khadra, Stuyfzand, & Khadra, 2017a).

directed to trenches or shafts to recharge underlying shallow aquifers); in-channel structures (subsurface dams, sand dams and gabions, which are usually built across ephemeral streams with the aim of building a new groundwater reservoir behind these barriers); infiltration galleries (covered trenches underground); recharge pits along roads (Jain, 2016); and subsurface storage facilities known as drywells (they receive, store and then infiltrate stormwater: Sasidharan, Bradford, Šimůnek, DeJong, & Kraemer, 2018).

The success of managed aquifer recharge generally depends on the hydrogeological setting, water availability and quality, selection of the proper recharge type, and pretreatment of the source water. Different aspects (e.g., technical, economic and legal) have been thoroughly tackled by various authors (e.g., Maliva & Missimer, 2012; Megdal & Dillon, 2015; Pyne, 2005). Still, many challenges exist, for instance well clogging, mobilization of arsenic, behaviour of pollutants, and recovery efficiency, notably in karst aquifers. No managed recharge type has overall applicability, and each has some pros and cons, as summarized in Table 1.

Viability of managed recharge in karstic domains

Managed aquifer recharge applications in karst have not gained unanimous faith yet due to many problems, of which the most notable are high permeability, with limited purification and chemical attenuation; very high transmission rate and unpredictable tunnelled flow, which hampers recovery; low overall porosity; limited storage capacity; high dispersivity; non-uniform response to recharge,; and complex surface terrains, often inappropriate for recharge basins and characterized by high natural infiltration rates, limiting the availability of surface water suitable for recharge (dolines may offer opportunities, however, e.g., the arid karst systems in Saudi Arabia: Schulz et al., 2016).

Due to this complexity, Daher et al. (2011) developed a list of criteria for aquifer rechargeability assessment, which was applied to a case study in Lebanon (the Damour aquifer). Rolf (2017) subsequently developed a framework to assess the suitability of managed aquifer recharge, including karstic systems. It uses criteria with many themes, related to aquifer characteristics, source water, and recharge techniques, among others. It was used to assess the application of managed recharge at nine potential sites, but the results showed no ideal choice of aquifer storage recovery in karstic units. These two methods are considered useful cornerstones to extend a more comprehensive hazard scale based on two major hydrodynamic factors:

- Rechargeability of the aquifer, which is the ease with which water can infiltrate and become available as groundwater storage. It is assessed based on available physical data and field observations to account for surface karst landforms, type of geologic formation (marl vs. limestone and/or dolomite), infiltration potential as controlled by steepness of the slope, and degree of karstification. Daher et al. (2011) suggested a detailed list of criteria resulting in five levels indexed from 0 to 4, with 3 and 4 being the most suitable for managed recharge.
- Aquifer retention, natural attenuation and storage. These parameters are assessed based on groundwater flow velocity (and thus residence time), preferential flow paths (e.g., via conduits where recharged water is easily lost), aquifer confinement (lateral and vertical), and aquifer inclination. Slower flow and longer residence is better, but no specific thresholds for velocity and transit time are yet known.

These two factors are the main parameters to assess the dynamics of managed recharge in karst, to assure that enough water storage with sufficient recovery is available. Other general aspects are to be added as well (e.g., environmental, economic, governance, source waters, and costs), before a full scheme can be selected with multi-criteria analysis.

Table 1. Main managed aquifer recharge types with their pros, cons and infiltration water source.

Type	Description	Pros	Cons	Water source
Surface infiltration from ponds or basins	Water spills on surface, then infiltrates and recharges the phreatic aquifer	• Low energy costs • Relatively low clogging potential of basin • Water quality improvement during aquifer passage • Reversal of or protection against seawater intrusion	• Evaporation losses, algae blooms, and atmospheric fallout of pollutants • Land restrictions (e.g., in urban areas or karst terrains) • Aquifer reactivity and redox reactions may degrade water quality (e.g., Fe, As and Mn) • Unsuitable for confined aquifers and impervious surfaces • Relatively high clogging risk of recovery system	• Precipitation • Urban stormwater • Surface water • Desalinated water • Harvested rainwater • Treated wastewater
Well/borehole recharge (e.g., vadose zone infiltration, or aquifer storage recovery)	Water enters wells discharging into either the vadose zone or the saturated zone (often (semi)confined)	• Small land area (suitable in densely populated areas) • Temporary subsurface storage to meet dry period demand (no evaporation, no algae blooms, no atmospheric fallout) • Water quality improvement during aquifer passage • More economical than traditional dams • Can recharge confined aquifers or overcome impervious surfaces • Old and dry wells can be used	• Mixing between recharge and ambient water • Quality of recharged water is critical (pretreatment required) • Losses by lateral flow • Aquifer reactivity and redox reactions may degrade water quality (e.g., Fe, As and Mn) • High clogging potential • Energy costs • Low recharge rate	• Surface water • Urban stormwater • Desalinated water • Treated wastewater • Harvested rainwater
In-channel systems (e.g., gabions, sand dams, subsurface dams)	Surface water is intercepted, spread, and allowed to infiltrate to subsurface	• Captures part of otherwise lost surface water • Needs little construction work compared to typical dams	• Evaporation losses • Unsuitable for confined aquifers • Unsuitable in settings with occupied floodplains • May enhance local floods in extreme events	• Streams (perennial or intermittent)
River bank filtration	Aquifer recharge is induced from a hydraulically connected river	• Water quality improvement • No disruption of land • Suitable for fractured and karst aquifers	• Requires permanent surface water (e.g., perennial streams) • Residence time may not be enough for chemical attenuation	• Perennial streams
Rainwater harvesting	Rainfall collected at rooftops is directed to recharge underlying shallow aquifers	• Economically feasible • No disruption of land • Meets dry-period demand • Old and dry wells can be used	• May not cover water needs • Low water quality during first flush (first minutes of a rain event)	• Rooftop rainwater

Managed aquifer recharge experience in Lebanon

Lebanon, as a mesothermal Mediterranean climatic area, has dry summers, mild and moist winters, and abundant sunshine. The cumulative annual rainfall in the coastal areas is about 600–1000 mm, increases to 1400 mm on the western flanks of Mount Lebanon due to orographic effects, and drops to less than 600 mm in the central regions due to the rain shadow on the leeward side of the orographic barrier. Lebanon has a surplus water budget, where total precipitation exceeds natural losses by evapotranspiration and groundwater flow across boundaries (Figure S1, supplementary online material); however, groundwater reserves are significantly stressed, and surface water is mostly lost to the sea if not detained in reservoirs. Weak water management and governance are among the main causes of water scarcity facing different territories in the country, notably the highly populated coastal zones. It is anticipated that these conditions will worsen soon due to population growth, increasing urbanization, and climate change. Managed aquifer recharge is thought to be a viable choice for water buffering (Khadra, 2017).

Managed recharge in Lebanon is currently limited to a few incomplete and unsuccessful trials, which ended with the onset of civil war in the mid-1970s (Daoud, 1973; MoEW & UNDP, 2014). But the water authorities still see some potential in it, and managed recharge has been added to the national water sector strategy aiming to recharge up to 200 Mm^3 of water by 2020 (MoEW, 2010). Thirty-three locations across the country were nominated as suitable sites for recharge of surface or treated effluent water (Figure S2, supplementary online material; MoEW & UNDP, 2014). However, no (fully) functional pilot has yet been installed, and no policy regulations have been written except for limited general guidelines in the water code ratified in April 2018.

Artificial recharge was earlier tested by Daoud (1973) in the Hazmieh area (south of Beirut's eastern suburbs; Figure 2) to prevent saltwater intrusion in a major dolomitic limestone aquifer. Four recharge attempts were carried out between April 1968 and May 1971 by injecting water from a nearby irrigation canal carrying water from the Beirut River. They induced a slight rise in groundwater level and reduced salinity (technical details in the supplementary online material). Intermittent injection has been resumed during short periods of the wet season since 2000. Neither proper monitoring nor reliable information is available, and salinization in the surrounding area is continuing (MoEW & UNDP, 2014). Later efforts evaluated the Hazmieh site and the chances of rerunning the same recharge wells. They concluded that implementing managed recharge in Hazmieh or elsewhere in Lebanon is viable (Prinz, 2016), but these conclusions remained at the desk level.

Another potential site is the Damour dolomitic limestone aquifer, south of Beirut (Figure 2). Its recharge is possible from the neighbouring perennial Damour River. One pilot injection programme was initiated in the 1970s, but operations soon ceased, and all related information was lost (MoEW & UNDP, 2014). A rechargeability and feasibility assessment by Daher et al. (2011) showed that managed aquifer recharge via infiltration ponds or injection wells to the main aquifer is not a good choice, which is in line with the anticipated hurdles of managed recharge in karst.

Meanwhile a project funded by the Dutch government is in progress to run a full pilot test and assess its overall performance. It is part of a programme aiming at strengthening the Lebanese water and agriculture sector. An exploration well was drilled in Khirbet

Qanafâr in Bekaa (central Lebanon) targeting the Miocene conglomerates (technical details in Burger et al., 2019; Stuyfzand, Khadra, & Burger, 2019; summarized in the supplementary online material). The success of any pilot there, even though in a clastic aquifer, may trigger local interest in testing or installing more sites.

Suggested solutions in karstic-dominated areas

Based on the pros and cons of managed recharge types (Table 1), two major techniques seem to have potential in karstic areas: river bank filtration along major rivers to recharge the hydraulically connected karstic aquifers; and borehole or trench recharge (notably in the vadose zone) using stormwater runoff, especially in major urban areas. Both methods need little land, which is advantageous for karstic terrains and urban areas, besides other hydraulic preferences as mentioned later. Other options exist as well:

- Infiltration of desalinated seawater in coastal aquifers where enough water is stored during low-demand periods for use in times of high demand. However, the density difference between the desalinated water and the ambient, more saline groundwater may force the stored freshwater to float upwards and spread out, reducing the recovery and benefits. This necessitates the use of a 'fresh storage saline extraction' scheme, with continuous pumping of saline water at a limited rate from below the stored cone-shaped freshwater (Van Ginkel, Olsthoorn, Smidt, Darwish, & Rashwan, 2010), or 'multiple partially penetrating wells', where deeper wells operate for recharge and shallower ones for recovery (Zuurbier, Zaadnoordijk, & Stuyfzand, 2014).
- A set-up of aquifer storage transfer recovery wells in which the injection wells are up-gradient of the recovery wells, thus capturing the bubble when drifted down-gradient, or aquifer storage transfer recovery with a target storage volume (volume required for recovery plus a buffer-zone volume). This technique had success at a pilot limestone aquifer in Hilton Head Island, South Carolina, where high recovery was achieved (David & Pyne, 2015).
- Combined managed aquifer recharge methods, such as beach filtration followed by reverse osmosis and deep well injection.
- Injection and storage of water in very deep aquifers that have not yet been considered for water supply. This needs exploration first, since the water there could be brackish. Still, it is an interesting option because the hydraulic gradient (responsible for bubble drift) could be low.
- Recharge of aquifers with a lithological, natural subterranean barrier that prevents the water being directly lost to the sea, e.g., tectonic blocks of impermeable rock or intruded magmatic dikes. No such setting is recognized along the Lebanese coast, but an inland example appears in Khraibe, south Lebanon (Figure 4).
- Artificial recharge of the alluvial quaternary unit that covers different spots across Lebanon and other eastern Mediterranean countries, including some coastal areas (e.g., in the Damour coastal plain). This unit has higher porosity (characteristic of uncemented clastic sediments), a more homogeneous character and lower permeability, together facilitating higher recovery efficiency than karst. Targeting such shallow units overlying karstic aquifers has been assessed positively elsewhere, for instance in Algarve, Portugal (Leitão et al., 2017).

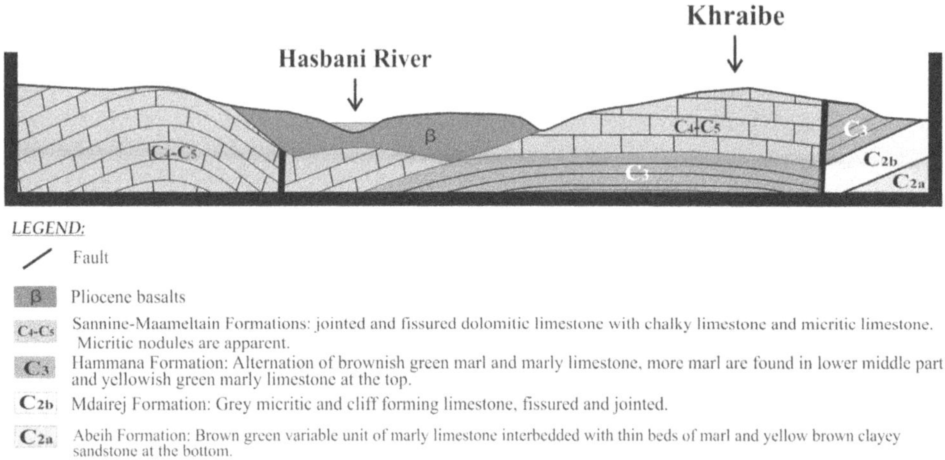

Figure 4. Geological cross section across Khraibe in south Lebanon. (The location of Khraibe in Lebanon is shown in Figure 2.)

Assessment of two managed recharge systems in Lebanon

River bank filtration

River bank filtration is associated with induced subsurface infiltration from nearby groundwater extraction, which creates a hydraulic preference for groundwater flow towards pumping wells. This constrains the flow and reduces the random losses expected in karst. Bank filtration is recommended nowadays as a superior alternative to surface water abstraction, to avoid the problems of turbidity, pathogens and pollution. It is widely spread in the world, and has had high success, for instance in Germany, Netherlands, Slovakia, Hungary, Switzerland, India, Egypt and the US (Dillon & Stuyfzand, 2019). This option can easily gain public support because of its easy implementation, low cost, good investment of surface water that is lost otherwise, and natural attenuation of river water during aquifer passage, resulting in lower concentrations of suspended material, pathogens and various chemical pollutants (Dillon, Miller, Fallowfield, & Hutson, 2002; Medema & Stuyfzand, 2002; Stuyfzand, 1998).

An example of a successful river bank filtration in karst occurs along the Damour River in Lebanon (Figure 2). Surface water infiltration is induced from the river bed (with a thin alluvial cover) before it flows and is then extracted from vertical wells tapping a karstified dolomitic limestone aquifer in the vicinity. According to a 2011 survey, wells within about 350 m from the channel (Figure S3, supplementary online material) intercept bank-filtered water, as confirmed by $\delta^{18}O$ (the ratio of oxygen-18 and oxygen-16) and chloride for a mixture of three end-members (river, local rain and infiltrated ocean water). These wells receive over 50% river water contribution (Khadra & Stuyfzand, 2014), summing up to a total discharge of about 3300 m^3/d, a value that was corroborated by 3-D numerical modelling (Khadra & Stuyfzand, 2018).

It is anticipated that installing additional wells close to the Damour River will increase the induced river recharge, provided that exacerbation of saltwater intrusion is avoided.

The closer the wells are to the river, the more surface water they intercept, but after a shorter subterranean detention. It is not clear whether this shorter detention will have a significant effect on water quality via filtration, adsorption and elimination of heavy metals, organic micropollutants, and bacteria and viruses (Stuyfzand, Juhàsz-Holterman, & de Lange, 2006).

Management of urban stormwater

Stormwater in the Lebanese coastal cities (including the capital, Beirut) flows in drains and as surface runoff, and ultimately discharges into the Mediterranean Sea. Capturing this water (or at least part of it) and using it to recharge the underlying groundwater reservoirs could be one way to alleviate water shortages (Figure 5). This could be done even in karstic aquifers via widespread networks of infiltration galleries, pits along roads, drywells, or urban infrastructure already having preferential flow paths (Bonneau, Fletcher, Costelloe, & Burns, 2017). In fact, the superposed effect of spreading out a network of infiltration sites with a significant amount of recharge might eventually lead to a general rise of the water table, despite the local losses or random flow responses expected in karst. This overall gain would improve many hydrological aspects, for instance augmenting groundwater storage, saving precious land surface, and reducing saltwater intrusion in coastal aquifers.

Previous trials (e.g., at the Andrews Farm in South Australia) showed success in using aquifer storage recovery with urban stormwater in brackish limestone (Pavelic, Dillon, Barry, & Gerges, 2006). Drywells are also very effective vadose zone facilities to recharge groundwater. In Greater Beirut (the Lebanese capital and its northern and southern suburbs), the same network was previously used for both stormwater and wastewater discharge to the sea. A recent government project rehabilitated Beirut's sewers and constructed a separate network for stormwater, which is expected to cover the whole city soon (CDR, 2017). No official reports have recorded the volumes of collected stormwater. The average rainfall in Greater Beirut is ~190 Mm^3/year, which exceeds annual water needs (about 124 Mm^3/year, assuming 170 litres per person per day for 2 million people). The infiltration rate in the city is expected to be nearly nil due to high urbanization (Safi et al., 2018). Therefore, capturing any of the urban

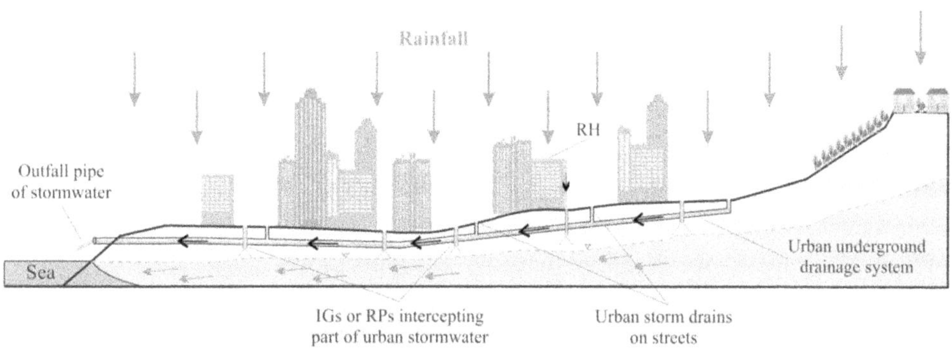

Figure 5. A coastal city where urban runoff is lost to the sea via drainage networks, or intercepted and forced to infiltrate via managed recharge systems.

stormwater, even if some is lost to the unpredictable tunnelled flow of karst, could be an excellent way to augment water storage and reduce the escalating salinization. It would improve the sustainability of the urban water cycle, and raise the Blue City Index. Assessing the overall feasibility of this technique and its influence on saltwater encroachment requires complex variable-density flow and solute transport simulations, coupled with some pilot sites, to evaluate the hydrogeochemical sustainability. The main hurdle is to have effective, very rapid pretreatment, removing suspended solids with very low maintenance, although karstic aquifers can handle high loads of suspended solids.

Water quality and hydrochemical analyses

Water quality concerns need to be seriously addressed when applying either river bank filtration, aquifer storage recovery or stormwater capture, not only in Lebanon but everywhere. First, there is the clogging potential of the source water. Clogging will reduce the efficiency of recharge or recovery facilities over time, and may incur high labour costs to regenerate the infiltration capacity of the river bed, basin or well. The clogging mechanism is mainly (bio)physical during infiltration (e.g., sedimentation due to high turbidity and total suspended solids, or the growth of biofilms and algae), and (bio)chemical during recovery (e.g., iron oxides, manganese oxides, calcite and biofilms caused by mixing of water from various environments and high fluxes near the well). Second, the chemical and microbiological quality of the source water may be problematic due to disposal of effluents and solid wastes, including nitrates, heavy metals, total organic carbon, organic micropollutants, bacteria and viruses, and radioactivity. Third, infiltrating river water, by interacting with deposited muds and then the aquifer matrix, can raise the levels of some geogenic chemical constituents (e.g., Fe, Mn, As, NH_4, Ca and HCO_3).

Available data on the Damour River (Table 2) show slightly polluted $CaHCO_3$ water characterized by low HCO_3, Ca and SiO_2, and relatively low Cl and SO_4 concentrations ($<$ 50 mg/L). Fe and Mn are under 0.04 mg/L, ammonium is under 0.1 mg/L, and nitrate averages 4 mg/L. The Damour river bank filtrate water mapped by Khadra and Stuyfzand (2014) mostly mimics the surface water source except for being much less polluted (Table 2). It is also characterized by pH stabilization, and attenuation of phosphate (34% removal) and some trace elements, e.g., Al, Cr, La, Ni, Pb, Sb and V, which recorded over 25% removal (Al recorded 98%). This testifies to the efficacy of the bank filtration process. However, Ba, Br and Sr are mobilized due to desorption and/or dissolution from hosting minerals, and Ba shows the highest mobilization, in line with the results of Khadra, Stuyfzand, and van Breukelen (2017b).

The increase of Na level in the bank filtrate points to Ca/Na exchange. Redox reactions are negligible, as indicated by the minor changes in O_2, SO_4, NO_3 and NH_4. Dissolution of Mg-calcite and SiO_2 is very limited, and hence only a small TDS increase is recorded. More concerns exist regarding the behaviour of potential pollutants in the river, and the chance that organic micropollutants, bacteria and viruses reach the river bank filtration wells. This could necessitate post-treatment of the raw extracted groundwater if used for drinking water supply, e.g., via coarse pre-filtration, sedimentation, rapid sand filtration, activated carbon filtration and chlorination.

Table 2. Median composition of the Damour River and the Damour river bank filtrate groundwater. Data acquired in 2011 (adapted from Khadra & Stuyfzand, 2014).

	Chemical water type	EC 25°C	Temp.	pH	Cl^-	SO_4^{2-}	HCO_3^-	NO_3^-	P	Na^+	K^+	Ca^{2+}	Mg^{2+}	Fe	Mn	NH_4^+	SiO_2	O_2
		µS/cm	°C	-	mg/L													
Detection limit					1	1	1	1	0.02	0.05	0.05	0.05	0.05	0.01	5×10^{-5}	0.1	0.04	0.01
Damour river bank Filtrate	F2CaHCO3	515	19.4	7.18	30.2	44.9	250	5.6	0.095	15.4	2.3	78.88	10.5	<0.01	0.000	<0.1	8.1	5.8
Damour River	g2CaHCO3	370	18.0	7.94	22.8	36.6	210	3.9	0.144	9.4	2.1	76.00	11.3	0.04	0.004	0.07	6.9	6.1

	Al	As	B	Ba	Br	Cr	Cu	Ge	La	Li	Mo	Ni	Pb	Pd	Rb	Sb	Sc	Sr	U	V	Zn
	µg/L																				
Detection limit	1	0.5	5	0.05	5	0.5	0.1	0.05	0.01	0.1	0.1	0.2	0.1	0.2	0.01	0.05	1	0.01	0.02	0.2	0.5
Damour river bank Filtrate	1	<0.5	27	34.7	120	0.7	0.7	<0.05	<0.01	2.0	0.3	0.3	0.3	<0.2	0.6	<0.05	<1	136	0.50	0.3	3.8
Damour River	53	<0.5	29	5.5	63.2	1.0	0.6	<0.05	0.03	2.1	0.5	0.5	0.4	<0.2	0.7	0.1	<1	115	0.5	0.6	1.9

As for drywells, limited research has tackled their performance and impact on the quantity and quality of groundwater (e.g., Sasidharan et al., 2018). Generally speaking, the stormwater influent should be properly managed to reduce suspended particles and avoid transport of surface pollutants, e.g., heavy metals and petroleum by-products, into the groundwater. So the water should be pretreated with sand, gravel and boulders or by self-cleaning strainers to reduce sediment accumulation from variable-sized loads, plant roots and wastes. Sufficient separation between the drywell or trench and the water table, for instance by the sedimentary material occasionally overlying karstic aquifers, is expected to provide passive treatment, acting as a natural filter to remove pollutants and impuritiesm although sometimes additional prefiltration (e.g., charcoal) is required. This way clogging is reduced, and improved water quality is expected to reach the groundwater.

Conclusions

Managed aquifer recharge, as a non-conventional method to augment water supply, is spreading worldwide because of its many advantages, including coping with water needs (safe and inexpensive temporal subterranean storage); improving water quality (often from polluted surface water to safe groundwater) via geo-purification or soil-aquifer treatment; mitigating saltwater intrusion in coastal areas; and adapting to climate change. There are still some obstacles to its application in karstic domains due to complex hydrodynamics and low chemical attenuation, although it also has advantages, including enhanced groundwater recharge and reduced clogging problems, even for turbid feed water. Therefore, the success of managed recharge in karstic settings is still questionable, and many doubts surround its application there.

In fact, managed aquifer recharge has a wide variety of applications and methods that could be adapted to fit into the different hydrogeological settings including karst. River bank filtration and artificial stormwater infiltration systems are among the main methods suggested in karstic domains, besides other alternatives, for instance the infiltration of desalinated seawater in coastal aquifers using fresh storage saline extraction, or multiple partially penetrating wells, or recharge of aquifers by natural subterranean lithological barriers that trap subsurface water.

River bank filtration is an easy, simple and cost-effective option that does not require technical sophistication. The assessment of one site in Lebanon (the Damour area, south of Beirut) confirms the efficacy of this technique in a dolomitic limestone karstic domain where Damour River water is captured and water quality improved via filtration, adsorption, and elimination of heavy metals and pollutants.

Management of stormwater in cities shows looks promising as well, even in karst where complete water storage is not possible. Capturing any portion of the stormwater that would otherwise be lost helps augment groundwater storage and reduce saltwater intrusion in coastal zones. This applies to the city of Beirut, where much water is lost through drains to the sea. Stringent regulations (e.g., on pretreatment) for safe, successful and sustainable installations are required in a country like Lebanon with random dumping and little governance of waste disposal. This necessitates further research with at least one pilot in a Lebanese urban environment on karstic grounds.

In conclusion, regardless of any previous perception of the inadequacy of managed aquifer recharge in karst, different available options can handle natural hydrodynamic complexities. This highlights the role of managed recharge as a main tool to cope with water needs, even in karstic-dominated areas. Like other karstic areas, Lebanon has a chance to step forward in managed aquifer recharge, initially with small pilots. The rule of thumb is to 'start small, learn as you go, and expand as needed' (Bouwer, 2002).

Acknowledgments

We would like to thank the editors and two anonymous reviewers for constructive comments that improved this manuscript.

Disclosure statement

No potential conflict of interest was reported by the authors.

ORCID

Wisam M. Khadra http://orcid.org/0000-0003-0590-8870

References

Asano, T. (1985). *Artificial recharge of groundwater*. Boston: Butterworth-Heinemann.

Bacchus, S. T., Bernardes, S., Xu, W., & Madden, M. (2015a). What Georgia can learn from aquifer storage and recovery (ASR) in Florida. In R. J. McDowell et al. (Ed.), *Proceedings 2015 Georgia Water Resources Conference*. Athens: University of Georgia.

Bacchus, S. T., Bernardes, S., Xu, W., & Madden, M. (2015b). Fractures as preferential flowpaths for aquifer storage and recovery (ASR) injections and withdrawals: Implications for environmentally sensitive near-shore waters, wetlands of the Greater Everglades Basin and the regional karst Floridan aquifer system. *Journal of Geography and Geology*, *7*(2), 117–155.

Bakalowicz, M. (2011). Management of Karst groundwater resources. In P. E. van Beynen (Ed.), *Karst management* (pp. 263–282). Dordrecht: Springer Science+Business Media.

Bonneau, J., Fletcher, T. D., Costelloe, J. F., & Burns, M. J. (2017). Stormwater infiltration and the 'urban karst': A review. *Journal of Hydrology*, *552*, 141–150.

Bouwer, H. (2002). Artificial recharge of groundwater: Hydrogeology and engineering. *Hydrogeology Journal*, *10*, 121–142.

Burek, P., Satoh, Y., Fischer, G., Kahil, M. T., Scherzer, A., Tramberend, S., … Wiberg, D. (2016). Water futures and solution: Fast track initiative (final report). Working Paper. Laxenburg, Austria: International Institute for Applied Systems Analysis. pure.iiasa.ac.at/13008/.

Burger, S., Brunke, M., Groen, K., Rolf, L., Metni, M., Matar, A., … Wiersma, A. (2019). *Managed aquifer recharge in Lebano: Pilot site selection for MAR in unconsolidated sediments in Lebanon*. Netherlands: Acacia Water report.

CDR. (2017). *Basic services of council for development and reconstruction in 2017*. Beirut: Council for Development and Reconstruction.

Daher, W., Pistre, S., Kneppers, A., Bakalowicz, M., & Najem, W. (2011). Karst and artificial recharge: Theoretical and practical problems: A preliminary approach to artificial recharge assessment. *Journal of Hydrology*, *408*(3–4), 189–202.

Daoud, A. (1973). *Groundwater recharge in the Beirut Area*. MS Thesis, American University of Beirut.

David, R., & Pyne, G. (2015). Aquifer storage recovery: An ASR solution to saltwater intrusion at Hilton Head Island, South Carolina, USA. *Environmental Earth Sciences, 73*, 7851–7859.
Dillon, P., & Stuyfzand, P. J.; 28 others. (2019). Sixty years of global progress in managed aquifer recharge. *Hydrogeology Journal, 27*, 1–30.
Dillon, P. J. (2005). Future management of aquifer recharge. *Hydrogeology Journal, 13*(1), 313–316.
Dillon, P. J., Miller, M., Fallowfield, H., & Hutson, J. (2002). The potential of riverbank filtration for drinking water supplies in relation to microsystin removal in brackish aquifers. *Journal of Hydrology, 266*, 209–221.
Hartog, N., & Stuyfzand, P. J. (2017). Water quality considerations on the rise as the use of managed aquifer recharge systems widens. *Water, 9*, 808.
Jain, S. K. (2016). Urban storm water management through artificial recharge and improvement in groundwater regime. *International Journal of Environmental Science, 1*, 385–391.
Khadra, W. M. (2017). *Analysis and remediation of the salinized, Damour coastal (Dolomitic) limestone aquifer in Lebanon*. PhD thesis, Delft University of Technology. doi:10.4233/uuid:6d4208be-65c1-43e8-afa0-5019f22c6167.
Khadra, W. M., & Stuyfzand, P. J. (2014). Separating baseline conditions from anthropogenic impacts: Example of the Damour coastal aquifer (Lebanon). *Hydrological Sciences Journal, 59*, 1872–1893.
Khadra, W. M., & Stuyfzand, P. J. (2018). Simulation of saltwater intrusion in a poorly karstified coastal aquifer in Lebanon (Eastern Mediterranean). *Hydrogeology Journal, 26*, 1839–1856. doi:10.1007/s10040-018-1752-z
Khadra, W. M., Stuyfzand, P. J., & Khadra, I. M. (2017a). Mitigation of saltwater intrusion by 'integrated fresh-keeper' wells combined with high recovery reverse osmosis. *Science of the Total Environment, 574*, 796–805.
Khadra, W. M., Stuyfzand, P. J., & van Breukelen, B. M. (2017b). Hydrochemical effects of saltwater intrusion in a limestone and dolomitic limestone aquifer in Lebanon. *Applied Geochemistry, 79*, 36–51.
Leitão, T. E., Ferreira, J. P. L., Martins, T., Oliveira, M. M., Henriques, M. J., Carvalho, T., … Rosa, M. J. (2017). Demonstrating managed aquifer recharge as a solution to water scarcity and drought: MAR to improve the groundwater status in south Portugal (MARSOL) project, final report, project ID 619120). Germany: European Union.
Maliva, R. G. (2020). *Anthropogenic aquifer recharge: WSP methods in water resources evaluation series no. 5*. Switzerland: Springer.
Maliva, R. G., & Missimer, T. M. (2012). *Aquifer storage and recovery and managed aquifer recharge using wells: Planning, hydrogeology, design, and operation*. Sugarland, TX: Schlumberger Water Services.
Masciopinto, C., & Carrieri, C. (2002). Assessment of water quality after 10 years of reclaimed water injection: The Nardò fractured aquifer (Southern Italy). *Groundwater Monitoring & Remediation, 22*(1), 88–97.
Medema, G. J., & Stuyfzand, P. J. (2002). Removal of micro-organisms upon basin recharge, deep well injection and river bank filtration in the Netherlands. In P. Dillon (Ed.), *Management of aquifer recharge for sustainability* (pp. 125–131). Adelaide, South Australia: Proceedings, 4th International Symposium on artificial recharge.
Megdal, S. B., & Dillon, P. (2015). Policy and economics of managed aquifer recharge and water banking. *Water, 7*, 592–598.
Missimer, T. M., Guo, W., Woolschlager, J., & Maliva, R. G. (2017). Long-term managed aquifer recharge in a saline-water aquifer as a critical component of an integrated water scheme in Southwestern Florida, USA. *Water, 9*, 774.
MoE/EU/UNDP. 2014. *Lebanon environmental assessment of the Syrian conflict & priority interventions*. Beirut: Ministry of Environment, Lebanon. Retrieved from http://www.lb.undp.org/content/lebanon/en/home/library/environment_energy/lebanon-environmental-assessment-of-the-syrian-conflict.html
MoEW. (2010). *National water sector strategy*. Beirut: Ministry of Energy and Water, Lebanon.

MoEW, & UNDP. (2014). *Assessment of groundwater resources of Lebanon.* Beirut: Ministry of Energy and Water, Lebanon.

Pavelic, P., Dillon, P. J., Barry, K. E., & Gerges, N. Z. (2006). Hydraulic evaluation of aquifer storage and recovery (ASR) with urban stormwater in a brackish limestone aquifer. *Hydrogeology Journal, 14*, 1544–1555.

Prinz, D. (2016). *Managed aquifer recharge in the Hazmieh area to prevent seawater intrusion in Beirut, Lebanon.* Bonn: Deutsche Gesellschaft für internationale Zusammenarbeit.

Pyne, R. D. G. (2005). *Aquifer storage recovery: A guide to groundwater recharge through wells* (2nd ed.). Florida: ASR Press.

Rolf, L. (2017). *Assessing the site suitability of Managed Aquifer Recharge (MAR) projects in karst aquifers in Lebanon, a multi criteria analysis.* MS thesis, Utrecht University, Netherlands.

Safi, A., Rachid, G., El-Fadel, M., Doummar, J., Abou Najm, M., & Alameddine, I. (2018). Synergy of climate change and local pressures on saltwater intrusion in coastal urban areas: Effective adaptation for policy planning. *Water International, 43*(2), 145–164.

Sasidharan, S., Bradford, S. A., Šimůnek, J., DeJong, B., & Kraemer, S. R. (2018). Evaluating drywells for stormwater management and enhanced aquifer recharge. *Advances in Water Resources, 116*, 167–177.

Schulz, S., de Rooij, G. H., Michelsen, N., Rausch, R., Siebert, C., Schüth, C., . . . Merz, R. (2016). Estimating groundwater recharge for an arid karst system using a combined approach of time-lapse camera monitoring and water balance modelling. *Hydrological Processes, 30*, 771–782.

Sprenger, C., Hartog, N., Hernández, M., Vilanova, E., Grützmacher, G., Scheibler, F., & Hannappel, S. (2017). Inventory of managed aquifer recharge sites in Europe: Historical development, current situation and perspectives. *Hydrogeology Journal, 25*(6), 1909–1922.

Stefan, C., & Ansems, N. (2018). Web-based global inventory of managed aquifer recharge applications. *Sustainable Water Resources Management, 4*(2), 153–162.

Stuyfzand, P. J. (1989a). Hydrology and water quality aspects of Rhine bank groundwater in the Netherlands. *Journal of Hydrology, 106*, 341–363.

Stuyfzand, P. J. (1989b). Quality changes of river Rhine and Meuse water upon basin recharge in the Netherlands' coastal dunes: 30 years of experience. In A. I. Johnson & D. J. Finlayson (Eds.), *Artificial recharge of Groundwater* (pp. 233–245). New York: Proceedings of the International Symposium, Anaheim,, American Society of Civil Engineers.

Stuyfzand, P. J. (1998). Simple models for reactive transport of pollutants and main constituents during artificial recharge and bank filtration. In J. S. Peters et al. (Ed.), *Artificial recharge of groundwater* (pp. 427–434). Amsterdam: Proceedings, 3rd International symposium on artificial recharge.

Stuyfzand, P. J. (2002). Quantifying the environmental impact and sustainability of artificial recharge systems. In Dillon (Ed.), *Management of aquifer recharge for sustainability* (pp. 77–82). Adelaide, South Australia: Proceedings, 4th International symposium on artificial recharge.

Stuyfzand, P. J., Juhàsz-Holterman, M. H. A., & de Lange, W. J. (2006). Riverbank filtration in the Netherlands: Well fields, clogging and geochemical reactions. In S. A. Hubbs (Ed.), *Riverbank filtration hydrology* (pp. 119–153). Netherlands: Springer.

Stuyfzand, P. J., Khadra, W. M., & Burger, S. (2019). *Evaluation of managed aquifer recharge in Lebanon, with focus on opportunities, problems with karstic aquifers, and preliminary results of an ASR pilot in the Beqaa Valley* (KWR report, KWR 2019.069). Netherlands.

Tröger, U. (2010). Managing the Thermal Aquifer System of Caldas Novas, Brazil. In E. Escalante (Ed.), *Achieving ground water supply sustainability & reliability through managed aquifer recharge* (pp. 591–596). Abu Dhabi, UAE: Proceedings, 7th International symposium on management of aquifer recharge.

UNESCO-IHP. 2011. *The impact of global change on water resources: The response of UNESCO's International Hydrological Programme.* Paris: UNESCO. Retrieved from: http://unesdoc.unesco.org/images/0019/001922/192216e.pdf

UN-Water. 2018. *The United Nations World Water Development Report 2018: Nature-based solutions for water*. Paris: UNESCO. Retrieved from: http://creativecommons.org/licenses/by-sa/3.0/igo/

Van Ginkel, M., Olsthoorn, T. N., Smidt, E., Darwish, R., & Rashwan, S. (2010). *Fresh storage saline extraction (FSSE) wells: Feasibility of fresh water storage in saline aquifer with a focus on the Red Sea coast, Egypt* (pp. 188–195). Abu Dhabi: Proceedings, 7th International symposium on management of aquifer recharge.

Vanderzalm, J., Page, D., Dillon, P., Lawson, J., Grey, N., Sexton, D., & Williamson, D. (2014). *A risk-based management plan for Mount Gambier stormwater recharge system: Stormwater recharge to the Gambier Limestone aquifer*. Adelaide, South Australia: Goyder Institute for Water Research, Technical Report 14/7.

Xanke, J., Goeppert, N., Sawarieh, A., Liesch, T., Kinger, J., Ali, W., … Goldscheider, N. (2015). Impact of managed aquifer recharge on the chemical and isotopic composition of a karst aquifer, Wala reservoir, Jordan. *Hydrogeology Journal*, *23*, 1027–1040.

Xanke, J., Jourde, H., Liesch, T., & Goldscheider, N. (2016). Numerical long-term assessment of managed aquifer recharge from a reservoir into a karst aquifer in Jordan. *Journal of Hydrology*, *540*, 603–614.

Zuurbier, K. G., Zaadnoordijk, W. J., & Stuyfzand, P. J. (2014). How multiple partially penetrating wells improve the freshwater recovery of coastal aquifer storage and recovery (ASR) systems: A field and modeling study. *Journal of Hydrology*, *509*, 430–441.

A multifaceted quantitative index for sustainability assessment of groundwater management: application for aquifers around Iran

Bahador Zarei, Esmaeel Parizi, Seiyed Mossa Hosseini and Behzad Ataie-Ashtiani

ABSTRACT

This study developed a groundwater sustainable management index including components of environmental, social, economic and institutional responsibility. Four significant indicators: integrated groundwater footprint, population density, percentage of groundwater used for agricultural and industrial sectors, and density of observation wells were adopted and executed for 443 of Iran's aquifers. The results indicated that the groundwater sustainable management for 32% of Iran's aquifers is poor–very poor, and this index is more sensitive to economic and social indicators. Finally, an inverse correlation was observed between the groundwater sustainable management index values and the coefficient of variation of normalized difference vegetation index ($R^2 = -0.437$).

Introduction

Groundwater is a valuable and ubiquitous source of freshwater that supports human health, socio-economic development and functioning of ecosystems in all climatic regions, including both urban and rural areas of developed and developing countries (Ferguson & Gleeson, 2012; Fraser et al., 2020; Javadi et al., 2011). Groundwater utilization is a major factor in maintaining the livelihood of rural farmers, achieving food security, national security and stability (Kelley et al., 2015). Groundwater resources fulfil the drinking water needs of more than half of the global population and 43% of all irrigation water requirements (Ayers & Westcot, 1994). On the other hand, the daily water needs of about 2.5 billion people around the world are solely met by groundwater resources (UNESCO, 2007).

Motivated by the accessibility and affordability of pumping technology, continuing increases in water demands along with decreases in precipitation and surface flows and an overreliance on groundwater resources have led to a groundwater depletion problem with adverse repercussions on dependent communities (Doell et al., 2014; Ghadimi &

Ketabchi, 2019). Major hotspots of depletion were observed in arid and semi-arid parts of the world, including north-west India, north-east China, north-east Pakistan, the High Plains and California Central Valley aquifers in the United States, central Iran, central Yemen, and southern Spain (Jakeman et al., 2016). About 67% of the groundwater used for global consumption (about 686 km^3/year) is extracted in countries characterized by climatic aridity, such as India (about 251 km^3/year, or 36.6%), China (about 112 km^3/year, or 16.3%), the United States (about 112 km^3/year, or 16.3%), Pakistan (about 64 km^3/year, or 9.3%), Iran (about 60 km^3/year, or 8.7%), Bangladesh (about 35 km^3/year, or 5.1%), Mexico (about 29 km^3/year, or 4.2%) and Saudi Arabia (about 23 km^3/year, or 3.4%) (Gleeson et al., 2012; Trichakis et al., 2016).

The socio-economic impacts of groundwater depletion are often translated into the problems of economic and social costs (Lezzaik & Milewski, 2018; World Bank, 2010). Economic costs arise due to greater capital investments in well construction and pumping plants associated with pumping from greater depths and reduced yield of pumping wells (Dalin et al., 2017; Konikow & Kendy, 2005). This imposes economic costs (i.e., electricity or fuel costs) associated with pumping water from greater depths. Land subsidence related to groundwater depletion damages public infrastructure and private property and imposes economic costs related to damage repairs born by official organizations and individuals (Mirzadeh et al., 2021; Nabavi, 2018). Social costs arise, especially for poor urbanites and rural dwellers who are forced to invest more time in finding alternative sources of water (Wegmann & Mußhoff, 2019). The continued degradation of and heightened competition over groundwater resources generates societal instability by forcing migration and creating underemployed and restive demographics (Castilla-Rho et al., 2019; Richey et al., 2015).

The sustainability of groundwater resources is threatened by various anthropogenic (e.g., overexploitation and rapid population growth) and climate factors (e.g., the severity and duration of droughts) (Foster & van der Gun, 2016; Goodarzi et al., 2016; Maiolo & Pantusa, 2019). The paradigm of weak sustainability considers that natural capital can be substituted with anthropogenic capital instead of the ecological–economic framework (Maiolo & Pantusa, 2019). After the definition of sustainable development by Brundtland (1987), Alley and Leake (2004, p. 13) introduced groundwater sustainability management as 'the development and use of these resources in a manner that can be maintained for an indefinite time without causing unacceptable environmental or socio-economic consequences'. Achieving groundwater sustainable management that meets all the aspects of the sustainability perspective for groundwater development is a challenge and vital for the sustainable development of countries (Senent-Aparicio et al., 2015; United Nations, 2015). The four main pillars of sustainability are environmental, economic, social and institutional elements (Bui et al., 2019).

Groundwater sustainable management is also affected by the support and suitability of governance framework (Pires et al., 2017), which refers to the provision for the authority to look after groundwater development and management, legal framework and control mechanism, and adequate capacity (availability of human resources, information and education) to carry out its duty (Pandey et al., 2011). The relationship between groundwater endowment and governance is explained by Collier's (2010) conceptual model. Equilibrium between the two systems is reached when relationships are resolved simultaneously. This model indicates that if the groundwater reserves are discovered after effective political systems have been established, then they may be more robustly managed by these institutions. In the regions with higher groundwater endowment (i.e., the high social value of groundwater), however, the quality of the groundwater governance will be weakened and hindered under the socio-economic developments. Here, the main question is how to provide enough quantified information for decision-makers to assist management decisions for improving groundwater sustainability. Efforts have been dedicated to quantifying groundwater sustainable management through indices (Lambán et al., 2011). The quantified index system would provide useful information for decision-makers to evaluate current policies and formulate and implement suitable strategies for the better management of these resources (Aayog, 2018; Pandey et al., 2011). For example, the groundwater sustainability indicator by UNESCO/IAEA/IAH (Vrba et al., 2007), the groundwater resource index by Mendicino et al. (2008), the groundwater sustainability infrastructure index by Pandey et al. (2011), the groundwater health index by Korbel and Hose (2011) and the environmentally sustainable groundwater management index by Hosseini et al. (2019) are some quantified index system for groundwater management.

Loucks (1997) proposed an approach for identifying sustainability indices relative to a weighted combination of reliability, resilience and vulnerability measures of various economic, environmental, ecological and social criteria. Despite the development of a dozen groundwater sustainable management indicators, a commonly integrated index based on internationally accepted scientific standard indicators that encompass all sustainability aspects and applicability for different aquifers must be defined (Pires et al., 2017). Due to spatiotemporal variation of quality and quantity data from an aquifer to an aquifer and the limited amount of data available, relating such index merits to the geographical and economic conditions of where they are being evaluated is challenging (Russo et al., 2014).

The unsustainable rates of groundwater abstraction in Iran are taking this country to a point where socio-economic development, political stability, ecosystem integrity, the health and welfare of natural systems, and human communities may seriously be threatened (Barati et al., 2019). After a long history of overdraft in Iran, almost no studies have been conducted to assess Iran's groundwater sustainable management while considering all infrastructures of sustainability (i.e., social, economic, environmental and institutional) (Sadeghi-Tabas et al., 2016). In a previous study, Hosseini et al. (2019) adopted 13 environmentally related indicators for assessing groundwater sustainable management in 30 key aquifers across Iran. Their results revealed the unsustainable status of groundwater management in this country from an environmental viewpoint. To date, no comprehensive analysis in support of developing a quantified index that includes all components of sustainability (environmental, social, economic and institutional) has been developed. This study evaluates groundwater sustainable management through the development of a sound quantified index that encompasses the four mentioned components of sustainability on the aquifer scale. For each component, quantified indicators have been selected based on the meeting of internationally scientific standards and the availability of data on the aquifer, and it was aggregated to the integrated index of groundwater sustainable management index by the analytical hierarchy process. To demonstrate the applicability of the adopted indicators, all were tested for 443 of Iran's aquifers (83% of Iran's total aquifers) based on the data availability. These aquifers are classified and ranked based on their groundwater sustainable management values and multicriteria decision-making method of the Preference Ranking Organization Method for Enrichment Evaluations (PROMETHEE) to evaluate the previous and current actions, plans, policies, and strategies on groundwater resources in this country. Ranking of the aquifers according to their groundwater sustainable management index values will help administrators and decision-makers to prioritize the areas that need new/revised policies in order to achieve sustainability in groundwater management.

Material and methods

Groundwater sustainable management is a manifold concept encompassing the dimensions of environmental, socio-economic and institutional responsibility. The process of assessing the means of achieving groundwater sustainable management in Iran translates into the analysis of the components and quantitative indicators listed in Table 1 and is also shown schematically in Figure 1: integrated groundwater footprint ($IGWF_T$) for the environmental dimension; population density (D_{Pop}) for the social dimension; percentage of groundwater used for the agricultural and industrial sectors (PGW_{ai}) for the economic dimension; and density of observation wells (D_{OW}) for the institutional dimensions. The w_k values assigned to the four indicators ($w_k, k = 1, 2, \ldots, 4$) were considered as one of the inputs of the PROMETHEE II method for ranking the aquifers (Figure 1). Determining the precise values of w_k through pairwise comparisons by experts is often a delicate task and involves the biased perceptions of the questioner (Brans & De Smet, 2016). To understand the relative influences of the w_i values on the variability in the multicriteria decision-making model output (i.e., ranking of aquifers), the local sensitivity analysis of w_i was adopted. During local sensitivity analysis, both equal and unequal weights were reasonably assigned to the four criteria to obtain their importance in the final groundwater sustainable management index. Because the geographical information system is a powerful tool for storing, manipulating, analysing and spatial data mapping (Assaf & Saadeh, 2009; Nampak et al., 2014), thus all data layers and outputs of the multicriteria decision-making model were prepared and mapped in ArcGIS Desktop 10.7.1.

Table 1. Components and indicators considered for the assessment of groundwater sustainable management in Iran's aquifers.

No.	Component	Indicator	Classification under the DPSIR framework	Description
1	Environmental	Integrated groundwater footprint ($IGWF_T$)	P	The $IGWF_T$ represents a water balance between recharge and outflows of groundwater, and also the role of groundwater quality
2	Social	Population density (D_{Pop})	P	Number of people living per km^2 of the aquifer
3	Economic	Percentage of groundwater used for agricultural and industrial sectors (PGW_{ai})	P, S, I	Annual amount of groundwater used by the agricultural and industrial sectors
4	Institutional	Density of observation wells (D_{OW})	S, R	Number of observing wells per km^2 of the aquifer

Note: DPSIR, drive forces, pressure, state, and impact and response.

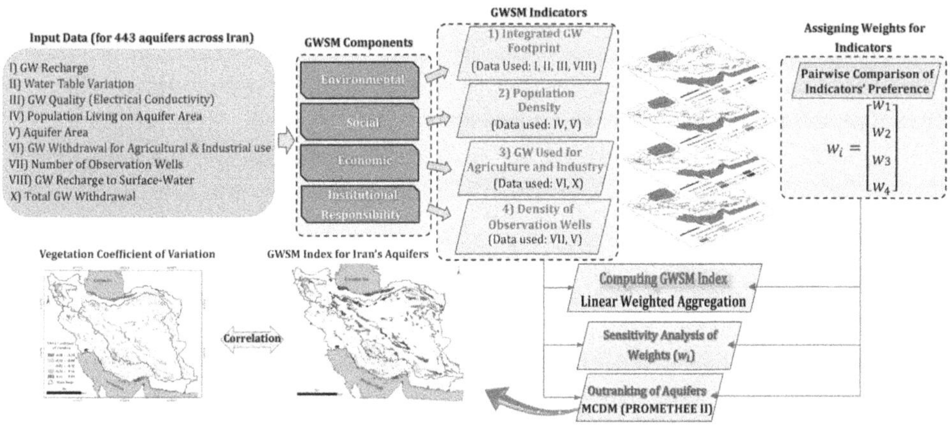

Figure 1. Schematic representation of the methodology adopted in this study to assess the groundwater sustainable management for 443 aquifers across Iran. Note: GWSM, groundwater sustainable management; GW, groundwater; and MCDM, multicriteria decision-making.

Indicators

The appropriate indicators for the assessment of groundwater sustainability through four sustainability components (environmental, social, economic and institutional) were selected based on the following criteria: availability of the required data regarding the groundwater data scarcity in the country, practicability, their ability to effectively represent the issues for which they are designed (effectiveness), and meeting universally scientific standards based on the review of the literature (e.g., Kumar et al., 2008; Pandey et al., 2011; Shah et al., 2003; UNESCO, 2007). Pires et al. (2017) evaluated an extensive list of 170 indicators related to water use and management based on scoring by an international panel of 14 experts. They concluded that 24 indicators complied with the majority of the sustainability criteria. The indicators chosen in the current study are consistent with those reported by Pires et al. (2017) and are described here under as well as in Table 1.

Integrated groundwater footprint, $IGWF_T$

This indicator expresses the status of the environmental component of the groundwater sustainability categorized in the pressure and response in the drive forces, pressure, state, and impact and response (DPSIR) framework. The DPSIR framework organizes the indicators according to the cause–effect schema under four categories of drive forces, pressure, state, and impact and response (WWAP, 2012). The groundwater footprint was first introduced by Gleeson et al. (2012) as the area required for sustainable use of groundwater for an aquifer and can be estimated using:

$$GWF = A_{aq} \times \left(\frac{\Delta H}{R - BF}\right) \tag{1}$$

where *GWF* or *GF* is the groundwater footprint; ΔH is the average annual variation of groundwater level (m); *R* is average annual recharge rate to aquifer (m); *BF* is groundwater contribution to surface flow (m) (which is very important during the low-streamflow period); and A_{aq} is the aquifer area (m^2). According to Smakhtin et al. (2004) and Requena et al. (2018), the *BF* can be considered equal to the lower decade of streamflow (Q_{10}), the monthly streamflow that is exceeded 90% of the time.

One challenge to computing the *GWF* indicator may be attributed to the average annual variation of groundwater level (variable ΔH in equation 1). Previous studies that used the *GWF* indicator for assessing the groundwater sustainable management on the aquifer scale (e.g., Gleeson et al., 2012; Hosseini et al., 2019; Kourgialas et al., 2018) did not encounter this challenge because the aquifers considered in their studies indicated groundwater abstraction and adopted the positive sign of ΔH. As groundwater abstraction increased, the value of ΔH was heightened. Therefore, a greater value of *GWF* represents the non-sustainable management of groundwater sustainable resource. In the current study, the dual behaviours of aquifers in the groundwater level (decreasing and increasing) were resolved by transforming the calculated values of *GWF* for the studied aquifers using equation (2). The strategy adopted is shown in Figure S1 in the supplemental data online, schematically:

$$GWF_T = \frac{-GWF + GWF_{min}}{GWF_{max}} \tag{2}$$

The *GWF* indicator was promoted to the *IGWF* by Kourgialas et al. (2018) by incorporating the crucial role of groundwater quality (i.e., electrical conductivity) into equation (1) to make it a sound decision-making tool for groundwater management:

$$IGWF = GWF \times \left(1 + \frac{A_{crit}}{A_{aq}}\right) \tag{3}$$

where A_{crit} is the area of an aquifer with a groundwater pollution problem. In this study, a threshold value of electrical conductivity = 1000 µS/cm groundwater that needs to be treated (EPA, 2001) was considered for the delineation of the area of aquifers that have a pollution problem. In this study, equations (1) to (3) were used to calculate $IGWF_T$:

$$IGWF_T = GWF_T \times \left(1 + \frac{A_{crit}}{A_{aq}}\right) \tag{4}$$

A greater $IGWF_T$ value indicates the non-sustainable management of groundwater resource and/or evidence of contamination, while aquifers with a lower $IGWF_T$ value suggests that groundwater resources are sustainably used and/or of appropriate quality. In the current study, mean long-term (2000–16) values of ΔH for 443 aquifers were calculated by the analysis of monthly groundwater level in observation wells using the Thiessen polygon method (Tatalovich et al., 2006) in different years. On the other hand, mean long-term (2000–16) values of *R* across Iran have previously been calculated by Iran's WRM Company (2017). Indeed, *R* was calculated using the model of lumped water balance, which was adopted by Iran's WRM Company. The total natural R to an aquifer using the general groundwater balance equation can be defined as (Parizi et al., 2020):

$$R = D_g + S_r + O_g + \Delta V \tag{5}$$

where R is the total natural R from river seepage, rainfall, groundwater inflow from other basins and return flows from water used for irrigation, industrial and domestic sectors; D_g is a draft from groundwater by springs, pumping wells and qanats; S_r is groundwater drainage into surface water; O_g is groundwater outflow to other basins; and ΔV is the change in groundwater storage. Moreover, to calculate the BF we used the monthly streamflow data of hydrometric stations from 2000 to 2016, which are obtained from the database of Iran's WRM Company. Also, A_{crit} is identified by interpolating the annual average values of electrical conductivity of sampled groundwater through a quality monitoring network for each aquifer during the observation period of 2000–16.

Population density over the aquifer area, D_{Pop} [people/km^2]

This pressure indicator shows the status of the social component of the groundwater sustainable management and can be computed by dividing the number of people who live in the aquifer area (N_{pop}) by the aquifer area (A_{aq}):

$$D_{Pop} = \frac{N_{pop}}{A_{aq}} \tag{6}$$

A greater value of D_{Pop} represents populated aquifers that lead to the exploitation of groundwater resources to meet the demand for more water in different sectors (e.g., the domestic consumption and goods production; Ayers & Westcot, 1994). To calculate the number of people who live in the aquifer area, in the first step the cities and villages located over the aquifer area were extracted. In the next step, the population data of the cities and villages for each aquifer were collected from Iran's Statistics Center (ISC) (http://www.amar.org.ir/default.aspx) based on mean census results in 2006, 2011 and 2016.

Percentage of groundwater withdrawal used for the agricultural and industrial sectors, PGW_{ai} (%)

This indicator reflects the status of the economic component of groundwater sustainable management classified the pressure, state and impact indicators. Agriculture is a dominant component of the global economy (Narany et al., 2016), and in agricultural water consumption-dominant regions such as Iran, agricultural activities are a crucial component of the economy (World Bank, 2010). Therefore, PGW_{ai} signifies the importance of groundwater in food security and social and economic rural and urban development, where the groundwater resource is consumed for irrigation or industry (Shah et al., 2003). Thus, a greater PGW_{ai} value indicates the more economic activities and the more efficient groundwater sustainable management. PGW_{ai} is defined as the fraction of total groundwater withdrawn (GW_{total}) used for the agricultural and industrial sectors (GW_{ai}):

$$PGW_{ai} = \frac{GW_{ai}}{GW_{total}} \times 100 \tag{7}$$

The groundwater is withdrawn through wells, springs and qanats in Iran (Hosseini et al., 2019). On the other hand, water consumption type in different sectors (agricultural, domestic and industrial) and the withdrawal volume have been determined by Iran's WRM Company (2017) in all of Iran's aquifers (Parizi et al., 2019). Thus, in this study, PGW_{ai} was calculated by collecting and analysing the water consumption type and the withdrawals volume data of wells, springs and qanats for 443 aquifers from Iran's WRM Company (2017).

Density of observation wells across the aquifer, D_{OW} (wells/km^2)

This state and response indicator is adopted to quantify the intuitional responsibility infrastructure of the groundwater sustainable management and can be computed by dividing the number of active observation wells (N_{ows}) by the aquifer area (A_{aq}):

$$D_{OW} = \frac{N_{ows}}{A_{aq}} \tag{8}$$

Satisfactory spatial data collection and monitoring of water table fluctuation over the aquifer as an institutional responsibility enable a long-term understanding of groundwater availability and anthropogenic effects on the groundwater resource (Hosseini et al., 2019). Monitoring data, if used in parallel with mapping and modelling, reflect a complete picture of the aquifer conditions needed for a proper and full assessment of the groundwater resource and benefit the decision-making process for protecting these resources (Pandey et al., 2011). Groundwater monitoring must be made with a network of observation wells and needs the continued financial support and commitment of the authority and stakeholders (i.e., Iran's WRM Company has the mandate for coordinating this issue in Iran). In developing countries, monitoring groundwater level fluctuations is likely expensive and a burdensome cost to the governance; therefore, the D_{OW} can be considered as a reasonable indicator for showing the economic ability of the authority in groundwater management (Abiye et al., 2018). To calculate D_{OW}, we collected mean long-term (2000–16) data of the number of active observation wells for 443 aquifers from Iran's WRM Company (2017).

Ranking aquifers using the multicriteria decision-making method

After defining the indicators, the PROMETHEE II technique developed by Brans (1982) and then extended by Brans and Vincke (1985) was used to rank 443 aquifers with respect to their groundwater sustainable management index values. This method is based on pairwise comparisons of indicators to obtain the relative preference of one indicator (i.e., groundwater sustainable management values of the aquifer) over others (Betrie et al., 2013) and uses three steps to the n alternatives (here, 443 groundwater sustainable management values) $A = \{a_1, a_2 \ldots, a_n\}$ based on m criteria, I_k (here, four indicators: environmental, social, economic and institutional; Jothibasu & Anbazhagan, 2016): (1) to define a preference function for each indicator; (2) to calculate the multicriteria preference index; and (3) to calculate the net flow score (Φ) using (Brans, 1982):

$$\Phi(a_i) = \frac{1}{n-1} \times \sum_{k=1}^{m} \sum_{a_j \in A} w_k \times \left[P_k(a_i, a_j) - P_k(a_j, a_i)\right] \tag{9}$$

where $P_k(a_i, a_j)$ is the preference function that gives the degree of preference expressed by decision-makers for alternative a_i over a_j ($i = 12 \ldots n$; $(a_i, a_j) \in A$) and produces a normalized value, $0 \leq P_k(a_i, a_j) \leq 1$. Six types of preference functions: usual, 'U'-shape, 'V'-shape, level, linear and Gaussian, were used in the literature (e.g., Brans & Mareschal, 2005). w_k is the relative weights of criterion I_k. The ranking of alternatives by PROMETHEE II is based on values of Φ, which are in the interval of $[-1 + 1]$. The best aquifer is that with the highest net flow score (Φ). To obtain the appropriate relative weights of indicators (w_k), a sequence of pairwise comparison judgements according to the analytic hierarchy process (Saaty, 2008) was adopted.

Calculating the vegetation coefficient of variation

Recently, studies have used the normalized difference vegetation index (NDVI) as an efficient indicator to assess groundwater levels and storage for naturally vegetation-covered areas (e.g., forests, grasslands and open shrublands) (Bhanja et al., 2019; Rigge et al., 2019). This index has been widely used as a suitable index for determining the long-term spatiotemporal changes of vegetation (Parizi et al., 2021). In the current study, the long-term coefficient of variation (CV) in the growing season (summer) was used to determine the relative extent of agricultural land in the studied areas. Summer imagery was used because this time period is less variable in terms of weather patterns and degree of cloudiness, and precipitation is typically low in this time period; therefore, it provides a clearer representation of vegetation (Aguilar et al., 2012; Xie et al., 2019). Moreover, it was assumed that any change in the NDVI during the growing season in successive years was mainly affected by human activities, that is, agriculture (Nhamo et al., 2019). For this purpose, the NDVI time series were first extracted using Landsat ETM+ satellite imagery from 2000 to 2016 with 16-day intervals and 30 m resolution on the Google Earth Engine platform (Gorelick et al., 2017). Then, the NDVI CV for each aquifer in the growing season was calculated.

Study aquifers

Iran is divided into six main basins (also called first-order basins), 30 secondary catchments and 609 study areas (also called 'plain'), and has 536 aquifers (IBWWOPS, Iran's Bureau of Water and Wastewater Operation and Protection Systems, 2015). Totally, 443 aquifers (83% of Iran's total aquifers) across the country include the appropriate quantity- and quality-monitoring networks. Figure 2 indicates the locations of the 443 study aquifers across Iran. This study incorporates all these aquifers to represent a comprehensive assessment of groundwater sustainable management across the country. According to Figure S2a in the supplemental data online,

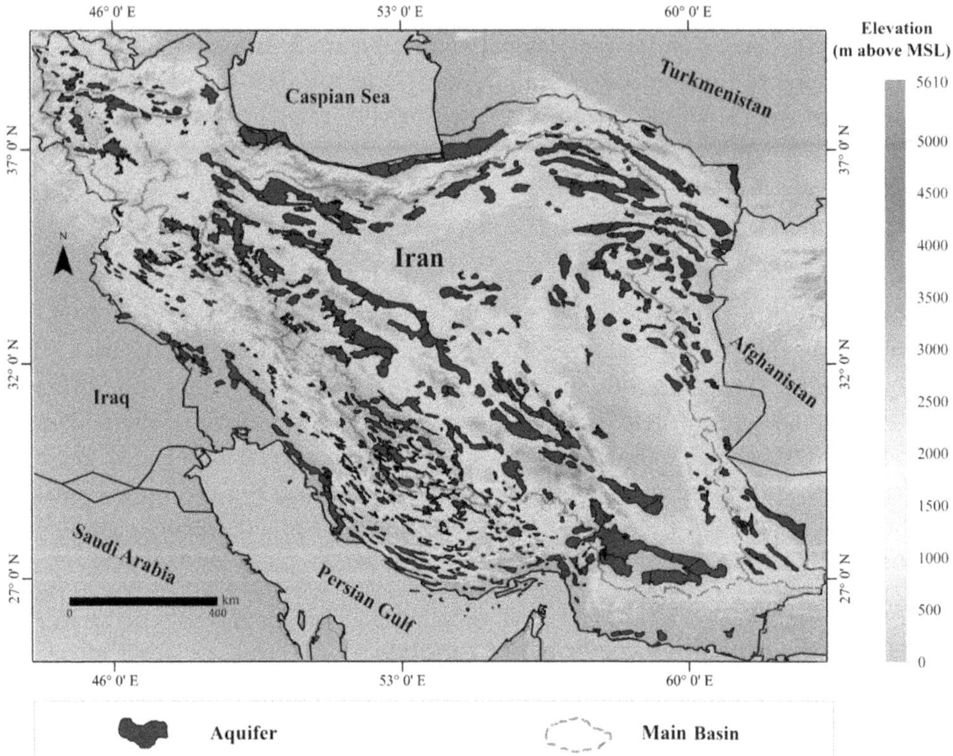

Figure 2. Location of the 443 study aquifers across Iran for assessing groundwater sustainable management.

9.4% and 67% of total aquifers in Iran receive annual precipitation of less than 100 mm and between 100 and 400 mm, respectively. Only 3.5% of aquifers receive annual precipitation of more than 1000 mm (http://www.irimo.ir). The trend of the number of drilled pumping wells (including deep and semi-deep wells) and the trend of groundwater withdrawal from Iran's 443 aquifers from 2003 to 2016 are shown in Figure S2b online. The total number of drilling pumping wells (only licenced wells) increased 1.7-fold from 434,535 to 742,433 wells (an increase of 23,700 wells/year on average), but the total withdrawal volume from groundwater resources by the pumping wells kept constant from 2003 to 2016 (43.7×10^9 m^3/year). This is due to the decrease in groundwater level in most aquifers of Iran and, thus, the decrease in the aquifer's transmissivity and yield of the pumping wells. As seen in Figure S2c online, population growth and, especially, the dramatic trend of urban population is a key driver of Iran's groundwater withdrawal by pumping wells during the same period. Pro-agricultural policies and agribusiness development evolved during the last 40 years in Iran, leading to the agricultural sector becoming responsible for consuming 89.14% of the groundwater withdrawal in the country. Potable water and industries consumed only 8.70% and 2.16% of the groundwater withdrawal, respectively (Iran's Water Statistical Yearbook, 2014). Although the agricultural sector consumes around 90% of groundwater, average irrigation systems efficiency in Iran is less than 35% (Madani, 2014), which is much lower than the irrigation systems efficiency of most developed countries, which lies in the range of 70–90% (FAO, 2016; Nazari et al., 2018). The steady increase in the exploitation of groundwater resources in Iran led to a declining level of groundwater equal to 0.36 m/year on average in many aquifers across the country (see Figure S2d online) and a reservoir deficit of 4220×10^6 m^3/year across 443 aquifers (see Figure S2e online).

As shown in Figure S2f online, prominent aquifers characterized by very significant long-term groundwater level declines are located mostly in arid and semi-arid zones (mostly in the south, south-west and north-east of Iran) due to less precipitation and increased pumping in those areas. Of the 609 total study areas (plains) across Iran, 31.4% are categorized by Iran's WRM Company as 'allowable' (groundwater depletion < safe yield), 44% are 'normal-prohibited' (groundwater depletion > safe yield), 22.4% are 'critical-prohibited' (groundwater depletion > safe yield; in addition, land subsidence, salt-water intrusion and deterioration of groundwater quality are also observed), and 2.13% are proposed to be critical-prohibited by the 2018 (IBWWOPS, Iran's Bureau of Water and Wastewater Operation and Protection Systems, 2019). The trend in the number of prohibited groundwater plains (including both normal- and critical-prohibited) in Iran over a 48-year period (1970–2018) is shown in Figure S2g online. During this period, the number of prohibited plains increased to around 2000% (from 20 plains in 1970 to 408 plains in 2018).

In 2014, Iran's Supreme Council on Water approved a new plan entitled 'Restoration and Rehabilitation of Groundwater Resources Plan' (IBWWOPS, Iran's Bureau of Water and Wastewater Operation and Protection Systems, 2015), which included 15 executive projects aimed at reducing the rate of groundwater decline, but not reversing it or backfilling the depleted volume. The plan mandated installing 'smart metering' systems in wells, putting Global Positioning System (GPS) on drilling machines to control their location during drilling, engaging groups for patrolling and inspection, creating a local water market, creating water-user associations, and replacing irrigation wells in prohibited areas with recycled water. The details of this national plan and its subprojects are well explained by Hosseini et al. (2019). During the most recent decade, the orientation of functions and activities of Iran's WRM Company, which has the mandate of coordinating water exploration, development and supply, has shifted from the groundwater developer to the integrated groundwater resources evaluator, planner and manager. The current study can assist government administrators in their development of effective groundwater resource management strategies for maintaining or adapting current policies or for establishing new ones.

Results and discussion

The spatial distribution and histogram of the initial data required for computing the four adopted indicators of environmental ($IGWF_T$), social (D_{Pop}), economic (PGW_{ai}) and institutional (D_{OW}) across 443 aquifers are shown in Figure S3a–p in the supplemental data online, and are summarized below.

Water table variation (ΔH)

Water table fluctuations in the study aquifers varied in the range of −2.37 to 0.51 m/year (average = −0.47 m/year) based on monthly data for the period 2000–16. As seen in Figure S3a in the supplemental data online, groundwater abstraction (negative ΔH) is observed in 93% of Iran's aquifers, mainly located in the central main basin, whereas increases in the groundwater table are observed only for 7% of total aquifers, averaging 0.08 m/year (see Figure S3b online). This variable contributes to the calculation of the environmental indicator (GWF).

Mean annual recharge (R)

The values of R in the studied aquifers were in the range of 0.00–2.93 m/year, with an average of 0.26 m/year (see Figure S3c,d in the supplemental data online). Aquifers located in the north-west, west and south-west of Iran received more recharge, and 52.6% of aquifers (233 aquifers) received recharge rates of less than 0.20 m annually.

Contribution of groundwater to surface flow (BF *or* $Q10$)

The spatial distribution and histogram of baseflow values (i.e., 10th percentile of streamflow at the hydrometric station located at the end of the groundwater basin) are shown in Figure S3e,f in the supplemental data online. The values of $Q10$ obtained for Iran's aquifers ranged from 0.0 to 6.71 m^3/s. The maximum contribution of groundwater to surface flow was observed in the few aquifers in the north of Iran (south of the Caspian Sea).

Aquifer area (A)

The area of the studied aquifers ranged from 12.3 to 5835.0 km^2 (average = 534 km^2) (see Figure S3g,h in the supplemental data online). Due to the wide plains with low slope and appropriate sediments (especially sand and gravel) (Iran's WRM Company, 2017), the largest aquifers are located in Kerman province (e.g., aquifers of Rudbar-e Jonubi, Rafsanjan and Bam-Narmashir).

Electrical conductivity of groundwater

The electrical conductivity of groundwater in Iran's aquifers varied from 304 to 29,041 μS/cm with an average value of 3164 μS/cm (see Figure S3i,j in the supplemental data online). Nearly 77% of all aquifers exceed the EPA's threshold value of groundwater quality (electrical conductivity = 1000 μS/cm) (EPA, 2001) and therefore needs to be treated. This problem is related to those aquifers that receive less recharge from the surface (see Figure S3c online). The areas of the aquifers where the electrical conductivity of groundwater exceeds the threshold value of 1000 μS/cm are considered as A_{crit} in equation (3). The variables ΔH, R, $Q10$ and electrical conductivity were considered when computing the $IGWF_T$ indicator for the 443 studied aquifers.

Number of piezometer or observation wells across aquifer (N_{ows})

As seen in Figure S3k,l in the supplemental data online, the number of observation wells for monitoring the groundwater level across aquifers are in the range of two wells (for aquifers located in the central main basin) to 250 wells (Gorgan aquifer in the south-east of the Caspian Sea), with 33 wells being the average. More than 60% of Iran's aquifers include a monitoring network with less than 30 observation wells. Dividing N_{ows} in each aquifer by the corresponding aquifer area reflects the institutional component (i.e., D_{OW}) for assessing groundwater sustainable management.

Population living on the aquifer area (N_{pop})

Based on mean census results in 2006, 2011 and 2016, the population living on the aquifer area ranged from no inhabitants (for 23 aquifers in the central main basin) to more than 12.3 million people (Tehran aquifer, in the northern part) (see Figure S3m,n in the supplemental data online). According to equation (6), dividing N_{pop} for each aquifer by the corresponding aquifer area gives the social component for assessing groundwater sustainable management in Iran's aquifers (i.e., D_{Pop}).

Groundwater use for agricultural and industrial sectors (GW_{ai})

In Iran's aquifers, groundwater resources provide from 33 to 1095 mcm/year of agricultural demands (see Figure S3o,p in the supplemental data online). Furthermore, groundwater extraction for the industrial sector ranges from 0.00 to 73.64 mcm/year, especially for megacities (e.g., Tehran, Mashhad and Esfahan) with an average of 2.38 mcm/year. This indicator serves to quantify the economic component of groundwater sustainable management.

Spatial distribution of $IGWF_T$, PGW_{ai}, D_{Pop} and D_{OW}

The spatial distributions of these indicators are shown in Figure 3a–d based on the Natural Breaks method (ESRI, 2013). According to Figure 3, $IGWF_T$ ranges from 1.00 to 2.76, and higher values of $IGWF_T$ were observed in the central main basin because of the more groundwater level decline, less recharge and higher electrical conductivity. PGW_{ai} ranges from 29% to 100%, D_{pop} ranges from 0.00 to 5137.00 people/km^2 and, eventually, D_{ow} ranges from 0.04 to 1.44 wells/km^2.

Ranking aquifers

The 443 aquifers were ranked with respect to the four developed indicators using the PROMETHEE II technique based on the corresponding net flow score (Φ) for each aquifer (equation 9), while the Gaussian type of preference function was adopted. The spatial distribution and histogram of the net flow score values (Φ) were calculated based on the PROMETHEE II method while considering equal relative weights for the four indicators (w_k = 0.25) are shown in Figure 4. Equal relative weights (w_k = 0.25) were assigned to the four indicators to avoid bias in the result because of unequal

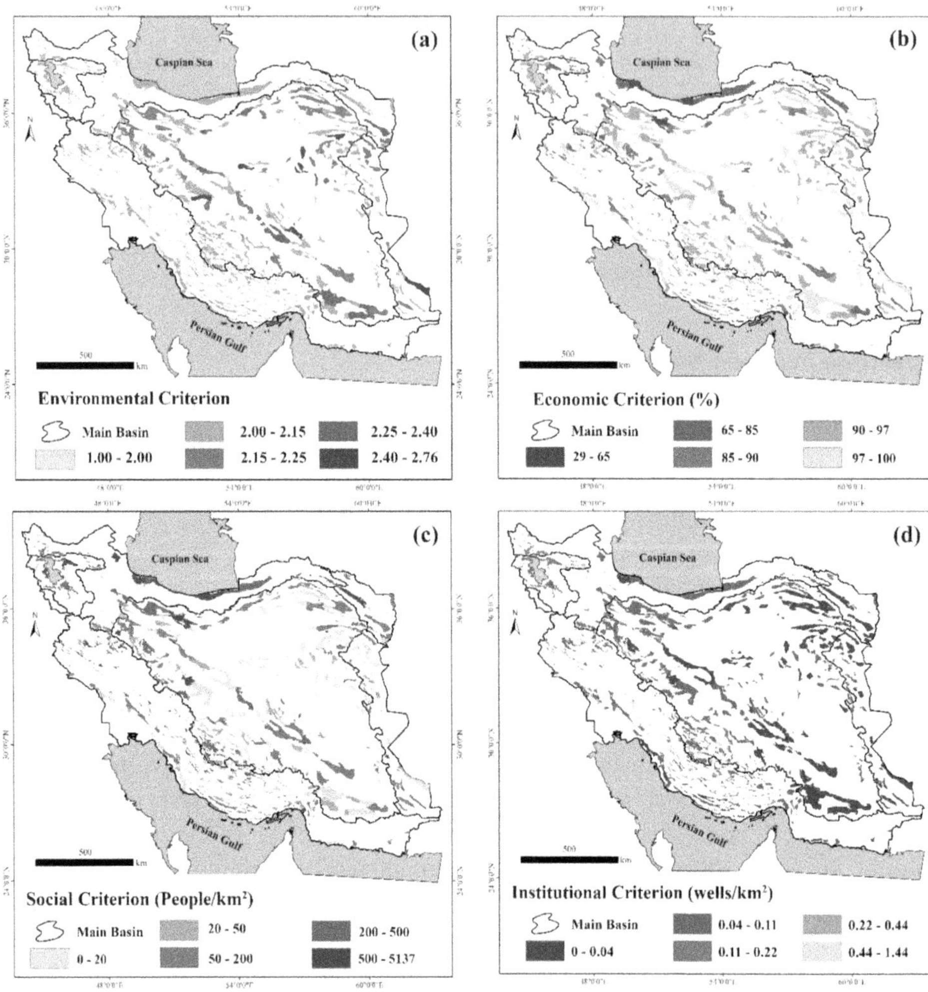

Figure 3. Spatial distribution of environmental ($IGWF_T$) (a), economic (PGW_{ai}) (b), social (D_{Pop}) (c) and institutional responsibility (D_{OW}) (d) components computed for 443 of Iran's aquifers.

weights and to give equal importance to the indicator for the calculation of the groundwater sustainable management index. The maximum and minimum values of Φ were 0.38 (for aquifers in central Iran) and −0.48 (for Tehran aquifer located in northern Iran), respectively. Lower Φ values indicate poor groundwater sustainable management in those aquifers in comparison with the others. The values of groundwater sustainable management index were calculated for Iran's aquifers (with equal weights assigned to the indicators), and they are categorized into five classes, as shown in Table 2. The negative and positive values indicate the unsustainable and sustainable status of groundwater sustainable management, respectively. The results indicated that 47.7% and 32.2% of Iran's aquifers are categorized as having good–very good and poor–very poor sustainability management, respectively. Relatively poor groundwater sustainable management was also observed, especially in the coastal aquifers and those close to metropolises (e.g., Tehran, Mashhad, Tabriz and Esfahan). The status of the

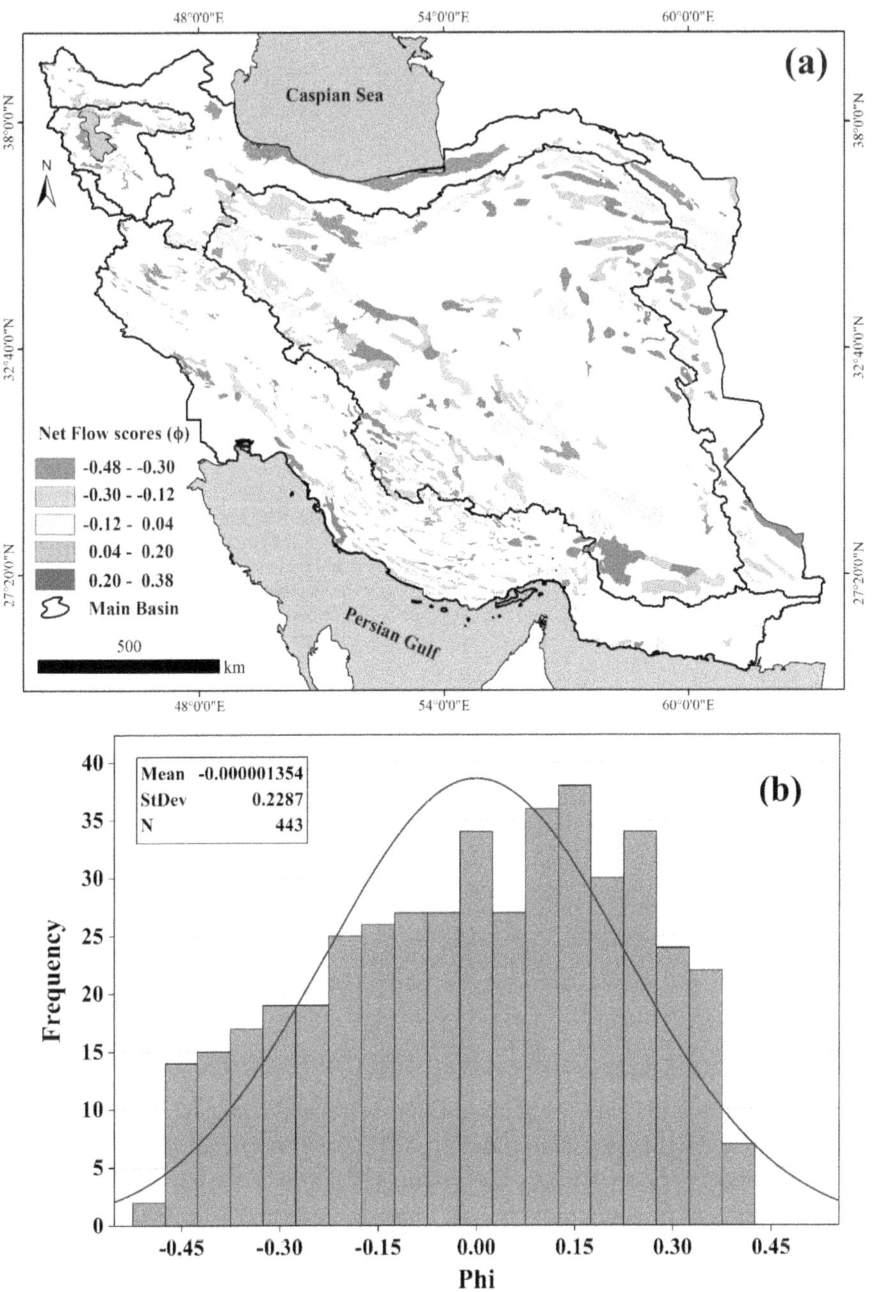

Figure 4. Spatial distribution (a) and histogram (b) of net flow score (Φ) computed through the PROMETHEE II method for the ranking of 443 of Iran's aquifers based on the indicators $IGWF_T$, PGW_{ai}, D_{Pop} and D_{OW}. Note: Negative and positive values indicate the unsustainable and sustainable status of groundwater sustainable management, respectively. StDev, standard deviation.

Table 2. Classification of Iran's aquifers based on groundwater sustainable management computed based on four indicators adopted in this study.

Status	Range of groundwater sustainable management index	No. of involved aquifers (% of total)	Status of indicators (component)*
Very poor	(−0.48, −0.30)	59 (13.32%)	Higher population density (social) Less exploitation of groundwater for irrigation and industry (economic) More efficient groundwater monitoring network (institutional) Less groundwater level declination and higher groundwater quality (environmental)
Poor	(−0.30, −0.12)	84 (18.96%)	Less exploitation of groundwater for irrigation and industry (economic) Higher population density (social) Less groundwater level declination and higher groundwater quality (environmental) More efficient groundwater monitoring network (institutional)
Relatively poor	(−0.12, 0.04)	90 (20.32%)	Higher groundwater level declination and lower groundwater quality (environmental) More efficient groundwater monitoring network (institutional) More withdrawal of groundwater for irrigation and industry (economic) Less population density (social)
Good	(0.04, 0.20)	108 (24.38%)	Less population density (social) Higher withdrawal for irrigation and industry (economic) Higher groundwater level declination and lower groundwater quality (environmental) Low efficiency of groundwater monitoring network (institutional)
Very good	(0.20, 0.38)	102 (23.32%)	Less population density (social) Lower groundwater level declination and higher groundwater quality (environmental) Higher withdrawal for irrigation and industry (economic) Low efficiency of groundwater monitoring network (institutional)

Note: *Equal weights of indicators ($w_k = 0.25$) is adopted to calculate the groundwater sustainable management index. (Negative and positive values indicate the unsustainable and sustainable status of groundwater sustainable management, respectively).

four indicators for each class of groundwater sustainable management index is also shown in Table 2. Higher population density (as the social indicator) and less exploitation of groundwater for agriculture and industry (as the economic indicator) are two main driving factors in aquifers with very poor groundwater sustainable management in comparison with the others.

To quantify the effect of changes in the values of w_k on the ranking of the 443 aquifers (i.e., values of Φ), the variations of w_k values and also those proposed by Pires et al. (2017) were investigated according to Table 3. The cumulative density function of net flow score values (Φ) obtained based on various values of w_k are shown in Figure S4 in the supplemental data online. The results highlight the crucial role of the social (D_{Pop}) and economic indicators (PGW_{ai}) on the values of Φ, while the minimum sensitivity of the Φ was observed for the indicators of environmental ($IGWF_T$) and institutional

Table 3. Weights (w_k) assigned to the four indicators considered for the assessment of groundwater sustainable management in 443 of Iran's aquifers and their sources.

			Assigned weight (w_k)					
No.	Component	Indicator	Equal weights	By Pires et al. (2017)	Variable weight			
1	Environmental	Integrated groundwater footprint ($IGWF_T$)	0.25	0.32	0.4	0.2	0.2	0.2
2	Social	Population density (D_{Pop})	0.25	0.26	0.2	0.4	0.2	0.2
3	Economic	Percentage of groundwater used for the agricultural and industrial sectors (PGW_{ai})	0.25	0.23	0.2	0.2	0.4	0.2
4	Institutional	Density of observation wells (D_{OW})	0.25	0.19	0.2	0.2	0.2	0.4

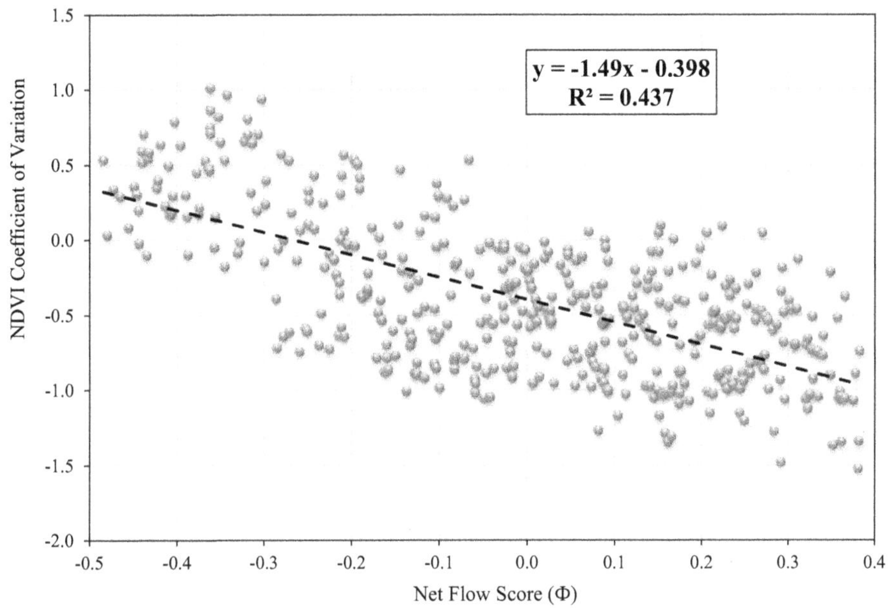

Figure 5. Normalized difference vegetation index (NDVI) coefficient of variation over 443 aquifer areas versus net flow score (Φ) obtained by the PROMETHEE II method considering the four groundwater sustainable management indicators.

responsibility (D_{OW}). It is noteworthy that these results are valid for the indicators adopted for each component of groundwater sustainability, and extending them to other indicators for the same component may not be allowable.

Relationship groundwater sustainable management index with vegetation CV

Figure S5 in the supplemental data online shows that the lowest and highest values of NDVI CV during the period 2000–16 are −4.58 and 0.94 and related to the central and northern (south of Caspian Sea) aquifers, respectively. The values of NDVI CV for the 443 studied aquifers versus the corresponding net flow scores (Φ) are plotted in Figure 5. The inverse relationship between the values of Φ and NDVI CV ($R^2 = -0.437$, $p < 0.01$) indicates that the vegetation CV in the growing season can be used as an indicator of groundwater sustainable management in the aquifers of Iran, which is known as an agricultural groundwater consumption-dominant region. Because groundwater in Iran is a primary source for supplying the water demands of dense vegetation (i.e., agricultural activities), vegetation cover quantified by the NDVI metric can be used as an indicator for groundwater withdrawal and sustainability.

Conclusions

To identify, reduce or eliminate threats to groundwater sustainability management, authorities need the assistance of sustainability components (i.e., social, economic, environmental and institutional responsibility) and quantitative tools for making more proactive, strategic and evidence-based management decisions. The current study aimed to facilitate the development of a sound, prioritized index to obtain and applicable on the aquifer scale to assess the sustainability of groundwater management in 443 aquifers across Iran. Application of the proposed index over 443 of Iran's aquifers may appropriately represent the current groundwater sustainable management situation in Iran and may assist in the evaluation of progress by Iran's governance in achieving groundwater sustainability and the refinement of adaptive groundwater management strategies. An important benefit of the proposed framework relies on the use of ground observations, thus avoiding complex model interpretations or water balance approaches to evaluating sustainable groundwater management.

The results of the current study revealed that the groundwater sustainable management in Iran's aquifers is relatively poor, especially in the coastal aquifers and those close to metropolises. The results also suggested that the NDVI CV can be used as a qualitative tool for assessing groundwater sustainable management in the region with human-influenced vegetation (i.e., agricultural activities). In Iran there is a need for improved groundwater governance for the monitoring and positive reinforcement of rules and regulations according to the local conditions of the groundwater basin while avoiding biased behaviour and reinforcing cooperative action. Incorporating other pertinent management options, including political, cultural and/or funding constraints, may reinforce the developed groundwater sustainable management index.

Acknowledgments

The authors thank Iran's Water Resources Management Company and the Iran Meteorological Organization, which helped to provide the required data of this paper.

Disclosure statement

No potential conflict of interest was reported by the authors.

Funding

The work was supported by The Organization for Researching and Composing University Textbooks in the Humanities (SAMT) as part of the 'Assessment and Analysis of Sustainable Water Resources Management in Arid Zones of Iran' project [grant number 100/12162].

ORCID

Bahador Zarei http://orcid.org/0000-0003-3343-6225
Esmaeel Parizi http://orcid.org/0000-0001-6458-1519
Seiyed Mossa Hosseini http://orcid.org/0000-0001-7161-8711
Behzad Ataie-Ashtiani http://orcid.org/0000-0002-1339-3734

References

Aayog, N. I. T. I. (2018). Composite water management index: A tool for water management. Government of India Publication, .180 http://www.niti.gov.in/writereaddata/files/document_publication/2018-05-18-Water-Index-Report_vS8-compressed.pdf

Abiye, T., Masindi, K., Mengistu, H., & Demlie, M. (2018). Understanding the groundwater-level fluctuations for better management of groundwater resource: A case in the Johannesburg region. *Groundwater for Sustainable Development*, *7*, 1–7. https://doi.org/10.1016/j.gsd.2018.02.004

Aguilar, C., Zinnert, J. C., Polo, M. J., & Young, D. R. (2012). NDVI as an indicator for changes in water availability to woody vegetation. *Ecological Indicators*, *23*, 290–300. https://doi.org/10.1016/j.ecolind.2012.04.008

Alley, W. M., & Leake, S. A. (2004). The journey from safe yield to sustainability. *Groundwater*, *42* (1), 12–16. https://doi.org/10.1111/j.1745-6584.2004.tb02446.x

Assaf, H., & Saadeh, M. (2009). Geostatistical assessment of groundwater nitrate contamination with reflection on DRASTIC vulnerability assessment: The case of the Upper Litani Basin, Lebanon. *Water Resources Management*, *23*(4), 775–796. https://doi.org/10.1007/s11269-008-9299-8

Ayers, R. S., & Westcot, D. W. (1994). Food, agriculture organization of the United Nations (FAO), water quality for agriculture. *Irrigation and Drainage, Rome*, *29* , 77044–2. https://www.fao.org/3/t0234e/t0234e00.htm

Barati, A. A., Azadi, H., & Scheffran, J. (2019). A system dynamics model of smart groundwater governance. *Agricultural Water Management*, *221*, 502–518. https://doi.org/10.1016/j.agwat.2019.03.047

Betrie, G. D., Sadiq, R., Morin, K. A., & Tesfamariam, S. (2013). Selection of remedial alternatives for mine sites: A multicriteria decision analysis approach. *Journal of Environmental Management*, *119*, 36–46. https://doi.org/10.1016/j.jenvman.2013.01.024

Bhanja, S. N., Malakar, P., Mukherjee, A., Rodell, M., Mitra, P., & Sarkar, S. (2019). Using satellite-based vegetation cover as indicator of groundwater storage in natural vegetation areas. *Geophysical Research Letters*, *46*(14), 8082–8092. https://doi.org/10.1029/2019GL083015

Brans, J. P. (1982). *L'ingénierie de la décision: L'élaboration d'instruments d'aide a la décision*. Université Laval, Faculté des sciences de l'administration.

Brans, J. P., & De Smet, Y. (2016). PROMETHEE methods. In S. Greco, M. Ehrgott, & J. R. Figueira (Eds.), *Multiple criteria decision analysis* (pp. 187–219). Springer.

Brans, J. P., & Mareschal, B. (2005). PROMETHEE methods. In S. Greco, M. Ehrgott, & J. R. Figueira (Eds.), *Multiple criteria decision analysis: State of the art surveys* (pp. 163–186). Springer. https://doi.org/10.1007/b100605

Brans, J. P., & Vincke, P. (1985). Note – A preference ranking organisation method: (The PROMETHEE method for multiple criteria decision-making). *Management Science*, *31*(6), 647–656. https://doi.org/10.1287/mnsc.31.6.647

Brundtland, G. H. (1987). *Report of the World Commission on Environment and Development: 'our common future'*. United Nations.

Bui, N. T., Kawamura, A., Du Bui, D., Amaguchi, H., Bui, D. D., Truong, N. T., Do, H. H. T., & Nguyen, C. T. (2019). Groundwater sustainability assessment framework: A demonstration of environmental sustainability index for Hanoi, Vietnam. *Journal of Environmental Management*, *241*, 479–487. https://doi.org/10.1016/j.jenvman.2019.02.117

Castilla-Rho, J. C., Rojas, R., Andersen, M. S., Holley, C., & Mariethoz, G. (2019). Sustainable groundwater management: How long and what will it take? *Global Environmental Change, 58*, 101972. https://doi.org/10.1016/j.gloenvcha.2019.101972

Collier, P. (2010). *The plundered planet: Why we must–and how we can–manage nature for global prosperity*. Oxford University Press. https://doi.org/10.25336/P6NG7C

Dalin, C., Wada, Y., Kastner, T., & Puma, M. J. (2017). Groundwater depletion embedded in international food trade. *Nature, 543*(7647), 700. https://doi.org/10.1038/nature21403

Doell, P., Mueller Schmied, H., Schuh, C., Portmann, F. T., & Eicker, A. (2014). Global scale assessment of groundwater depletion and related groundwater abstractions: Combining hydrological modeling with information from well observations and GRACE satellites. *Water Resources Research, 50*(7), 5698–5720. https://doi.org/10.1002/2014WR015595

EPA. (2001). Parameters of Water Quality Interpretation and Standards. *Ireland*, 133. https://www.epa.ie/pubs/advice/water/quality/Water_Quality.pdf

ESRI. (2013). *Arc Hydro: GIS for water resources*. https://doi.org/10.1109/CESCE.10.169

FAO. (2016). *Country water report*. Food and Agriculture Organization of the United Nations, Rome. http://www.fao.org

Ferguson, G., & Gleeson, T. (2012). Vulnerability of coastal aquifers to groundwater use and climate change. *Nature Climate Change, 2*(5), 342. https://doi.org/10.1038/nclimate1413

Foster, S., & van der Gun, J. (2016). Groundwater governance: Key challenges in applying the Global Framework for Action. *Hydrogeology Journal, 24*(4), 749–752. https://doi.org/10.1007/s10040-016-1376-0

Fraser, C. M., Kalin, R. M., Kanjaye, M., & Uka, Z. (2020). A methodology to identify vulnerable transboundary aquifer hotspots for multi-scale groundwater management. *Water International, 45*(7–8), 865–883. https://doi.org/10.1080/02508060.2020.1832747

Ghadimi, S., & Ketabchi, H. (2019). Possibility of cooperative management in groundwater resources using an evolutionary hydro-economic simulation-optimization model. *Journal of Hydrology, 578*, 124094. https://doi.org/10.1016/j.jhydrol.2019.124094

Gleeson, T., Wada, Y., Bierkens, M. F., & van Beek, L. P. (2012). Water balance of global aquifers revealed by groundwater footprint. *Nature, 488*(7410), 197. https://doi.org/10.1038/nature11295

Goodarzi, M., Abedi-Koupai, J., Heidarpour, M., & Safavi, H. R. (2016). Development of a new drought index for groundwater and its application in sustainable groundwater extraction. *Journal of Water Resources Planning and Management, 142*(9), 04016032. https://doi.org/10.1061/(ASCE)WR.1943-5452.0000673

Gorelick, N., Hancher, M., Dixon, M., Ilyushchenko, S., Thau, D., & Moore, R. (2017). Google earth engine: Planetary-scale geospatial analysis for everyone. *Remote Sensing of Environment, 202*, 18–27. https://doi.org/10.1016/j.rse.2017.06.031

Hosseini, S. M., Parizi, E., Ataie-Ashtiani, B., & Simmons, C. T. (2019). Assessment of sustainable groundwater resources management using integrated environmental index: Case studies across Iran. *Science of the Total Environment, 676*, 792–810. https://doi.org/10.1016/j.scitotenv.2019.04.257

IBWWOPS, Iran's Bureau of Water and Wastewater Operation and Protection Systems. (2015). *Water and Wastewater master plan studies*.

IBWWOPS, Iran's Bureau of Water and Wastewater Operation and Protection Systems. (2019). Water and wastewater master plan studies.

Iran's Water Statistical Yearbook. (2014) *Published by Iran's Bureau of Water and Wastewater planning*. Ministry of Energy, p. 302 http://isn.moe.gov.ir/

Iran's WRM Company. (2017). *Iran's Water Resources Management Company*. https://www.wrm.ir

Jakeman, A. J., Barreteau, O., Hunt, R. J., Rinaudo, J. D., Ross, A., Arshad, M., & Hamilton, S. (2016). Integrated groundwater management: An overview of concepts and challenges. In A. Jakeman (Ed.), *Integrated groundwater management* (pp. 3–20). Springer. https://doi.org/10.1007/978-3-319-23576-9

Javadi, S., Kavehkar, N., Mohammadi, K., Khodadadi, A., & Kahawita, R. (2011). Calibrating DRASTIC using field measurements, sensitivity analysis and statistical methods to assess groundwater vulnerability. *Water International*, *36*(6), 719–732. https://doi.org/10.1080/02508060.2011.610921

Jothibasu, A., & Anbazhagan, S. (2016). Modeling groundwater probability index in Ponnaiyar River basin of South India using analytic hierarchy process. *Modeling Earth Systems and Environment*, *2*(3), 109. https://doi.org/10.1007/s40808-016-0174-y

Kelley, C. P., Mohtadi, S., Cane, M. A., Seager, R., & Kushnir, Y. (2015). Climate change in the Fertile Crescent and implications of the recent Syrian drought. *Proceedings of the National Academy of Sciences*, *112*(11), 3241–3246. https://doi.org/10.1073/pnas.1421533112

Konikow, L. F., & Kendy, E. (2005). Groundwater depletion: A global problem. *Hydrogeology Journal*, *13*(1), 317–320. https://doi.org/10.1007/s10040-004-0411-8

Korbel, K. L., & Hose, G. C. (2011). A tiered framework for assessing groundwater ecosystem health. *Hydrobiologia*, *661*(1), 329–349. https://doi.org/10.1007/s10750-010-0541-z

Kourgialas, N. N., Karatzas, G. P., Dokou, Z., & Kokorogiannis, A. (2018). Groundwater footprint methodology as policy tool for balancing water needs (agriculture & tourism) in water scarce islands – The case of Crete, Greece. *Science of the Total Environment*, *615*, 381–389. https://doi.org/10.1016/j.scitotenv.2017.09.308

Kumar, M. D., Patel, A., Ravindranath, R., & Singh, O. P. (2008). Chasing a mirage: Water harvesting and artificial recharge in naturally water-scarce regions. *Economic and Political Weekly*, *43*(35), 61–71. https://doi.org/10.2307/40278725https://doi.org/10.2307/40278725

Lambán, L. J., Martos, S., Rodríguez-Rodríguez, M., & Rubio, J. C. (2011). Application of groundwater sustainability indicators to the carbonate aquifer of the Sierra de Becerrero (Southern Spain). *Environmental Earth Sciences*, *64*(7), 1835–1848. https://doi.org/10.1007/s12665-011-1016-x

Lezzaik, K., & Milewski, A. (2018). A quantitative assessment of groundwater resources in the Middle East and North Africa region. *Hydrogeology Journal*, *26*(1), 251–266. https://doi.org/10.13140/RG.2.2.13830.6816

Loucks, D. P. (1997). Quantifying trends in system sustainability. *Hydrological Sciences Journal*, *42*(4), 513–530. https://doi.org/10.1080/02626669709492051

Madani, K. (2014). Water management in Iran: What is causing the looming crisis? *Journal of Environmental Studies and Sciences*, *4*(4), 315–328. https://doi.org/10.1007/s13412-014-0182-z

Maiolo, M., & Pantusa, D. (2019). Sustainable Water Management Index, SWaM_Index. *Cogent Engineering*, *6*(1), 1603817. https://doi.org/10.1080/23311916.2019.1603817

Mendicino, G., Senatore, A., & Versace, P. (2008). A Groundwater Resource Index (GRI) for drought monitoring and forecasting in a Mediterranean climate. *Journal of Hydrology*, *357*(3–4), 282–302. https://doi.org/10.1016/j.jhydrol.2008.05.005

Mirzadeh, S. M. J., Jin, S., Parizi, E., Chaussard, E., Bürgmann, R., Delgado Blasco, J. M., & Mirzadeh, S. H. (2021). Characterization of Irreversible Land Subsidence in the Yazd-Ardakan Plain, Iran from 2003-2020 InSAR Time Series. *Journal of Geophysical Research: Solid Earth*, *126*, e2021JB022258. https://doi.org/10.1029/2021JB022258https://doi.org/10.1029/2021JB022258

Nabavi, E. (2018). Failed policies, falling aquifers: Unpacking groundwater overabstraction in Iran. *Water Alternatives*, *11*(3), 699–724. https://www.cabdirect.org/cabdirect/abstract/20203104921

Nampak, H., Pradhan, B., & Manap, M. A. (2014). Application of GIS based data driven evidential belief function model to predict groundwater potential zonation. *Journal of Hydrology*, *513*, 283–300. https://doi.org/10.1016/j.jhydrol.2014.02.053

Narany, T. S., Ramli, M. F., Fakharian, K., & Aris, A. Z. (2016). A GIS-index integration approach to groundwater suitability zoning for irrigation purposes. *Arabian Journal of Geosciences*, *9*(7), 502. https://doi.org/10.1007/s12517-016-2520-9

Nazari, B., Liaghat, A., Akbari, M. R., & Keshavarz, M. (2018). Irrigation water management in Iran: Implications for water use efficiency improvement. *Agricultural Water Management*, *208*, 7–18. https://doi.org/10.1016/j.agwat.2018.06.003

Nhamo, L., Ebrahim, G. Y., Mabhaudhi, T., Mpandeli, S., Magombeyi, M., Chitakira, M., & Sibanda, M. (2019). An assessment of groundwater use in irrigated agriculture using multi-spectral remote sensing. *Physics and Chemistry of the Earth, Parts A/B/C, 115*, 102810. https://doi.org/10.1016/j.pce.2019.102810https://doi.org/10.1016/j.pce.2019.102810

Pandey, V. P., Shrestha, S., Chapagain, S. K., & Kazama, F. (2011). A framework for measuring groundwater sustainability. *Environmental Science & Policy, 14*(4), 396–407. https://doi.org/10.1016/j.envsci.2011.03.008

Parizi, E., Bagheri-Gavkosh, M., Hosseini, S. M., & Geravand, F. (2021). Linkage of geographically weighted regression with spatial cluster analyses for regionalization of flood peak discharges drivers: Case studies across Iran. *Journal of Cleaner Production, 310*, 127526. https://doi.org/10.1016/j.jclepro.2021.127526

Parizi, E., Hosseini, S. M., Ataie-Ashtiani, B., & Simmons, C. T. (2019). Representative pumping wells network to estimate groundwater withdrawal from aquifers: Lessons from a developing country, Iran. *Journal of Hydrology, 578*, 124090. https://doi.org/10.1016/j.jhydrol.2019.124090

Parizi, E., Hosseini, S. M., Ataie-Ashtiani, B., & Simmons, C. T. (2020). Normalized difference vegetation index as the dominant predicting factor of groundwater recharge in phreatic aquifers: Case studies across Iran. *Scientific Reports, 10*(1), 1–19. https://doi.org/10.1038/s41598-020-74561-4

Pires, A., Morato, J., Peixoto, H., Botero, V., Zuluaga, L., & Figueroa, A. (2017). Sustainability Assessment of indicators for integrated water resources management. *Science of the Total Environment, 578*, 139–147. https://doi.org/10.1016/j.scitotenv.2016.10.217

Requena, A. I., Ouarda, T. B., & Chebana, F. (2018). Low-flow frequency analysis at ungauged sites based on regionally estimated streamflows. *Journal of Hydrology, 563*, 523–532. https://doi.org/10.1016/j.jhydrol.2018.06.016

Richey, A. S., Thomas, B. F., Lo, M. H., Reager, J. T., Famiglietti, J. S., Voss, K., & Rodell, M. (2015). Quantifying renewable groundwater stress with GRACE. *Water Resources Research, 51*(7), 5217–5238. https://doi.org/10.1002/2015WR017349

Rigge, M., Homer, C., Wylie, B., Gu, Y., Shi, H., Xian, G., & Bunde, B. (2019). Using remote sensing to quantify ecosystem site potential community structure and deviation in the Great Basin, United States. *Ecological Indicators, 96*, 516–531. https://doi.org/10.1016/j.ecolind.2018.09.037

Russo, T., Alfredo, K., & Fisher, J. (2014). Sustainable water management in urban, agricultural, and natural systems. *Water, 6*(12), 3934–3956. https://doi.org/10.3390/w6123934

Saaty, T. L. (2008). Decision making with the analytic hierarchy process. *International Journal of Services Sciences, 1*(1), 83–98. https://doi.org/10.1504/IJSSCI.2008.017590

Sadeghi-Tabas, S., Samadi, S. Z., Akbarpour, A., & Pourreza-Bilondi, M. (2016). Sustainable groundwater modeling using single- and multi-objective optimization algorithms. *Journal of Hydroinformatics, 19*(1), 97–114. https://doi.org/10.2166/hydro.2016.006

Senent-Aparicio, J., Pérez-Sánchez, J., García-Aróstegui, J., Bielsa-Artero, A., & Domingo-Pinillos, J. (2015). Evaluating groundwater management sustainability under limited data availability in semiarid zones. *Water, 7*(8), 4305–4322. https://doi.org/10.1007/s10040-017-1601-5

Shah, T., Roy, A. D., Qureshi, A. S., & Wang, J. (2003, May). Sustaining Asia's groundwater boom: An overview of issues and evidence. In *Natural Resources Forum* (Vol. 27, No. 2, pp. 130–141). Blackwell Publishing Ltd. https://doi.org/10.1111/1477-8947.00048

Smakhtin, V., Revenga, C., & Döll, P. (2004). A pilot global assessment of environmental water requirements and scarcity. *Water International, 29*(3), 307–317. https://doi.org/10.1080/02508060408691785

Tatalovich, Z., Wilson, J. P., & Cockburn, M. (2006). A comparison of Thiessen polygon, kriging, and spline models of potential UV exposure. *Cartography and Geographic Information Science, 33*(3), 217–231. https://doi.org/10.1559/152304006779077318

Trichakis, I., Burek, P., de Roo, A., & Pistocchi, A. (2016). *Development of an integrated groundwater and surface water model in a pan-European scale.* Paper presented at 2nd EWaS International Conference. Chania, Greece.

UNESCO. (2007). *GW Resources Sustainability Indicators.* (IHP-VI Series on GW No. 14).
United Nations. (2015). *The Millennium Development Goals Report 2015.* http://www.undp.org/content/dam/undp/library/MDG/english/UNDP_MDG_Report_2015.pdf
Vrba, J., Girman, J., van der Gun, J., Haie, N., Hirata, R., Lopez-Gunn, E., & Wallin, B. (2007). Groundwater resources sustainability indicators. In A. Lipponen (Ed.), *Groundwater resources sustainability indicators* (p. 114). UNESCO.
Wegmann, J., & Mußhoff, O. (2019). Groundwater management institutions in the face of rapid urbanization–Results of a framed field experiment in Bengaluru, India. *Ecological Economics, 166*, 106432. https://doi.org/10.1016/j.ecolecon.2019.106432
World Bank. (2010). *World development indicators database.*
WWAP (World Water Assessment Programme). (2012). *World water assessment programme: The united nations world water development report 4. Managing Water under Uncertainty and Risk.*
Xie, S., Liu, L., Zhang, X., Yang, J., Chen, X., & Gao, Y. (2019). Automatic land-cover mapping using Landsat time-series data based on google earth engine. *Remote Sensing, 11*(24), 3023. https://doi.org/10.3390/rs11243023

Part III

Groundwater users

Whither collective action? Upscaling collective actions, politics and basin management in the process of 'legitimizing' an informal groundwater economy

Marta Rica, Aurélien Dumont, Fermín Villarroya and Elena López-Gunn

This article examines different forms and levels of collective action by aquifer users in securing access to over-allocated groundwater resources using a case study of La Loma, Úbeda (Jaén, Spain), one of the largest olive-growing areas in the world. It shows how opportunities for collective water management increase at the basin level as bargaining spaces increase but also how political rent influences the institutional designs that emerge. The article identifies an opportunity to redesign the organizational and institutional configurations by both securing access to water and strengthening collaborative spaces at the basin level.

Introduction

Groundwater is a common pool resource with the classic characteristics of substractability and non-excludability (Feeny, Berkes, McCay, & Acheson, 1990; Ostrom, Burger, Field, Norgaard, & Policansky, 1999). Each of these characteristics is important for the analysis of collective action, because they offer both constraints and opportunities for resource users. Resource users must decide whether it is worthwhile engaging in a collective process to address the problems that confront them, given the transaction costs of coordination (Lopez-Gunn & Martínez Cortina, 2006). Two key questions for management are: (1) whether the initiative for groundwater use regulation can come from groundwater users or instead needs an external, higher-scale agent such as a water authority (Lopez-Gunn & Martínez Cortina, 2006); and (2) whether collective-action rules can be devised by users or also require an external agent.

La Loma, in Úbeda, Spain, is one of the largest olive-growing areas in the world. It was extensively converted to irrigation starting in the mid-1990s and represents a prime example of entrepreneurial user initiative to exploit groundwater resources. The formal groundwater regulator has been a follower and spectator in the actions led by users. This is contextualized in the Spanish legal framework, where groundwater became a public good in the 1985 water law and fell under the responsibility of the river-basin authorities, who grant water rights for its use (Fornés, de La Hera, Llamas, & Martínez-Santos, 2007). Yet this boom in groundwater use was often not acknowledged by public authorities and constitutes an example of the "silent revolution" (Llamas & Martínez-Santos, 2005) that has taken place in many parts of the world. Due to the nature of groundwater, users often have relative autonomy, which lends itself to user initiative in developing groundwater resources. In the particular case of arid and semi-arid areas, irrigation can make a substantial difference in the productivity of agriculture, often doubling or tripling productivity (López-Gunn, Rica, & van Cauwenbergh, 2012). In a context where the current groundwater use is mainly informal, the process of water rights allocation is an action arena where cross-scale interplay occurs. Water rights allocation is a good starting-point to analyze the crystallization of both institutions and power dynamics, where the political influence on resource use and user relationships can be analyzed carefully (Agrawal, 2003).

In this complex context of resource use, what are the organizational forms and objectives of collective action by users? How can they be typified or characterized? How do they emerge and why? This article will reflect on these questions in relation to La Loma, particularly looking at opportunities for self-governance and co-management as a possibility to tame groundwater chaos (Blomquist, 1992; López-Gunn et al., 2012; Shah, 2009). The article introduces cross-scale collective action, including aspects related to coordination with the water authorities (the regulator) and to the upscaling of action by groundwater user groups. Analyzing issues of cross-scale interaction, especially with respect to "ecological sustainability, social welfare, equity, and cultural autonomy", is critical in redesigning or restructuring regimes to produce more desirable outcomes (Young, 2006, p. 19).

In particular, the article looks at the role that water users have played and how they contribute or could contribute to a development model that internalizes and operates within its resource boundaries, while seeking to maintain existing socio-economic benefits. It also reflects on the relationship model with other key actors, mainly the regulator, the Guadalquivir Basin Water Authority (hereafter CHG, from Confederación Hidrográfica del Guadalquivir), and also the role political factors have played in either supporting or hindering collective action, e.g. supporting or weakening user collectives. The question is whether users act collectively to resolve a common pool resource (CPR) problem or rather that it is other motivations or objectives, not directly related to water conservation, that explain their coordination, in a context where the use of the resource is challenged by the water authorities.

The article is structured as follows. The next section documents the exponential growth in intensive groundwater use in the region of La Loma over a relatively short 12-year period. The third section focuses on the emergence of collective entities for groundwater use, their typology and their main claims, and then analyzes some collective initiatives and whether these helped users realize their objectives. This success or failure is then related to the political context and political ecology of the region and basin. The fourth section discusses what attempts to sustain the existing model through the regulation of informal use mean for upscaling management to other levels and coordinating at the regional and basin levels. This means engaging with the existing tensions between formal and informal water use, and the potential arbitrating or mediating role of the water authority to regulate the situation. The fifth section addresses the recent attempts of aquifer users to get access to surface water as a solution to aquifer overuse in a context where all the water resources have been allocated at the river-basin scale. The last section therefore explores whether collective action or co-management can be done at a larger scale, taking into account the increased options to share benefits and the current vulnerability of the existing socio-economic system, in particular the lack of economic diversification. Thus, the question is whether upscaling can in fact lead to increased resilience to avoid eventual collapse from intensive resource use through the conjunctive use of surface

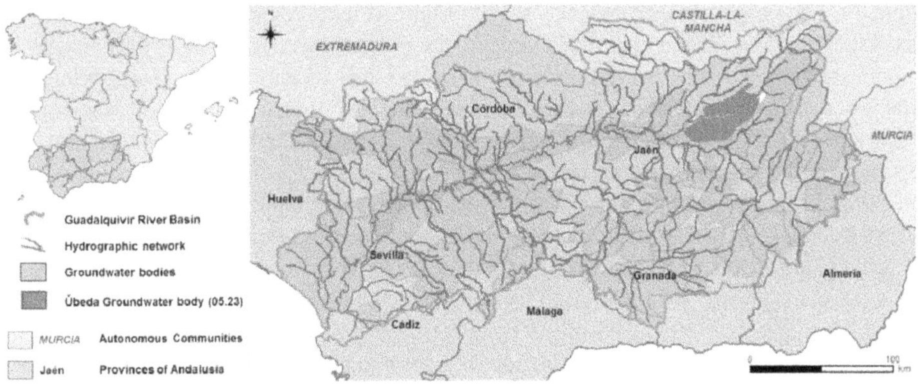

Figure 1. Localization of the Úbeda groundwater body within the Guadalquivir River basin.

and groundwater at the basin level or if instead it is a way to externalize and upscale the problem of intensive aquifer overuse.

Case-study description of La Loma

La Loma of Úbeda is located in the Upper Guadalquivir Basin, in Southwest Spain. Groundwater resources come from groundwater body 05.23 Úbeda (Figure 1).[1] Both the Guadalquivir and Guadalimar Rivers flow close to this area. These rivers are not available to La Loma irrigators, because surface water is committed to other uses further downstream. As will be seen below, one of the consequences of intensive groundwater use has been a new claim by farmers in La Loma for access rights to take water from these surface-water flows in winter.

From the hydrogeological point of view, there are two superimposed aquifers: the Tertiary Miocene aquifer and the Jurassic carbonate aquifer (Gollonet, Rubio, Martín, González-Ramón, & Peinado, 2002). The Jurassic aquifer is composed of calcareous and dolomite materials about 100 m thick (González-Ramón, Rubio, & López Geta, 2007). In its northern part, around the Guadalimar River, it behaves as an unconfined aquifer, with a surface area of 252 km^2; in the south, the aquifer submerges under Miocene sediments and behaves like a confined aquifer. The total extension of the aquifer is about 626 km^2. It is recharged by about 50 MCM/y (million cubic metres per year) of rainwater, according to the CHG (2012), and also by the Guadalimar River. Aquifer reserves are estimated at approximately 220–330 MCM. Test drillings provide yields of about 19 L/s. The Miocene aquifer has an area of about 170 km^2. It is composed mainly of impervious marls, with sands and conglomerate on the bottom and sandstone on the top. The aquifer behaves as unconfined and is recharged mainly by rainwater infiltration, estimated at 7.6 MCM/y (CHG, 2012), while the reserves are estimated to be 33–50 MCM. However, the exploitation rate of this aquifer is higher than its resources: 17.5 MCM/y (CHG, 2012). Test drillings produce yields lower than 10 L/s. Overall, the Jurassic and Miocene aquifers show an unsustainable exploitation rate: the CHG (2012) estimates abstractions around 80 MCM/y, compared to the 58 MCM/y renewable rainwater infiltration. In addition, this renewable resource contributes to the natural outflow of the Guadalimar River. Consequently, the appropriation of this resource through groundwater pumping has potential impacts downstream.

Initially, a large governance obstacle was the limited hydrological knowledge of the groundwater resources in the area. For example, the previous Guadalquivir Hydrological Plan, published in 1998 by the CHG, only contemplated the Miocene aquifer, disregarding the Jurassic aquifer as well as the possible connections with rivers in the area. It was not until 2001 that the CHG considered the Jurassic aquifer as part of the hydrogeological unit, taking into account its connections with the Miocene aquifer. By then, as will be seen below, spontaneous intensive groundwater use was already increasing rapidly. Some regulations were then established, mainly on the distance wells should be from each other and what area could be irrigated. These measures were decided by the CHG, with economic sanctions for users violating the norms. While the aquifers were already widely exploited, there was very limited information from the CHG on its dynamics and piezometric evolution. In some areas, perforations had reached almost 1000 m deep, reaching water that was estimated to be 25,000 years old (Nuñez et al., 2006). However, the perturbation generated by pumping in confined aquifers propagates rapidly (a few months or years in an aquifer of the size of the La Loma), meaning that the boundary conditions changed and that pumping resulted in the diversion of more surface-water flows to the aquifer.

In the Upper Guadalquivir Basin, with an average temperature of 16.2 ºC and 611 mm/y of rainfall, conditions are very favourable for olive groves. It is an area where people traditionally practised extensive agriculture, alternating dry-land olive groves with other crops such as vine and cereals. Over the last 30 years the agricultural area has experienced a dramatic transformation in two ways: first, in the shift to what is effectively a monoculture agricultural model of intensive olive-grove cultivation; and second, from dry-land farming to an intensively irrigated (although highly efficient) agricultural model through drip and subsurface irrigation of olive groves. This trend in mode of olive-grove production can be seen all around the Guadalquivir Basin, where olive groves occupy almost 60% of the agricultural surface. Between 1997 and 2008, the cultivated area increased by 50%, mainly in irrigated olive groves (Dumont, Salmoral, & Llamas, 2013). In La Loma itself, agriculture is the main source of activity, with approximately 40,000 ha of olive fields.

Factors such as drought events, inherent to Mediterranean climate variability, together with the EU's common agricultural policy, which subsidized olive production (Graaff, de, & Eppink, 1999; Sánchez Martínez, Gallego, & Araque, 2011), meant that farmers had the combined incentive to irrigate due to the increased productivity brought about by irrigation. In the late 1990s and early 2000s, farmers benefitted from higher EU subsidies based on production per hectare. Salmoral et al. (2011) find that dry olive-grove productivity is €1163/ha, while the productivity of an irrigated farm is almost double (€2080/ha). These factors, and the fact that surface-water resources were effectively 'closed' to farmers in the area, explain the groundwater drive and motivations for farmers in La Loma.

Figure 2 shows how the irrigated area and water demand have increased notably over the last 15 years. In 1999 there were approximately 12,000 ha, with estimated withdrawals of 23 MCM/y. Only three years later, by 2002, it is estimated that water abstraction in the area had doubled, with 25,000 ha consuming more than 46 MCM/y (WWF/Adena, 2006). Recently the situation has stabilized, with approximately 40,000 ha and 80 MCM/y groundwater abstractions (Cifuentes, 2012).

In both Spanish water planning and under the EU Water Framework Directive, basin plans are a normative requirement, and the current Guadalquivir Basin Plan establishes the 'available resource' for the groundwater body at 46 MCM/y, only half of the current level of abstraction. Thus, the issue of how to reduce abstraction, while ensuring the socio-economic viability of the existing system, heavily dependent on olive groves, represents a big challenge.

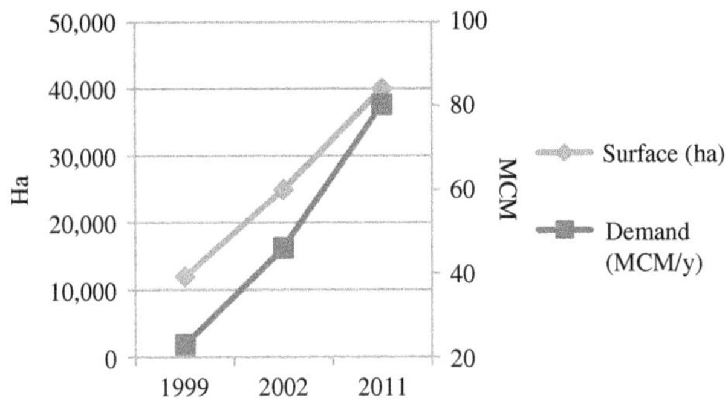

Figure 2. Groundwater use and irrigated land evolution. Data from WWF/Adena (2006) and Cifuentes, (2012).

Emergence of collective action and contested water rights: from well sharing to claims and collaborative spaces?

Drought in the early 1990s, combined with the EU's common agricultural policy subsidies based on production, started a race to find groundwater resources. Pioneer farmers in the region and neighbouring areas saw a good business opportunity, and the first wells were drilled in the central parts of the aquifer. Various drilling companies landed in the region, offering their services to farmers, in most cases with no prior hydrogeological studies. Most of these farmers, not wanting to miss the chance to irrigate, and seeing the relatively large investments required, organized into small groups to jointly undertake the capital-intensive investment needed to construct the well and irrigation systems. The first type of collective action seen in La Loma was thus based on the creation of informal water user associations (WUAs) as a form of collective financial investment, accompanied by norms on the repartition of groundwater shares. Little strategic planning was carried out, and distances between wells were not taken into account. Often, wells were drilled based on the *de minimis* right to 7000 m^3/y established under Spanish law (Hogson, 2013; López-Gun, Bea, Huertas, Vay, & Zorrilla, 2014), before permission was obtained from the water authorities, as required under the 1985 Water Act.[2]

The logic of the constitution of WUAs followed both territorial closeness and social ties. Each single farmer owns on average a relatively small 1–2 ha plot of land. There was not a homogeneous pattern in the WUAs' emergence. In some areas, like the municipality of Villacarrillo, a WUA included 50–60 ha. In others, such as Sabiote, the area covered was 300 ha.

These informal associations started a process of formalization, elaborating rules, choosing a representative or president and a board made up of members, hiring a *relojero* ('clock watcher') to operate the pump and irrigation system, and making joint decisions on how and when to distribute water, normally by consensus in member assemblies. The next step was the gradual formal constitution into irrigation communities. Under the Spanish regulatory framework, there are a range of organizational legal forms for WUAs, though the main ones are irrigation communities (public corporations recognized by the water authorities as water users) and civil societies (private associations often not recognized by the water authorities) (Rica, López-Gunn, & Llamas, 2012).

The second level of collective action was based on upscaling collective interests beyond the irrigation system (the common well) to defend collective access to abstraction rights. The emergence of this initiative was promoted in 2001 by some of the most active farmers, supported by the Asociación Agraria de Jóvenes Agricultores (ASAJA, agrarian association of young farmers) farming union, who realized that the situation should be stabilized to preserve the resource base and farmers' livelihoods and that legitimate access to and use of water resources should be formalized (i.e. water rights should be granted). The formalization of the initiative led to the creation of the Association of Wells in the Carbonated Aquifer of La Loma (hereafter, the Association). The Association was born as a result of meetings and conversations between leading farmers in the area. The number of members in the Association increased over time and represented approximately 24,000 ha out of the 40,000 ha irrigated olive groves over the groundwater body 0.23 Úbeda. It included 247 individual farmers and 179 WUAs or irrigation communities. An important aspect however is that a rival organization developed in parallel, which included farmers irrigating around 8000–9000 ha in the area who chose to be represented by an agricultural trade union, the Union de Pequeños Agricultores, (UPA, Small Farmers' Union), which operates at both the municipal level in Villacarrillo and at the provincial level in Jaén. What is interesting in this second level of collective action is the power element that became prevalent in the process.

Two groups emerged, both associated with farming unions, but of opposite political leanings. The interest of farmers in the support of the emergence and formal constitution of these second-order water user groups was to meet a strategic need to secure water access by the formalization of water use through the allocation of water rights. Technically, these WUAs were abstracting water informally (through *de minimis* rights), which could occur as long as the basin and aquifer plan were not closed. Farmers were aware that – as the 1985 Spanish Water Act contemplates – water rights would be awarded as a 'collective' right, i.e. titled to a WUA composed of a number of individuals as joint owners of the well through the granting of a state concession to use groundwater.

A few first-order WUAs that had applied early in the process succeeded in getting a formal water right from the CHG. The water authority was reluctant to officially recognize water use by granting water rights until better data on available resources were available. The outcome was that to this day, hundreds of WUAs (and the individual farmers in each WUA) remain in an illegal situation when using groundwater, generating an 'informal water economy'.

For farmers (and their first-order WUAs) considering upscaling collective action, tackling 'informality' was the major driver, and their strategy then turned to securing their water rights. This was inherently a political process, with individual farmers and WUAs seeking an organization that would be well positioned from a power perspective to secure state water rights concessions from the CHG (Table 1).

As mentioned earlier, the Association had close links with the ASAJA, a trade union traditionally associated with the conservative party; meanwhile, the UPA had close links with the labour party. This became a process of clientelistic politics, with farmers exchanging their individual interests in securing water rights for political rents (votes and membership) based on the organization they perceived as best aligned with the existing political constellation. These organizations expanded from the local trade unions to the province of Jaén, to Seville, where the CHG is located, and eventually to Madrid, which ultimately held the key to the decisions, because the senior posts in the CHG are politically appointed. The result is a multilevel political game to secure groundwater rights in exchange for political rents.

The third level then refers to power dynamics taking place at both regional and national levels. In 2005, the Association joined the AEUAS (Asociación Española de Usuarios de Aguas Subterráneas, or Spanish Association of Groundwater Users), a

Table 1. Rules proposed by the Confederación Hidrográfica del Guadalquivir. Adapted from CHG-ITGE (2001).

- All users who want to be included in the allocation of water rights must go through a formal open process of application for a water concession.
- All users with the same concession must constitute a water user association.
- Irrigation units must be larger than 1000 ha, with each individual landowner not having more than 150 ha of irrigated area.
- Wells must be built in such a way that only one aquifer is tapped (no connection between different aquifer layers).
- No more than 1000 m^3 will be assigned per hectare.
- The level of exploitation is conditional on an improved water-level monitoring network.
- Once the process of water rights distribution is finished, users will be required to constitute a General Community of La Loma de Úbeda.
- Water rights guarantee only groundwater, not surface water.
- The application period lasts until all the available groundwater resources are allocated.

national group. From the perspective of the Association, this had clear strategic advantages. First, it allowed them to share experiences and learn from other groundwater user associations in Spain with similar problems, particularly in relation to securing access to water rights. Second, it allowed them to seek financial support from national programmes that could increase their technical capacity, which is key in terms of having access to better data, information and knowledge on the resource base. This in turn could prove crucial in negotiations over available resources with the water authority. Third, and important from a power perspective, it gave them space and *de facto* legitimacy to hold conversations at the national level with the ministry. This was important because the CHG is hierarchically dependent on the central ministry. Thus, collaborative institutional spaces and power bargaining were intertwining at both regional and national levels and impacting collective action to secure access to water resources at the local level.

Cross-scale collective action and politics: the interplay between river-basin authorities, central and regional governments and aquifer associations in securing water rights

A few months before the establishment of the Association, the Guadalquivir water authority (CHG) approved some uses and management rules for the aquifer (CHG-IGME, 2001). As explained earlier, this froze the process of granting water rights concessions until there was enough knowledge on aquifer dynamics. By October 2007 a new regulatory process was opened after the CHG had received the results from a study on hydrology and water balance commissioned from the Geological and Mining Institute of Spain (IGME). In this newly opened regulatory process, the CHG established a basic set of rules for granting groundwater rights, disregarding surface–groundwater interaction, as shown in Table 1.

However, three months after opening the application process, the CHG closed it on the basis that water rights applied for already exceeded the available water identified in the hydrogeological study. From the perspective of the CHG, new groundwater user associations should be developed, by redesigning the existing ones and creating a ‘macro water user association’ or ‘macrocommunity’. Since most land holdings are small, the new size rules posed problems for farmers. There was also the problem of existing wells that had drilled into both upper and lower aquifer layers, which in theory would now not be permitted. A more fundamental problem came from the relatively large number of users who had not applied in the 3-month window of the regulatory process due to lack of information or awareness of the process[3] or simply because they were automatically excluded through lack of compliance with one or more of the requirements of the application process. This meant that approximately 10% of farmers were *de facto* excluded from the process of water rights formalization (Rubiales, Secretary of the Association, personal communication, 2011). Although this is a small percentage, it was enough to play a significant role in overall water allocation and the water balance.

In summary, after approximately 15 years of irrigation history in the region, approximately 65% of the water used is abstracted informally (Cifuentes, 2012). Thousands of users that had applied for water rights concessions were waiting either for their water rights to be formalized or to be informed of their exclusion. The failure to reach an agreement with the water authority was reflected in sanctions on farmers. The region became a battlefield between the water policing service of the CHG and the farmers defending their existing water use. Collective action here was the claim to the legitimacy of groundwater use.

For the Association and the UPA, the key objective therefore was to secure groundwater concessions for their members. In water management, where politics are inherent to the system, the question is how equitable water management is and whether decisions are taken clientelistically or through patronage. In the region of La Loma, the effects of clientelistic politics and rent seeking (Ramos & Merino, 1999) are well portrayed. The CHG erred on the side of caution and did not grant water rights due to a lack of knowledge of aquifer dynamics (Cifuentes, 2012). Thus, user organizations continued to state their claims, while at the same time changing the action arena, upscaling discourses, claims and strategies to the central-government (national) level. This was a strategy based on exercising pressure from the top down, since getting agreements from the CHG is dependent on the central government, which at the time (from 1996 to 2004) was led by the conservative party. Meanwhile, at the regional level in Andalusia, the Socialist Party had been governing since 1982. In the early period (2000–2004), the political constellation was favourable to the Association. In 2004 the Association held negotiations with the water authority to get water rights concessions for all those farmers who collaborated by providing accurate information on abstraction, which had been unavailable until that moment. Meanwhile, the UPA trade union representatives, closely associated with the regional Socialist Party, showed opposition to this agreement, using equity as an argument on the basis that only a few wealthy farmers would get water rights. In the second period (2004–2011), after the March 2004 national elections were held, the Socialist Party won at the national level. This gave additional power to the UPA at both regional and national levels. The agreements held by the previous leaders were dropped, and no concessions were granted.

The victory of the Socialist Party at the national level, and therefore a change in the political orientation of the CHG, meant that there was a change in the interlocutor with users, with the UPA instead of the Association now in conversations and negotiations on water rights concessions. In interviews, members of the Association stated that the UPA wanted to be instrumental in granting water rights, because it was a means to securing political support and votes. There is clear evidence that while for the period 1999–2004 the Association had good access to the CHG and the central government, this changed in 2004, with the elections, when the UPA had closer contacts with the water authorities and national government.

For the period of January 2009 to October 2011, the responsibility of granting water rights was delegated to the regional government of Andalusia, instead of the CHG (which was dependent on the central government).[4] During those years, both the regional authority and a newly created regional water agency agreed that approximately 10,000 ha would be legalized in La Loma, with water rights concessions arranged through the UPA. The Association opposed the agreement, arguing it was both illegal and unfair. The whole process went to court, and as a result, some of the concessions were stopped. Yet at the same time, approximately 600 ha (extendable to 1,400 ha with the same water quantity per ha) were granted water rights concessions for surface water from the Guadalquivir River's surplus winter flow.

Under Spanish water law, once an aquifer is declared overexploited, the creation of a groundwater users general community is required at the scale of the aquifer, along with a management plan to regulate withdrawals. In December 2008, a process to declare the Úbeda groundwater body overexploited was opened. In relation to the first requirement, the Association is a bottom-up initiative from the year 2001 that has now applied to be legitimized and constituted as the general community by the CHG. However, as explained earlier, the Association does not include all users, particularly those whose water rights were recognized after mediation by the UPA agricultural trade union. Thus, at present, there is a potential conflict for the CHG, since both the Association and the UPA have contributed to the creation of these 'macrocommunities', and they are split along political lines, both having applied for the legitimation of users' water rights.

Opportunities for the actual institutional and organizational design: a balance with alternative resources?

Meanwhile, in view of the difficulties of securing water rights, farmers' strategies started to also look at the resource base – in other words, the technical options for bringing in alternative surface-water resources – and in turn, the institutional and organizational challenges and opportunities this would present for users and collective action. Groundwater users, who are aware of the threat posed to their livelihoods from diminishing aquifer reserves and/or the increased pumping costs from the lowering of aquifer levels, turned their attention first to the potential for aquifer recharge and second to use of surface water pumped directly from the Guadalquivir and Guadalimar Rivers. The limitation is that the basin, as a whole, has a deficit of approximately 430 MCM/y (Cifuentes, 2012). Therefore, the approach suggested at the moment is technical in nature, based on opportunities to recharge the aquifer by coordinating uses in summer and winter and storing the so-called 'surplus' water of wet years in the aquifer. It would eventually have to engage in a process of regulating the informal irrigation sector with formal water rights concessions. This could violate the environmental objectives of the Water Framework Directive and would need to be integrated into the whole basin balance (Hernández Herrero, 2010).

Water in the upper part of the basin is at the moment promised for other users further downstream. In addition, there is an environmental flow requirement. However, and most important, this has also put on the table discussions of the allocation of water at the catchment scale, and of water efficiency, productive use and equitable allocations. The technocratic discussion now becomes a political one: is reallocation of water possible in this basin? Who is responsible for the investment needed for this alternative scenario, of e.g. aquifer recharge? The CHG is taking a cautious position, in view of conflicts that might be caused with existing water rights downstream and economic aspects regarding investment. Meanwhile, upstream and downstream users are aware of these implications. At present, most water rights are granted to users downstream, for traditional surface-water irrigation systems. For the same crops of olive groves, water concessions downstream are 6000 m^3/ha per year, while in La Loma the maximum is 1500 m^3/ha per year. Since Spain does not operate under a system of 'first in time, first in right', potential reallocations are dependent on issues related to potential trading of water rights, requirements for rational use, effective monitoring and compensation. Groundwater users in La Loma are engaging in equity discourses, stating, "This is not fair: downstream users have cheaper water and use it less efficiently." Thus, in an interesting twist, technical solutions are also on the table that go across scales, where farmers upstream argue for making more water available upstream through the modernization of water use downstream to redistribute the water 'savings' in La Loma. However, the idea of modernizing irrigation systems, raising irrigation-water application efficiency to obtain 'new resources', is an inaccurate view; in many settings, it does not reduce water consumptive use ('dry water savings') (Perry, 2011; Seckler, 1996).

All user groups in La Loma think that it is necessary to bring more water sources into the game in order to keep the regional economy alive. This strategy of finding resilience in their system by looking beyond the resource boundaries of their system to other areas and irrigation sectors is common in other cases when resource boundaries are hit with intensive groundwater use (Rica & López Gunn, 2011). It could be argued, however, whether this strategy looks at the whole socio-ecological resilience of the system, including aquifer recovery, or whether it is mainly focused on social resilience and the pathology of intense resource use (Holling & Meffe, 1996). In addition, irrigation has increased the production of olive oil to the point where overproduction has been reached, and the price of olives has fallen to a point that rainfed growing is hardly profitable. Granting water rights contributes to the validation of this extractive intensive groundwater model for the regional economy.

Another challenge that conjunctive use of surface water and groundwater brings is in terms of organizational design. According to the CHG (2012), the long-term survival of olive-grove production in La Loma depends on the constitution and effective performance of both individual water user associations and also a strong and effective general association (the second level of associations). For equitable water use based on a conjunctive-use scenario, the CHG would prefer a single general association at the groundwater-body scale, representing all water user communities and individual users. The challenge that remains is to create a structure that is able to control and monitor abstraction as well as having a clear sanctioning regime for non-compliance, backed by the CHG and user associations. In general, the current trend for groundwater users in the area is to modify the actual existence of several small WUAs into larger ones which look at overall resource efficiency, not just with respect to water but also increasingly considering where water and energy use are most efficient. However, in contrast to this technical, rational and resource-efficient approach, there is the reality of a messy, bottom-up, heavily politicized local WUA distribution. Here the trends go in the opposite direction from 'technocratic rationality'.

As discussed earlier, the real challenge at the aquifer level and furthermore at the basin scale is how to allocate and potentially reassign and/or make water rights more flexible. A general water community would be an ideal solution, as specified – and required – under the current Spanish water law, and further reinforced under Andalusian water law. This is the solution sought currently by the CHG, the Association and some users. However, it leaves out water communities linked to the UPA agricultural trade union. A facilitation process may be needed based on consensual information on hydrological data for the creation of an organization where legitimate collective action can be carried out for the management of the aquifer nestled in a larger basin and which includes representatives from both organizations to ensure representativeness.

Conclusion: collective solutions for collective problems, but at what scale?

The La Loma case study highlights some relevant challenges, opportunities and key issues for institutional design in facilitating collective action to manage intensively used groundwater. The first issue is to adapt decision making towards a hydrological optimum, taking into account that the resource system is not homogeneous and presents certain hydrological dynamics once intensive resource use is a *fait accompli*. The second issue relates to the scale at which management, organization and problems occur. These scales do not always match, and collective action at the local level, once the problem has changed scale, may not be sufficient to solve collective problems. The solution has now evolved; it needs to take into account variables that escape the scope of the (local) collective organization. Thus, a challenge and an opportunity for the resilience of the system, both social and ecological, lies in devising institutional designs (currently absent in La Loma) which can upscale the boundaries of the current collective decision-making institutions and the arenas for conflict resolution. This could be done in cooperation with the water authorities, following a 'co-management' model which is better suited to e.g. redistributing wells according to sound hydrogeological criteria (Berkes, 2009; Plummer, 2009).

According to collective action theory, user groups could potentially solve the problem of difficulty of exclusion through self-regulation and mutual restrictions of access (Ostrom, 1990). Systems of exploitation of natural resources are in a constant dynamic process, with alternating phases of crisis and equilibrium. The resilience of these systems depends heavily on the adaptive capacities of groups like, in our case, groundwater users (Folke et al., 2002). Sometimes, however, local institutions cannot deal with the conservation problem because the resource's scope is beyond their ability to manage, bringing us to issues of complexity and scale (Ostrom et al., 1999, qtd. in Berkes, 2004).

These community groundwater organizations may not yet have the technical and institutional capacity, or indeed the complex institutional arrangements, needed to solve intrinsic problems in common pool resources, which would strengthen the overall resilience of the socio-ecological system. For collective-action design and performance in La Loma, one could thus question whether users are free-riding or cooperating. Are these organizations somehow limited to administrative advice, focused mainly on a bureaucratic role to process or secure water rights applications? As was highlighted, La Loma appropriators do value the common pool resource; indeed, their whole economy is based on intensive groundwater use. There is an acknowledgement of the common pool resource problem, which in fact triggered the emergence of collective action. However, so far rules have not been commonly designed or indeed agreed on. There is not enough knowledge, or at least officially sanctioned knowledge, of the aquifer dynamics, which has been a crucial factor in not regulating the situation. Both the Association and the UPA, shown on paper as a second level of collective action, represent emergent self-regulation attempts to devise a larger-scale entity for institutional arrangements. However, these cannot yet be considered successful. Not all users are represented, and no decision making regarding use or management is carried out. A fundamental problem is that the starting-point is to have as far as feasible clear access to both the volume of water resources and the aquifer boundaries. The lack of this resource and institutional baselines thus hinders the evolution of further institutionalization of groundwater use and collective action at a higher level.

This sheds light on the theoretical potential that polycentric governance systems have to cope with tragedies of the commons (Ostrom, 2005), distorted in our case study. The main challenges to collective action or co-management (or even polycentric governance) are currently how to address or incorporate the reality of water politics, including coping with potential rent-seeking games. If properly devised, this same threat can become a strength, if water politics is played out as a deliberative process, with an open and transparent argumentative turn, which incorporates good water governance criteria and practice (Rogers & Hall, 2002; Lautze, de Silva, Giordano, & Sanford, 2011). The political interests of different organizations are in fact part and parcel of the democratic system, confronted first by the issue of water rights, which have to give farmers a level of security, while also guaranteeing the resource base, to be 'on the safe side' before negotiating with the water authorities over which water is to be used and when. Both the Association and the UPA work independently on similar issues, creating competition for water rights acquisition. As we have seen, political colour determines the development of the granting process.

Therefore, a good starting-point for discussion would be to generate an agreed baseline on resource use. Until this information is available to support the water rights granting process, clientelistic behaviour from users, matched by the rent seeking of certain groups, is a determining factor halting the establishment of common rules or institutions for resource management. Rent seeking is thus a disincentive for equitable collective action. As Schlager (2002) argues, appropriators must decide whether or not to engage in the process of changing the rules to address the condition of the resource. Their knowledge of the resource and their ability to control it directly affect their decisions. Ultimately, under the current power constellation, the CHG has the last word in technical terms on the route pursued (e.g. recharge, diversion, etc.). It also has the last word in water-allocation terms and has legal legitimacy to decide on the management and sustainable use of the resource, taking into account the whole basin and its interconnections. Unless more space is granted to users' collective action, effectively achieving more distributed power, it could be questioned whether any one agent (e.g. the CHG) will really have the power (or legitimacy) to execute decisions.

La Loma shows the complex dynamics in the recognition of informal rights or an informal (water) economy. Here, it is shown how informal users organize themselves in a context where theoretically no more water is available for allocation, or in fact the aquifer is over-allocated. This process of legalization is a real test for collective action at different levels. These levels highlight the potential for an upscaling of collective action: a first level of water user groups sharing a well, a second level of aquifer associations, and finally the potential coordination with other (downstream) water user groups at the catchment level, possibly mediated by the water authority. Yet the 'impossibility of satisfying everybody' has been mixed up with politics and vote-catching at the regional level, a competition between groups and politically opposed lobbies. This is highly conflictive because the economy has a highly specialized and very vulnerable model of production, with monoculture of intensive irrigated olive groves but with very little redundancy, and thus the risk and stakes are high for all actors concerned. Yet, the potential upscaling of emergent collective action at the local scale, together with a strengthening of collaboration between farmers and the regulator, offer some opportunities for co-management of conjunctive water use at the basin level.

Notes

1. The numbers correspond to the nomenclature officially used by the Spanish Hydrological Plans, adapted to the European Water Framework Directive.
2. Spanish Water Act, 29/1985 (1985).

3. It was not well advertised: the only publication was in the official bulletin of the Andalusian region.
4. This was the result of the claim by the Andalusian government that the special clause of the Spanish constitution covering any river basin entirely within the borders of an autonomous community (a Spanish region) should apply in the case of Guadalquivir, because more than 90% of the river basin is located in Andalusia. However, this was challenged in courts by other autonomous communities and management was returned to an agency depending on the central state at the end of 2011.

References

Agrawal, A. (2003). Sustainable governance of common-pool resources: Context, methods, and politics. *Annual Review of Anthropology, 32*(1), 243–262. Retrived from http://www.annualreviews.org/doi/abs/10.1146/annurev.anthro.32.061002.093112. doi:10.1146/annurev.anthro.32.061002.093112

Berkes, F. (2004). Rethinking community-based conservation. *Conservation Biology, 18*(3), 621–630. Retrived from http://www.blackwell-synergy.com/doi/abs/10.1111/j.1523-1739.2004.00077.x. doi:10.1111/j.1523-1739.2004.00077.x

Berkes, F. (2009). Evolution of co-management: Role of knowledge generation, bridging organizations and social learning. *Journal of Environmental Management, 90*(5), 1692–1702. doi:10.1016/j.jenvman.2008.12.001

Blomquist, W. (1992). *Dividing the waters: Governing groundwater in Southern California.* San Francisco, CA: Institute for Contemporary Studies Press.

CHG. (2012). *Recursos de la MAS 05.23 Úbeda. Propuesta del Plan hidrológico de la cuenca del Guadalquivir.* Documento para la consulta pública 18 pp. Confederación Hidrográfica del Guadalquivir. Retrived from http://www.chguadalquivir.es

CHG-IGME. (2001). *Propuesta de Normas de Explotación.* U.H. 05.23 (Úbeda). Confederación Hidrográfica del Guadalquivir-Instituto Geológico Minero de España.

Cifuentes, V. (2012). Experiencias de los usuarios en el uso conjunto del agua, in *Jornadas de AEUAS*, 21 June 2012, Úbeda, Spain

Dumont, A., Salmoral, G., & Llamas, M. R. (2013). The water footprint of a river basin with a special focus on groundwater: The case of Guadalquivir basin (Spain). *Water Resources and Industry, 1–2*, 60–76. doi:10.1016/j.wri.2013.04.001

Feeny, D., Berkes, F., McCay, B., & Acheson, J. M. (1990). The tragedy of the commons: Twenty-two years later. *Human Ecology, 18*(1), 1–19. doi:10.1007/BF00889070

Folke, C., Carpenter, T., Elmqvist, L., Gunderson, C. S., Holling, C. S., & Walker, B. (2002). Resilience and sustainable development: Building adaptive capacity in a world of transformations. *Ambio, 31*(5), 437–440. And International Council for Science, *Series on Science for Sustainable Development* No. 3. Retrived from http://www.sou.gov.se/mvb/pdf/resiliens.pdf

Fornés, J. M., de La Hera, Á., Llamas, R., & Martínez-Santos, P. (2007). Legal aspects of groundwater ownership in Spain. *Water International, 32*(4), 676–684. doi:10.1080/02508060.2007.9709698

Gollonet, J., Rubio, J. C., Martín, C., González-Ramón, A., & Peinado, T. (2002). *Los acuíferos de la Loma de Úbeda, reserva estratégica para el desarrollo del olivar. Presente y futuro de las aguas subterráneas en la provincia de Jaén.* IGME ISBN-84-7840472–4

González-Ramón, A., Rubio, J. C., & López Geta, J. A. (2007). *Los acuíferos de la Loma de Úbeda (Jaén).* Ministerio de Educación y Ciencia IGME 56 pp.

Graaff, J., & de, Eppink, A. J. (1999). Olive oil production and soil conservation in southern Spain, in relation to EU subsidy policies. *Land Use Policy, 16*(4), 259–267. doi:10.1016/S0264-8377(99)00022-8. Retrived from http://www.sciencedirect.com/science/article/pii/S0264837799000228.ISSN 0264–8377.October 1999.

Hernández Herrero, E. (2010). *Directiva Marco del Agua y aspectos cuantitativos del Plan Hidrológico de Demarcación del Guadalquivir.* In Botín Foundation´s Water Observatory seminars.

Hogson, S. (2013). Water governance and the role of tenure and rights in coping with agricultural water scarcity'; Expert Consultation, Jan 2013

Holling, C. S., & Meffe, G. (1996). Command and control and the pathology of natural resource management. *Conservation Biology, 10*(2), 328–337.

Lautze, J., de Silva, S., Giordano, M., & Sanford, L. (2011). Putting the cart before the horse: Water governance and IWRM. *Natural Resources Forum, 35*(1), 1–8. doi:10.1111/j.1477-8947.2010.01339.x

Llamas, M. R., & Martínez-Santos, P. (2005). Intensive groundwater use: Silent revolution and potential source of social conflicts. *Journal of Water Resources Planning and Management, 131* (5), 337–341. Retrived from http://link.aip.org/link/JWRMD5/v131/i5/p337/s1&Agg=doi. doi:10.1061/(ASCE)0733-9496(2005)131:5(337)

López-Gunn, E., Bea, M., Huertas, R., Vay, L., & Zorrilla, P. (2014). *Water Tenure in Spain*; Report commissioned by FAO Water governance and water scarcity programe (January 2014).

Lopez-Gunn, E., & Martínez Cortina, L. (2006). Is self-regulation a myth? Case study on Spanish groundwater user associations and the role of higher-level authorities. *Hydrogeology Journal, 14*(3), 361–379. doi:10.1007/s10040-005-0014-z

López-Gunn, E., Rica, M., & van Cauwenbergh, N. (2012). *Taming the groundwater chaos*. In L. De Stefano & M. R. Llamas (Eds.), *Water, agriculture and the environment: Can we square the circle?* London, UK: Taylor and Francis.

Nuñez, I., Moreno, L., Rubio, J. C., Pérez Zabaleta, E., Gollonet, J., Araguás, L., & González, A. (2006). Evolución hidrogeoquímica del Acuífero carbonatado profundo de la Loma de Úbeda (Jaén). *Boletín del IGME N°, 117*, 613–616. Número monográfico especial.

Ostrom, E. (1990). *Governing the commons: The evolution of institutions for collective action.* Cambridge, UK: Cambridge University Press.

Ostrom, E. (2005). *Understanding Institutional Diversity* I. University, ed., Princeton, NJ: Princeton University Press. Retrived from http://www.amazon.com/dp/0691122385.

Ostrom, E., Burger, J., Field, C. B., Norgaard, R. B., & Policansky, D. (1999). Revisiting the commons: Local lessons, global challenges. *Science, 284*, 278–282. doi:10.1126/science.284.5412.278

Perry, C. (2011). Accounting for water use: Terminology and implications for saving water and increasing production. *Agricultural Water Management, 98*(12), 1840–1846. doi:10.1016/j.agwat.2010.10.002

Plummer, R. (2009). The adaptive co-management process: An initial synthesis of representative models and influential variables. *Ecology and Society, 14*(2), 24. Retrived from URL: http://www.ecologyandsociety.org/vol14/iss2/art24/

Ramos, J. L., & Merino, A. (1999). *Las comunidades de regantes y la nueva política del agua: los problemas de la acción colectiva*. Paper presented at the Iberic Congress on Water Management, Zaragoza, Spain.

Rica, M., & López Gunn, E. (2011). Institutional Diversity and incentives in collective action: groundwater user groups and their role in SES resilience. In *14th IWRA World Water Congress*, Porto de Galinhas, Brasil.

Rica, M., López-Gunn, E., & Llamas, R. (2012). Analysis of the emergence and evolution of collective action: An empirical case of Spanish groundwater user associations. *Irrigation & Drainage, 61*(Suplement S-1), 115–125. doi:10.1002/ird.1663

Rogers, P., & Hall, A. (2002). *Introducing effective water governance*. Stockholm: Global Water Partnership.

Salmoral, G., Dumont, A., Aldaya, M. M., Rodríguez Casado, R., Garrido, A., & Llamas, M. R. (2011). *Análisis de la huella hídrica extendida de la cuenca del Guadalquivir.* Madrid. Retrieved from http://www.fundacionbotin.org/monografias_observatorio-del-agua_publicaciones.htm

Sánchez Martínez, J. D., Gallego, V. J., & Araque, E. (2011). El olivar andaluz y sus transformaciones recientes. *EstudiosGeográficos, 72*(270), 203–229. doi:10.3989/estgeogr.201109

Schlager, E. (2002). Rationality, cooperation, and common pool resources. *American Behavioral Scientist, 45*(5), 801–819. Retrived from http://abs.sagepub.com/cgi/doi/10.1177/0002764202045005005. doi:10.1177/0002764202045005005

Seckler, D. (1996). *The new era of water resources management: From "dry" to "wet" water savings*. Colombo.

Shah, T. (2009). *Taming the Anarchy. Groundwater Governance in South Asia.* Washington, DC: Resources for the Future, IWMI.

WWF/Adena. (2006). *Uso ilegal del agua en España*. Madrid.

Young, O. (2006). Vertical interplay among scale-dependent environmental and resource regimes. *Ecology and Society, 11*(1): 27. [online] http://www.ecologyandsociety.org/vol11/iss1/art27/.

OPEN ACCESS

Participatory rural appraisal to assess groundwater resources in Al-Mujaylis, Tihama Coastal Plain, Yemen

Wahib Al-Qubatee, Henk Ritzema, Adel Al-Weshali, Frank van Steenbergen and Petra J. G. J. Hellegers

ABSTRACT

A participatory rural appraisal (PRA) conducted in the Al-Mujaylis area, Tihama Coastal Plain, Yemen provided a contribution, as a bottom-up approach, to the assessment of the needs of communities and their views on how to avoid groundwater degradation. It was found that PRA tools could be applied usefully in an area with data scarcity and a culturally different context. It is concluded that adopting a research approach between top down and bottom up is most valuable and effective.

Introduction

Tihama plain, located along Yemen's Red Sea coast and also known as Yemen's breadbasket, represents a fertile agricultural area of great importance for the food security of Yemen. In the 1970s, a large proportion of Yemen's population (nearly 90%) relied on agriculture and animal husbandry as a source of income. The dominant crop cultivated along the coast was date palm, and more inland date palm, cotton, cereal (sorghum, millet, maize and sesame), fodder, citrus, vegetable crops (onion, tomato and okra), mango and banana were cultivated (Irrigation Improvement Project, 2002). Because of the low prices of imported products compared with domestically produced products, the economic profitability was low (Cohen & Lewis, 1979). As a result, emigration to Saudi Arabia attracted a large segment of the labour force and led to a decline in agricultural production compared with the population growth. By 2010, the percentage of the population who relied economically on agriculture as a source of income was reduced to 54% (Ministry of Agriculture and Irrigation, 2012). Agriculture in the midstream areas mainly depends on the conjunctive use of spate and groundwater for irrigation, while in the downstream areas it depends only on groundwater in addition to scarce rainfall (Al-Qubatee, Ibrahim, From Dalseng, Al-Weshali, & van Steenbergen, 2013).

To improve agricultural production, the Food and Agricultural Organization (FAO) of the United Nations recommended improving irrigation systems in the Tihama region (Tipton & Kalmbach, 1974). The government initiated irrigation-improvement projects through the construction of reservoirs and water-harvesting structures. It also supported the agricultural sector by banning fruit imports to enhance the profitability of locally produced fruits, subsidizing diesel fuel, and easing the import of drilling and pumping equipment (Ward, 2009). A study in the area of Wadi Zabid (*wadi* in Arabic means a valley) concluded that the best way forward was to build 10 diversion weirs (five were actually built) along the wadi to divert spate water into the existing canals and to build structures in the irrigation system to facilitate the control and distribution of water (Tipton & Kalmbach, 1974). This type of spate irrigation requires the strong control and participation of the users in the water distribution, thus a participatory approach is required. Other Middle Eastern governments also tried to overcome high water demand by building similar diversion structures in combination with desalination plants and deepening groundwater wells, but these water supply solutions are costly and often do not reduce the increasing water demands (Zeitoun, Allan, Al Aulaqi, Jabarin, & Laamrani, 2012).

As a result of these investment programmes, agricultural production has increased considerably, e.g., the banana cultivation areas in the midstream of Wadi Zabid increased from 20 ha in 1980 to 3500 ha in 2000 (Van Steenbergen et al., 2010). This, however, happened at the expense of the sustainable use of groundwater resources as the number of drilling wells in Wadi Zabid and Wadi Rima increased by more than five times between 1987 and 2008, from about 2421 to 12,339 wells (National Water Resources Authority (NWRA), 2008). The groundwater degradation is a result of the change in the balance between abstraction and recharge. Negative impacts of the construction of upstream dams on the downstream areas include reduced surface flow/runoff, dried-up wells and water being lost from the dams' reservoirs through evaporation (Al-Qubatee, 2009). In addition, in some areas where people invested in small dam construction, they consider it to be their own property, which resulted in giving themselves priority water rights and restricting others from becoming shareholders (Vermillion & Al-Shaybani, 2004). Groundwater degradation is also a result of the combined effects of (1) the intensification of agriculture activities (El Ayni, Cherif, Jrad, & Trabelsi-Ayadi, 2012); (2) the decrease of the recharge because of the decrease in precipitation and the reduction in river flows caused by the construction of dams in the upstream parts of the catchment (El Ayni, Manoli, et al., 2012); and (3) economic incentives such as the diesel subsidies, which encourage groundwater abstraction instead of groundwater conservation (Burke, Moench, & Sauveplane, 1999; Hellegers, Perry, & Al-Aulaqi, 2011). The declining access to groundwater affects not only food production but also the socio-economic situation of the population, e.g., an degeneration of the health conditions caused by reduced access to drinking water and a reduction in the cultivation and use of medicinal plants (Sharma, 2009). Along the Red Sea coast, the drop in groundwater levels has also resulted in seawater intrusion (NWRA, 2009).

Participatory rural appraisal (PRA) is a term used to describe a growing range of approaches and methods that can be used to encourage stakeholders to participate in analysing and assessing measurements in land planning and to bring their own knowledge to the dialogue (Chambers, 1994). This approach was developed in the early 1990s to initiate a paradigm shift from the top-down to a bottom-up approach, and from blueprint planning to an interactive learning process. It is a shift from extractive survey questionnaires to experience sharing with local people (Cavestro, 2003; Sijbesma & Postma, 2008). In the participatory approach it is important to combine the implicit knowledge of the community with the distinct scientific knowledge of the researchers and decision-makers in order to overcome lack of data on the hydrogeological and the often complex societal ecosystems (Ritzema, Quang Anh, & Thi Kim, 2011), to increase understanding of the complexity of problems, to reach solutions that are acceptable for stakeholders and that can be implemented (Burke et al., 1999; Zanetell & Knuth, 2002), and to achieve effective water governance (Barreira, 2003). Moreover, participatory methods have been used successfully to provide sufficient data for participatory groundwater modelling in areas with lack of continuous data recording, and at the same time lead to an agreement between the stakeholders on the water resources management plan (action plan) to sustain both the livelihood and the ecosystem (Ritzema, Froebrich, Raju, Sreenivas, & Kselik, 2010).

The government started to apply an integrated water resources management (IWRM) approach to managing basins that were facing a critical depletion of their groundwater resources. Taher, Ward, Fadl, Saleh, and Sultan (2013) assessed the first phase of such a project in the Sana'a Basin. They concluded that by implementing IWRM in alleviating water degradation, its advantages were being flexible, adaptable, and responsive to local conditions and needs. However, they suggested that for follow-up phases, more emphasis should be placed on the development of a decentralized participatory approach and to improve water legislation. In Tihama plain, numerous groundwater resources studies have been conducted (e.g., Tesco-Viziterv-Vituki, 1971b; Al-Eryani, 1979; DHV, 1988; Al-Kebsi, 2000; Abu-Lohom, 2002; Nasher, Al-Sayyaghi, & Al-Matary, 2013). These, however, focused mainly on technical aspects such as hydrology, hydrogeology, hydrogeochemistry, geophysics, geochemistry and modelling and lack an IWRM approach. Numerous authors have indicated that complex water-management problems require a fully integrated research approach (social, economic, institutional, political and technical) with a focus on engaging the different and sometimes contradictory interests of all stakeholders (Burke et al., 1999; Hussein, 2016; Margerum, 2007; Pahl-Wostl, 2007; Raadgever, Mostert, & Van de Giesen, 2012; Von Korff, Daniell, Moellenkamp, Bots, & Bijlsma, 2012). Participatory approaches are gaining importance to make information available to researchers, environmental managers and decision-makers (Zvoleff & An, 2014), not only to discuss interventions but also through collective simulation and modelling (Bots, Bijlsma, Von Korff, Van der Fluit, & Wolters, 2011; Dionnet et al., 2013). Although participatory research in the field of natural resources and environmental management is increasing, some aspects are not receiving adequate attention (Trimble & Lázaro, 2014). Especially, the extent to which stakeholders are interested in, and capable of, being involved effectively in governance varies widely, and is often not addressed satisfactorily (Smiley, De Loë, & Kreutzwiser, 2010). In this study, we argue that adopting a research approach between top down and bottom up is more valuable and effective, especially in the areas of high illiteracy. The objective of this paper is to evaluate the use of such a mixed approach to PRA in areas with data scarcity and in a cultural context where Islamic religious values encourage people to apply *Shura* (Islamic term for consultation) among themselves, with the experts and specialists, and to be more cooperative and altruistic. The usefulness of PRA as a bottom-up approach in assessing the degradation of groundwater resources in the Al-Mujaylis area is evaluated with the aim to obtain a better understanding of the effects on the livelihoods of the local population and of the local stakeholder's views on the problems and their views on appropriate solutions.

Study area

Yemen has an ancient civilization and historical heritage known as 'Arabia Felix' (Jungfer, 1987). The water harvesting and diverting techniques are the most famous examples (Brunner & Haefner, 1986; Harrower, 2010), i.e., the great and famous dam of Ma'rib with its antique irrigation system that has been in operation for more than two millennia (Brunner & Haefner, 1986). Nowadays, the continuous increase in the construction of harvesting and diversion structures in the upstream/midstream areas without adequate water resources management is resulting in negative externalities such as groundwater resources degradation in downstream areas. The degradation of groundwater resources has both environmental and socio-economic impacts in the affected areas. At the end of 2012, field research was conducted in the downstream (Al-Mujaylis village and surrounding areas) and midstream areas of Wadi Zabid and Wadi Rima (Figure 1) to study the cause–effect relationship between land-use changes and groundwater degradation and the socio-economic impacts.

Wadi Zabid and Wadi Rima, with catchment areas of 4450 km^2 (Qaid, 2007) and 2900 km^2 (Land Resources Division, 1977) respectively, originate from the Yemen Highlands (Western slopes). The mean annual precipitation varies from 100 mm at the coast to 350 mm in the foothills and to more than 550 mm in the Eastern mountains, with temperatures ranging between 18 and 40°C (NWRA, 2009). The mean annual runoff (water flow) in Wadi Zabid is 136.73 million m^3 (from 1970 to

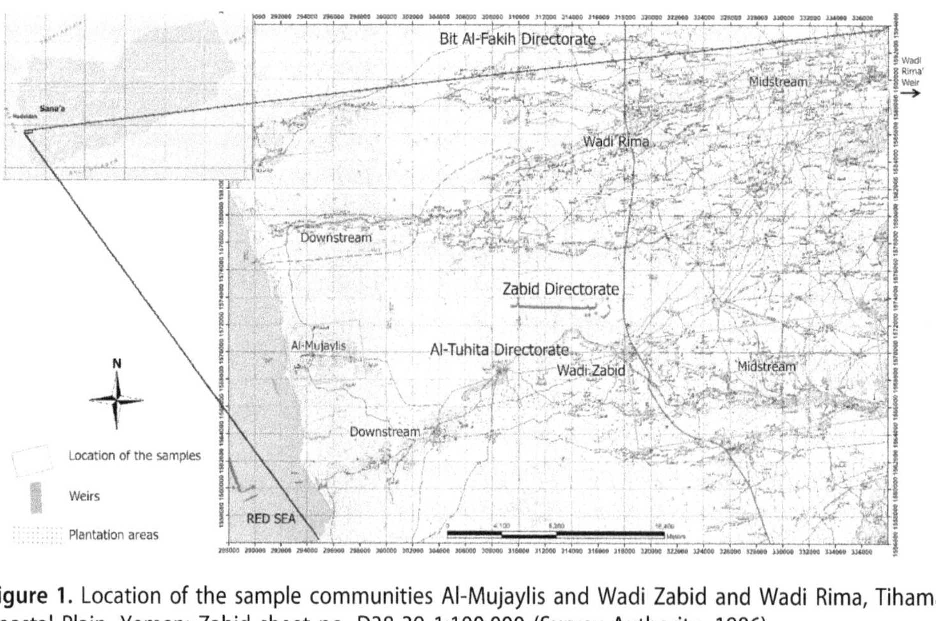

Figure 1. Location of the sample communities Al-Mujaylis and Wadi Zabid and Wadi Rima, Tihama Coastal Plain, Yemen; Zabid sheet no. D38-39 1:100,000 (Survey Authority, 1986).

1987) and in Wadi Rima is 85.68 million m^3 (from 1976 to 1987) (DHV, as cited in NWRA, 2008). Only after exceptional heavy rainstorms does the runoff reach the sea (Tesco-Viziterv-Vituki, 1971a).

The population in the middle and downstream part of Wadi Zabid (Zabid and Al-Tuhita directorates) and Wadi Rima (Bit Al-Fakih directorate) was 464,545 in 2004 (Central Statistical Organization, 2005). The population mainly depends on agricultural activities and ranching for its income. The total cultivated area in the three directorates is about 66,223 ha (Tihama Development Authority (TDA), 2007).

Materials and methods

A combination of PRA techniques was used, as described in detail by Mascarenhas et al. (1991), Cavestro (2003), Bhandari (2003) and Van der Schans and Lemperiere (2006). In order to ensure accuracy, to avoid bias, to find points of consensus/contradiction, to facilitate mutual learning, to integrate knowledge and to reach the reliability and validity desired, we used the following approach:

- A combination of six PRA tools was selected, namely semi-structured interviews (individual and group discussion), resources sketch mapping, daily calendars, time lines, transect walks and direct observations, and the preparation of problem and solution trees (Table 1). Each tool was used to address a specific topic (Table 2), and verification of the data was achieved by methods triangulation of at least three tools to cross-check the results of the different tools. In addition, the triangulation of the data sources was achieved by involving different respondents (farmers, key informants and emigrants), from different areas (downstream and midstream) and including secondary sources of information (previous studies, available data records, maps, Google Earth etc.). Moreover, the investigator triangulation was achieved by involving an interdisciplinary team from different local institutions (the observations of different investigators

Table 1. Participatory rural appraisal (PRA) tools and number of times tools were applied in the study.

	Semi-structured interview		
Area	Individual interviews	Group discussions	Other PRA tools
Al-Mujaylis	25 (four with key informants and 10 with people who migrated from the area)	Four (one group with women)	One resources map One daily calendar One time line Two transect walks and direct observations Two problem and solution trees with ranking
Wadi Zabid	22 (six with key informants)	Three (one with women)	One daily calendar One time line Two problem and solution trees with ranking
Wadi Rima	22 (five with key informants)	One group	One resources map One time line One transect walk and direct observation Three problem and solution trees with ranking

Sources: Al-Qubatee et al. (2013, 2015).

Table 2. Participatory rural appraisal (PRA) tools applied in the study area.

PRA tools		Description of the tools	NTTA[a]
Semi-structured interview	Individual Group discussion Key informant Emigrant	To know the situation in the area, the problems and the suggested solutions. Open questions were used that allowed the interviewer and respondents to raise new issues and topics during the discussion (Butler, Monroe, & McCaffrey, 2015). Stakeholders were interviewed individually to avoid the influence of or on others, especially in areas where there is mistrust as a result of unsolved conflicts over water resources. Group discussion was used to reach consensus and variety. Key informant opinion is important from the technical, institutional, legislation and political prospective. Some who migrated were interviewed as they are victims of the environmental degradation in the area	44 8 15 10
Resource sketch map		To know the socio-economic status and activities of people. We explained to respondents the purpose of this tool. From ideally the highest location in the area, respondents used local materials (chalk and coal) to draw the sketch map. The resources sketch map included land use, land-use change, and the features and infrastructures in the area	2
Daily calendar		To know farm and household activities in the area, the economic situation and working hours in the fields. Farmers were asked for their opinions on working hours, their main income sources, a household's activities and constraints	2
Time line		To know the situation in the area over the last 50 years. Respondents were asked (especially elderly people) to define major historical events that occurred in the past. Then they can make links between certain water and environmental issues and those events (easy to remember). This tool indicated the changes over time in the land use and agriculture productivity, groundwater situation, desertification, sand dune movement, and migration in the study area. Flip charts were used for this activity	3
Transect walk and direct observation		To know the current situation and to validate the information that was extracted by the other PRA tools. Simple measurements were carried out	3
Problem and solution tree		To know the causes and suggested solutions of the problems in the area from the part stakeholders' point of view. Groups of people sat together to define the water resources problems in their areas using a tree diagram. It was used as a problem analysis technique. Respondents were asked to rank the causes and suggested solutions of the problems according to their point of view and their preferences. A flip chart matrix was used	7

Notes: [a]NTTA, number of times tools are applied.

Al-Mujaylis: 25 semi-structured interviews, four group discussions, two transect walks, one resources sketch map, one time line, one daily calendar, and two problem and solution trees with ranking in addition to direct observation.

Wadi Zabid: 22 semi-structured interviews, three group discussions, one time line, one daily calendar, and two problem and solution trees with ranking in addition to direct observation.

Wadi Rima: 22 semi-structured interviews, one group discussion, one transect walk, one resources sketch map, one time line, and three problem and solution trees with ranking in addition to direct observation.

Sources: Adapted from Al-Qubatee et al. (2013, 2015).

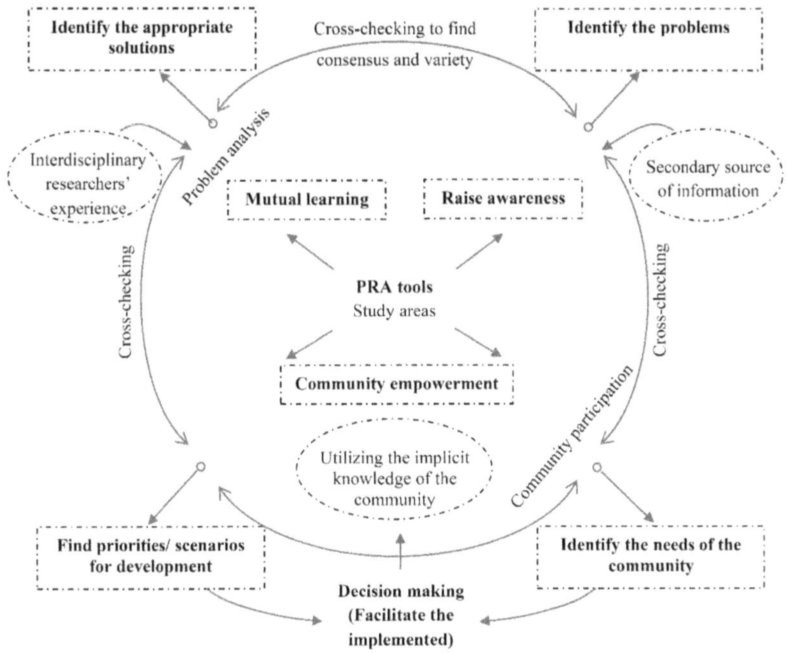

Figure 2. Interaction cycle among the different participatory rural appraisal (PRA) components (inputs) used for problem analysis, determination of priorities for development and community empowerment (outcomes).

in the same study area). Local materials, chalk, coal, stone and trees sticks were used to achieve the PRA activities in addition to flip charts, marker pens, maps, global position system (GPS) and cameras.

- The sampling method was 'stratified random sampling' in combination with 'purposive sampling'. Stakeholders from the areas downstream and midstream of Wadi Zabid and Wadi Rima were included. The sampling included different local communities and decision-makers such as farmers and most key informants (parliamentary representatives, managers, engineers, head of the Water User Association (WUA), head of the Irrigation Council, sheikhs, imams and teachers). The group discussions with the different stakeholder groups (farmers and key informants) included groups from the areas downstream and midstream of Wadi Zabid and Wadi Rima (Table 1).
- Cross-checking was performed between the results of different PRA tools to study the points of consensus and/or contradiction.
- The results obtained with the PRA tools were compared with available secondary sources of information.
- The study was conducted by an interdisciplinary team of four members from the Water and Environment Center (WEC), the Geological Survey and Mineral Resources Board (GSMBR), the National Water Resources Authority (NWRA) and the Tihama Development Authority (TDA) respectively. One team member was a female researcher who interviewed the women and conducted discussion groups with

them. This facilitated the communication process according to local customs and culture (a conservative community with a high percentage of illiteracy).

For data analysis, respondents defined and ranked the problems and solutions by themselves in the field. We used memos, flip charts (scanned with a digital camera) and/or audio and video to record all PRA tools applied in the field. Data coding was done manually for the semi-structured interviews and the results of discussion groups according to the objectives of the study (semi-structure questionnaires) and for any other emerging themes. Validity and reliability were achieved through the above-mentioned steps.

The interrelationship between the PRA components, inputs for problem analysis, determination of priorities for development and community empowerment, and the outputs are depicted in Figure 2. It shows how the implicit knowledge of the community, the interdisciplinary researchers' experience and the available secondary sources of the information were integrated (1) to identify the problems; (2) to identify appropriate solutions; (3) to find priorities/scenarios for development; and (4) to identify other needs of the community.

Results

Groundwater status

From the semi-structured interviews in Al-Mujaylis (Table 1), it was found that in this area groundwater is the major source for drinking, domestic, agriculture and livestock water use. People rely heavily on agriculture for their income in addition to fishing and remittances from their migrant relatives. The dominant agricultural crop is palm tree with some fodder for livestock. From semi-structured interviews, especially from the elderly men and women, we learned that the groundwater level in Al-Mujaylis was at a depth of less than 0.5 m below the soil surface 50–60 years ago. In 2013, the groundwater level had dropped to more than 12 m below the soil surface. From the interviews we also learned that there have been changes in well type and depth, drilling and water-pumping techniques.

Respondents expressed their assessment of water quality based on the taste of the water and, in some cases, how it compared with the taste of bottled mineral water. They indicated that the water quality in most wells in Al-Mujaylis is still fresh while other areas have higher water salinity that results in dying of the date palms in Al-Fazah village downstream of Wadi Zabid. This was confirmed by information from secondary sources: electric conductivity (the indicator of total dissolved solids) in most wells in Al-Mujaylis is between 0.7 and 1.6 dS/m, whereas water salinity in Wadi Zabid is between 0.8 and 17.8 dS/m and in Wadi Rima is between 0.6 and 10.9 dS/m (NWRA, 2009).

With one-group discussions (more than 10 respondents) in Al-Mujaylis, time-line tools were used to characterize the changes in groundwater status, agriculture activities and production, desertification and emigration changes since 1962 (Table 3). During the interviews, respondents were asked to connect the previous status with distinct past events, and during the discussion attempts were made to connect changes in the status of groundwater, agricultural activities and production, desertification and emigration

Table 3. Time line in the Al-Mujaylis area.

		Groundwater status			Agriculture activities				
Year	Spat water	Total depth	Water level	Water quality	Area	Crop type	Yield (%)	Desertification	Emigration
1962	Floods come from heavy rainfall in the area	0.5 m, no drilling till 1975	Near the surface (< 0.5 m)	Fresh	All agriculture lands	Palm	100	None	None
1979	Same as above	As above	0.5 m	Fresh	All agriculture lands	Palm	100	None	Few (10 people)
1985	Same as above	6–8 m	5 m	Fresh	Start of agriculture lands shrinking	Palm	100	Start to cover palms	
1990	Same as above	12 m	7–8 m	Fresh	As above	Palm	25	As above	30%
2000	Rainfall shortage	16–17 m	5–9 m	Fresh	Continuous decrease in agriculture lands	Palm	25	Continuous cover of palms	60%
2011	Same as above	30–50 m	12	Fresh	Most decreased in agriculture lands	Palm	10	It remains at 15% of palms	

Sources: Al-Qubatee et al. (2013, 2015).

changes with distinct events. For example, there was agreement between participants that after 1985 the groundwater level started to fall, the total agriculture area started to decline, the desertification areas started to increase and people's emigration from the area increased. The participants also agreed that after 2000, rainfall became more scarce and this was associated with the continuous drop in groundwater levels, the reduction of agricultural areas, a decrease in yields and an increase in desertification, a result being that about 60% of the population in Al-Mujaylis village move away from the area (Table 3). In an interview, the head of the TDA indicated that the dams built in the mountain areas of Wadi Zabid decreased the water flow to downstream areas. That was confirmed in other interviews with the head of the Environmental Department of the TDA, Zabid Branch, and the head of the Irrigation Council in Wadi Zabid. The current situation in the area was assessed by transect walks (involving researchers and farmers) and direct observations. The cross-checking between the outcomes of the PRA activities and the data collected from external sources of information confirmed the results related to the changes in groundwater levels, wells types and depths. Triangulation was used not only to validate findings but also for deepening and expanding researchers' understanding (Yeasmin & Rahman, 2012). However, there is a critique of using this method (Blaikie, 1991). Each method has positive and negatives aspects, but we found that this method helped to get more accuracy and validity.

Impact of groundwater degradation on livelihoods

Negative externalities have emerged in the study area, e.g., desertification has increased and date palms are dying. A resource sketch map, which was drawn by local people in Al-Mujaylis, gives information about the socio-economic situation and people's activities in the area (Figure 3). The resource sketch map includes livelihood status, types of houses and infrastructure in the area. This tool allowed us to obtain a better understanding about the overall area because it was drawn when standing on the roof of a high building (a school). By combining direct observations with a Google Earth map, a high level of precision and validity was achieved. The photographs in Figure 3 show palm farms invaded by sand dunes, the heavy spread of *Prosopis juliflora* trees. *Prosopis juliflora* is an evergreen tree native from South America; it is fast growing, nitrogen fixing, and tolerant to arid conditions and saline soils (El-Keblawy & Al-Rawai, 2005). It survives where other tree species have failed and in many cases has become a major nuisance. In 2000, it was rated one of the world's top 100 worst invasive alien species (Lowe, Browne, Boudjelas, & De Poorter, 2000). Date palm trees are dying as a result of rainfall shortages, groundwater decline and spread of *P. juliflora*. The results of the PRA activities show that after 2000 sand dunes have covered about 60% of the agricultural lands (Table 3).

The increased depth of the groundwater is impairing the economic situation of the people as well. In the past, agriculture did not need extensive irrigation technology because the groundwater was shallow (< 1 m below the soil surface). Today's cost of drilling a borehole to a depth of 40–50 m with pumping equipment is high and can reach US$8400. Most people in the area cannot afford to do this; only those few who can obtain financial support from emigrated relatives or other sources can afford borehole wells.

The production cost of palm tree agriculture has increased due to the harsh natural conditions, and the yield of the palm trees has gone down from 30 kg per palm tree per

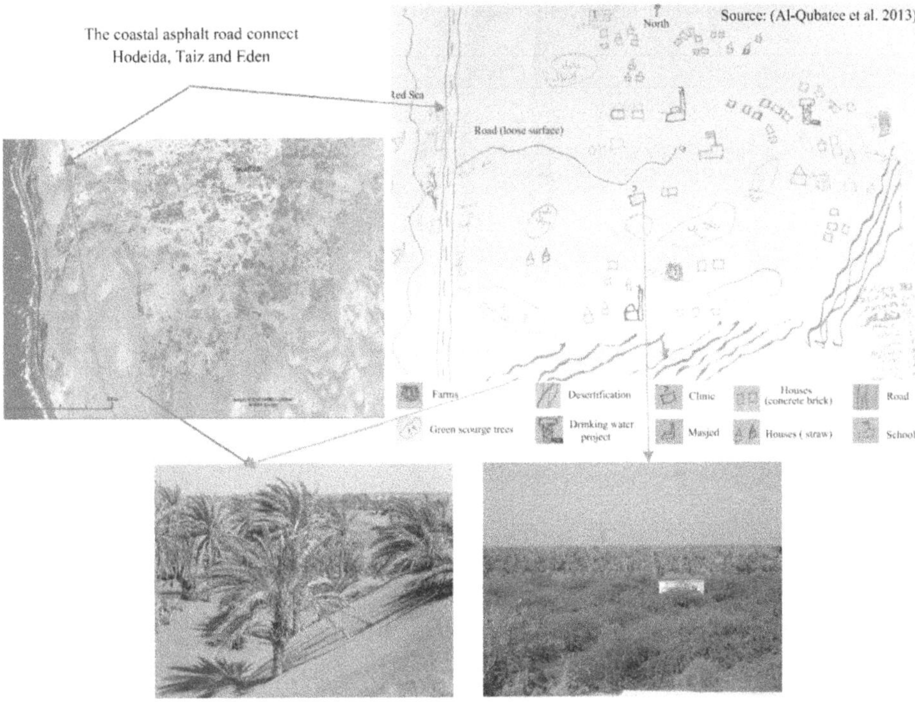

Figure 3. Resources sketch map of Al-Mujaylis with connection to Google Earth and direct observation (photographs) for precision and validity: (left) palm trees in Al-Zakham hamlet completely covered and in the process of being covered by sand dunes and (right) the heavy spread of *Prosopis juliflora* surrounding the palm trees and clinic centre. Photographic sources: Al-Qubatee et al. (2013, 2015).

year in the years before 1990 to only 5 kg per palm tree in 2011. Lack of rain, falling water tables, drought, sand dune movements and the spread of *P. juliflora* have impacted heavily on agricultural activities. Thus, palm tree agriculture is no longer the profitable business that it was previously.

According to local people, the overall decline in agricultural and livestock activities has led to increased poverty and migration from the area. This is reflected in daily farm activities: farmers spent just three hours per day in the field, and in the numbers of migration and current poverty levels. Al-Mujaylis, which consists of 10 hamlets, has a total population of 2642 (1328 male, 1314 female) according to the last census in 2004 (Central Statistical Organization, 2005). According to a key informant (a teacher), about 85% of the people are living in extreme economic poverty and the other 15% in moderate economic poverty (personal communication with Mr Ibrahim, 2013). Migration has increased after 2000, reaching 50–60% in 2015 (Table 3).

Stakeholder views on the causes and solutions of water resources problems

Eight different discussions groups, with 5 to 10 respondents in each group (including farmers and key informants), were formed to identify their problems and to propose solutions (Table 1). For example, the groups of people in the downstream areas close to the Red Sea (four different groups) identified and ranked the causes of the water

Table 4. Priorities of the water resources problems and the suggested solutions, downstream of the wadis.

Ranking	Al-Mujaylis	Al-Tefaf	Al-Gah (1)	Al-Gah (2)[a]
Causes of the water resources problems				
1	Rainfall shortage	Poverty	Rainfall shortage	Rainfall shortage
2	Random drilling wells (groundwater depletion)	Rainfall shortage	Spate of irrigation practices	Over abstraction, groundwater depletion
3	*Prosopis juliflora* trees	*P. juliflora* trees	Over abstraction and groundwater depletion	Increased agriculture areas
4	Spate irrigation	Construction of diversion structures	Farming crops consume a lot of water, such as banana	Constructed dams in the mountain catchment areas
5	Construction of weirs and dams	Crop pattern change upstream	Constructed dams in the mountain catchment areas	Crop pattern change, mango and banana instead of cereals
6		Groundwater depletion because of high abstraction upstream	Spread of *P. juliflora* trees	Absence of high-efficiency irrigation techniques
7		Random drilling of wells without the legal distance		
Suggested solutions				
1	Support farmers by modern irrigation etc.	Support farmers (marketing etc.)	*Zakat* giving	Implement high-efficiency irrigation
2	Cope with *P. juliflora* trees, desertification (reactivate the International Fund for Agricultural Development [IFAD] project)	Cope with *P. juliflora* trees and desertification	Implement high-efficiency irrigation techniques	Build dams and weirs in suitable places
3	Stop random well drilling and implement distance between wells	Build dams at the end of the wadis before water discharge into the sea	Regulate well drilling	Cultivate cereal crops
4	Water harvesting (at the end of the wadis before water discharge into the sea)	Solve the problem of diversion weirs to serve all people	Farm crops that can withstand drought	Use of natural fertilizers
5		Regulate well drilling	Select a suitable location for dams and weirs	

Note: [a]For this group no ranking was done in the field for the causes of the problem.
Sources: Al-Qubatee et al. (2013, 2015).

resource problems in the area, proposed solutions for these problems and ranked them according to their priority and preferences (Table 4).

People in the midstream of Wadi Zabid and Wadi Rima have their own interests and conflicts about the distribution of spate water. For example, people in midstream of Wadi Zabid depend on traditional water rights called in Arabic '*arf*' (common law) that were introduced by Sheikh Al-Jabarti more than 600 years ago based on the general rule '*Al-a'la fi-l-a'la*' (priority for the upstream riparian residents). These rules give priority to spate water rights to three groups of users in the midstream of Wadi Zabid within the 20 km downstream of the weirs. The water that exceeds their needs flows further downstream and (only in exceptional cases) reaches the Red Sea, 50 km from the

Table 5. Priorities of the problems in the midstream of Wadi Zabid and Wadi Rima.

Ranking	Al-Gerbah area, within the first group of Wadi Zabid	Al-Gerbah area, within the first group of Wadi Zabid (2)	Al-Mawi canal, within the second group of Wadi Zabid	Al-Jarubah area, Wadi Rima
Causes of the water resources problems				
1	Rainfall shortage	Rainfall shortage	Rainfall and spate water shortage	Convert the spate water of Wadi Rima to the southern area
2	Groundwater depletion because of banana farming and an increase in agriculture areas	Random drilling of wells, the water law is not implemented	Banana farming, which consumes a lot of water	No fair spate water distribution
3	Constructed dams in the mountain catchment areas of Wadi Zabid	Spate irrigation	Constructed weirs and dams in the mountain catchment areas of Wadi Zabid	Rainfall shortage
4	Water law not implemented fairly	Banana farming, which consumes a lot of water	Random drilling of wells	Crop pattern change, banana farming instead of cereals
5			Absence of high-efficiency irrigation techniques	
Suggested solutions				
1	Stop building harvesting dams in the mountain catchment areas of Wadi Zabid	Implement water law and stop the random drilling of wells	Stop or decrease banana farming	Justice in spate water distribution and implement the law
2	Implement the regulation for drilling of wells	Construct water-harvesting structures	Support farmers by high-efficiency irrigation techniques	Support farmers with diesel, marketing and technical
3	Implement new spate water-distribution systems as the system in the southern part of Yemen (Lahj & Abyan[a])	Support farmers by high-efficiency irrigation techniques	Regulate the drilling of wells	Educate farmers on how to conserve water
4	Replace banana crops with other crops like cereals	Decrease banana farming by farming another crops like cereals	Regulate the construction of dams and manage the water of constructed dams	Reduce crops that consume a lot of water, such as banana
5			Raise the awareness of farmers about the importance of water	

Note: [a]Spate water rights in those areas: farmers upstream use the spate for irrigation only once, thereafter letting the water pass to the next farm, and so on until the water reaches the last beneficiary who depends on the amount of the spate. The second spate would then begin from the farmer after the last one to receive the previous spate, and so on.
Sources: Al-Qubatee et al. (2013, 2015).

foothills. The outcome of the problem and solution tree exercise for the midstream stakeholders shows the most important causes of the problems and the priority for the suggested solutions in midstream Wadi Zabid and Rima (Table 5). Although there is neither cooperation nor conflict between downstream and midstream residents of the wadis, there were clear similarities in their views about the problems and possible solutions. The leading causes of the problems mentioned in both areas are: (1) rainfall shortage; (2) the effect of the construction of dams and diversion structures; (3) changes in cropping pattern and the low efficiency of irrigation techniques; and (4) random

Table 6. Usefulness of participatory rural appraisal (PRA) tools to achieve the objectives of the study, arranged from the most to the least useful.

	Objectives			
	To assess the degradation of groundwater resources			
PRA tools and other sources of information	Groundwater status	Livelihoods	To obtain stakeholders' views on the problems and their solutions	To overcome lack of data[a]
Semi-structured interview	+ + + + +	+ + + + +	+ + +	+ + + +
Transect walk and direct observation	+ + + + +	+ + + +	+ + +	+ + + +
Groups discussion	+ + + +	+ + +	+ + + +	+ + + +
Time line	+ + + + +	+ + + +	+	+ + + + +
Problem and solution trees	+ + +	+ +	+ + + + +	+ + +
Key informant interview	+ + + +	+ + +	+ + +	+ + +
Resource sketch map	+	+ + + +	+ + +	+ + +
Daily calendar	–	+ + +	–	+

Notes: Many plus signs (+) mean the strong contribution of the method to the achievement of the objectives; a negative sign (–) means no contribution at all).
[a]Groundwater status, spate water, traditional rights of spate water distribution, irrigation system, land use, livelihoods status etc.

drilling of wells and over-abstraction of groundwater. The common solutions suggested are (1) to support farmers to introduce high-efficiency irrigation techniques and assistance to market their crops; (2) the selection of suitable locations for future dams and diversion structures and better management of the water in the existing harvesting structures; (3) the regulation of well drilling and implementation of the water laws; and (4) the cultivation of crops that can withstand drought and have lower water requirements, e.g., cereals.

Key informants were asked about their role in the solutions for all the above-mentioned problems. The response of one parliamentary representative in the region was that each institution should take its responsibility to improve the situation. Suggestions included ideas such as to improve cooperation between government and related institutions and to improve legislation related to water resources to overcome the weak points in the laws (both formal and tradition laws). These results confirmed our expectation that each PRA tool applied was well suited to the various objectives of this study (Tables 2 and 6).

Discussion

There is strong support for more change from the current management practices towards a more adaptive and flexible approach to get more attention paid to tackle water management from an integrated approach, including human, physical, biological and biogeochemical components (Pahl-Wostl, 2007). The trend in the latest literature is towards considering water as interrelated to other sectors, within the broader context, through a problemshed rather than a watershed approach (Hussein & Grandi, 2017). Many arguments exist regarding the involvement of stakeholders in information production, which is regarded as strategic in shaping policies. In fact, the critical hydro-politics literature argues that discourses are constructed and deployed by powerful groups in a society or by powerful riparian countries in order to drive towards certain (favourable) solutions – while closing less favourable policy solutions (Zeitoun & Warner, 2006). The construction of discourses is linked to public opinion on how they absorb and reproduce through daily practices these discourses, and on how they may challenge and contest dominant discourses. Korfmacher (2001) stated that there is an ongoing debate on the subject of public involvement in the processes of knowledge production and decision making. He mentions several issues that justify the involvement of society, such as the democratic, substantive and pragmatic aspects of the process. The argument he made is that citizen involvement in watershed modelling will support decision-making and the implementation of those decisions. This perspective also supports the idea of educating people with the expectation that better understanding will lead to significant support of policy output. Baldwin and Ross (2012), based on a case study of the water-stressed Lockyer Catchment in Australia, stated that applying an action research methodology and consensus-building techniques in the early stage of the regulation process reduces the number of conflicts, leads to mutual learning and builds trust between parties (government and local communities).

However, there are also arguments against stakeholder involvement. Bureaucratic theory takes the view that some aspects of the decision-making process need technical expertise and are, therefore, best left to agencies with expertise. Korfmacher (2001) argues against public participation in watershed modelling, because they may lack expertise. He argues that there can be the risk of biased input, de-legitimization and/or over-legitimization, misrepresentation of consensus and insufficient influence. Moreover, Moote, McClaran, and Chickering (1997) stated that providing a forum for social discussion (participatory democracy approaches) on its own does not ensure successful collaboration, consensus and/or that decisions will be acceptable to all stakeholders, especially where different interests emerge.

Raadgever et al. (2012) argues that effective collaborative research will only work where the stakeholders are highly motivated (an accurate stakeholder analysis is required) and have equal input in and influence on the research process. This takes time and calls for an intensive collaborative research process.

In our study, we found that PRA tools are an effective method to benefit from the implicit knowledge of the community in which to fill gaps in our knowledge and data records. The PRA technique facilitated the generation of information over the groundwater status (in the past and at the moment), livelihoods of the local community, problems/constraints and views on appropriate solutions that this study aims to address (Table 6). This approach helped us to formulate appropriate solutions and to obtain a better understanding of people's preferences. The application of a combination of PRA tools helps us to reach a better understanding of the current groundwater situation and its effects on the livelihoods of people in the area.

The reliability of the generated information was assessed via triangulation and cross-checking between the outputs of the PRA process with external sources of information. The groundwater table contour map prepared by Tesco-Viziterv-Vituki (as cited in Tipton & Kalambach, 1980) confirms our findings that in areas close to the sea the groundwater table was at a depth of 10 m $^{+}$MSL (mean sea level), less than 1 m below the soil surface (Tipton & Kalambach, 1980, based on Tesco-Viziterv-Vituki, 1971b).

Based on our results, the involvement of stakeholders in problem assessment and the identification of proposed solutions is both important and desirable. This approach gave us insight into the community's perspectives on the issues under discussion, which is virtually impossible to do with just a quantitative approach. The communities suggested solutions that they felt were possible for effective implementation. However, we suggest that adopting a research approach between top down and bottom up is more valuable and effective, especially in the areas with high illiteracy. Lack of knowledge of the community on some issues can lead to unrealistic suggestions that may create conflicts and/or create other problems in future. In this respect, we had to assess whether the suggestions were realistic or unrealistic from a specialist point of view, including the various social, technical and economic aspects.

The disadvantages of applying the PRA method include the greater effort required to analyse the qualitative data generated by six different tools and because of the different emerging themes and, therefore, the difficulty of applying statistical methods for generalization compared with the analysis of quantitative data (structured and closed-ended questionnaire). Also, the difficulty in getting representation from the whole community required much time and resources. We overcome this by applying as many of the different PRA tools along the wadi as possible (Table 1) and to concentrate more on the results that were ranked by the stakeholders in the field. Another disadvantage of this method is the risk that the discussions are dominated by certain group(s) of people (especially where illiteracy is high). We overcame this by use of individual and group interviews. Individual interviews avoided influence from and on others, while group discussion led to increased mutual understanding and conjunctive learning between researchers and stakeholders.

Conclusions and recommendations

This paper suggests that PRA is a useful method for assessing the degradation of groundwater and its effects on the livelihoods of the inhabitants of the Al-Mujaylis area, overcoming gaps in the available data and getting a better understanding of the local stakeholders' views on the problems, their causes and appropriate solutions. The method gave a clear picture about groundwater degradation with regard to the groundwater drawdown and the effect this had on the livelihood of the community (a higher cost of farming, increased poverty and migration from the area). The challenge was to select the most appropriate and useful combination of tools to address a specific topic. The strongest tools are the semi-structured interviews, the transect walk in combination with direct observations and groups discussion, time lines, and problem and solution trees (Tables 2 and 6). The result of group discussion and the use of problem and solution trees reflect the stakeholders' views on the problems, the causes of the problems and appropriate solutions. By applying PRA techniques, stakeholders were able to analyse the causes of the problems and to propose appropriate solutions for mitigating or resolving the water problems in the area. The results of all the groups' input were compared with found areas of consensus (common solutions between the groups living downstream and midstream of the wadi).

The application of the PRA method contributed to obtaining an appropriate amount of data regarding the status of groundwater, land use, livelihood situations and the economic situation of the community. The method fits with our context as less attention has been paid to the social aspects of water studies. However, applying a complete bottom-up approach is not useful, especially in areas of high illiteracy. For example, respondents suggested that the government should subsidize the installation of high-efficiency irrigation techniques. From an experts point of view, however, it can be argued that high-efficiency irrigation techniques in closed water basins will not result in a high water saving because part of the water used in irrigation will return (non-consumptive water) to the aquifer over the short and/or long terms (Frederiksen & Allen, 2011; Gleick, Christian-Smith, & Cooley, 2011). Respondents also suggested that the government should continue the diesel subsidy. From an expert's point of view, however, economic incentives such as irrigation techniques and/or a diesel subsidy can be better oriented to water-saving activities (Hellegers, Perry, Al-Aulaqi, Al-Eryani, & Al-Hebshi, 2008). This is especially true for areas threatened by environmental catastrophes, such as groundwater depletion, seawater intrusion and desertification, as in our case study area. On the other hand, applying a pure top-down approach is not fully successful and faces some limitations. For example, the government implemented a project to cope with the desertification, but that project was not completely successful because people in the area preferred a water drinking project. Therefore, they were later allowed to convert some of those wells for drinking water. Furthermore, between 2004 and 2006, the government increased the height of two weirs in the midstream area by 1 m because sediments had accumulated upstream of it. Stakeholders downstream were unsatisfied with this, but on the other hand stakeholders on both sides of the weirs suffer from sediment accumulation in front of the weir. Participatory study and consensus-building was required along with the technical study.

From our study we conclude that adapting a research approach that is balanced between a totally top-down and a bottom-up approach and between delegitimization and over-legitimization will be more integrated and useful. The community views and suggestions have to be taken into account and studied by the concerned government institutions in a multidisciplinary approach, including technical, social and economic factors. Only then can appropriate decisions be made about solutions to be implemented.

The next step in our study will be participatory groundwater modelling to determine the most appropriate solutions suggested by stakeholders. The implicit knowledge of the local communities helped us to fill the gaps in our data recorded that will be used as inputs in this groundwater modelling. In addition, attention needs to be given to examining the wider impacts of the suggested solutions; the production value of the water; and to find the most commonly preferred solutions by stakeholders. These further developments will help in the policy and decision-making process and facilitate effective implementations.

Acknowledgement

This paper is based on a study carried out by the Water and Environment Centre, Sana'a University, Yemen, and MetaMeta, the Netherlands, in 2013 under the activity of Groundwater in the Political Domain, CoCooN project (a knowledge, research and innovation programme on conflict and cooperation over natural resources in developing countries). This paper is part of PhD research study conducted under the Water Resource Management Group, Wageningen University.

Disclosure statement

No potential conflict of interest was reported by the authors.

Funding

This work was supported by the Netherlands Organization for Scientific Research (NWO), Ministry of Foreign Affairs of the Netherlands. Project Number: W 07.68.307.00.

References

Abu-Lohom, N. M. (2002). *Hydrogeological, geophysical and geochemical investigation of the aquifers in Wadi Siham, Tihama area, Republic of Yemen* (Unpublished doctoral dissertation). Pune: University of Pune.

Al-Eryani, M. L. (1979). *Hydrology and ground water potential of the Tihama-Yemen Arab Republic* (Unpublished doctoral dissertation). Tempe: University of Arizona.

Al-Kebsi, A. Y. (2000). *Assessment of saltwater intrusion in a costal aquifer, Wadi Surdud, Yemen Republic* (Unpublished doctoral dissertation). Mumbai: Indian Institute of Technology.

Al-Qubatee, W. (2009). *Artificial recharge assessment of groundwater through Mekhtan Dam Bani Husheish, Sana'a basin - Yemen and ways to improve it out of the perspective of IWRM* (Unpublished master's thesis). Jordan University & Cologne University.

Al-Qubatee, W., Ibrahim, A., From Dalseng, C., Al-Weshali, A., & Van Steenbergen, F. (2013). *Groundwater depletion, Sand dune Formation and Local Responses: Case study Al-Mujaylis. CoCoon: Groundwater in the Political Domain.* Sana'a: Water and Environment Center, Sana'a University & 's Hertogenbosch: MetaMeta Research.

Al-Qubatee, W., Ibrahim, A., From Dalseng, C., Al-Weshali, A., Van Steenbergen, F., & Ritzema, H. (2015). Assessing the effectiveness of water governance in combating groundwater degradation in Al-Mujaylis, Tihama Coastal Plain, Yemen. In M. Fayyad, S. Sandri, M. Weiter, & D. Zikos (Eds.), *Social water studies in the Arab Region* (pp. 197–216). Berlin: Humboldt-Universität zu Berlin.

Baldwin, C., & Ross, H. (2012). Bridging troubled waters: Applying consensus-building techniques to water planning. *Society & Natural Resources, 25*(3), 217–234. doi:10.1080/08941920.2011.578120

Barreira, A. (2003). The participatory regime of water governance in the Iberian Peninsula. *Water International, 28*(3), 350–357. doi:10.1080/02508060308691709

Bhandari, B. B. (2003). *Participatory rural appraisal (PRA) Virgina.* Japan: Institute for Global Environmental Strategies.

Blaikie, N. W. H. (1991). A critique of the use of triangulation in social research. *Quality & Quantity, 23*, 115–136. doi:10.1007/BF00145701

Bots, P. W. G., Bijlsma, R., von Korff, Y., Van der Fluit, N., & Wolters, H. (2011). Supporting the constructive use of existing hydrological models in participatory settings: A set of 'rules of the game'. *Ecology and Society, 16*(2), 16. doi:10.5751/ES-03643-160216

Brunner, U., & Haefner, H. (1986). The successful floodwater farming system of the Sabeans, Yemen Arab Republic. *Applied Geography, 6*, 77–86. doi:10.1016/0143-6228(86)90030-5

Burke, J., Moench, M., & Sauveplane, C. (1999). *Groundwater and society: Problems in variability and points of engagement.* World Bank Technical Paper: 31–52.

Butler, W. H., Monroe, A., & McCaffrey, S. (2015). Collaborative implementation for ecological restoration on US Public Lands: Implications for legal context, accountability, and adaptive management. *Environmental Management, 55*(3), 564–577. doi:10.1007/s00267-014-0430-8

Cavestro, L. (2003). *PRA-participatory rural appraisal concepts methodologies and techniques.* Padua: University of Padua, Faculty of Agriculture. Retrieved 14 January, 2007, from http:// wwwagrariaunipdit/agraria/master/02-03/participatory%20rural%20appraisalpdf

Central Statistical Organization. (2005). *Statistical year book 2005.* Yemen: Central Statistical Organization, Ministry of Planning and International Cooperation.

Chambers, R. (1994). Participatory rural appraisal (PRA): Challenges, potentials and paradigm. *World Development, 22*(10), 1437–1454. doi:10.1016/0305-750X(94)90030-2

Cohen, J. M., & Lewis, D. B. (1979). Capital-surplus, labor-short economies: Yemen as a challenge to rural development strategies. *American Journal of Agricultural Economics, 61* (3), 523–528. doi:10.2307/1239440

DHV. (1988). *Tihama basin water resources study, Groundwater development in the Tihama coastal plain.* Sana'a: Ministry of Agriculture and Fisheries Resources, Tihama Development Authority (TDA).

Dionnet, M., Daniell, K. A., Imache, A., von Korff, Y., Bouarfa, S., Garin, P., … Rougier, J.-E. (2013). Improving participatory processes through collective simulation: Use of a community of practice. *Ecology and Society, 18*(1), 36. doi:10.5751/ES-05244-180136

El Ayni, F., Cherif, S., Jrad, A., & Trabelsi-Ayadi, M. (2012). Impact of agriculture activities on a coastal aquifer in Tunisia and options for a better water management. *Water International, 37* (7), 871–883. doi:10.1080/02508060.2012.733671

El Ayni, F., Manoli, E., Cherif, S., Jrad, A., Assimacopoulos, D., & Trabelsi-Ayadi, M. (2012). Deterioration of a Tunisian coastal aquifer due to agricultural activities and possible approaches for better water management. *Water and Environment Journal, 27*(3), 348–361. doi:10.1111/j.1747-6593.2012.00354.x

El-Keblawy, A., & Al-Rawai, A. (2005). Effects of salinity, temperature and light on germination of invasive *Prosopis juliflora* (Sw.) D.C. *Journal of Arid Environments, 61*, 555–565. doi:10.1016/j.jaridenv.2004.10.007

Frederiksen, H. D., & Allen, R. G. (2011). A common basis for analysis, evaluation and comparison of offstream water uses. *Water International, 36*(3), 266–282. doi:10.1080/ 02508060.2011.580449

Gleick, P. H., Christian-Smith, J., & Cooley, H. (2011). Water-use efficiency and productivity: Rethinking the basin approach. *Water International, 36*(7), 784–798. doi:10.1080/ 02508060.2011.631873

Harrower, M. J. (2010). Geographic Information Systems (GIS) hydrological modeling in archaeology: An example from the origins of irrigation in Southwest Arabia (Yemen). *Journal of Archaeological Science, 37*, 1447–1452. doi:10.1016/j.jas.2010.01.004

Hellegers, P. J. G. J., Perry, C., Al-Aulaqi, N., Al-Eryani, A., & Al-Hebshi, M. (2008). *Incentives to reduce groundwater extraction in Yemen.* The Hague: Wageningen University.

Hellegers, P. J. G. J., Perry, C. J., & Al-Aulaqi, N. (2011). Incentives to reduce groundwater consumption in Yemen. *Irrigation and Drainage, 60*, 93–102. doi:10.1002/ird.538

Hussein, H. (2016). *An analysis of the discourse of water scarcity and hydropolitical dynamics in the case of Jordan* (Unpublished doctoral dissertation). Norwich: University of East Anglia.

Hussein, H., & Grandi, M. (2017). Dynamic political contexts and power asymmetries: The cases of the Blue Nile and the Yarmouk Rivers. *International Environmental Agreements: Politics, Law and Economics.* doi:10.1007/s10784-017-9364-y

Irrigation Improvement Project. (2002). *Water management in Wadi Tuban & Wadi Zabid, water Management plans & spate Management models.* Yemen: Ministry of Agriculture and Irrigation/The World Bank.

Jungfer, E. (1987). Grundwassererganzung und Grundwassernutzung in den Wusten des Yemen [Supplementing groundwater and using groundwater in the deserts of Yemen]. *Geographische Rundschau, 39*(7–8), 408–410, 417.

Korfmacher, K. S. (2001). The politics of participation in watershed modeling. *Environmental Management, 27*, 161–176. doi:10.1007/s002670010141

Land Resources Division. (1977). *Mountain plains and wadi Rima' project, a land and water resources survey, an investment and development plan for wadi Rima'*. London: British Ministry of Overseas Development.

Lowe, S., Browne, M., Boudjelas, S., & De Poorter, M. (2000). *100 of the World's Worst Invasive Alien Species A selection from the Global Invasive Species Database*. The Invasive Species Specialist Group, World Conservation Union (IUCN), 12pp.

Margerum, R. D. (2007). Overcoming locally based collaboration constraints. *Society & Natural Resources, 20*, 135–152. doi:10.1080/08941920601052404

Mascarenhas, J., Shah, P., Joseph, S., Jayakaran, R., Devavaram, J., Ramachandran, V., … Pretty, J. (1991). *Participatory rural appraisal: Proceedings of the February 1991 Bangalore PRA trainers workshop: RRA notes, No. 13.* London: IIED and Bangalore: MYRADA.

Ministry of Agriculture and Irrigation. (2012). *A promising sector for diversified economy in Yemen: National agriculture sector strategy 2012-2016.* Yemen: Author.

Moote, M. A., McClaran, M. P., & Chickering, D. K. (1997). Theory in practice: Applying participatory democracy theory to public land planning. *Environmental Management, 21*, 877–889. doi:10.1007/s002679900074

Nasher, G., Al-Sayyaghi, A., & Al-Matary, A. (2013). Identification and evaluation of the hydrogeochemical processes of the lower part of Wadi Siham catchment area, Tihama plain, Yemen. *Yemen Arabian Journal of Geosciences, 6*, 2131–2146. doi:10.1007/s12517-011-0471-8

National Water Resources Authority (NWRA). (2008). *Tihama water resources assessment, Zabid water resources management, district (Wadi Zabid & Wadi Rema'a), Well inventory results.* Sana'a: Ministry of Water and Environment.

National Water Resources Authority (NWRA). (2009). *Wadi Zabid and wadi Rima water quality.* Sana'a: Ministry of Water and Environment.

Pahl-Wostl, C. (2007). Transitions towards adaptive management of water facing climate and global change. *Water Resources Management, 21*, 49–62. doi:10.1007/s11269-006-9040-4

Qaid, K. Q. (2007). *Assessment study of the impact of agricultural demonstrations program on production and income in Zabid and Tuban valleys.* Sana'a: Irrigation Development Project.

Raadgever, G., Mostert, E., & van de Giesen, N. C. (2012). Learning from collaborative research in water management practice. *Water Resources Management, 26*, 3251–3266. doi:10.1007/s11269-012-0070-9

Ritzema, H., Froebrich, J., Raju, R., Sreenivas, C., & Kselik, R. (2010). Using participatory modelling to compensate for data scarcity in environmental planning: A case study from India. *Environmental Modelling & Software, 25*, 1450–1458. doi:10.1016/j.envsoft.2010.03.010

Ritzema, H. P., Quang Anh, L., & Thi Kim, B. (2011). Colloborative research to improve the water management in two polders in the Red River Delta in Vietnam. In A. Van Paassen, J. Van Den Berg, E. Steingröver, R. Werkman, & B. Pedroli (Eds.), *Knowledge in action* (pp. 57–84). Wageningen: Mansholt Publication Series, Wageningen Academic Publ.

Sharma, K. D. (2009). Groundwater management for food security. *Current Science. 96*, 1444–1447. Retrieved from http://re.indiaenvironmentportal.org.in/files/Groundwater%20 management%20for%20food%20security.pdf

Sijbesma, C., & Postma, L. (2008). Quantification of qualitative data in the water sector: The challenges. *Water International, 33*(2), 150–161. doi:10.1080/02508060802022209

Smiley, S., De Loë, R., & Kreutzwiser, R. (2010). Appropriate public involvement in local environmental governance: A framework and case study. *Society & Natural Resources, 23*, 1043–1059. doi:10.1080/08941920802627491

Survey Authority. (1986). *Topographic map, Zabid sheet No. D38-39 1:100,000.* Yemen: Author.

Taher, T., Ward, C., Fadl, N., Saleh, A., & Sultan, M. (2013). Planning for integrated water resources management: Case study Sana'a Basin, Yemen. *Iwtj 4*, Retrieved from http://iwtj.info/wp-content/uploads/2014/06/V3-N4-P61.pdf

Tesco-Viziterv-Vituki. (1971a). *Survey of the apicultural potential of the Wadi Zabid.* YAR; land tenuer and water right. Budapest: Author.

Tesco-Viziterv-Vituki. (1971b). *Survey of the apicultural potential of the Wadi Zabid. YAR; Technical Report 2 – hydrology and hydrogeology.* Budapest: Author.

Tihama Development Authority (TDA). (2007). *Cultivated area in different seasons and main crops in each district of Hodeidah Governorate, Monitoring and evaluation deportment, spreadsheet.* Hodeida: Author.

Tipton and Kalmbach. (1974). *Proposed plan for water allocation, Wadi Zabid irrigation project.* Hodeida, Yemen, and Denver, CO: Tihama Development Authority (TDA).

Tipton and Kalambach. (1980). *The Wadi Zabid groundwater potential.* Hodeida, Yemen, and Denver, CO: Tihama Development Authority (TDA).

Trimble, M., & Lázaro, M. (2014). Evaluation criteria for participatory research: Insights from coastal Uruguay. *Environmental Management, 54,* 122–137. doi:10.1007/s00267-014-0276-0

van der Schans, M. L., & Lemperiere, P. (2006). *Manual: Participatory rapid diagnosis and action planning for irrigated agricultural systems (PRDA).* Rome: Food and Agriculture Organization of the United Nations.

Van Steenbergen, F., Lawrence, P., Haile, A., Salman, M., Faures, J., Anderson, I., ... Ratsey, J. (2010). *Guidelines on spate irrigation.* Rome: Food and Agriculture Organization of the United Nations (FAO).

Vermillion, D. L., & Al-Shaybani, S. (2004). *Small dams and social capital in Yemen: How assistance strategies affect local investment and institutions.* Colombo: International Water Management Institute.

Von Korff, Y., Daniell, K. A., Moellenkamp, S., Bots, P., & Bijlsma, R. M. (2012). Implementing participatory water management: Recent advances in theory, practice, and evaluation. *Ecology and Society, 17*(1), 30. doi:10.5751/ES-04733-170130

Ward, C. (2009). Water conflict in Yemen: The case for strengthening local resolution mechanisms. In N. V. Jagannathan, A. S. Mohamed, & A. Kremer (Eds.), *Water in the Arab world* (pp. 233–267). Washington, DC: The World Bank.

Yeasmin, S., & Rahman, K. F. (2012). 'Triangulation' research method as the tool of social science research. *Bup Journal, 1,* 2219–4851. Retrieved from http://www.bup.edu.bd/journal/154-163.pdf

Zanetell, B. A., & Knuth, B. A. (2002). Knowledge partnerships: Rapid rural appraisal's role in catalyzing community- based management inVenezuela. *Society & Natural Resources, 15,* 805–825. doi:10.1080/08941920290107576

Zeitoun, M., Allan, T., Al Aulaqi, N., Jabarin, A., & Laamrani, H. (2012). Water demand management in Yemen and Jordan: Addressing power and interests. *The Geographical Journal, 178,* 54–66. doi:10.1111/j.1475-4959.2011.00420.x

Zeitoun, M., & Warner, J. (2006). Hydro-hegemony – a framework for analysis of transboundary water conflicts. *Water Policy, 8,* 435–460. doi:10.2166/wp.2006.054

Zvoleff, A., & An, L. (2014). Analyzing human–landscape interactions: Tools that integrate. *Environmental Management, 53,* 94–111. doi:10.1007/s00267-012-0009-1

Federal reserved rights and California's Groundwater Management Act: resolving groundwater rights tensions in California and the western United States

Stefanie Viktoria Caroline Schulte

ABSTRACT

California introduced its first comprehensive groundwater legislation (the Sustainable Groundwater Management Act) in 2014 to address the overuse of the state's groundwater resources and to establish a pathway for sustainable groundwater use. However, a recent decision by the US Court of Appeals for the Ninth Circuit declaring the existence of federal reserved rights to groundwater could complicate the future implementation of California's new groundwater legislation. This article explores the impact of the case and offers recommendations on how the tensions between the court's decision and the existing legal framework for groundwater management in California could be reduced.

Introduction

The diversity of California's climate, geography and population density, together with the inter-temporal scarcity of water supply, has caused conflict over water resources and led to extended legal disputes between competing water users (Moran & Cravens, 2015). To address some of these challenges, California amended its water management and governance arrangements in 2014, following the passage of a three-bill legislative package collectively referred to as the Sustainable Groundwater Management Act (SMGA). The act establishes California's first comprehensive groundwater management and governance framework under the direction of local public agencies with water supply, water management and land use responsibilities (California Legislature, 2016). While the act mandates local planning and state administrative reviews (Owen, 2017), the implementation of this relatively unique model of local–state groundwater governance could be complicated by a recent court adjudication – *Agua Caliente Band of Cahuilla Indians v Coachella Valley Water District*, No. 15-55896 (Ninth Circuit 2017) – which has raised a number of key issues at the intersection of tribal law and water law that have not been conclusively resolved by US courts in previous cases. The court decision in the Coachella Valley dispute has also highlighted several unresolved issues with California's groundwater regulation which will require resolution before a consistent long-term management framework for groundwater resources can be established.

The legal dispute concerns the questions of whether the Agua Caliente Band of Cahuilla Indians (the tribe) has federal reserved rights to the Coachella Valley Groundwater Basin, ownership rights to the basin's porous space, and a right to unimpaired water quality. In March 2017, the United States Court of Appeals for the Ninth Circuit resolved Stage 1 of the dispute by granting the tribe federal reserved rights to the Coachella Valley's groundwater aquifer. Stage 2 of the dispute, which was originally scheduled to commence in the US District Court on 22 January 2018 (Coachella Valley Water District [CVWD], 2018a), focuses on whether the tribe has ownership rights to the Coachella Valley Groundwater Basin's porous space and a right to unimpaired water quality in conjunction with its federal reserved right to groundwater.

As federal reserved rights are not subject to state-based legislation (Blumm, 2016), the Ninth Circuit's decision interferes with the implementation of the SGMA and raises questions about the scope of regulatory powers conferred by the state authorities on the public agencies responsible for its implementation. The case also highlights the existing problems of integrating federal reserved rights into California's water management framework and shows that this issue has not been resolved by the introduction of the SGMA. To the contrary, the case makes clear that California will continue to struggle with the incorporation of federal reserved rights into the new groundwater management and governance framework.

This article explores the likely impact of the Ninth Circuit's decision on the future implementation of California's SGMA. The first part describes the complexities of water resource management in California and more specifically in the Coachella Valley. The second part discusses California's existing common law water rights framework, while the third part discusses California's new regulatory approach to groundwater management, with a particular focus on the regulatory instruments made available through the new groundwater management legislation. The fourth part analyzes how the recent groundwater court case in the Coachella Valley could complicate the implementation of the SGMA and lead to protracted legal disputes between different groundwater rights holders who may be aggrieved by future augmentation to groundwater management decisions. The article concludes with four recommendations to amend the new groundwater management framework and address the emerging challenges between different groundwater rights holders.

Complexities of California's water resource management

Given California's variable precipitation and regular dry periods, surface water supply in the state can be highly variable (California Department of Water Resources [CDWR], 2015). This variability not only poses challenges for California's water management but also influences the interaction between surface water and groundwater use in the state (CDWR, 2018a). According to the CDWR (2018a), California's 515 groundwater basins and sub-basins contribute between 38% and 46% of the state's total water supply. However, some rural communities and larger regional centres depend entirely on groundwater (Langridge, 2017). In regions with both surface water and groundwater access, the variations in groundwater use are closely linked to weather conditions and the availability of surface water supplies. In dry periods, when surface water is scarce, water users often turn to groundwater as an alternative water source (US Geological Survey, 2018). This substitution effect from surface water to groundwater was observable during California's recent drought, in 2011–2017 (Farr, Jones, & Lui, 2016), when increased groundwater extraction lowered groundwater levels and reduced groundwater quality (US Geological Survey, 2018). The impact of the drought was most pronounced in central and southern California, where groundwater extraction levels far exceeded natural or managed groundwater recharge (CDWR, 2017a), leading to subsidence and groundwater access issues.

One of the groundwater basins affected by the drought was in the Coachella Valley of Southern California. The Coachella Valley Groundwater Basin has four groundwater sub-basins (CDWR, 2017a) that are relied on by urban and industrial water users due to the region's hot desert climate. The groundwater sub-basins are predominantly managed by two public water agencies, the CVWD and the Desert Water Agency (CVWD, 2018b). These two agencies were formed in 1918 and 1961, respectively, to deliver domestic and irrigation water, collect and recycle wastewater, provide stormwater protection, replenish the groundwater basin and promote water conservation (CVWD, 2018c). While both water agencies hold contracts with the State Water Project, the Coachella Valley Water District also has a federal contract for Colorado River water (CVWD, 2018c). Both water agencies also have contracts with the Metropolitan Water District to exchange State Water Project water for Colorado River water, which the agencies import via the All-American Canal to recharge their groundwater basins (CVWD, 2018c).

Since the 1970s and particularly during the recent drought (2011 to early 2017), water demand in the Coachella Valley has exceeded the water agencies' groundwater replenishment efforts, leading to a decline in local groundwater levels (US Geological Survey, 2017). Although the water agencies had groundwater management plans prior to the introduction of the SGMA (CVWD, 2017), in Bulletin 118 the CDWR (2017b) listed three of the four Coachella Valley groundwater sub-basins as medium-priority basins, thereby making the groundwater sub-basins subject to the newly introduced SGMA.

The Coachella Valley Groundwater Basin is also subject to a legal dispute between the two public water agencies and a resident Native American tribe, the Agua Caliente Band of Cahuilla Indians (Agua Caliente Band of Cahuilla Indians v Coachella Valley Water District, et al. and Desert Water Agency et al., 2017a). The 400-member tribe is a federally recognized Native American tribe that has occupied land in the Coachella Valley for over 5000 years (Agua Caliente Band of Cahuilla Indians, 2018a). The tribe is the single largest land owner in the area due to an executive order which established a 30,000-acre reservation for the tribe in 1976–77 (Agua Caliente Band of Cahuilla Indians, 2018a). The tribe also holds surface water access rights to the Whitewater River and its tributaries (Whitewater River Decree of 1938) and has a correlative right to the Coachella Valley Groundwater Basin.

The tribe has raised concerns about the management of the Coachella Valley's Groundwater Basin for over two decades (Agua Caliente Band of Cahuilla Indians, 2015). In several letters to the water agencies, the tribe has emphasized the importance of protecting and preserving Coachella Valley's water resources and demanded the cessation of the water agencies' groundwater replenishment programme due to concerns about water quality degradation (Agua Caliente Band of Cahuilla Indians, 2018b). Despite the requests, no resolution on this matter was reached. In response, the tribe filed an action for declaratory and injunctive relief with the US District Court for the Eastern District of California in 2013. The action contained three legal issues (Agua Caliente Band of Cahuilla Indians v Coachella Valley Water District & Desert Water Agency, 2013) that each could challenge California's existing common law water rights framework and further complicate the implementation of the recently introduced SGMA, as discussed further below.

California's common law water rights framework

The legal framework that underpins California's water use is complex, as surface water and groundwater resources are governed by two different legal frameworks. While surface water rights are based on a prior appropriation system, groundwater rights are built on a framework known as correlative rights (California Environmental Protection Agency, 2018).

California's correlative rights system for groundwater resources is closely related to riparianism in that overlying landowners have an inherent right to withdraw all the groundwater they can beneficially use on their land, subject only to the correlative rights of other landowners. Thus, water rights are not subject to a priority-based allocation system, and groundwater right holders are not inherently guaranteed a specific volume of water. Instead, every landowner enjoys the right to make reasonable use of the water adjoining their property. In times of scarcity, correlative rights are apportioned among overlying landholders, whereby the rights of each landowner are determined based on his/her current reasonable and beneficial need for water (Ratliff, 2016). Although groundwater use reporting has been in place in some Californian basins (California State Water Resource Control Board, 2019), many groundwater users had not been required to monitor, meter, or record their annual groundwater extraction until the introduction of the SGMA (CDWR, 2019a). This lack of comprehensive monitoring and metering has made it difficult to enforce the implied restrictions on groundwater use during droughts. Irrespective of any restrictions, an overlying landowner cannot lose a correlative right through non-use, nor is he/she subject to temporary appropriation by other landowners (Ratliff, 2016).

With the introduction of the SGMA in 2014, another layer of groundwater regulation was added to the existing common law water rights framework (CDWR, 2018e). While the new legislation requires changes to the management and governance of groundwater resources, it does not amend the existing correlative rights system. To be more specific, the legislation explicitly states that neither the act nor the actions of the agencies that will assume responsibility over the management of groundwater resources shall 'determine or alter common law water rights or water rights priorities' (California Legislature, 2017).

This provision will likely complicate the implementation of the SGMA and hinder the achievement of its objectives. This is because the SGMA's objective – to achieve a sustainable use of California's groundwater resources within 20 years – will likely require curtailments in individual or cumulative groundwater extraction, which will be resisted by existing correlative rights holders, who may insist that their existing common law water rights be respected. The potential conflict between the common law water rights framework and the SGMA has yet to be fully explored. However, the inherent tension between these two systems could lead to additional legal challenges brought by correlative rights holders, who may be aggrieved by any changes to the existing groundwater management arrangements (as discussed further in later sections). Further, the recent court decision on the question of federal reserved rights to the Coachella Valley Groundwater Basin highlights that California has yet to resolve the long-standing issue of how to integrate federal reserved rights into its groundwater management and governance framework.

California's new regulatory approach to groundwater management

Before the SGMA, groundwater management was a voluntary process (CDWR, 2003) unless there was an adjudication or a special act. In 2014, these diverse and often voluntary groundwater management arrangements were replaced by a state-wide groundwater management and governance framework (California Legislature, 2014). This new framework is supported by a state-wide programme, California Statewide Groundwater Elevation Monitoring (CDWR, 2018b) and a programme to classify California's groundwater basins *Bulletin 118* (CDWR, 2018c).

Combined with these two resources, the SGMA establishes an unusual model of local–state groundwater governance, which requires local public agencies with water supply, water management and land use responsibilities to form groundwater sustainability agencies (GSAs) and to prepare groundwater sustainability plans (GSPs) for 127 designated groundwater basins and sub-basins in California that are classified as 'medium priority' or 'high priority' in Bulletin 118 (CDWR, 2016). In cases where a GSA cannot be formed or is unable to prepare a GSP, the state government has the authority to develop and implement an interim plan until a local GSA can assume the management functions of the groundwater basin or sub-basin (California Legislature, 2014). The SGMA requires all medium- and high-priority groundwater basins listed in Bulletin 118 to be managed sustainably within 20 years (CDWR, 2018e). For all remaining basins in California, GSPs are encouraged but are not mandatory under the SGMA (California Legislature, 2015a).

For the development of a GSP, a GSA is required to determine the local groundwater basin's sustainable yield, which is defined as the 'maximum quantity of water... that can be withdrawn annually from a groundwater basin without causing an undesirable result' (California Legislature, 2019). An undesirable result is any of the following (Langridge, 2017):

- Chronic lowering of groundwater levels resulting in a significant and unreasonable depletion of supply;
- Significant and unreasonable reduction of groundwater storage;
- Significant and unreasonable seawater intrusion;
- Significant and unreasonable degraded water quality;
- Significant and unreasonable land subsidence; and
- Depletion of interconnected surface waters that has significant and unreasonable adverse impacts.

Notably, the SGMA does not specify any quantitative groundwater usage targets or explicitly define key terms like 'overdraft' or 'water quality', putting the legislation in the category of performance-based rather than prescriptive regulation. The lack of prescriptive targets gives the GSAs greater flexibility in tailoring their future groundwater management approaches to the particular local conditions and permits more innovative solutions to groundwater governance. However, there are a number of practical challenges that need to be overcome.

Prior to the introduction of SGMA, approximately 2300 local public agencies were involved in California's groundwater management (Nelson R., 2012). As of 1 July 2017, 250 GSAs had formed to cover the 127 high- and medium-priority groundwater basins. This suggests that some groundwater basins are managed by more than one GSA (Milman, Galindo, & Blomquist, 2018). For the Coachella groundwater sub-basins, five agencies have applied to be GSAs: CVWD, Desert Water Agency, Mission Springs Water, Indio Water Authority and Coachella Water Authority (CDWR, 2019b). Of these five agencies, the CVWD and Desert Water Agency have the largest supply areas and encapsulate the entire Coachella Valley groundwater basin (CVWD, 2019).

While the consolidation of California's local public agencies and the formation of GSAs were challenging, the future co-management of groundwater basins by multiple GSAs will be particularly difficult and will require significant cooperation and coordination on highly technical issues such as water budgets and water resource assessments (Milman et al., 2018).

GSAs also retain their existing responsibilities as water supply, water management and land use authorities, raising concerns about the increased regulatory burden on these agencies and future resource constraints in the absence of additional state funding (Moran & Cravens, 2015). Resource constraints will likely become apparent when GSAs have to report on their GSP implementation progress, or when they have to conduct extensive outreach and consultation sessions with all beneficial users of their designated groundwater basin, including Native American tribes (CDWR, 2018d). As Native American tribes are considered sovereign entities, engagement with tribes can often be more challenging than standard stakeholder consultation, as tribes often have different expectations around adequate consultation processes in the context of their specific regulatory powers over trust lands (CDWR, 2017c). Tribes' status as domestic dependent nations also place them under federal law jurisdiction, with 'government-to-government' consultation requirements to discuss matters concerning tribal rights or resources (CDWR, 2017c). Future consultation on the new local–state groundwater management and governance framework may therefore require GSAs to adopt different consultation frameworks to address the diversity in stakeholder requirements. Also, it appears that while GSAs have a requirement to consult with Native American tribes if these are beneficial users of the designated groundwater basins, the SGMA does not provide a direct pathway for tribes to form a GSA (CDWR, 2017c).

Regulatory instruments and enforcement under the Sustainable Groundwater Management Act

The SGMA has vested significant regulatory powers in local public agencies when acting in their capacity as GSAs (Nelson & Perrone, 2016). Under the SGMA, GSAs are authorized to apply command-and-control mechanisms (e.g., penalties and sanctions for non-compliance with groundwater extraction limits); cap and trade market-based instruments (e.g., permits and approvals of groundwater allocation transfers), and communication tools (e.g., groundwater extraction data collection and well registration), as well as more innovative approaches to conservation management (e.g., groundwater recharge projects, water imports and voluntary fallowing) (Nelson & Perrone, 2016). Of these regulatory instruments, the communication tools will be critical for the GSAs to understand groundwater usage patterns, set appropriate water extraction levels, monitor groundwater extraction, and enforce compliance with the SGMA's sustainable yield requirement or other groundwater extraction limits.

The SGMA also empowers GSAs to directly regulate groundwater pumping by restricting or suspending well production, prohibiting new well construction and requiring measurement or metering of groundwater extraction (California Legislature, 2015b). While these restrictions may be necessary to achieve the SGMA's sustainable yield requirements or to follow any court orders relating to federal reserved rights, it is likely that these direct restrictions on groundwater extraction will be challenged by existing correlative rights holders, who see them as an infringement on their rights under the existing common law water rights framework. In anticipation of these legal challenges, it is questionable as to whether the GSAs are incentivized to apply these stricter regulatory instruments. Also, as GSAs retain their primary responsibilities as water supply authorities and will continue to derive revenue from water users (i.e., customers) via fees for their water delivery functions (Milman et al., 2018), the conflicting dual water supply and water conservation responsibilities might limit the GSAs' willingness to impose onerous and restrictive regulation (California Legislature, 2015c). Additional investigation into the incentives of GSAs to implement certain regulatory tools and their potential inherent conflict of interest should be considered to ensure that the overarching state policy goal of sustainable groundwater management is achieved.

Another related problem is that the SGMA provides little guidance on whether or when particular regulatory tools should be deployed. Under the theory of 'really responsive regulation' introduced by Baldwin and Black in 2008, regulators should account for a broad range of factors before deploying regulatory instruments, including the interaction effects of various instruments, the institutional setting and the behavioural responses of those impacted by the regulation. Yet the SGMA does not indicate if and whether these kinds of factors should be considered by GSAs in the use of their regulatory tools.

With respect to behavioural responses, for example, the SGMA appears to be silent on an appropriate strategy to deal with the heterogeneity of groundwater users, including Native American tribes. This is disappointing because an understanding of the values, norms and behavioural responses of groundwater users with respect to the new regulatory approach and the various regulatory instruments will be critical to steering groundwater users towards more sustainable groundwater extraction levels. Since Elinor Ostrom's (1999) famous work *Governing the Commons*, there has been widespread recognition that social norms are important in the context of collective-action problems like groundwater management. This point was also raised in other recent work (Castilla-Rho, Rojas, Anderson, Holley, & Mariethoz, 2017), where it was suggested that compliance with groundwater extraction limits is highly sensitive to social norms of groundwater users, and that nudging (rather than a more prescriptive or intrusive regulatory measure) could be sufficient to achieve a desired regulatory outcome (Castilla-Rho et al., 2017). This work could provide some useful insights for GSAs when deciding which regulatory instrument to deploy, because failure to design appropriate groundwater management approaches that address local conditions could have unintended consequences, including further depletion of groundwater resources or unjustified costs to local economies.

Lastly, while the assortment of regulatory instruments under the SGMA provides greater flexibility for GSAs to manage their designated groundwater basins, the SGMA provides little guidance on how the regulatory instruments could be deployed effectively in combination to achieve its sustainable yield requirement. This lack of guidance might create some challenges for smaller and less experienced GSAs, who may not have the technical expertise or resources to assess an appropriate regulatory instrument mix or to evaluate which regulatory tools may be incompatible with each other. The choice of an optimal regulatory instrument mix might be further complicated and constrained by the recent Ninth Circuit decision regarding federal reserved rights in the Coachella Valley, as will be argued in the discussion and recommendations section.

Federal reserved rights dispute and the Sustainable Groundwater Management Act

The Agua Caliente Band of Cahuilla Indians filed a lawsuit against the CVWD and the Desert Water Agency in 2013 to seek declaratory and injunctive relief concerning the following matters (Agua Caliente Band of Cahuilla Indians v Coachella Valley Water District & Desert Water Agency, 2013):

- That the court recognizes the tribe's federal reserved rights to groundwater underneath the Coachella Valley;
- That the court prevents the CVWD and the Desert Water Agency from withdrawing groundwater in amounts that would interfere with the tribe's rights; and
- That the court enjoins the two public water authorities from continuing to augment the groundwater basin's water supplies with imported water from the Colorado River.

At the commencement of the litigation, the parties agreed to trifurcate the case and commence the proceedings on the legal question of whether the tribe had federally reserved rights to groundwater in the Coachella Valley Groundwater Basin (Agua Caliente Band of Cahuilla Indians v Coachella Valley Water District & Desert Water Agency, 2013). Stage 2 of the litigation was scheduled to commence in the US District Court in January 2018 and concerns the questions of whether the tribe has rights to the Coachella Valley Groundwater Basin's porous space and a right to unimpaired water quality. It is anticipated that Stage 3 of the case, which concerns the quantification of the tribe's federal reserved rights, will follow after the resolution of the Stage 2 issues.

The origin of the federal reserved rights can be found in the US Supreme Court case *Winters v. United States* (1908), 207 U.S. 564. In that case, the court established the principle that when the US government sets aside public land for a particular purpose (such as an Indian reservation), it may reserve an implied right to water. In his judgement, Justice McKenna stated: 'The power of the Government to reserve the waters and exempt them from appropriation under the state law is not denied, and could not be. That the Government did reserve them we have decided, and for a use which would be necessarily continued through the years.' While the Supreme Court determined that the establishment of a reservation consequentially reserved surface water necessary to fulfil the federal government's purpose of the reservation, the court failed to define the purpose of a reservation or to explain how this principle may apply to a riparian water management system, like California's correlative rights framework.

In a subsequent US Supreme Court case, *Cappaert v United States* (1976), 426 U.S 128, 138, the court stated:

> When the Federal Government withdraws its land from the public domain and reserves it for a federal purpose, the Government, by implication, reserves appurtenant water then unappropriated to the extent needed to accomplish the purpose of the reservation. In so doing the United States acquires a reserved right in unappropriated water which vests on the date of the reservation and is superior to the rights of future appropriators.

The court's language in *Cappaert* illustrates that the court placed the federal reserved rights firmly in a prior-appropriation system of water management, effectively suggesting that if the water had not been reserved, then it would have been subject to prior appropriation under state law (Ratliff, 2016). The legal dispute in the Agua Caliente case differs from *Cappaert* in that it asked whether the federal reserved rights doctrine applies equally to a riparian water management system where a tribe already has access to surface water and groundwater.

It should be noted that the court in Cappaert had the opportunity to address the issue of federal reserved right to groundwater but chose not to resolve the issue fully. The court did state that 'since the implied-reservation-of-water-doctrine is based on the necessity of water for the purpose of the Federal reservation, we hold that the United States can protect its water from subsequent diversion, whether the diversion is of surface or groundwater'.

In comparison, the court in In re Gen. Adjudication of all Rights to Use Water in the Big Horn river Sys. (1988), 753 P.2d 76, 99, refused to extend the federal reserved rights doctrine to groundwater (Petition for Writ of Certiorari, 2017). In its judgement, the court concluded that while the Indian reservation has a federal reserved right to surface water, the 'reserved water doctrine does not extend to groundwater' (In re Gen. Adjudication of all Rights to Use Water in the Big Horn river Sys., 1988).

Finally, the Arizona Supreme Court is the only court that has explicitly found federal reserved rights to groundwater in a riparian water management system, thereby producing a ruling that is the most closely related to the Agua Caliente case. Relying on the reasoning provided in *Cappaert*, the Arizona Supreme court held in In re Gen. Adjudication of All Rights to Use Water in Gila River Sys. & Source (1999), 989 P.2d 739, 751 n.6, that federal law does not differentiate between surface water and groundwater 'when identifying the water to be protected'. Thus, the court held that 'the significant question for the purpose of the [federal] reserved rights doctrine is not whether the water runs above or below the ground but whether it is necessary to accomplish the purpose of the reservation'. As the quote indicates, the Arizona

Supreme Court extended federal reserved rights to groundwater but applied two restrictions. First, the court held that federal reserved rights should only be granted to fulfil the purpose of the reservation and no more; and second, the holder of a federal reserved right should enjoy greater protection from groundwater pumping than other water rights holders, 'to the extent that greater protection may be necessary to maintain sufficient water to accomplish the purpose of the reservation'. While the Arizona Supreme Court gave no reason for these conditional provisions, it can be stipulated that the court offered a reasonable accommodation to the fact that federal reserved rights do not easily fit in a riparian water management system.

In the Agua Caliente case, the US District Court for the Eastern District of California (March 2015) and the US Court of Appeals for the Ninth Circuit (March 2017) held that the tribe possessed a federal reserved right to the Coachella Valley Groundwater Basin (Agua Caliente Band of Cahuilla Indians v Coachella Valley Water District, 2017b). In its judgement, the Ninth Circuit did not elaborate on why federal reserved rights to groundwater were granted in the context of the tribe's other water rights, including its correlative right to the Coachella Valley Groundwater Basin (Joint Brief of Appellants Coachella Valley Water District, et. al. and Desert Water Agency, et. al., 2015). The court simply stated that the tribe's correlative right under the California state law was 'irrelevant for the analysis' (Agua Caliente Band of Cahuilla Indians v Coachella Valley Water District, 2017b). The Ninth Circuit has thereby provided the broadest application of federal reserved rights to riparian groundwater management systems to date. The judgement will likely generate future tensions between tribal and non-tribal groundwater users around future access to the Coachella Valley Groundwater Basin.

As a consequence, the CVWD and the Desert Water Agency filed a Petition for Writ of Certiorari (2017) with the US Supreme Court on the question of 'whether, when and to what extent the Federal reserved right doctrine applies to groundwater'. This appeal by the two water agencies was rejected by the Supreme Court in November 2017.

Conflict with the Sustainable Groundwater Management Act

The decision of the Ninth Circuit can be considered a significant victory for the Agua Caliente Band of Cahuilla Indians. Although the quantification of the tribe's federal reserved rights has yet to be resolved, the decision of the Ninth Circuit has given the tribe greater leverage in having its concerns over the future use and management of groundwater resources in the Coachella Valley heard and acted on, and in securing access for tribal members (although thc water agencies have never restricted water access by the tribe). The court decision has also triggered a rethink about future stakeholder engagement, as the existence of federal reserved rights will give the tribe a special status in influencing future decisions on the management and governance of the Coachella Valley's Groundwater Basin.

However, the decision of the Ninth Circuit has also illustrated the long-standing issues with integrating federal reserved rights into the Californian water management and governance framework. Given the characteristics of federal reserved rights, the decision will change the future groundwater management arrangements in the

Coachella Valley, impact the implementation of SGMA across California and set an important precedent for future tribal litigations in the western US.

On a valley-specific level, the Ninth Circuit's decision will likely complicate the future management of the Coachella Valley Groundwater Basin because federal reserved rights fall outside the regulatory enforcement powers of the state or the local public agencies (Blumm, 2016). As the tribe would not be obliged to comply with any of the GSA's future extraction limits, or to adhere to any groundwater sharing principles, this could make it extremely difficult for the GSAs to develop an effective groundwater management plan. It would also make it more difficult for the GSAs to introduce some new regulatory instruments, because they may infringe on the tribe's federal reserved rights. In effect, the GSAs have to manage the Coachella Valley Groundwater Basin in the presence of a potentially unconstrained and unmonitored groundwater user with as yet indeterminate rights to the groundwater resource.

Furthermore, if the GSAs are restricted in their ability to monitor the tribes' groundwater use, it may not be possible to determine appropriate groundwater allocations for all non-tribal users (unless the quantity of the tribe's federal reserved rights is known). If the GSAs have to estimate the tribe's groundwater use, this could lead to over-extraction or under-utilization of the groundwater resource. Both could have either a long-term environmental impact or a socio-economic impact, as non-tribal groundwater users may have made investment decisions on the basis of previous groundwater access arrangements.

The existence of the federal reserved water rights to groundwater could also have an impact on other groundwater users in dry years because water access may be curtailed due to the priority of federal reserved groundwater rights, depending on the quantification of the tribe's right (Agua Caliente Band of Cahuilla Indians v Coachella Vally Water District & Desert Water Agency, 2017c). This might be particularly contentious in times of drought and could spark further legal challenges by other groundwater users who are aggrieved by the introduction of a priority system that does not respect the existing correlative rights framework.

At a state level, the recognition of federal reserved rights to groundwater in the Coachella Valley could have wider implications for the management of groundwater resources in California, because every tribal reservation in California could have, by extension of the Ninth Circuit's decision, a federal reserved right to groundwater basins, because every reservation needs water to exist. This fact could constrain other GSAs in their attempts to implement the new groundwater management and governance framework.

In an even broader perspective, the decision in the Agua Caliente case has an impact beyond California. It is the current precedent throughout the Ninth Circuit, and thus, any future cases in California or the rest of the Ninth Circuit that present the same issue as the ones presented in the Agua Caliente case should be decided in the same way by any federal courts in the Ninth Circuit. This is important, as a vast number of reservations are outside California, including Nevada and Arizona. These states are also part of the Ninth Circuit and have individual state-based legislation and regulation on groundwater management (Languages, 2015). As an *amicus curiae* brief by the states of Nevada, Arizona, et al. highlighted (Statement of the States of Nevada et al. as Amicus Curiae, 2017), there is concern that an extension of federal reserved rights to groundwater could mean that state governments are restricted in the management of their own groundwater resources, or they could be subject to extensive legal challenges by other Native American tribes on the basis of the Agua Caliente case (Statement of the States of Nevada et al. as Amicus Curiae, 2017). It will also be interesting to see how courts outside the Ninth Circuit, but with similar geographical and climatic conditions as well as extensive tribal reservations, will treat the Agua Caliente decision when faced with a similar case. While the Ninth Circuit's decision will not be binding on equal courts in other states, it is expected to be given some consideration in future rulings on federal reserved rights.

In all, it appears that the decision will increase the complexity of future groundwater management in California and might necessitate a reconsideration of the existing correlative rights system or lead to further amendments to the SGMA.

Discussion and recommendations

The legal dispute between the two water agencies and the Agua Caliente Band of Cahuilla Indians has provided an opportunity to revisit the existing conflicts between federal reserved rights and California's previous and new approaches to the groundwater management and governance framework. It has also provided an opportunity to consider the role of Native American tribes in California's water resource management.

Irrespective of the outcome of the two remaining stages of the litigation, the decision in the Agua Caliente case could disrupt the implementation of the new groundwater management framework in California, because any (yet to be determined) federal reserved rights would fall outside the jurisdiction of the SGMA. The effectiveness and broad-scale application of the new legislation could then be significantly diminished. However, whether the decision will ultimately impact California's new state-wide groundwater management framework will depend on how the decision manifests itself in a change to the water management practices in California and in the western US more broadly. In other cases involving water rights for Native American tribes, the existence of a right to water resources has not necessarily translated into tangible benefits for these tribes (Fleck, 2016).

While it is too early to assess whether the decision will set off a cascade of other regulatory amendments, the litigation between the Agua Caliente Band of Cahuilla Indians and the two water agencies in Southern California has illustrated that further work on the integration of federal reserved rights into the existing water management framework is necessary to successfully implement a state-wide groundwater management and governance framework in California. The following four aspects should be considered ahead of a resolution to Stages 2 and 3 of the litigation to mitigate further conflict between Native American tribes and California's water agencies.

First, the litigation between the tribe and the water agencies demonstrates that groundwater management in California will continue to be a delicate balancing act between the state's common law water rights system, federal reserved rights and the new groundwater management approach under the SGMA. Prior to the full implementation of the SGMA, further consideration should be given as to whether there is an inherent conflict between California's common law water rights framework, federal reserved rights and the new groundwater management framework, and if so, whether the SGMA and its regulatory instruments could be refined to mitigate further legal challenges between competing water rights holders. As highlighted earlier, further litigation is likely if the GSAs impose access restrictions for correlative groundwater rights holders as a result of water scarcity and the priority of federal reserved rights. Thus, state authorities should consider how an equitable and efficient allocation system between different water rights holders can be established that will not result in an ongoing legal challenge between current tribal and non-tribal groundwater users.

Second, it should be assessed whether any of the new authorized regulatory instruments are incompatible with federal reserved rights, and how the GSPs or other negotiated arrangements could address this incompatibility. It is evident that the SGMA considered federal and tribal law issues in the drafting of the legislation (California Legislature, 2015d). As outlined in the California Water Code, in the 'adjudication of groundwater rights or in the management of groundwater basins or sub-basins by a GSA, Federal reserved water rights to groundwater shall be respected in full' (California Legislature, 2015d). In case of a conflict between federal and state law, federal law shall prevail. While this suggests that the recently granted federal reserved rights to groundwater unequivocally rank higher in priority than state-based correlative rights, it is yet unclear how this clause will operate in practice, particularly without a wide-scale groundwater monitoring network that can assess individual water users' groundwater extractions and impose penalties in case of over-extraction. Also, in future drought scenarios, it is yet unclear how any groundwater access restrictions would be distributed among the various water rights holders. Whether such issues will arise in the Coachella Valley will depend on the next stages of the litigation and the following protracted adjudication process, which could take decades.

Third, the legal dispute between the tribe and the water agencies made it clear that an improved dialogue between Native American tribes, water agencies and non-tribal water users is critical to understand each group's needs and to assess whether a negotiated outcome is feasible. Given that tribes have a unique status as a domestic dependent nation with sovereign authority, culturally respectful engagement will be an important first step to working collaboratively with tribes and to create an open forum where information can be shared, and trust can be built (Lucero, DuPraw, Di Vittorio, & Ceppos, 2017). This level of trust and sharing of information has clearly not existed between the tribe and the two water agencies. As highlighted in Agua Caliente's response to Stage 2 issues, the tribe has complained about the water agencies' lack of engagement: 'the defendants continue to exclude Agua Caliente from meaningful, substantive participation in groundwater management that affects Federal reserved rights' (Agua Caliente Band of Cahuilla Indians v Coachella Valley Water District & Desert Water Agency, 2017b). Although the two agencies have established a tribal GSA working group to share information and encourage discussion, further engagement through these advisory committees or appointments of tribal members to the GSA boards will be critical to create mechanisms for tribes to more actively engage in groundwater management.

Fourth, depending on the results of the case, the SGMA may still provide an avenue for the GSAs to accommodate both the federal reserved rights and the existing correlative rights of other landholders. If the water agencies can establish annual water allocation announcements and transfer systems within the GSA's designated groundwater areas (CDWR, 2016), it could provide all tribal and non-tribal groundwater users with another management tool to deal with their own unique water needs, including an option to trade any unused allocations by either group within a particular location or time period, creating income for the tribe and enabling other non-tribal groundwater users to access unused groundwater allocations. While the establishment of an effective groundwater market will require comprehensive data on resource availability, usage patterns and recharge behaviour, as well as effective groundwater extraction monitoring, a local, decentralized groundwater market could be an important tool for allowing groundwater users in the Coachella Valley to manage the shift in water rights in the Coachella Valley.

With the final court decision on the other phases still outstanding, the feasibility and practical application of any of these proposed approaches have yet to be tested. But it is likely that an opportunity to try any of these possible tools may only arise after the adjudication process has been completed.

Conclusion

The legal dispute in the Coachella Valley has highlighted the complexities and uncertainties surrounding future groundwater management arrangements in California. The Ninth Circuit's decision on the Agua Caliente Band of Cahuilla Indian's federal reserved rights to groundwater has left many questions unanswered, including whether non-tribal water rights holders will be able to legally challenge the decision of the court or the new water management regulation under the SGMA.

It is clear, however, that ongoing engagement and consultation between the GSAs, groundwater users and tribes will be critical for a successful implementation of the new local–state groundwater management and governance framework without further legal challenges – a process that is time-consuming and resource-intensive and undermines the consistent application of a groundwater management framework in California.

Further assessment of the institutional setup of GSAs and their regulatory tools (including monitoring and enforcement mechanisms) will also be essential to refine regulatory strategies and instruments under the SGMA and to ensure compliance with the new framework. Also, an ongoing assessment of how individual groundwater users and GSAs respond to the various regulatory tools will be important to determine the effectiveness of the legislation.

Finally, competition for scarce and variable water supplies (groundwater and surface water) will require well-resourced institutions that can establish shared rules and norms around how to allocate water across competing users, and that are able to manage conflict and to adapt in the face of future climate stresses and regulatory changes (Garrick, 2015). As the future funding of GSAs is uncertain, significant risk remains as to whether the policy goal of sustainable groundwater in the next 20 years is achievable in California under current conditions.

Acknowledgments

I thank Dr Belinda Reeve for her help and guidance in writing this article. Her ongoing encouragement and detailed advice have been invaluable.

Disclosure statement

No potential conflict of interest was reported by the author.

References

Agua Caliente Band of Cahuilla Indians. (2015). News & events. Retrieved from http://www.aguacaliente.org/content/News%20and%20Events/?showStoryID=84&siteSearch=water&highlight=water

Agua Caliente Band of Cahuilla Indians. (2018a). Cultural history. Retrieved from http://www.aguacaliente.org/content/History%20and%20Culture/

Agua Caliente Band of Cahuilla Indians. (2018b). Frequently asked questions regarding the Agua Caliente water lawsuit. Retrieved from www.aguacaliente.org.au:http://www.aguacaliente.org/downloads/AguaCaliente_WaterRights_FAQ.pdf

Agua Caliente Band of Cahuilla Indians v Coachella Valley Water District, No 15-55895 (Ninth Circuit Court March 2017b).

Agua Caliente Band of Cahuilla Indians v Coachella Valley Water District & Desert Water Agency, ED CV 13-00883 JGB SPx (United States District Court, Central District of California, Eastern Division, May 14, 2013).

Agua Caliente Band of Cahuilla Indians v Coachella Valley Water District, et al. and Desert Water Agency et al., 15-55896: Joint Brief of Appellants Coachella Valley Water, et al. and Desert Water Agency (Ninth Circuit Court, March 2017a).

Baldwin, R., & Black, J. (2008). Really responsive regulation. *Modern Law Review*, *71*, 59–94.

Blumm, M. C. (2016). Federal reserved water rights as a rule of law. *Idaho Law Review*, *52*, 369.

California Department of Water Resources. (2003). Groundwater management in California. Retrieved from www.water.ca.gov:https://water.ca.gov/LegacyFiles/pubs/groundwater/bulletin_118/california's_groundwater__bulletin_118_-_update_2003_/bulletin118-chapter2.pdf

California Department of Water Resources. (2015). *California's groundwater update 2013*. Sacramento: Author.

California Department of Water Resources. (2016). Bulletin 118 interim update 2016. Retrieved from www.water.ca.gov:https://water.ca.gov/LegacyFiles/groundwater/bulletin118/docs/Bulletin_118_Interim_Update_2016.pdf

California Department of Water Resources. (2017a). Maps and report: groundwater level change (2011 to fall 2016). Retrieved from www.water.ca.gov:https://www.water.ca.gov/LegacyFiles/groundwater/maps_and_reports/MAPS_CHANGE/DOTMAP_F2016-F2011.pdf

California Department of Water Resources. (2017b). Appendix 1-13: CASGEM Compliance. Retrieved from https://www.water.ca.gov/LegacyFiles/irwm/grants/docs/Archives/Prop84/Submitted_Applications/P84_2014Drought/Indio%20Water%20Authority%20(201498760026)/Attachment%201%20-%20Att1_DG_Eligible_1of12.pdf

California Department of Water Resources. (2017c). *Guidance document for the sustainable management with tribal governments*. Sacramento: Author.

California Department of Water Resources. (2018a). Groundwater. Retrieved from https://www.water.ca.gov/Water-Basics/Groundwater

California Department of Water Resources. (2018b). Groundwater monitoring (CASGEM). Retrieved from https://www.water.ca.gov/Programs/Groundwater-Management/Groundwater-Elevation-Monitoring–CASGEM

California Department of Water Resources. (2018c). Bulletin 118: Groundwater Information Center. Retrieved from https://www.water.ca.gov/Programs/Groundwater-Management/Bulletin-118

California Department of Water Resources. (2018d). Guidance document for the sustainable management of groundwater: engagement with tribal governments. Retrieved from https://water.ca.gov/-/media/DWR-Website/Web-Pages/Programs/Groundwater-Management/Sustainable-Groundwater-Management/Best-Management-Practices-and-Guidance-

Documents/Files/Guidance-Document-for-Sustainable-Management-of-Groundwater—Engagement-with-Trib

California Department of Water Resources. (2018e). SGMA groundwater management. Retrieved from https://www.water.ca.gov/Programs/Groundwater-Management/SGMA-Groundwater-Management

California Department of Water Resources. (2019a). Groundwater update 2013a. Retrieved from https://water.ca.gov/-/media/DWR-Website/Web-Pages/Programs/California-Water-Plan/Docs/Update2013/GroundwaterUpdate/Californias-Groundwater-Update-2013–A-Compilation-of-Enhanced-Content-for-California-Water-Plan-Upd.pdf

California Department of Water Resources. (2019b). Groundwater sustainability agencies. Retrieved from https://water.ca.gov/-/media/DWR-Website/Web-Pages/Programs/Groundwater-Management/Sustainable-Groundwater-Management/Groundwater-Sustainability-Agencies/Files/Comparison-of-GSAs-Formed-Using-2003-Basin-Boundaries-vs-2016-Basin-Boundaries.pdf

California Environmental Protection Agency. (2018). State water resources control board. Retrieved from Water Rights Process: http://www.waterboards.ca.gov/waterrights/board_info/water_rights_process.shtml#law

California Legislature. (2014). *Assembly Bill No 1739; Senate Bill No 1168; Senate Bill No 1319. Groundwater Management.* Sacramento: California Legislative Information.

California Legislature. (2015a). *California water code. Division 6, Part 2.74, Chapter 1, s.10720.7 (b).* Sacramento, CA: Author.

California Legislature. (2015b). *California water code. Division 6, Part 2.74, Chapter 5 s 10725.8 and 10826.4(a).* Sacramento, CA: Author.

California Legislature. (2015c). *California water code. Division 6, Part 2.74, Chapter 1, s10720.8.* Sacramento: California Legislature.

California Legislature. (2016). *California Water Code Division 6, Part 2.74, Chapter 2, s1721(n).* Sacramento: California Legislature.

California Legislature. (2015d). *California water code. Division 6, Part 2.74, Chapter 1, s10720.3.* Sacramento, CA: Author.

California Legislature. (2017). *California Water Code. 10720.5(b), 10726.4(a)(2), 10726.8(b), 10735.8(i), 10720.5(a) and 10736.4.* Sacramento: California Legislature.

California Legislature. (2019). California legislative information. Retrieved from http://leginfo.legislature.ca.gov/faces/codes_displaySection.xhtml?sectionNum=10721.&lawCode=WAT

California State Water Resource Control Board. (2019, February 16). Groundwater recordation program. Retrieved from https://www.waterboards.ca.gov/waterrights/water_issues/programs/groundwater_recordation/

Cappaert v United States, 426 U.S 128,138 (US Supreme Court 1976).

Castilla-Rho, J. C., Rojas, R., Anderson, M., Holley, C., & Mariethoz, G. (2017). Social tipping points in global groundwater management. *Nature Human Behaviour, 1*, 640–647.

Coachella Valley Water District. (2017). Decades of efforts pay off in successful groundwater management. Retrieved from http://www.cvwd.org/DocumentCenter/View/811/Successful-Groundwater-Management-PDF

Coachella Valley Water District. (2018a). Agua Caliente fact sheet updated. Retrieved from http://www.cvwd.org/DocumentCenter/View/3670/Fact-Sheet-Tribal-Litigation-February-2018-PDF-

Coachella Valley Water District. (2018b). About CVWD. Retrieved from http://www.cvwd.org/288/About-CVWD

Coachella Valley Water District. (2018c). *District Where does my water come from?* Retrieved from http://www.cvwd.org/154/Where-does-my-water-come-from

Coachella Valley Water District. (2019). Stakeholder meeting presentation. Retrieved from http://www.cvwd.org/DocumentCenter/View/3236/September-7-2016-SGMA-Stakeholder-Meeting-Presentation?bidId=

Farr, T., Jones, C., & Lui, Z. (2016). *Progress report: Subsidence in California March 2015.* Pasadena, CA: California Institute of Technology, Jet Propulsion Laboratory.

Fleck, J. (2016). *Water is for fighting over.* Washington: Island Press.

Garrick, D. (2015). *Water allocation in river under pressure: water trading, transaction costs and transboundary governance in the Western US and Australia.* Cheltenham, UK: Edward Elgar.

In re Gen. Adjudication of all Rights to Use Water in Gila River Sys. & Sources, 989 P2d 739,751 n.7 (Supreme Court of Arizona 1999).

In re Gen. Adjudication of all Rights to Use Water in the Big Horn river Sys., 753 P.2d. 76, 99 (Supreme Court of Wyoming, May 6, 1988).

Joint Brief of Appellants Coachella Valley Water District, et. al. and Desert Water Agency, et. al., 15-55896 (US Court of Appeals For the Ninth Circuit 2015).

Langridge, R. (2017). Accounting for climate change and drought in implementing sustainable groundwater management. *Water Resource Management, 31*(2).

Languages, A. I. (2015). Maps of United States Indians by States. Retrieved from http://www.native-languages.org/states.html

Lucero, S., DuPraw, M., Di Vittorio, S., & Ceppos, D. (2017, September). Engaging with tribes: Implementing California's Sustainable Groundwater Management Act: Requirements, rights, options and opportunities as concerns tribes. *Water Report*, p. 163.

Milman, A., Galindo, L., & Blomquist, W. A. (2018). Establishment of agencies for local groundwater governance under California's Sustainable Groundwater Management Act. *Water Alternatives, 11*(3), 474.

Moran, T., & Cravens, A. (2015). *California's Sustainable Groundwater Management Act of 2013, Recommendations for preventing and resolving groundwater conflicts.* Stanford, CA: Stanford Law School.

Nelson, R. (2012). Assessing local planning to control groundwater depletion: California as a microcosm of global issues. *Water Resource Research, 48,*1–14.

Nelson, R., & Perrone, D. (2016). Local groundwater withdrawal permitting laws in the South-Western U.S: California in comparative context. *Groundwater, 6*, 747–753.

Ostrom, E. (1999). *Governing the commons: The evolution of institutions for collective action.* Cambridge: Cambridge University Press.

Owen, D. (2017, November). Cooperative subfederalism. Retrieved from https://papers.ssrn.com/sol3/papers.cfm?abstract_id=3071907

Petition for Writ of Certiorari, 17-40, –42 (Supreme Court of the United States, July 5, 2017).

Ratliff, D. (2016). A proper seat at the table: Affirming a broad winters right to groundwater. *University of Denver Water Law Review, 19*, 239–260.

Statement of the States of Nevada et al. as Amicus Curiae, 17-40, –42 (Supreme Court of the US, August 8, 2017).

US Geological Survey. (2017). California water science centre. Retrieved from Land Subsidence in the Coachella Valley: https://ca.water.usgs.gov/land_subsidence/coachella-valley-subsidence.html

US Geological Survey. (2018). Drought impacts. Retrieved from California Water Science Center: https://ca.water.usgs.gov/data/drought/drought-impact.html

Winters v United States, 207 U.S 564 (US Supreme Court 1908).

Part IV

Groundwater for irrigation

Learning from the past to build the future governance of groundwater use in agriculture

Olivier Petit, Aurélien Dumont, Stéphanie Leyronas, Quentin Ballin, Sami Bouarfa, Nicolas Faysse, Marcel Kuper, François Molle, Charlotte Alcazar, Emmanuel Durand, Ridha Ghoudi, Aline Hubert, Selin Le Visage, Imane Messaoudi, Marielle Montginoul, Seyni Ndao, Audrey Richard Ferroudji, Jean-Daniel Rinaudo, Julie Trottier, Olivia Aubriot, Mohamed Elloumi, Marc Boisson, Rhoda Fofack-Garcia, Frédéric Maurel, Dominique Rojat, Bruno Romagny and Emmanuelle Salgues

ABSTRACT

The use of groundwater is increasing worldwide, particularly in agriculture. This leads to pumping races (from which the poorest farmers are often excluded), environmental disasters and the degradation of groundwater quality. Based on discussions between scientists and operational experts in two workshops held in 2018 and 2020, this paper, after taking stock of the dynamics and motivations of groundwater use in agriculture, reviews the solutions most commonly proposed, in particular by public authorities, to regulate the access to and use of this resource and to limit its over-exploitation. These (often optimistic) solutions generally combine regulatory or economic instruments, or indirect measures linking water to other issues, and mechanisms based on the participation of all users. However, they rarely question the intensive agricultural systems driving groundwater demand and fail to recognize the multifunctional nature of groundwater. To overcome these hurdles, we draw up possible ways forward for policymakers and resource users to develop negotiated solutions.

Introduction

Groundwater use for agriculture, which soared in the second half of the 20th century (Siebert et al., 2010), often competes with other uses. From an environmental viewpoint, groundwater feeds springs and rivers and helps to preserve wetlands, which are reservoirs of biodiversity. Furthermore, it is massively and increasingly used to supply agriculture, industry and drinking water for cities. Thus, its governance demands serious consideration to protect groundwater and establish pathways for its sustainable use.

This paper contributes to the debate on the preservation of groundwater, focusing on agricultural use, since a growing proportion of the world's irrigated land area – currently at around 40% – relies on this resource (FAO, 2016). Furthermore, increased groundwater use is often linked to the intensification of agriculture, which explains how agriculture can account for around 70% of global groundwater abstraction (Siebert et al., 2010), making it a strategic resource.

The dynamics of agricultural groundwater use are driven by the economic return it provides along with comparatively easy and secure access to water resources. There is a trade-off between these benefits and the current and future consequences of depleting flows and reserves, with the added complexity of the invisibility of groundwater sources. This can result in the loss of groundwater-dependent ecosystems or farmers engaging in a pumping race that excludes the poorest among them, thereby deepening economic and social inequalities (Ameur et al., 2017; Aubriot, 2013; Hoogesteger & Wester, 2015; Jouili et al., 2013; Mukherji, 2006).

The classical solutions proposed to control and limit groundwater abstraction at the institutional level generally combine: (1) regulatory instruments (permits, bans, quotas, zoning and well closures), economic instruments (fees and subsidies) or indirect measures linking water to other issues (energy and food security) and (2) participatory mechanisms involving all water users through community-based management schemes or other contractual arrangements (Molle & Closas, 2020a, 2020b; Villholth et al., 2018). Such solutions tend to face operational, financial, social, cultural and often political hurdles. More importantly, they rarely question the intensive agricultural systems that underlie the growing demand for groundwater or recognize the multifunctional nature of the resource. It would seem, therefore, that these solutions are mired in public policy contradictions by, on the one hand, aiming to reduce groundwater pumping to limit the environmental and social consequences of intensive use, while, on the other, maintaining the current level of withdrawal to 'buy' social peace, and even encouraging the development of new sources of water supply to strengthen the capitalistic and entrepreneurial modes of agriculture.

Academic work on groundwater governance is now well developed, and the topic has seen a revival of interest in the past few years with the publication of numerous studies and review articles (Closas & Villholth, 2020; Megdal et al., 2015; Molle et al., 2018; Rinaudo et al., 2020; Rouillard et al., 2021; Villholth et al., 2018). Groundwater governance is understood in this paper as 'the overarching framework of groundwater-use laws, regulations and customs, as well as the processes of engaging the public sector, the private sector, and civil society' (Megdal et al., 2015, p. 2). What is often missing from the academic literature, however, is a dialogue between scientists, stakeholders and decision-makers that would lead to credible options for future action and policy recommendations. The purpose of this paper is threefold: (1) to present the drivers of the current groundwater crisis, (2) to review the solutions generally proposed; and (3) to discuss possible ways forward.

The analyses that follow build on discussions between scientists and operational experts working in the field of groundwater governance brought together by the Agence Française de Développement (AFD – French Development Agency) and the Comité scientifique et technique pour l'eau agricole (COSTEA – Scientific and Technical Committee for Water in Agriculture). Two workshops took place – in November 2018 and January 2020 – to evaluate the various experiences of groundwater governance in agriculture and find ways to improve them. The workshops were built on a sharing of experiences on the basis of case studies presented during the first workshop, followed by a comparison and discussion of the observable trends, the limitations of the current modes of governance and the challenges faced in overcoming them. The second workshop was more specifically spent designing potential ways forward, on the basis of thematic groups and multiple points of view, focusing on the collective dimension of groundwater governance and going beyond the specificities of each case study. They resulted in the publication of a policy paper on which this article is largely based (Dumont et al., 2021).

The researchers, practitioners and experts drew on their knowledge of context-specific arrangements during the workshops, and these points of reference are important in illustrating our argument. The findings of this working group were also discussed in a webinar in March 2021, which allowed us to run our diagnosis and progress paths by a much larger audience.

In this article we focus on the quantitative aspects of groundwater (over)abstraction and only address qualitative issues (e.g., pollution and salinization) in terms of the former. Neither do we look at transboundary aquifers as these are more related to international relations. Selected case studies are based on the authors' fieldwork, meaning identified cases of over-exploitation, as in North America, Australia and China, for example, are not discussed as such.

The remainder of the paper is structured as follows. Following an overview of the use of groundwater for agriculture around the world we highlight several often-underemphasized issues to stress the actors' motivations in increasing pumping. These include a lack of knowledge of the hydrogeological dynamics and social aspects, and political inertia in the institutions responsible for groundwater governance. We then analyse some classical solutions that have been proposed and already largely implemented, particularly by public authorities, to regulate the access to and use of this resource, while highlighting their shortcomings. We then delineate some ways forward that deal with the collective aspects of groundwater governance while reconsidering the modalities of knowledge production and sharing in an attempt to grasp the socio-economic and environmental dimensions. The suggestions are intended to overcome the deadlock faced by policies focused on simply increasing the supply of water by advocating for a vision of groundwater that acknowledges and preserves its multifunctional nature.

The dynamics of groundwater use in agriculture: from economic motivations to institutional inertia

In this section, in addition to taking stock of the situation regarding groundwater use in agriculture, we also consider certain aspects sometimes underemphasized in the literature when looking at the inherent complexity of the dynamics of aquifers and their use in agriculture.

The social, economic and environmental consequences of intensive groundwater use

The pressure on groundwater is increasing constantly, in both quantitative (withdrawals) and qualitative terms (pollution and salinization), and there is a risk that this will increase with climate change, especially in semi-arid and arid areas (Taylor et al., 2013).

In many parts of the world, in the North and South, groundwater use has allowed farmers to secure and increase their income. Indeed, some regions owe the majority of their development to groundwater. For example, the 'green revolution' in India, which reduced famine, was largely enabled by the intensive use of groundwater combined with fertilizers (Aubriot, 2006). At the global scale, food security depends on international trade in agricultural produce, a significant proportion of which comes from groundwater-based irrigation, including crops that are also widely produced under rain-fed conditions, such as cereals (Dalin et al., 2017).

While many regions already use groundwater intensively, elsewhere (particularly in Sub-Saharan Africa) an inability to invest in the development potential of the resource, whether for economic, political or institutional reasons, means it is yet to be exhausted (Cobbing & Hiller, 2019).

The intensive use of groundwater has environmental consequences that are well identified but not always recognized or documented. The most obvious are the lowering of water tables, the degradation of the resource (pollution or marine intrusion into coastal aquifers) and land subsidence issues. Local groundwater withdrawals also have impacts at other scales on groundwater uses and resources: water that would have naturally discharged into springs, rivers or wetlands is instead captured, modifying the water exchanges between the surface and underground. This is particularly problematic in periods of drought, when groundwater naturally has a regulating effect by supporting rivers at low flow or wetlands connected to water tables. With the prospect of climate change, these issues are likely to become increasingly important and it is becoming urgent to prevent groundwater over-exploitation where the dynamics of intensive use can still be reversed.

Aquifers can also be the host or vector of various water-quality problems. The natural quality of groundwater may be unsuitable for its various uses. Sometimes, exploiting groundwater, which is itself of adequate quality, can dislodge contaminants naturally present in a geological layer in contact with the aquifer, such as arsenic. The inadequate management of boreholes (in terms of their design or maintenance), linking water tables of different qualities, has also been widely documented (Santi et al., 2006). Furthermore, qualitative aspects are paramount when active aquifer management techniques are being considered, such as managed artificial recharge using treated wastewater or surface water. Finally, groundwater pollution should be taken into consideration, whether diffuse (nitrates and pesticides), accidental or due to spatially localized activities, and whether or not they depend on groundwater use. Irrigating with low-quality water, combined with poor drainage, can raise the level of the water table and thus its salt content, leading to soil salinization.

All these quantitative and qualitative phenomena generate cascades of consequences for which there is no known short-term remedy, even if, for example, the source of pollution is eliminated. They unfold across various spatial and temporal scales, affecting not only those using the aquifer but also downstream surface water users or the ecosystems.

Environmental impacts tend not to be adequately recognized in processes of water resource allocation. De Graaf et al. (2019) estimate that by 2050, in 42–79% of basins worldwide, excessive groundwater abstraction will disrupt ecological flows (defined as the quantity, seasonality and quality of watercourse flows required to sustain freshwater and estuarine ecosystems, as well as the needs and well-being of the people who depend on them). They point out that the alteration of surface flows may prove to be far more problematic than the loss of aquifer reserves and increased pumping costs.

Intensive abstraction also has economic and social impacts, particularly in terms of growing inequalities. Irrigators who cannot afford to invest in deeper wells, or pay the higher cost of water abstraction caused by lower groundwater levels, can become excluded from access to the resource (Ameur et al., 2017; Mukherji, 2006). In the southern Indian state of Tamil Nadu power relations have been observed between farmers considering themselves as 'water masters', who have the means to drill a borehole, and those who depend on the water made available to them by these water masters in exchange for money or a portion of their own harvest (Aubriot, 2013). In Tunisia the implementation of a neoliberal policy in the mid-1990s favoured private investment, and the conditions of access to land and water resources profoundly changed, weakening those small-scale family farms that are unable to compete and become marginalized, as in the case of Sidi Bouzid (Fautras, 2021; Jouili et al., 2013).

Determinants and motivations of intensive groundwater use: from local actors to national policies

Groundwater exploitation for agriculture is driven by multiple factors at different levels that add up to dictate the intensity of abstraction. Here we discuss individual dynamics (in relation to networks of local actors), national agricultural and energy policies and, more broadly, macroeconomic factors (such as agricultural markets).

For farmers, groundwater abstraction provides a solution to restricted or insufficient access to surface water. It allows them to avoid the limitations of collectively managed surface water (community-based or public irrigation schemes), such as having to take turns in access or choose their crop based on the availability of water. It can arise from individual initiative or collective efforts to develop new territories, which, often promoted by the state, have fuelled agricultural intensification and expansion in arid or even desert areas, such as the Sahara in Algeria (Amichi et al., 2020).

The level of abstraction is closely tied to drilling techniques. In line with technological progress and according to their financial capability, farmers are gradually abandoning traditional wells that draw water from alluvial or shallow aquifers in favour of tapping deeper ones. This is facilitated by the availability of different sources of energy (electricity, diesel and gas). Easy access to deep aquifers prompts farmers to release themselves from collective surface irrigation systems, as observed in southern India (Aubriot, 2013) and Palestine (Trottier & Perrier, 2018), for instance. This shift in aquifer use, along with interactions between neighbouring wells and the lowering of the water table, causes traditional wells to fail, leading to their abandonment. Similarly, in the case of the Saïss aquifer system in Morocco, the number of farmers tapping the deep aquifer through tubewells significantly increased in the 2000s. This led to the gradual abandonment of traditional wells with limited access to water due to the geological characteristics of the aquifer and interactions between neighbouring wells, thus shifting the farmers from a world of water scarcity to one of apparent abundance (Fofack et al., 2018). The same was observed in Tunisia, where shallow wells are progressively transformed into boreholes to tap deeper aquifers (Elloumi, 2016). In the Niayes region of Senegal we still observe the use of traditional wells (using pulley systems) alongside motor pumps.

Advanced drilling techniques aid farmers in accessing very deep aquifers. In extreme cases farmers can drill boreholes of several hundred metres to irrigate high added-value crops, such as the olive groves of the Loma de Úbeda aquifer in Spain. Their high cost means that several farms must join together to cultivate them, each gaining access to water in proportion to their investment, recalling the 'tubewell companies' of Gujarat in India (Shah & Bhattacharya, 1993). However, for those without hydrogeological knowledge there is a high risk of 'dry drilling', which, along with the high cost of drilling, may discourage farmers from undertaking such a project (Fofack et al., 2018).

Furthermore, public policies (particularly those concerning energy) directly or indirectly encourage this exploitation. Many farmers use engines, and these are powered either by subsidized fossil fuels (petrol, diesel or gas) connected to the power grid or by solar energy, which is developing rapidly. When the state subsidizes energy, farmers have an indirect incentive to increase their abstraction. The consequences are worrying in some already vulnerable aquifers, as illustrated by a recent study carried out by Gupta (2019) in the state of Rajasthan in India.

In addition to energy subsidies, agricultural subsidies for certain crops can further encourage groundwater abstraction (Rouillard & Rinaudo, 2020). In Spain, European subsidies for the production of olive oil from La Loma, or wine from the Mancha Occidental aquifer, have greatly increased pressure on groundwater.

The intensive use of groundwater is often linked to the production of export crops and the role of international agro-industries. These could be cereals from the Great Plains of the United States, fruits and vegetables from the Mediterranean region, asparagus from Peru, avocados from Chile, pistachios or almonds from California or wine, mangoes and other tropical fruits from Australia. The liberalization of international trade and access to new markets (access to the European market for Spain in the 1980s, then for Morocco in the 2000s) play a decisive role in increasing abstraction. While these phenomena involving the exportation of 'virtual groundwater' (Dalin et al., 2017) generate local development, they are often accompanied by a monopolization of resources by agribusiness and exacerbate water crises at the local level, leading to social and environmental crises, as observed, for instance, in Peru and Chile (Rinaudo & Donoso, 2019). This also raises questions about contemporary modes of consumption, whereby consumers increasingly rely on off-season, water-intensive and imported food. A clear example is the consumption of tomatoes all year round, implying their production in far-flung places.

Over-allocation of rights and widespread illegal use

Even when farmers are aware of the risk of the depletion of the resource on which their livelihood depends, they often remain in a pumping race comparable with the 'tragedy of open access' described by Hardin (1968), whereby the pursuit of individual interests in the exploitation of a common resource inevitably leads to its over-exploitation if no control mechanism is set.

State regulation of abstraction through legal frameworks that integrate groundwater is increasingly common. However, over-exploitation and its consequences have been tackled only belatedly, meaning that public authorities have allocated excessive rights to the resource. This can be partly explained by a lack of knowledge of the environmental consequences described above as well as by a lack of political will to take long-term issues into account.

It also results from an understanding of existing uses based on land ownership alone. In the valleys of northern Chile (Copiapo and La Ligua Petorca), for example, the total water rights allocated are four times higher than the available renewable resource, compounded by the promotion of water markets (Box 1).

The over-allocation of rights is almost systematically compounded by the problem of illegal abstraction. This can involve unauthorized well drilling, unauthorized irrigation of agricultural land or abstraction exceeding the allocated right. This has been described in Spain, where, according to a report released by WWF (2006, p. 20), the Ministry of the Environment counted no fewer than 510,000 illegal boreholes in the early 2000s, each of which was abstracting more than 7000 m^3 per year. This 'illegal' abstraction is motivated by the profitability of irrigated crops and is often indirectly encouraged by public subsidies, as described above, as well as crops that generate income and employment (Novo et al., 2015) for the local economy, or even as a means to preserve social order. The need to regulate illegal use and adjust rights where they have been over-allocated are therefore crucial in many contexts around the world, in developing and developed countries alike (Schulte & Cuadrado Quesada, 2020).

BOX 1.

Overallocation of groundwater in northern Chile.

In the Copiapo Valley of northern Chile water rights have been over-allocated. Several factors can explain this situation of over-allocation. From a technical point of view, the available resource was evaluated following a relatively wet period, which resulted in its overestimation. The state was also under political pressure, which led to the continued allocation of temporary water rights (subsequently 'regularized') once the over-exploitation was recognized. The existence of water markets exacerbated the situation by allowing users whose wells were dry to sell their rights to other users in better locations, and facilitating the transfer of water rights from users with less intensive practices to those with more regular consumption (Rinaudo & Donoso, 2019). The model used to calculate the distribution of water rights (surface and groundwater) in Chile was based not only on a lack of data regarding the dynamics of the aquifer but also on a calculation that took the allocated water rights and not the amounts used as the quantity of water extracted. The model proved to be inappropriate and reinforced already existing inequalities in access to water (Budds, 2009).

BOX 2.
Multifunctional nature of aquifers: the example of the Crau plain in France.

In the Crau plain in the south of France (550 km^2), the water table is located at highly variable depths (from 6 to 60 m, depending on the area), and gravity irrigation has helped sculpt the landscape through the restitution of irrigation water to the shallow aquifer. It supplies a vast agricultural and market-gardening area (the Crau wetlands), as well as a nature reserve and a Natura 2000 site (in a more arid area comprised of steppes: the dry Crau). Apart from agriculture, diverse economic activities use the resource. Groundwater supplies drinking water not only to the inhabitants who live within its perimeter (approximately 100,000), but also to a great many users (170,000) who live nearby. This situation explains why considering the perimeter of the aquifer alone is insufficient to identify all relevant actors affected by its management. The problems of the different areas located within the hydrogeological perimeter of the aquifer (dry Crau and Crau wetlands), as well as those concerning the areas located beyond it, differ significantly. From the preservation of the environment to support for a quality agricultural sector and drinking-water supply, the challenges are all interlinked in different ways depending on the scales concerned and should be approached together with a view to preserving the common patrimony of the groundwater of the Crau region.

Source: *Syndicat mixte de gestion de la nappe phréatique de la Crau* (Joint Association for the Management of the Crau aquifer). https://www.symcrau.com/.

Lack of knowledge and status quo policy

The invisibility of aquifers makes the scarcity of the resource difficult to grasp – a central factor in over-exploitation. General awareness is further hampered by a lack of hydrogeological surveys and, when they exist, meagre public availability of the data. As such, knowledge remains confined to expert circles.

The interactions of aquifers with ecosystems and surface water, and the interdependence of scales, while often insufficiently documented, are crucial in understanding how issues are embedded in territories. There is much debate around water withdrawals for gravity irrigation, which are considered as excessive, yet local actors and scientists emphasize that it is return flows to the aquifer that are crucial for domestic use as well as to sustain downstream wetlands, as illustrated by the case of the Crau plain in France (Box 2).

The lack of knowledge or uncertainty surrounding hydrogeological science is sometimes used as an excuse for inaction. For too long the claim that exhaustive knowledge must be acquired before anything can be done has justified the status quo and significantly delayed the adoption of measures to limit abstraction, or has extended consultation periods, as has been observed in the case of the Beauce aquifer in France (Petit, 2009).

Challenges in implementing solutions

When the sustainability of groundwater use is threatened, solutions are proposed alone or in combination: (1) schemes to increase the resource base or to save water, (2) regulatory instruments implemented by the authorities and (3) community initiatives.

Supply augmentation and water-saving schemes: ill-adapted and counterproductive solutions

A drop in the groundwater level can often prompt the exploitation of additional water resources, that is, water transfers (sometimes over long distances) or unconventional sources, such as desalinated water or treated wastewater. Such supply augmentation options do not challenge the intensive agricultural systems at the origin of such a drop and tend to be costly both economically and environmentally. Transfers shift pressure onto the donor basin: wastewater often has already been used, and treating it does not 'add water' (however desirable this might be), while desalination comes with energy consumption and environmental impacts. In addition, while pressure may be relieved in the short term, boosting supply only encourages further shortages.

In the case of Morocco, substituting groundwater by surface water for urban supply is openly advocated so that farmers access more resources (Del Vecchio, 2020).

This move towards excluding non-agricultural users from groundwater resources can be also observed in Campo de Dalias in southern Spain where seawater desalination is a new source of supply to urban users, who bear its high cost (Dumont, 2015). In contrast, in Tunisia, groundwater is brought from inland plains to supply drinking water to urban areas on the coast. These transfers compete with local uses. In the plains of Sidi Bouzid or Kairouan, populations living near the aquifers request an increase of the allocation of these resources for local irrigation (Elloumi, 2016).

Supply augmentation solutions are often implemented alongside moves to save water based on technical innovations such as drip irrigation. Yet, while intended to save water, such innovations often result in agricultural intensification (the introduction of crops with high added value but often higher water requirements) and an increase in the irrigated area, meaning a hike in water consumption (Birkenholtz, 2017; Dumont et al., 2013; Trottier et al., 2020; Ward & Pulido-Velazquez, 2008). In Morocco, for example, the arrival of drip irrigation on the Saïss plain was accompanied by a 50% increase in the irrigated area between 2005 and 2014 and the doubling of groundwater abstraction (Kuper, Ameur et al., 2017).

Moreover, when areas irrigated with surface water undergo a change in irrigation technique it can lead to a drastic decrease in recharge due to irrigation 'losses', jeopardizing the available recharge/abstraction balance in areas irrigated with both surface and groundwater. This has been observed in several Indian states and is also illustrated by the Chilean Copiapo valley and French Crau case studies mentioned above.

Among the technical solutions, managed aquifer recharge is in a category of its own by reinforcing the natural capacities of aquifers. This may include facilitating surface water infiltration during flood events or intercepting some subsurface flows through detention devices (Smith et al., 2016). The proper functioning of these devices may be the responsibility of local communities as illustrated by the flood spreading system in Ghardaïa in Algeria (Jadeja et al., 2018).

However, the risk associated with such solutions must also be recognized, particularly when they involve regular maintenance, such as the re-infiltration of treated wastewater or the use of infiltration wells. In India, for example, managed aquifer recharge is a consensual policy, but its implementation is controversial (Box 3).

Long-term monitoring appears to be necessary to assess the quantitative and qualitative impacts of the recharge structures.

BOX 3.

Managed aquifer recharge in India.

In India, groundwater accounts for more than half of the water used for irrigation and most domestic water. In the union territory of Puducherry and the state of Tamil Nadu, where groundwater is over-exploited and subject to pollution and seawater intrusion in the coastal area, managed aquifer recharge structures are built or planned under national programmes with the support of international donors. Long-term monitoring appears to be necessary in both quantitative and qualitative terms (risk of pollution). There is a lack of user participation in their design and implementation and of maintenance, and in some cases they are diverted from their initial purpose: from recharge to abstraction. Artificial recharge systems are often seen as a means to increase water supply without addressing demand. Thus, the purpose of recharge is not challenged while the means are. For example, farmers and non-governmental organizations (NGOs) actively defend traditional 'tanks' – seasonal lake–reservoirs that contribute to recharge – against artificial recharge wells, which they consider to be too expensive (Richard-Ferroudji et al., 2018).

Regulatory actions taken by public authorities faced with their own dilemmas

Public authorities play a pivotal role in groundwater management. This is primarily due to the state's control of the management of land, the environment and natural resources, particularly water, which is often defined as being in the public domain. Formally, the state is also the guarantor of the general interest and of fair access to the resource for the various types of user. Moreover, public policy in domains such as agriculture, energy and public works may increase or reduce the demand for groundwater.

Faced with over-exploitation, or simply to exercise their control of the resource, public authorities (central government, federal states, local authorities or even basin agencies) may implement a variety of regulatory and economic instruments (Figure 1). These can include permits, bans, well closures, quotas, zoning by irrigated area or land use, etc. Public authorities may introduce economic measures, such as direct or indirect taxes and subsidies. These are designed to send a financial signal to farmers to limit their abstraction but are usually too small to have a significant effect. Some tools can appear contradictory or even encourage abstraction, whether directly (subsidized agriculture or electricity) or indirectly (subsidized water-saving technology).

Furthermore, the spatial extent of aquifers and the invisibility of the resource mean monitoring and controlling abstraction are difficult and expensive: abstraction can be carried out from any location, without its effect being directly noticed. The distribution of water volumes among the various users is complex, and the means of controlling abstraction are limited, especially in the absence of collective action and/or when the state faces legal constraints. Even if significant resources are allocated, public authorities are often helpless against the growth of (often illegal) boreholes (millions in India). They tend to respond with ill-adapted punitive systems, complicated and costly procedures to legalize illegal wells and unenforceable volumetric permit systems for groundwater abstraction. The common result is counterproductive behaviour, such as false declarations, metre tampering or even corruption (Molle & Closas, 2020a).

However, some experiments aim to counter the individualistic nature of groundwater exploitation (Box 4).

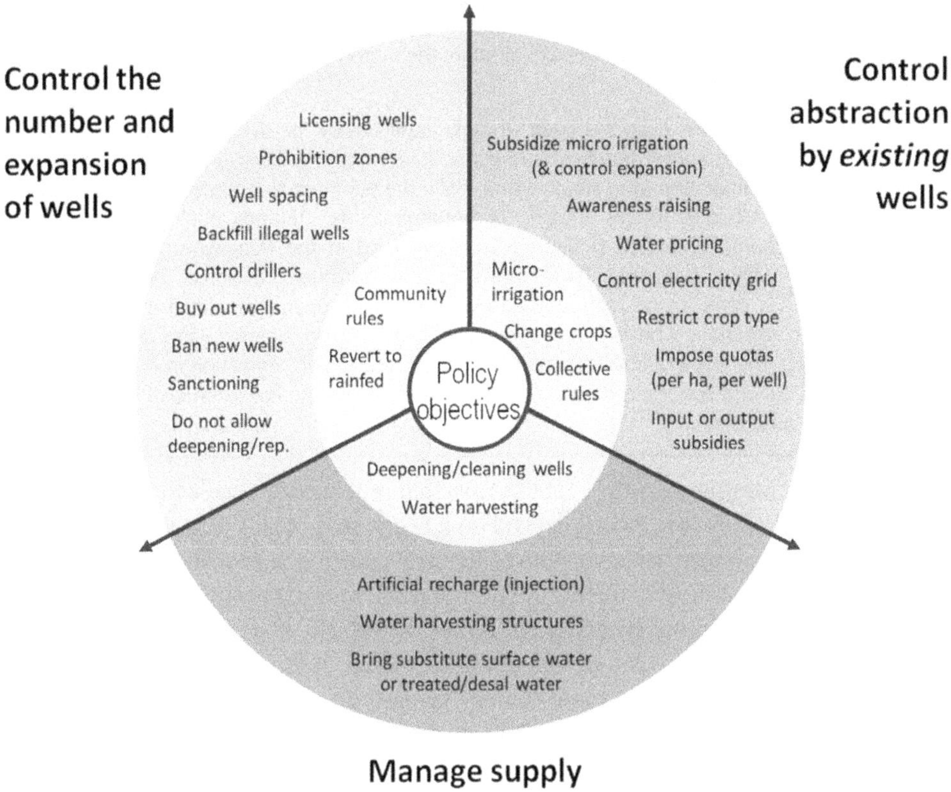

Figure 1. Regulating groundwater access and use: the public authority's toolbox. Source: Molle and Closas (2020a).

BOX 4.
Role of irrigation cooperatives in the Izmir region of Turkey.

In the Izmir region of Turkey, the size of the area irrigated with groundwater grew from 100,000 ha in 1978 to over 700,000 ha in 2014 (Le Visage & Kuper, 2019). Despite a failure (or lack of will) to 'combat illegal drillers' (Apaydın, 2011, p. 326), the water administration has sought to remain involved in monitoring this irrigation, particularly through the creation of irrigation cooperatives. Responsible for the management of collective boreholes at village level, these cooperatives irrigated almost 480,000 ha in 2014. Thus, the state (partially) supervises abstraction through collective boreholes authorized by the administration and equipped under its technical oversight. The cooperatives play a key role in local agricultural development by facilitating access to groundwater without the farmers having to individually invest in private boreholes. However, this model should not be idealized. Cooperatives often face difficulties, including the depth of the water table, poorly drilled boreholes, the incomplete or late collection of fees depending on fluctuations in agricultural income and high energy costs, and are financially unsustainable.

The obstacles to the effective public management of groundwater resources are often of a political nature. The political cost of restricting access to resources encourages public authorities to favour short-term economic development to the detriment of sustainable resource management. A certain 'social tolerance' of illegal use is thus commonly observed

BOX 5.
When the state supports groundwater exploitation: the example of the Saïss plain in north-west Morocco.

Since the 1980s, the Saïss plain has been strongly impacted by the dynamics of groundwater irrigation. Having 'one's well, borehole and water' had become a question of modernity and dignity – a sign of liberation from state control – and the pumping race turned into an identity race. Yet, state subsidies for the digging of boreholes, drip irrigation equipment and the planting of orchards, linked to facilitated access to land in the framework of the 'Green Morocco Plan', have benefited large farms to the detriment of small local farmers. Groundwater has come to be monopolized by a few major operators with financial means and political connections who feel free to use the resource in an intensive (and uncontrolled) manner. Hence, there is a deliberate absence of state control and regulation in order to stifle growing political and social tensions in rural areas and ensure the coexistence of a mosaic of farms (Messaoudi, 2022).

(López-Gunn & Martinez-Cortina, 2006). Any attempt to restrict access to the resource is opposed by those whose livelihoods depend on it (Loch et al., 2020). The political costs are closely related to the financial costs, both to irrigators and to the people depending on the resource at the local scale (Novo et al., 2015). The deadlock is compounded by the fact that the state is not a homogeneous entity but rather the vehicle of several visions, expressed through policies from different sectors at various levels, which sometimes prove to be contradictory. A national agricultural strategy favouring food self-sufficiency or production for export will result, as mentioned above, in greater extraction (Box 5).

Attempts by governments to reduce water rights have been documented throughout the world, revealing common difficulties. To address the short-term cost to the local economy and the loss of farmers' income caused by restricted water rights, 'gradual restrictions' can be proposed either to balance dry and wet years, as in Beauce in France (Petit, 2009), or to reduce abstraction in the long term, as in Spain (Sanchis-Ibor et al., 2017). The state may also buy back water rights, either temporarily or permanently. For example, the buyback included in the Spanish government's Plan Especial del Alto Guadiana aimed to allocate rights to vine irrigators who had none, as well as to restore natural outflows from the aquifer to the wetlands of the Tablas de Daimiel National Park, thereby justifying the cost of the measure to public finances. However, detailed accounting reveals that the anticipated results were overly optimistic, in particular because the water rights for vines (irrigated in summer) are fully used each year, unlike for cereals whose level of irrigation depends on rainfall in spring. A right is not equivalent to actual use (Dumont, 2015).

From community to co-management: actors, legitimacy, limitations

While the role of public authorities is critical, their action alone is not sufficient. The difficulties in enforcement and the mixed results of regulatory and economic tools can be linked to a lack of user involvement in both their design and implementation. Users can be seen as responsible for the deterioration of the resource and as the direct beneficiaries of its preservation.

The problem of groundwater over-exploitation is relatively recent, yet some local agricultural organizations have existed for a long time, sometimes focused on water management and sometimes on groundwater, as with the qanats of North Africa and the Near and Middle East. In some cases, robust, age-old institutions for surface water management can inspire the design of specific institutions for groundwater, as in the Huerta de Valencia in Spain or in the oases of Algeria. These organizations are very rare and, where they do exist, are not adapted to the scale of the problem caused by the intensive extraction made possible by boreholes. Therefore, it is not usually possible to rely on them to generate collective action for groundwater management. As stressed by Landy et al. (2021, p. 29), 'there is nothing in the basic nature of water that makes it fundamentally prone to be a commons. It needs a social collective agreement for it to be managed "commonly" and in a sustainable way'.

Cases are reported, or even promoted, where aquifer users are deeply involved in management without it constituting 'self-regulation' per se (López-Gunn, 2003). The state often remains a key actor in regulating the use of water resources. Often it invites consultation with users by establishing rules or taking action. This can be illustrated by the confiscation of drills used for illegal wells in Morocco, by the Tunisian state's clampdown in Bsissi in the late 1990s, which led to a discussion that resulted in the creation of an Agricultural Development Group (ADG) for groundwater management (Molle & Closas, 2020b), or by Spain's official 'declaration of over-exploitation' that made the creation of user associations compulsory. A similar process has taken place in France with the establishment of water user associations. Collective dynamics to ensure that users adhere to allocation schemes or to create or strengthen joint practices are therefore generally co-management solutions in which public authorities remain fully involved (Rouillard & Rinaudo, 2020). However, the diversity of institutional arrangements is large, and the relative power of public authorities and groundwater users varies from one case to another, as stressed by a recent comparison between groundwater management institutions in France, Spain and the western United States (Rouillard et al., 2021).

There are many drawbacks to the establishment and legitimacy of co-management groups (Molle & Closas, 2020b). These include a lack of precedent in cooperation, the prioritizing of individual interests or a desire to maintain the social status linked to access to water, and a perception of the resource as a private object (due to its legal status in some countries) that precludes any collective action or intervention by the state. The cohesion of these groups is hampered by the diversity of profiles (from small farmers to agribusiness representatives) and the legitimacy of their leaders (sometimes self-appointed). Sometimes user groups fail to win the support of all their members; sometimes several user groups coexist; and in some cases these associations are not strongly supported by the state.

The co-management model, whether observed or promoted, can turn into a negotiation behind closed doors between the public authorities and the (representatives of powerful) direct users of the aquifer (Del Vecchio, 2020). There is often no real space for debate or decision-making that involves all stakeholders to address the diversity of issues, including environmental challenges and future generations, which are difficult to represent in the debate.

Possible ways forward

Turning to groundwater for agriculture is motivated by various factors and reflects a view of the resource as abundant and cheap to access. As exemplified by Kuper, Amichi et al. (2017, p. 726) in the case of North Africa, 'the availability of pumped groundwater 24 hours per day, 7 days a week provided a sense of abundance, at least in the short term'. This perception can lead to over-exploitation – a trend that cannot be reversed by traditional solutions, such as the enforcement of regulatory tools by the state, often poorly equipped and with little legitimacy at the local level, or by increased supply (dams, reservoirs, water transfers with artificial recharge, etc.). The possible ways forward presented in this section are not intended to be a complete overview of all groundwater governance issues but are based on collective aspects relating to the necessity (1) to negotiate knowledges and visions, (2) to broaden the debate beyond the volumetric dimension and (3) to balance the roles of user communities and public authorities – elements that are barely addressed in groundwater governance issues but are, we argue, decisive.

Negotiating knowledges and visions

Sharing knowledge on the functioning of aquifers is crucial. The effect of abstraction on the natural interactions of aquifers with surface water and ecosystems must be accurately assessed in order to quantify the available resource while taking into consideration all relevant variables. It is then necessary to have an estimate of effective abstraction. A clear distinction must be made between allocated water rights, effective abstraction and net consumption (considering the possible infiltration of drainage water). Moreover, the intricate interconnections of aquifers with ecosystems and surface water must be taken into account. For example, favouring pressurized irrigation to 'save water' where surface irrigation has helped, for many decades or centuries, to provide drinking water or sustain ecosystems reveals only a partial understanding of the water cycle. Finally, the resource/abstraction dynamic must be known and shared. The temporal response of aquifer systems in relation to qualitative and quantitative pressures and to their possible reductions must also be clearly established. Any action taken could take several decades to bear fruit, and efforts could be jeopardized if this dynamic is not adequately acknowledged. This is particularly the case for marine intrusion into coastal aquifers. Intermediate indicators can be formulated to reveal possible improvements.

When public authorities model aquifer function and abstraction estimates the results are often questioned by users who compare them with their daily experience of the scarcity of the resource. Hence, scientific knowledge should be open to resource users' empirical knowledge, rather than creating a hierarchy, with 'expert' knowledge ranked above 'local' knowledge. A dialogue is needed between various forms of knowledge, the distinctions between which are often artificial (Agrawal, 1995). At the same time empirical evidence is not sufficient to understand the functioning of deep aquifers, and local knowledge related to this 'new' resource is often constructed by bore-owners and biased towards their own interests (Aubriot, 2013).

Having an institutional space for dialogue is crucial to confront and negotiate knowledge and visions based on tools and indicators that can be easily understood. Avenues worth exploring include tools derived from citizen science, the establishment of local observatories by managers (Box 6) and education, awareness and training programmes. An analysis of 11 case studies in France highlights the effectiveness of a variety of communication methods in making groundwater visible, from scientific reports to artistic representations, including maps, photography and digital media, and recruiting spokespeople to advocate for groundwater (Richard-Ferroudji & Lassaube, 2020).

BOX 6.
Observatories: important forums for dialogue.

Observatories generally focus on long-term monitoring requirements, and their scope depends on the actors concerned. In the case of groundwater, they must enable boreholes to be located, monitor water levels and the main qualitative variables, maintain test sites, facilitate aquifer modelling work, and so on. The Indo-French Centre for Groundwater Research based in Hyderabad, India, fits this model (Maréchal et al., 2018). Beyond strictly hydrological aspects, these systems provide detailed information on the various types of ecosystems that exist in the territory and their relations with groundwater, as well as on groundwater use and user profiles, particularly from a socioeconomic point of view. The range of tools includes indicators, mapped information, and socio-anthropological and economic analyses. Observatories must also stimulate ongoing reflection by users and support information-exchange and consultation mechanisms. This is rarely the case, however, and observatories often remain focused on the production of scientific knowledge.

This institutional space would allow the integration of new knowledge and adapt to changing conditions (climate issues for instance). Social learning will help to create and strengthen a collective approach to groundwater governance. One challenge in creating a functional space will be to identify the key actors (visible or invisible) and their interest in formalizing their often-informal practices.

A negotiated vision and collectively established regulations can also encourage compliance. In the Chaouïa region of Morocco, for example, decision-making was informed by the development of different reference scenarios shared among stakeholders (Faysse et al., 2014).

It is often difficult to establish a common vision, since stakeholders will each have their own legitimate viewpoint. More generally, as mentioned above, developing an institutional space for dialogue is a delicate process, as existing hierarchies and individual/collective interests in groundwater use will certainly be revealed, requiring specific negotiation approaches and tools. Also, dealing with over-exploitation often demands challenging existing water use, and beyond that, some of the intensive agricultural systems put in place.

Nevertheless, achieving a meaningful dialogue requires the actors involved first to agree to step out of their various positions and engage in a real negotiation, and while such techniques are helpful, strong political support will be required to make this happen.

Broadening the debate beyond the volumetric dimension

A concerted decision-making process should find a context-specific solution rather than imposing an apparently universal strategy. The histories of the different users and uses of the resource must be taken into account, particularly those who have been deprived of access due to a drop in water table levels and find themselves marginalized (cf. interviews in Coste et al., 2007). The development of different baseline scenarios shared among actors helps to inform decision-making. In particular, it is important to understand the dynamics of entrepreneurial agriculture compared to those of traditional groundwater uses by developing alternative scenarios (Petit et al., 2018).

Shared diagnoses should be able to focus on the notions of 'risk' and 'crisis'. Some over-exploited aquifers have such a deficit that it seems improbable, even if pumping were discontinued, that they could return to their former water levels in the medium term. It is therefore advisable to intervene as quickly as possible, as soon as the early-warning signs of a future crisis are detected (Petit et al., 2017).

BOX 7.
Local planning for integrated water resource management: the case of Niayes in Senegal.

In the Niayes region of Senegal, which forms a coastal strip between Dakar and Saint-Louis, the state, with support from the non-governmental organization (NGO) Gret, is setting up local water platforms in municipalities to address the deterioration of groundwater. Following two years of consultation, local actors have agreed a diagnosis and now share a vision for sustainable and equitable water resource management. They have also chosen their mode of governance and drawn up local plans for integrated water resource management. Memoranda of understanding for the plans have been signed by the platforms, ministerial management and mayors of the relevant municipalities, and approved by the sub-prefects. This territorial and democratic approach stems from a political will to test local integrated water resource management through a process that connects the different levels rather than being top-down. However, the prerogatives of these platforms still need to be guaranteed through a reform of the legislative framework, currently underway, to assign the roles that are as yet only set out in their articles of association. Whether the platforms are truly representative of local actors and how power is balanced within them warrants examination.
Source: https://www.gret.org/projet/recherche-action-sur-la-gestion-integree-des-ressources-en-eau-dans-les-niayes/.

Broadening the debate beyond the establishment of an abstraction limit can generate change based on a set of socioeconomic factors. Indeed, when public authorities focus solely on the preservation of a common resource, such as groundwater, improvements are rarely seen. Diverging interests come into conflict, requiring negotiation processes that ensure the integration of all actors, including right-holder communities (Box 7). Demonstrating the environmental and social impacts of groundwater exploitation in relation to the actors concerned can help define a negotiated pathway to reduce abstraction. It is therefore political, economic and social stakes, as much as environmental issues, that motivate the search for sustainable management.

Balancing the roles of user communities and public authorities

The formation of user communities (self-organization by irrigators or externally imposed) is generally a prerequisite to managing the resource. However, it is insufficient in itself, and many groundwater-related issues are debated in discussion forums at other levels. Integrating citizens into these spaces, as well as users who are excluded from the resource, is also important. Thus, the state should reform its approach and regulatory tools to open up these negotiation spaces, although this often goes against deep-seated practices of non-transparency and centralized decision-making. It becomes crucial then to strengthen human resources (staff and skills) with, for example, the intervention of legitimate mediators with recognized skills.

Depending on the maturity of the user communities, they may be given certain responsibilities but also positive incentives in exchange for commitments (see the example of La Mancha Oriental, Spain) (Box 8). If it is to be effective, this concerted management requires leadership which is seen to be legitimate, making it possible to

BOX 8.
Organization and 'self-governance' of the irrigators of La Mancha Oriental in Spain.

The aquifer of La Mancha Oriental is used to irrigate approximately 100,000 ha of cereals, vegetables and vines. Contributing naturally to the flow of the River Júcar, which crosses it, the volumes extracted have an adverse impact on the aquatic ecosystems and the sustainability of traditional irrigation downstream (the area surrounding Valencia). It is noteworthy that the creation of a user association of irrigators to address the situation has reduced abstraction by approximately 25%. This is mainly due to cooperative regulation by the irrigators and a penalty system established in collaboration with the authorities. As the latter have a legitimate user representative with which to engage, they can impose restrictive measures during droughts more easily, such as financial compensation in exchange for reducing abstraction. This relative success (the adverse environmental impacts are still significant) can be explained by the existing social organization of the irrigators and the local 'social capital', and by a change in certain agricultural practices, as well as by several 'external' factors that could call into question the 'spontaneous' aspect of this organization, such as the additional supply of surface water (Molle & Closas, 2020b).

put in place specific arrangements and economic incentives, as well as to take full advantage of the entire range of regulatory instruments and render them effective on the ground.

A responsibility that could be handed to users is deciding how the resource is allocated once the overall level of abstraction has been established, as has happened in France, for instance (Rouillard & Rinaudo, 2020). It could also be the case that any request for a new abstraction right must receive the user group's approval before being submitted to the public authorities. This autonomy can strengthen adherence and compliance with the level of abstraction but is unlikely to be achieved if users do not receive some 'carrots' (e.g., additional supply from other sources, subsidies, etc.) in exchange for accepting the logic of the 'stick' (Molle & Closas, 2020b). More 'flexible' water distribution instruments, such as water markets or the allocation of a collective right to an irrigator community, can result in a higher rate of water right use: irrigators who would not have used their individual water right for a variety of reasons may transfer it to others. In addition, control may no longer be solely the responsibility of the public authorities and could thus involve users, or at least create a common sense of responsibility.

Finally, the inclusion of users in the sphere of decision-making and compromise can be based on a set of solutions integrating groundwater management and the mobilization of additional resources (surface water, re-use or desalination) while supporting agricultural development. This comprehensive approach can make it possible both to restore the water balance in the short term and ensure that management mechanisms are in place for the long term.

Conclusions: strengthen the understanding of groundwater as a common patrimony and develop a territorial project

We conclude that it is imperative to move away from the dominant vision of groundwater as a natural capital to be exploited for the sole purpose of agricultural intensification and economic development. Rather, defining what constitutes a common patrimony (Calvo-Mendieta et al., 2017) among heterogeneous actors whose interests may sometimes clash is helpful in negotiating the sharing and management of groundwater. Technical and economic solutions must emerge from the agreements reached in social collective arenas at different levels.

Initiatives should derive from an approach in which groundwater is considered a common patrimony (e.g., prioritizing the preservation of wetlands to take account of long-term issues). This way, problems are more likely to be discussed collectively and solutions found that depart from the conception of water as a commodity (Calvo-Mendieta et al., 2017). This also means being aware of the crucial aspect of access to groundwater for vulnerable groups as a means of livelihood and poverty alleviation. Indeed, the development of agribusiness, often with state support, denies certain users access to groundwater and neglects the environmental consequences of its intensive exploitation.

The rules and instruments arising from the consultation process must be legitimized and politically backed at the local level to improve the chances of sustainable groundwater management. According to Calvo-Mendieta et al. (2017, p. 127), 'common patrimony becomes meaningful only in the identification of a community of holders who recognize themselves as being the guarantors for the preservation and transmission of the common resources in a given territory'. This means that local public actors and 'indirect' users should be involved in decisions concerning groundwater management. Hence, analysing the stakes of this management for the territory and the local economy is an effective way to achieve the adherence of local representatives and users, and, therefore, to widen the debate beyond the hydrogeological unit to address the multifunctional nature of groundwater. Communication and awareness-raising should be directed at the general public, especially when decisions by local representatives risk being misunderstood.

Finally, initiatives to promote groundwater preservation should ensure that its management is fully integrated into discussions on territorial development and basin management so that all local actors value it and consider it in future planning. Territorial development refers here to the design of a socially constructed space where actors can project themselves and build activities and policies that allow this territory to reveal its specific resources (Caron, 2015). Appreciating the interconnectedness of surface and groundwater, and the various spaces delineated by overlapping aquifers, helps to identify which individual actors, and which formal and informal organizations, should be involved for each specific situation or issue. It is therefore important to renew development frameworks and models at a territorial scale, building on existing collectives. This is helpful in collectively determining which territorial project 'makes sense', particularly where the potential remains to increase groundwater exploitation, as in Sub-Saharan Africa.

Acknowledgments

This paper is the result of insights gained from the discussions of a joint working group of the Agence Française de Développement (AFD – French Development Agency) and the Comité scientifique et technique pour l'eau agricole (COSTEA – Scientific and Technical Committee for Water in Agriculture) project (www.comite-costea.fr).

Disclosure statement

No potential conflict of interest was reported by the authors.

Funding

This work was supported by the Agence Française de Développement [CZZ2162].

ORCID

Olivier Petit http://orcid.org/0000-0002-7987-0767
Aurélien Dumont http://orcid.org/0000-0002-7223-5939
Sami Bouarfa http://orcid.org/0000-0003-2965-6663
Nicolas Faysse http://orcid.org/0000-0002-5683-8473
Marcel Kuper http://orcid.org/0000-0002-1240-0592
François Molle http://orcid.org/0000-0001-5748-8770
Selin Le Visage http://orcid.org/0000-0002-5765-9688
Marielle Montginoul http://orcid.org/0000-0002-8029-6475
Audrey Richard Ferroudji http://orcid.org/0000-0001-8437-6766
Jean-Daniel Rinaudo http://orcid.org/0000-0001-8539-574X
Julie Trottier http://orcid.org/0000-0002-4899-8215
Olivia Aubriot http://orcid.org/0000-0001-9539-1806
Mohamed Elloumi http://orcid.org/0000-0001-9979-8988
Rhoda Fofack-Garcia http://orcid.org/0000-0002-2882-0399
Bruno Romagny http://orcid.org/0000-0002-7764-5785

References

Agrawal, A. (1995). Dismantling the divide between indigenous and scientific knowledge. *Development and Change*, *26*(3), 413–439. https://doi.org/10.1111/j.1467-7660.1995.tb00560.x

Ameur, F., Amichi, H., Kuper, M., & Hammani, A. (2017). Specifying the differentiated contribution of farmers to groundwater depletion in two irrigated areas in North Africa. *Hydrogeology Journal*, *25*(6), 1579–1591. https://doi.org/10.1007/s10040-017-1569-1

Amichi, F., Bouarfa, S., Kuper, M., & Caron, P. (2020). From oasis archipelago to pioneering Eldorado in Algeria's Sahara. *Irrigation and Drainage*, *69*(Suppl. 1), 168–176. https://doi.org/10.1002/ird.2308

Apaydın, A. (2011). Groundwater legislation in Turkey: Problems of conception and application. *Water International*, *36*(3), 314–327. https://doi.org/10.1080/02508060.2011.586750

Aubriot, O. (Ed.). (2013). *Tank and well irrigation crisis: Spatial, environmental and social issues* (p. 405). Cases in Puducherry and Villupuram Districts (South India). Concept Publishing Company.

Aubriot, O. (2006). Baisse des nappes d'eau souterraine en Inde du Sud: Forte demande sociale et absence de gestion de la ressource. *Géocarrefour*, *81*(1), 83–90. https://doi.org/10.4000/geocarrefour.1805

Birkenholtz, T. (2017). Assessing India's drip-irrigation boom: Efficiency, climate change and groundwater policy. *Water International*, *42*(6), 663–677. https://doi.org/10.1080/02508060.2017.1351910

Budds, J. (2009). Contested H_2O: Science, policy and politics in water resources management in Chile. *Geoforum*, *40*(3), 418–430. https://doi.org/10.1016/j.geoforum.2008.12.008

Calvo-Mendieta, I., Petit, O., & Vivien, F.-D. (2017). Common patrimony: A concept to analyze collective natural resource management. The case of water management in France. *Ecological Economics*, *137*, 126–132. https://doi.org/10.1016/j.ecolecon.2017.02.028

Caron, P. (2015). Territory: With government and market, a major institutional component to achieve resilience. *Natures Sciences Sociétés*, *23*(2), 175–182. https://doi.org/10.1051/nss/2015038

Closas, A., & Villholth, K. G. (2020). Groundwater governance: Addressing core concepts and challenges. *WIREs Water*, *7*(1), e1392. https://doi.org/10.1002/wat2.1392

Cobbing, J., & Hiller, B. (2019). Waking a sleeping giant: Realizing the potential of groundwater in Sub-Saharan Africa. *World Development*, *122*, 597–613. https://doi.org/10.1016/j.worlddev.2019.06.024

Coste, N., & Ploumpidis, N., (with Anupama, K., Aubriot, O. & Gunnell, Y.). (2007). *Bittersweet waters: Irrigation practices and modern challenges in South India*. Trilingual DVD (French, English, Tamil), Publications Hors Série 5. French Institute of Pondicherry. https://vimeo.com/223447091

Dalin, C., Wada, Y., Kastner, T., & Puma, M. J. (2017). Groundwater depletion embedded in international food trade. *Nature*, *543*(7647), 700–704. https://doi.org/10.1038/nature21403

De Graaf, I. E. M., Gleeson, T., Van Beek, L. P. H., Sutanudjaja, E. H., & Bierkens, M. F. P. (2019). Environmental flow limits to global groundwater pumping. *Nature*, *574*(7776), 90–94. https://doi.org/10.1038/s41586-019-1594-4

Del Vecchio, K. (2020, July). Gestion des eaux souterraines au Maroc: Entre priorités du développement agricole et préoccupations environnementales. Les apports de la science politique. *Note COSTEA*, 7.

Dumont, A. (2015). *Flows, Footprints and Values: Visions and Decisions on Groundwater in Spain* (p. 311). Doctorado en Geología e Ingeniería Geológica. Universidad Complutense de Madrid.

Dumont, A., Leyronas, S., Petit, O., & Ballin, Q. (2021). Acting together for the sustainable use of water in agriculture. Proposals to prevent the deterioration and overexploitation of groundwater. *AFD Policy Paper*, 8. https://www.afd.fr/en/ressources/acting-together-sustainable-use-water-agriculture

Dumont, A., Mayor, B., & López-Gunn, E. (2013). Is the rebound effect or Jevons paradox a useful concept for better management of water resources? Insights from the irrigation modernisation process in Spain. *Aquatic Procedia*, *1*, 64–76. https://doi.org/10.1016/j.aqpro.2013.07.006

Elloumi, M. (2016). *La gouvernance des eaux souterraines en Tunisie* (p. 121). IWMI Project Report, n°7, Groundwater governance in the Arab World.

FAO. (2016). *Global diagnostic on groundwater governance*.

Fautras, M. (2021). *Paysans dans la révolution. Un défi tunisien* (p. 494). IRMC-Karthala.

Faysse, N., Errahj, M., Imache, A., Kemmoun, H., & Labbaci, T. (2014). Paving the way for social learning when governance is weak: Supporting dialogue between stakeholders to face a groundwater crisis in Morocco. *Society & Natural Resources*, *27*(3), 249–264. https://doi.org/10.1080/08941920.2013.847998

Fofack, R., Billaud, J.-P., Kuper, M., & Petit, O. (2018). Analyse du basculement des modes d'extraction des eaux souterraines dans le Saïss (Maroc). Vers une reconfiguration du monde des eaux cachées? *Développement durable et territoires*, *9*(2). https://doi.org/10.4000/developpementdurable.12197

Gupta, E. (2019). The impact of solar water pumps on energy–water–food nexus: Evidence from Rajasthan, India. *Energy Policy*, *129*, 598–609. https://doi.org/10.1016/j.enpol.2019.02.008

Hardin, G. (1968). The tragedy of the commons. *Science*, *162*(3859), 1243–1248. https://doi.org/10.1126/science.162.3859.1243

Hoogesteger, J., & Wester, P. (2015). Intensive groundwater use and (in)equity: Processes and governance challenges. *Environmental Science & Policy*, *51*, 117–124. https://doi.org/10.1016/j.envsci.2015.04.004

Jadeja, Y., Maheshwari, B., Packham, R., Bohra, H., Purohit, R., Thaker, B., Dillon, P., Oza, S., Dave, S., Soni, P., Dashora, Y., Dashora, R., Shah, T., Gorsiya, J., Katara, P., Ward, J., Kookana, R., Singh, P. K., Chinnasamy, P., Goradiya, V., . . . Chew, M. (2018). Managing aquifer recharge and sustaining groundwater use: Developing a capacity building program for creating local groundwater champions. *Sustainable Water Resources Management*, *4*(2), 317–329. https://doi.org/10.1007/s40899-018-0228-6

Jouili, M., Kahouli, I., & Elloumi, M. (2013). Appropriation des ressources hydrauliques et processus d'exclusion dans la région de Sidi Bouzid (Tunisie centrale). *Études rurales*, *192* (192), 117–134. https://doi.org/10.4000/etudesrurales.9929

Kuper, M., Ameur, F., & Hammani, A. (2017). Unravelling the enduring paradox of increased pressure on groundwater through efficient drip irrigation. In J.-P. Venot, M. Kuper, & M. Zwarteveen (Eds.), *Drip irrigation for agriculture: Untold stories of efficiency, innovation and development* (pp. 85–104). Routledge.

Kuper, M., Amichi, H., & Mayaux, P. L. (2017). Groundwater use in North Africa as a cautionary tale for climate change adaptation. *Water International*, *42*(6), 725–740. https://doi.org/10.1080/02508060.2017.1351058

Landy, F., Ruiz, L., Jacquet, J., Richard-Ferroudji, A., Sekhar, M., Guétat-Bernard, H., Oger-Marengo, M., Venkatasubramanian, G., & Noûs, C. (2021). Commons as demanding social constructions: The case of aquifers in rural Karnataka. *International Journal of Rural Management*, *17*(1), 27–54. https://doi.org/10.1177/0973005220945428

Le Visage, S., & Kuper, M. (2019). Sous les gölet, les forages. Infrastructures d'irrigation et trajectoires des territoires de l'eau dans la région d'Izmir (Turquie). *Développement durable et territoires*, *10*(3). https://doi.org/10.4000/developpementdurable.15839

Loch, A., Perez-Blanco, C. D., Carmody, E., Felbab-Brown, V., Adamson, D., & Seidl, C. (2020). Grand theft water and the calculus of compliance. *Nature Sustainability*, *3*(12), 1012–1018. https://doi.org/10.1038/s41893-020-0589-3

López-Gunn, E., & Martinez-Cortina, L. (2006). Is self-regulation a myth? Case study on Spanish groundwater user associations and the role of higher-level authorities. *Hydrogeology Journal*, *14* (3), 361–379. https://doi.org/10.1007/s10040-005-0014-z

López-Gunn, E. (2003). The role of collective action in water governance: A comparative study of groundwater user associations in La Mancha Aquifers in Spain. *Water International*, *28*(3), 367–378. https://doi.org/10.1080/02508060308691711

Maréchal, J.-C., Selles, A., Dewandel, B., Boisson, A., Perrin, J., & Ahmed, S. (2018). An observatory of groundwater in crystalline rock aquifers exposed to a changing environment: Hyderabad, India. *Vadose Zone Journal*, *17*(1), 180076. https://doi.org/doi:10.2136/vzj2018.04.0076

Megdal, S. B., Gerlak, A. K., Varady, R. G., & Huang, L. Y. (2015). Groundwater governance in the United States: Common priorities and challenges. *Groundwater*, *53*(5), 677–684. https://doi.org/10.1111/gwat.12294

Messaoudi, I. (2022). Quelle modernisation rurale pour quel développement? Appropriations foncières, recompositions de l'espace et dualisme agraire au Maroc. *Annales de géographie*, forthcoming.

Molle, F., & Closas, A. (2020a). Why is state-centered groundwater governance largely ineffective? A review. *WIREs Water*, *7*(1), 1. https://doi.org/10.1002/wat2.1395

Molle, F., & Closas, A. (2020b). Comanagement of groundwater: A review. *WIREs Water*, *7*(1), 1. https://doi.org/10.1002/wat2.1394

Molle, F., López-Gunn, E., & van Steenbergen, F. (Eds.). (2018). The local and national politics of groundwater overexploitation. *Water Alternatives*, *11*(3), 445–846. https://www.water-alternatives.org/index.php/tp1-2/1903-vol11/334-issue11-3

Mukherji, A. (2006). Political ecology of groundwater: The contrasting case of water-abundant West Bengal and water-scarce Gujarat, India. *Hydrogeology Journal*, *14*(3), 392–406. https://doi.org/10.1007/s10040-005-0007-y

Novo, P., Dumont, A., Willaarts, B. A., & López-Gunn, E. (2015). More cash and jobs per illegal drop? The legal and illegal water footprint of the Western Mancha Aquifer (Spain). *Environmental Science & Policy, 51*, 256–266. https://doi.org/10.1016/j.envsci.2015.04.013

Petit, O., Kuper, M., & Ameur, F. (2018). From worker to peasant and then to entrepreneur? Land reform and agrarian change in the Saiss (Morocco). *World Development, 105*, 119–131. https://doi.org/10.1016/j.worlddev.2017.12.031

Petit, O., Kuper, M., López-Gunn, E., Rinaudo, J. D., Daoudi, A., & Lejars, C. (2017). Can agricultural groundwater economies collapse? An inquiry into the pathways of four groundwater economies under threat. *Hydrogeology Journal, 25*(6), 1549–1564. https://doi.org/10.1007/s10040-017-1567-3

Petit, O. (2009). Un regard rétrospectif sur l'évolution de la gouvernance de l'irrigation en Beauce (1993–2008). *Science et changements planétaires/Sécheresse, 20*(3), 262–270. https://doi.org/10.1684/sec.2009.0194

Richard-Ferroudji, A., & Lassaube, G. (2020). The challenge of making groundwater visible: A review of communication approaches and tools in France. In J.-D. Rinaudo, C. Holley, S. Barnett, & M. Montginoul (Eds.), *Sustainable groundwater management. Global issues in water policy, vol 24* (pp. 191–209). Springer.

Richard-Ferroudji, A., Raghunath, T. P., & Venkatasubramanian, G. (2018). Managed aquifer recharge in India: Consensual policy but controversial implementation. *Water Alternatives, 11*(3), 749–769. https://www.water-alternatives.org/index.php/alldoc/articles/vol11/v11issue3/463-a11-3-16/file

Rinaudo, J.-D., & Donoso, G. (2019). State, market or community failure? Untangling the determinants of groundwater depletion in Copiapó (Chile). *International Journal of Water Resources Development, 35*(2), 283–304. https://doi.org/10.1080/07900627.2017.1417116

Rinaudo, J.-D., Holley, C., Barnett, S., & Montginoul, M. (Eds.). (2020). *Sustainable groundwater management. Global issues in water policy* (Vol. 24). Springer.

Rouillard, J., Babbitt, C., Pulido-Velazquez, M., & Rinaudo, J.-D. (2021). Transitioning out of open access: A closer look at Institutions for Management of Groundwater Rights in France, California, and Spain. *Water Resources Research, 57*(4), e2020WR028951. https://doi.org/10.1029/2020WR028951

Rouillard, J., & Rinaudo, J.-D. (2020). From State to user-based water allocations: An empirical analysis of institutions developed by agricultural user associations in France. *Agricultural Water Management, 239*, 106269. https://doi.org/10.1016/j.agwat.2020.106269

Sanchis-Ibor, C., García-Mollá, M., & Avellà-Reus, L. (2017). Effects of drip irrigation promotion policies on water use and irrigation costs in Valencia, Spain. *Water Policy, 19*(1), 165–180. https://doi.org/10.2166/wp.2016.025

Santi, P. M., McCray, J. E., & Martens, J. L. (2006). Investigating cross-contamination of aquifers. *Hydrogeology Journal, 14*(1–2), 51–68. https://doi.org/10.1007/s10040-004-0403-8

Schulte, S., & Cuadrado Quesada, G. (2020). Reducing entitlements when groundwater has been over-allocated: Policy issues and options. In J. D. Rinaudo, C. Holley, S. Barnett, & M. Montginoul (Eds.), *Sustainable groundwater management. Global issues in water policy* (Vol. 24, pp. 315–332). Springer.

Shah, T., & Bhattacharya, S. (1993). Farmer organizations for lift irrigation: Irrigation companies and tubewell cooperatives of Gujarat. Irrigation Management Network, Overseas Development Institute, London, Network paper, n°26.

Siebert, S., Burke, J., Faures, J. M., Frenken, K., Hoogeveen, J., Doll, P., & Portmann, F. T. (2010). Groundwater use for irrigation – A global inventory. *Hydrology and Earth System Sciences, 14* (10), 1863–1880. https://doi.org/10.5194/hess-14-1863-2010

Smith, M., Cross, K., Paden, M., & Laban, P. (2016). *Spring – Managing groundwater sustainably* (p. 133). IUCN.

Taylor, R., Scanlon, B., Döll, P., Rodell, M., van Beek, R., Wada, Y., Longuevergne, L., Leblanc, M., Famiglietti, J. S., Edmunds, M., Konikow, L., Green, T. R., Chen, J., Taniguchi, M., Bierkens, M. F. P., MacDonald, A., Fan, Y., Maxwell, R. M., Yechieli, Y., Gurdak, J. J., & Treidel, H. (2013). Ground water and climate change. *Nature Climate Change*, *3*(4), 322–329. https://doi.org/10.1038/nclimate1744

Trottier, J., Leblond, N., & Garb, Y. (2020). The political role of date palm trees in the Jordan Valley: The transformation of Palestinian land and water tenure in agriculture made invisible by epistemic violence. *Environment and Planning E: Nature and Space*, *3*(1), 114–140. https://doi.org/10.1177/2514848619876546

Trottier, J., & Perrier, J. (2018). Water driven Palestinian agricultural frontiers: The global ramifications of transforming local irrigation. *Journal of Political Ecology*, *25*(1), 292–311. https://doi.org/10.2458/v25i1.22759

Villholth, K., López-Gunn, E., Conti, K., Garrido, A., & van der Gun, J. (Eds.). (2018). *Advances in groundwater governance*. CRC Press/Balkema.

Ward, F. A., & Pulido-Velazquez, M. (2008). Water conservation in irrigation can increase water use. *Proceedings of the National Academy of Sciences*, *105*(47), 18215–18220. https://doi.org/10.1073/pnas.0805554105

WWF. (2006). *Illegal water use in Spain. Causes, effects and solutions*. WWF/Adena.

OPEN ACCESS

Drivers of groundwater utilization in water-limited rice production systems in Nepal

Anton Urfels, Andrew J. McDonald, Timothy J. Krupnik and Pieter R. van Oel

ABSTRACT

Most rice farmers in Nepal's Terai region do not fully utilize irrigation during breaks in monsoon rainfall. This leads to yield losses despite abundant groundwater resources and ongoing expansion of diesel pumps and tubewell infrastructure. We investigate this puzzle by characterizing delay factors governing tubewell irrigation across wealth and precipitation gradients. After the decision to irrigate, different factors delay irrigation by roughly one week. While more sustainable and inexpensive energy for pumping may eventually catalyze transformative change, we identify near-term interventions that may increase rice farmers' resilience to water stress in smallholder-dominated farming communities based on prevailing types of irrigation infrastructure.

Introduction

Feeding a projected global population of 9 billion in 2050 will require focused efforts to address trade-offs and capitalize on synergies between natural resources management and food security objectives, necessitating broad-based transitions to sustainable intensification technologies and management approaches (Pretty & Bharucha, 2014). The Eastern Indo-Gangetic Plains of South Asia host the world's highest density of rural poor, pervasive yield gaps, and relatively abundant water resources (Bharati, Sharma, & Smakthin, 2016; Jain et al., 2017). They are therefore a global priority for sustainably increasing food production while ensuring that yield gains are accompanied by acceptable social and environmental costs and that long-term viability of the resource base is maintained.

Despite an overall abundance of water resources, water stress is one of the main factors limiting staple crop productivity in the Eastern Indo-Gangetic Plains. Increasingly erratic monsoon-season (*kharif*) precipitation poses a particular threat to the rice-based production systems that dominate the landscape (Turner & Annamalai, 2012). Many farmers in the Eastern Indo-Gangetic Plains apply supplementary 'life-saving' irrigation to overcome rainfall deficits during *kharif*, which is enabled by shallow water tables and broad coverage of groundwater irrigation infrastructure. Nevertheless, several studies suggest that irrigation use is typically 'too little, too late' and that even resource-poor farmers may improve yield, profitability and resilience with more judicious water use (Kishore, Sharma, & Joshi, 2014). Helping more farmers transition from 'life-saving' to 'productivity-enhancing' irrigation strategies will therefore be crucial for establishing pro-poor sustainable intensification pathways in the Eastern Indo-Gangetic Plains, particularly in view of contemporary climate variability and progressive change.

In contrast to the Western Indo-Gangetic Plains in India and Pakistan, where groundwater decline jeopardizes sustainability, abstraction in many parts of the Eastern Indo-Gangetic Plains is judged to be well within the limits of safe operating space, with significant scope for increased use (Bharati et al., 2016). Government of Nepal estimates suggest that groundwater use in the region could increase more than fivefold before breeching sustainable abstraction thresholds (Shrestha, Tripathi, & Laudari, 2018). Meanwhile, diesel-powered shallow tubewells are quickly growing in number, whereas investment in large canal infrastructure is diminishing (Bharati et al., 2016). This presents an opportunity to increase productivity-enhancing irrigation use in the Eastern Indo-Gangetic Plains, with the proviso that recurrent monitoring and adaptive management are essential to detect and mitigate early signs of receding groundwater levels. However, studies increasingly recognize that infrastructure development alone is insufficient to achieve optimal use of water resources (Qureshi, Ahmad, & Krupnik, 2015). Shallow tubewell-based irrigation is encouraged or discouraged by farm- and community-scale decisions, along with broader economic and policy incentives (Kishore et al., 2014). Understanding and harnessing these socio-hydraulic dynamics is particularly important in regions like the Eastern Indo-Gangetic Plains that are dominated by smallholder agricultural systems in which farmers face seasonal cash liquidity constraints, the costs of pumping are high, farmers are risk averse, and tenancy arrangements often discourage investments in intensification (Jain et al., 2017; Sugden, 2014).

This study characterizes drivers of shallow tubewell irrigation use to identify development support pathways that are likely to increase irrigation use and climate resilience among rice farmers in Nepal's component of the Eastern Indo-Gangetic Plains, the Terai – the country's 'breadbasket' lowland region, which runs parallel to its southern border with India. To understand the complexities of socio-hydraulic systems, an interdisciplinary perspective is essential (Massuel et al., 2018). We use a mixed-methods approach to characterize shallow tubewell irrigation infrastructure and farmers' decision processes around its use for *kharif* rice. Emphasis was placed on diesel-pump-based systems, the dominant form of irrigation in much of the Terai. The Terai is similar in physiography and cropping patterns to the neighbouring Indian states of Bihar and eastern Uttar Pradesh, and also north-eastern Bangladesh, although in Nepal agricultural policies and support programmes are less generous and rural infrastructure is generally underdeveloped (Shah et al., 2009).

Study area

This study was conducted in three districts of Nepal's Terai: Rupandehi, Banke and Kailali. These districts span a co-varying gradient of precipitation and wealth (Figure 1).

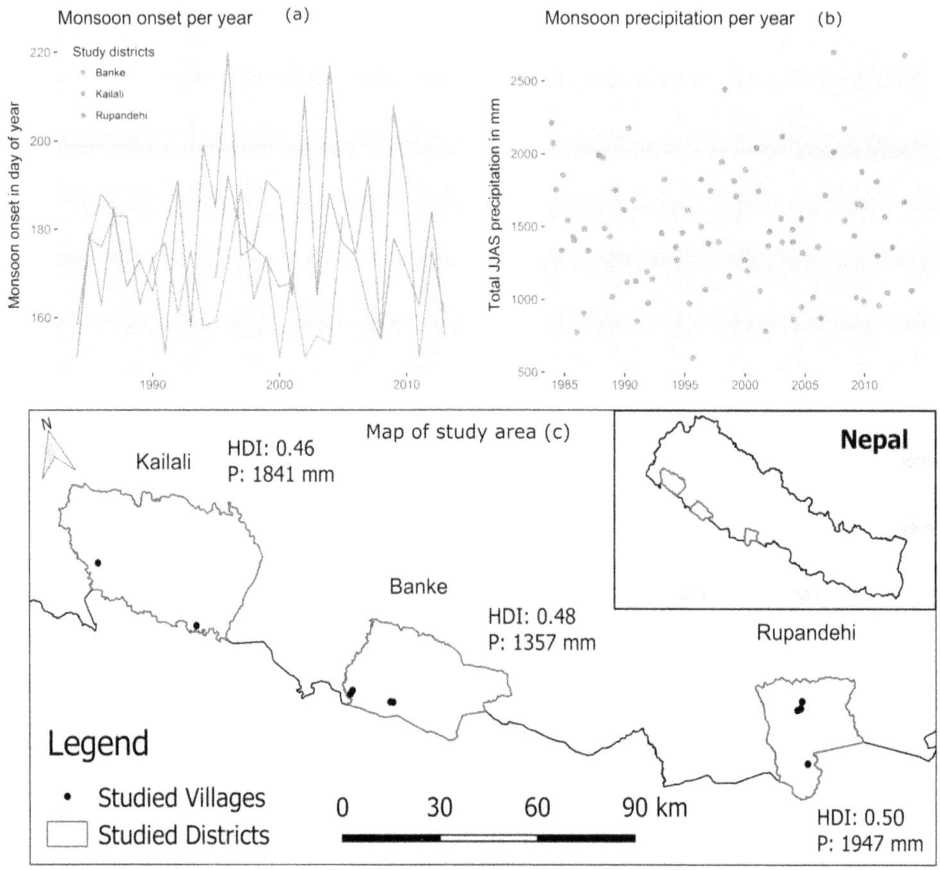

Figure 1. Study area: (a) monsoon onset by district; (b) variability in total June–September precipitation; (c) locations in the Terai of Nepal where household interviews and surveys were conducted.

They were chosen in consultation with senior irrigation policymakers who verified the district's representativeness for the Terai. In each of these districts, around 80% of the precipitation and river discharge occurs from July to October, during the monsoon (Shrestha et al., 2018). Flooding is common during this period. River flows diminish significantly shortly thereafter, hindering the use of canal-based surface water irrigation on a year-round basis (Bharati et al., 2016). A large part of Nepal's cultivated land area (ca. 40%) is reportedly reached by canal irrigation schemes, but considerably less is actually used. But an estimated 42% of the Terai's farmers do have access to shallow tubewells, according to the Nepal National Sample Census of Agriculture 2011–2012 (Central Bureau of Statistics, 2012).

The aquifers in the Terai are complex and consist of poorly ordered alluvial sediments that generally become finer towards the south. An unconfined aquifer extends to a depth of around 50 metres below ground level but varies in thickness, sometimes even at the village scale. These unconstrained aquifers may be underlain by a confined aquifer that can extend to a depth of 200 m. (Bonsor et al., 2017). Well yields vary from a few litres per second to more than 10 L/s. Bharati et al. (2016) report an average annual water table depth of 4.6 m across our study areas, with a seasonal fluctuation of about 3 m and a peak in August following recharge from monsoon precipitation. However, no reliable long-term water table data are publicly available, and levels may vary between locations.

Methods and data sets

Mixed-methods approach

To identify drivers of rice irrigation practices in areas where shallow tubewells predominate in the Terai, we employed a mixed-methods approach including semi-structured interviews with farmers, policy makers, and farming communities. These were coupled with household surveys and ethnographic decision-tree models (Roth & Botha, 2009). Interview and survey data were used to define the variables and sequences of decisions represented in the decision-tree models. An overview of data sets, figures and tables, and analytical approach used in their development is presented in Figure 2.

Site selection

We used time-series estimates of net primary productivity from remote-sensing imagery to identify villages using a range of irrigation practices. Net primary productivity values correlate with crop water uptake (Ciais et al., 2005). Therefore, areas with high temporal variation in agricultural net primary productivity generally have unreliable access to or suboptimally applied irrigation, while areas with low net primary productivity variation can be understood as having less moisture limitation of agricultural productivity.

As a proxy for irrigation, we used MODIS17A3H annual net primary productivity raster data from 2000 to 2014 (500 m^2 grids) to calculate the standard deviation of net primary productivity for each cell over a 14-year period. Then, we identified areas with a large range of irrigation use surrounding them by assigning each cell the standard deviation of its 3×3 neighbourhood. We identified the villages in the upper 10% of variability in access to irrigation surrounding them using a village location database based on a village data shapefile obtained from the Survey Department of the Ministry of Agriculture, Land Management and Cooperative. These villages (four in Rupandehi, three in Banke) were earmarked for our study. Selected locations were cross-checked with agricultural experts in each district to verify the presence of *kharif* rice production and shallow tubewells. This method produced results for Rupandehi and Banke, but the results for Kailali were problematic due to high occurrence of non-cropped areas. Thus, we omitted the neighbourhood calculation for Kailali and selected one village in the upper 10% and one village in the lower 10% of interannual net primary productivity variability.

Interview and survey data

Data were collected between September and December in 2016. Key informants (27) were chosen from government organizations with a mandate related to irrigation. These included senior officers of the Department of Irrigation, Department of Agriculture and Department of Electricity at the district level. These officers were approached at least

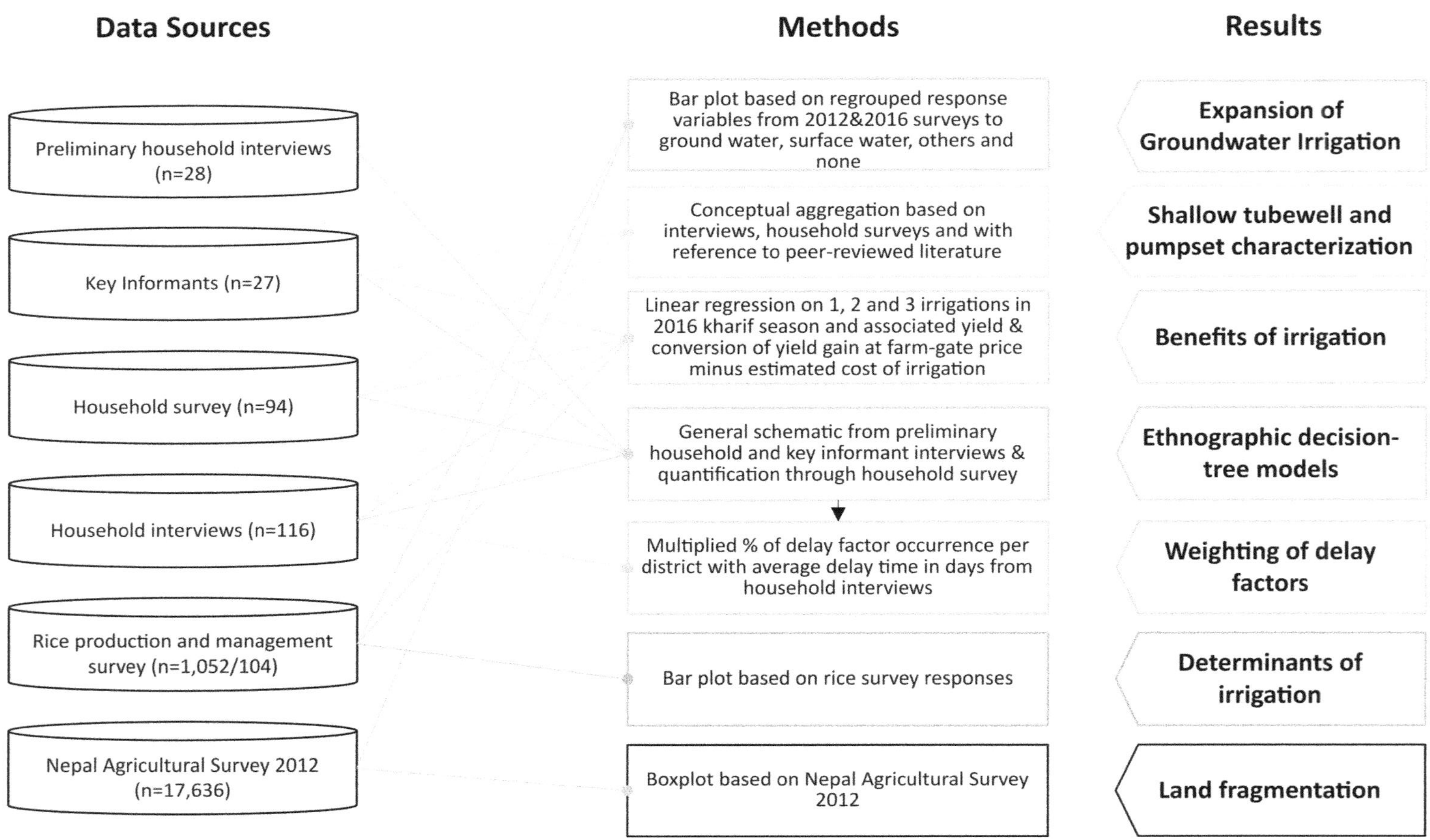

Figure 2. Overview of the data collected, methods employed and results utilized in this study.

twice, once at the beginning of the study and once at the end, to reconfirm responses. At the local level, Village Development Committee secretaries and other local leaders were interviewed on a recurrent basis as questions arose.

At the farm level, semi-structured household interviews (116) and more formal household surveys (94) were conducted. At least 12 household heads were randomly interviewed in each village, covering all sub-village administrative units (blocks). Following interviews, short ad hoc group discussions were often had with neighbours. Data from these sources were used to cross-check previous results and were added to the household interview data set. We also used secondary data from the Nepal Agriculture Survey and the National Census of Agriculture, collected in 2011–12 (Central Bureau of Statistics, 2012), along with data from rice production practice surveys (1052 households) in the Terai collected by the Cereal Systems Initiative for South Asia (CSISA) in 2016 (Paudel, Maharjan, Guerena, Rai, & McDonald, 2017).

Shallow tubewell irrigation characteristics

To develop the context in which decisions are made and aid interpretation, we used several auxiliary data sets to complement the ethnographic decision-tree models. The spread of shallow tubewells was estimated with data from the Nepal Agriculture Survey (2012) and the CSISA production practice survey, as described earlier. Pricing data and technical specifications for pumps and tubewells were collected from machinery dealers and well drillers. Irrigation costs paid by farmers were estimated through surveys and household interviews. Interview data were triangulated with several key informants and farmers.

Potential profits from supplementary irrigation applied to *kharif* rice were estimated by combining operational cost data for irrigation from our own surveys with 'farm gate' price data for rice from the FAO and the incremental yield benefits of additional irrigation estimated by linear regression from the CSISA production practice survey. An asymptote regression fit would be appropriate, but we lacked sufficient observations of farmers applying more than three irrigations. So we focused on the yield increase associated with one, two and three irrigations, which can be described by a linear fit (Figure 3). The benefit is expected to level off for higher numbers of irrigations, as other inputs such as seed or fertilizer may become limiting factors; it will also vary significantly by year, depending on rainfall patterns. We take the increased profit from additional irrigation (in USD per hectare) to be

$$(\mathrm{FGP} \times \mathrm{YI}) - (\mathrm{IP} \times \mathrm{ID} \times 10,000 \times \mathrm{EF}) \tag{1}$$

where FGP is the farm grate price of rice (USD/kg), YI is the yield increase per irrigation (kg/ha), IP is the irrigation water price (USD/m^3) translated from hourly rates charged in the field at an assumed discharge of 10 L/s, ID is irrigation depth (m) (Sudhir-Yadav, Humphreys, Kukal, Gill, & Rangarajan, 2011), and EF is water losses of 33%, commonly found in basin irrigation. The SD in yield per hectare translates to USD 18 per hectare using the same assumed irrigation costs. We assumed USD 1 per litre of diesel, USD 3 per hour of pumping as a pump rental price (based on key informants), USD 260 per ton of rice produced (based on FAO information), and 368 kg rice yield increase irrigation per

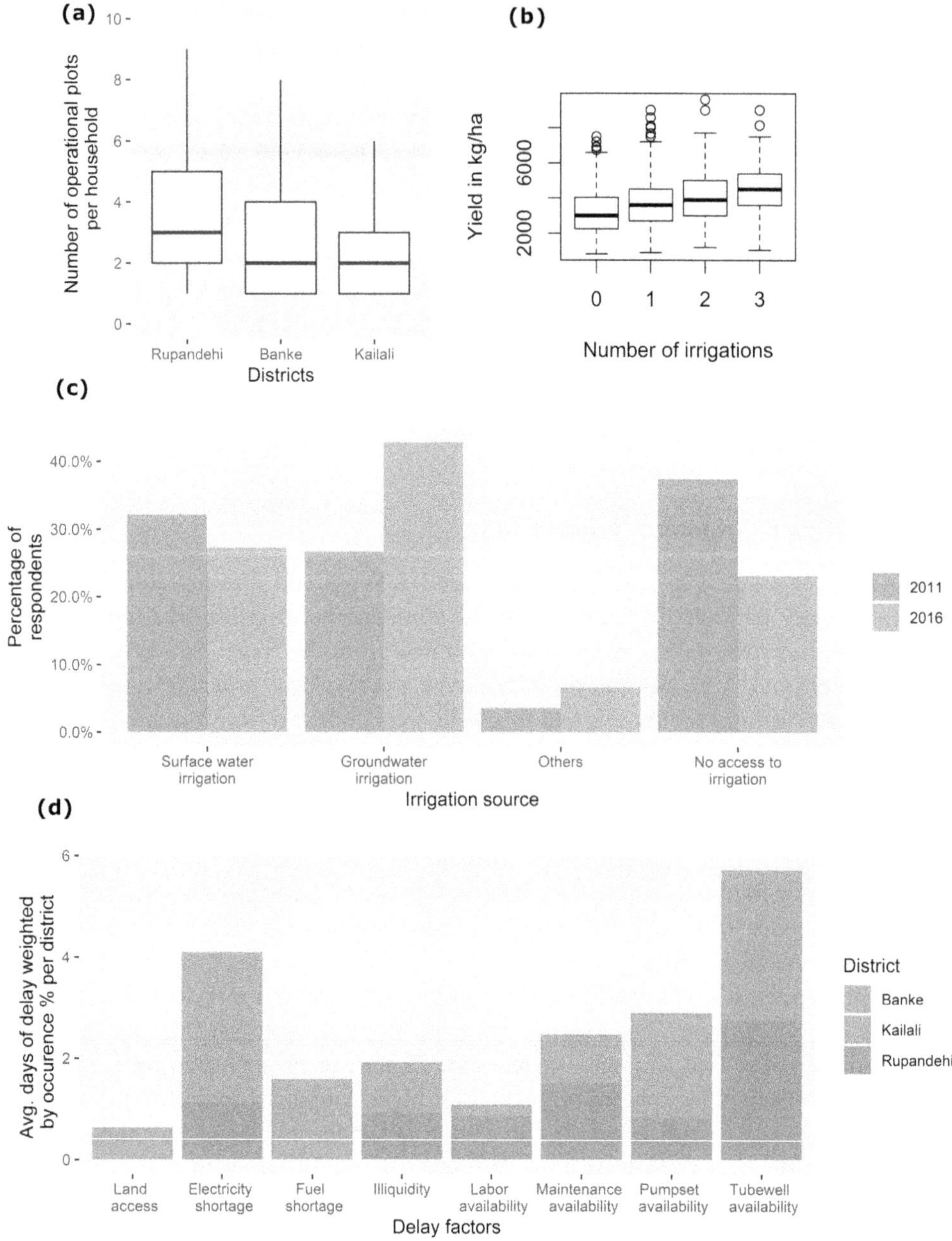

Figure 3. Shallow tubewell irrigation characteristics: (a) land fragmentation in western Terai; (b) yield increase per irrigation (Paudel et al., 2017); (d) relative importance of supplementary irrigation delay factors reported by *kharif*-season rice.

irrigation (based on linear model fit to rice crop cut survey data; coefficient 368.46, standard error 36.34, $p \leq 2e\text{-}16$).

Precipitation information (1984–2013, Figure 1) was drawn from government data from stations in Bhairawa (Rupandehi), Nepalgunj (Banke) and Sandepani (Kailali). The monsoon onset dates used in Figure 1(a) were calculated using the methods of Fitzpatrick, Parker, and Willetts (2016). Human Development Index (HDI) is based on United Nations Development Programme (2014). The land fragmentation figures in Figure 3(a) are based on Central Bureau of Statistics (2012) data.

Ethnographic decision-tree model construction

Ethnographic decision-tree models are flow charts that aim to capture decision-making processes from the language and concepts respondents use rather than a predefined conceptual framework. They are based on interviews and subsequent cross-validation (Roth & Botha, 2009). They are suitable for representing decision processes of rather homogeneous groups, such as farmers within an agro-ecosystem or entrepreneurs within a city (Roth & Botha, 2009). Our model was based on preliminary unstructured interviews with farmers to identify factors that influence the choice to provide supplementary irrigation to rice. Factors that negatively influence or delay farmers from choosing to provide supplementary irrigation were also identified. Our approach followed Roth and Botha (2009) and was based on evidence that farmers' general perceptions are sufficiently reliable indicators of local trends and patterns (Banerjee, Chandrasekhar, Duflo, & Jackson, 2018). We first built a generalized decision schematic through key informant interviews with farmers and their perceptions of decision points and factors that influence them. To quantify these relationships, results from the household survey were used to assign each branch of the decision tree a weight reflecting the proportion of farmers making that decision. We then compared the quantifications across districts.

Ethnographic decision-tree models were only constructed with farmers that have access to shallow tubewells for supplementary irrigation. The decision process is initiated by farmer assessments of the need for irrigation and may be reset at any point that sufficient rainfall is received before irrigation has been applied. Similarly, to highlight bottlenecks for the use of supplementary irrigation for *kharif* rice, we represent changes in decision preferences only in the case of irrigation delay, not for situations in which farmers choose to irrigate unencumbered.

The ethnographic decision-tree models have two components: irrigation triggers and delay factors. The first refers to when and under what circumstances farmers decide to apply supplemental irrigation. The CSISA production practice survey (Paudel et al., 2017) indicates that plant (59%), soil (56%), weather (43%) and neighbouring farmers' practices (31%) play a role. Farmers indicated that disagreements among them on the status of their crops or soil were relatively rare. The terminology used to describe crop and weather tended to be subjective and idiomatic (e.g., 'plants smile', 'clouds look fatter'). Soil moisture was most commonly described in terms of the width of cracks that form as rice fields dry over successive days without floodwater. These were taken as 'no cracks/hairline cracks', 'small cracks' (up to 2.5 cm wide), and 'large cracks' (over 2.5 cm). Since the soil criteria are readily observed by farmers and fairly consistently applied as a management rule, we decided to limit ethnographic decision-tree model quantification to soil conditions to maintain parsimony, reduce translation error and ease comparison between farmers.

Irrigation delay factors are the second component. Shallow tubewell delays are irrigation lags associated with turn-taking as farmers wait to use shared well infrastructure. Energy delays happen when electricity or fuel (depending on the type of pump) is needed but unavailable. Liquidity delays refer to the inability of farmers to pay for energy or rental of a pumpset when required. Pumpset delays are caused when a farmer must borrow or rent a pump. Transport delay refers to the transport of pumps to boreholes in command areas lacking road access. (Moving engines across established fields can damage crops; it is generally only permitted if neighbouring farmers have already irrigated.) Labour delays refer to a dearth of farm labour to oversee irrigation. As irrigation is generally considered a task for men, some households engage relatives or friends if male members are not present. Finally, maintenance delays refer to the time it takes to source spare parts or engage a technician to repair a broken pumpset.

Results: shallow tubewell irrigation characteristics and use patterns

Key shallow tubewell irrigation characteristics

Pumpsets deployed for shallow tubewells in the Terai are described in Table 1. Using standards described by Bom, van Raalten, Majundar, Duali, and Majumder (2001), oversized and inefficient pumpsets of 5 horsepower or more are common. Farmers and agricultural machinery dealers differentiate between large pumps (5–10 HP, not easily transportable) and smaller and more portable pumps (3–5 HP). These might seem more suitable for vegetable cultivation. But at a constant discharge, the horsepower requirement of a pump depends on system head (the pressure required to move water through the system). With well yields of ca. 10 L/s, and only a few metres of system head, the smaller pumps are adequate. With increasing system head (e.g., because of several hundred metres of delivery pipes) higher horsepower is warranted, or discharge may decrease.

The distribution of small and large pumpsets is nonuniform: 72% of small pumpsets owned by surveyed households were found in Banke District, but only 18% of large pumpsets. The reasons for this disparity are not clear, but household survey respondents said that high mobility of the pumps, a lack of financial liquidity and a relatively high irrigation frequency with lower energy costs are the main drivers of the preference for small equipment. The smaller pumps, primarily imported from China, are perceived as less durable, while the larger pumps, primarily from India, are considered strong and more reliable, although their higher cost and power make them a worse fit for most shallow tubewell infrastructure.

Between 2011 and 2016, groundwater abstraction in the Terai using shallow tubewells increased by approximately 15%, while surface water irrigation decreased by roughly 5% (Figure 3). Data from the CSISA production practices survey also suggest future growth in shallow tubewell infrastructure, with nearly half of the respondents planning to provide supplementary irrigation to *kharif* rice more frequently. In addition, 34% of farmers surveyed indicated their intention to purchase shallow tubewell technologies (e.g., boreholes, pumps, piping) in the next year. This is not surprising given that yield benefits may bring substantial economic benefits (Table 1). Even in a 'good' monsoon, clear yield benefits can be expected from increasing supplementary irrigation intensity. The 2016 CSISA survey shows that farmers produced 368 kg/ha more rice, on average, when irrigating a first, second or third time (SE 34 kg); very few farmers irrigated four or more times. The impact of additional irrigations, as well as the timing of irrigation and growing-season weather, are important but beyond the purview of this analysis. Based on current

Table 1. Overview of locally available pumpsets.

	Capital expenditure (USD)	Operational expenditure (USD/h)	Power (HP)	Mobility	Popularity of *kharif* rice irrigation	Potential profit from extra irrigation (USD/ha)
Large diesel pumpsets	350–650	1–2 (0.003/m^3)	5–10	Immobile, transported by bullock cart	High. Regarded as strong and durable. Status symbol.	57
Small diesel pumpsets	250–450	0.3–0.7 (0.014/m^3)	3–5	Mobile, transported by bicycle	Low. Regarded as weak and easily damaged.	83
Electric pumpsets	150–250	0.07–0.12 (0.042/m^3)	1.5–2.5	Mobile, transported by bicycle	Medium. Inexpensive and efficient, but depends on unreliable and hard-to-get electricity.	93
Rented pumpsets	0	3–4.5 (0.083/m^3)	3–10	Depends on rented pumpset	High, but depends on availability, social capital and cash/credit availability.	18

Source: Household interviews and surveys, key informants.

practices in the Terai of Nepal, applying additional irrigation is generally lucrative for pump owners and potentially for pump renters, if rental prices can be reduced.

From the farmer and household survey, as well as key informant interviews, operational aspects emerged as another factor of considerable, and perhaps intensifying, importance to groundwater irrigation. Land fragmentation, measured as the number of operational fields maintained by a farm household, decreases from east to west, with an interquartile range of 2–5 non-adjacent plots per household in Rupandehi, decreasing to 1–3 in the far-western Kailali District (Figure 3). These data indicate that transaction costs for shallow tubewell use are much higher in Rupandehi than in Kailali.

Shallow tubewell irrigation use patterns

Most surveyed farmers decide to apply supplementary irrigation to rice in the *kharif* season based on visual observation of soil cracking as drying occurs; they wait until large cracks (over 2.5 cm) are present before irrigating (Figure 4). But criteria differ between districts, with 47% of the farmers in Banke waiting for large cracks, whereas 68% and 71% of farmers wait for large cracks in Rupandehi and Kailali, respectively. Key informant interviews indicated that Banke farmers tend to irrigate earlier because of access to electric pumps and prevalence of small pumpsets, which are less expensive to operate than the large pumpsets mostly found in Rupandehi and Kailali. Our data suggest that most farmers do not recognize the link between delayed irrigation and yield outcomes, as irrigation is often considered as a mechanism to save the crop rather than as a productivity-enhancing investment, and farmers often attribute yield penalties to inferior quality of groundwater compared to rainwater.

Factors influencing farmers' choice to delay *kharif*-season supplementary irrigation of rice, weighted by their percentage of occurrence and average time of delay, are presented in Figure 4 and Table 2. Pumpset rental and queuing at shallow tubewells are major delay factors in all districts. In Kailali and Rupandehi, maintenance and liquidity constraints cause delays, pointing to problems with the availability of overly expensive and high-horsepower diesel pumpsets, primarily from India. Differences between land access and labour availability reflect higher rates of male labour out-migration in Rupandehi than in other districts (Ministry of Labor and Employment, 2017). Land access also reflects older and harder-to-transport pumpsets being used in Kailali. In Banke, electricity shortages and poor maintenance of pumps are important constraints. Based on these results, we discuss potential and complementary policies and development interventions that can be applied at relevant spatial scales and that could help optimize the use of supplementary irrigation for *kharif*-season rice in the Terai in the face of increasing precipitation uncertainty.

Nearly half (47%) of all delay pathways pertain to one of the following four pathways: only electricity supply constraints (23%); only shallow tubewell queuing (8%); only pumpset delays (6.7%); and all of the three previous factors together (9%). Once a decision has been made to irrigate, on average farmers reported that total delay (after the decision) amounts to one week, and that irrigation takes place 13 days after the last rainfall event (12.9 days for electricity delays, 10.4 days for queuing delays, 11.3 days for pumpset rental delays, and 15.25 days for all three occurring together).

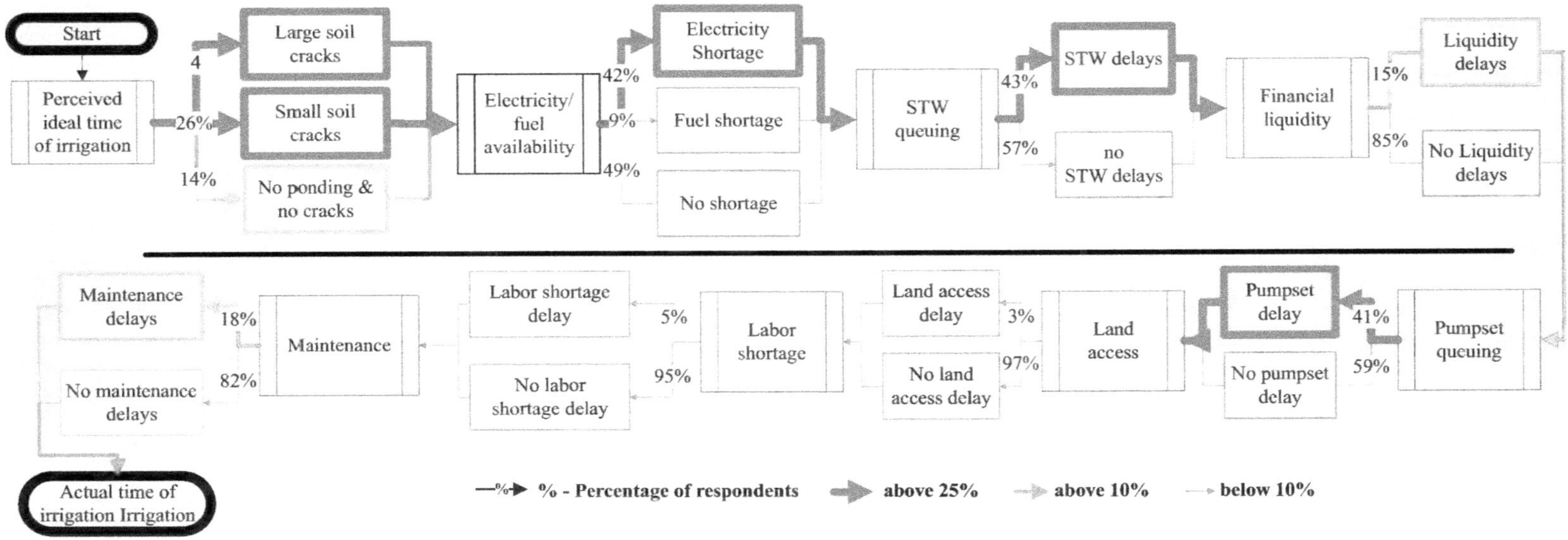

Figure 4. Quantification of ethnographic decision-tree models regarding the decision to use shallow tubewell irrigation, based on household survey results in all three study districts. The decision process can be restarted any time rain occurs.

Table 2. Irrigation delay factors reported by monsoon-season rice farmers, with average delay and percentage of occurrence in each district.

		Occurrence (%)		
Delay factor	Average delay (days)	Kailali	Banke	Rupandehi
Transportation	7	9	0	0
Electricity shortage	3	27	71	38
Fuel shortage	5	32	0	0
Illiquidity	3	27	6	31
Labour availability	5	5	0	17
Maintenance availability	4	18	6	38
Pumpset rental	3	32	37	28
Queuing at shallow tubewell	4	45	29	69

Discussion

All three districts seem to follow a similar overall development trajectory of increasingly erratic monsoon rainfall to which farmers respond through shallow tubewell expansion. But late irrigation scheduling and delay factors reduce productivity and production stability. This is reflected in generally long delays for queuing for wells and pumpsets in all districts. On closer inspection, however, precipitation and wealth differences between the districts arguably attune them to differing configurations of the same overall development trajectory. Most notably, Banke receives considerably less rainfall, leading to coordinated government investments in agricultural electrification and thereby incentivizing farmers to invest in electric pumpsets. This is reflected in the finding that unreliability of electricity is more critical in Banke than in the other two districts. Higher moisture stress and familiarity with smaller-horsepower electric pumpsets may also explain the prevalence of smaller-horsepower diesel pumpsets in Banke than in Kailali and Rupandehi.

Kailali and Rupandehi receive similar amounts of precipitation, but Rupandehi is wealthier. Increasing labour shortages (due to structural changes in the economy away from agriculture) and less fuel shortages are arguably a sign of this disparity. Similarly, land access issues persist in Kailali but not in Rupandehi, pointing at a larger use of bullock carts for transporting pumpsets to the field. Financial liquidity appears equally important in Kailali and Rupandehi but less so in Banke. This indicates that more frequent pump rental in Rupandehi and large inefficient pumpsets operated by resource-poorer farmers in Kailali remain a key issue in diesel pump irrigation, in contrast to areas where electric and small and efficient pumpsets are more common.

A single overall strategy to support farmers in responding effectively to monsoon breaks is therefore unlikely to suffice, because sustainable irrigation development needs to cater to contextual factors that vary geographically and even between households within the same community. Furthermore, groundwater development programmes risk widening existing social inequality between users (Wilson, 2002). This calls for pro-poor governance frameworks to ensure that different strata of water users benefit equitably from groundwater development. Recognizing these challenges, we propose three modular support pathways that correspond to nested spatial scales and can be prioritized for investment according to the socio-hydraulic characteristics of the target region. The main goal is to reduce the socio-economic barriers to groundwater use in an equitable and sustainable manner. Lastly, we briefly discuss the current state and potential of electrically powered shallow tubewell use and highlight the need to improve groundwater governance along with increasing use patterns.

As a general caveat, our findings should be contextualized within the longer-term change and development processes in this region. A recent drop in prices for solar panels has convinced many scientists, organizations and governments that solar-powered irrigation systems will be a key transformative technology for rural development in South Asia (Hartung & Pluschke, 2018). Similarly, policy imperatives for improved rural electrification, such as the ambition that Nepal could be the Battery for South Asia and India's initiative to bring electricity to 'the last village', suggest that grid-powered irrigation could plausibly increase in the nearer term, although large uncertainties remain about the pace and extent of rural electrification. The support pathways put forward in this article seek to identify options that are likely to encourage improved water use for diesel-powered irrigation until reliable grid- and solar-powered irrigation become widely available. Improved diesel systems could also be of use where electrically powered irrigation systems remain unfeasible. Understanding farmers' decision making with regard to current technologies could also contribute to better delineating problems and identifying target groups for scaling up solar and grid-powered irrigation systems. But such analyses remain outside the scope of this article.

Support pathway 1: efficiency gains at the farm level

Our findings point at two key interventions that could improve the water and energy efficiency of pumpset irrigation at the farm level to encourage farmers to move from 'life-saving' to 'productivity-enhancing' irrigation of *kharif* rice, one regarding irrigation scheduling and the other regarding operational efficiency.

Irrigation scheduling

Rice farmers in the Terai generally use the formation of large soil cracks as a cue for irrigation scheduling. This strategy can dramatically increase the costs of irrigation due to percolation losses from the root zone. Furthermore, allowing soils to dry past 40 kPa water tension can reduce rice yield by 50% in some soil types (Sudhir-Yadav et al., 2011); and large cracks typically form in even drier conditions. In addition to limited knowledge of the relationship between water stress and yield loss, late irrigation scheduling may also reflect farmers' tendency to be cash-investment averse rather than yield-or-profit-loss averse. In the case of marginal farmers, investment aversion in the Eastern Indo-Gangetic Plains is often linked to unfavourable tenancy agreements (Sugden, 2014). Risk aversion has been observed elsewhere as a limiting factor to the adoption of climate-smart agricultural practices intended to increase resilience, a situation in which both awareness raising and policy measures may be needed to boost uptake (Rai, Bhatta, Acharya, & Bhatta, 2018). Changing farmer's perceptions of the value and profitability of timely irrigation of *kharif* rice, in tandem with lowering irrigation costs to reduce investment aversion, have been observed to raise crop production and farmers' income (Qureshi et al., 2015).

Another factor in farmers' willingness to implement timely scheduling is uncertainty about the occurrence of the next rainfall. In general, the surveyed farmers are understandably reluctant to irrigate if rainfall could occur. Unfortunately, weather forecast information and climate-informed advisories for irrigation have not yet been widely deployed in South Asia. The World Meteorological Organization has acknowledged the need to improve the availability of climate information to better inform decision makers, with specific attention to agriculture. The Global Framework for Climate Services and a number of pilot efforts have developed out of this (Hewitt, Mason, & Walland, 2012). Another approach emerging in agricultural climate services is the sharing of crucial climate data made available by global producing centres with regional climate centres and national hydro-meteorological services. But the challenge is to build scalable decision frameworks that effectively leverage climate services and communicate them successfully to farmers to encourage behavioural change towards cost-effective and risk-reducing irrigation management.

Lastly, farmers mistakenly believe that groundwater is inferior in quality to rainwater. This can be addressed through clearer extension messaging. Scientific knowledge on irrigation scheduling is mainly anchored in assessing crop water requirements (e.g., Sudhir-Yadav et al., 2011), and advisory and decision-support systems may benefit from addressing these findings in their trainings and messaging.

Improving operational efficiency

Late irrigation scheduling practices in the Terai are primarily influenced by cash-investment aversion, so lowering operation costs may be an additional strategy to encourage farmers to move from 'life-saving' to 'productivity-enhancing' irrigation of *kharif* rice. Our results suggest that the use of inappropriately sized pumps reduces operational inefficiency through wasteful energy consumption. Mismatching pumps is a common problem in developing countries, where information on pumpset efficiency is often lacking and the choice of pumps is limited (Bom & van Steenbergen, 1997; G.J. Bom et al., 2001). For example, a crucial efficiency issue is the problem of friction losses in water delivery pipes. After vertical pumping, many farmers only need to use delivery pipes for short horizontal distances, while others may have to convey water over hundreds of metres using lay-flat pipes. Pump power requirements are much lower for the former group, especially with larger-diameter lay-flat pipes.

Several measures can address the prevailing scenario of low pump efficiency. First, simply adjusting engine speed can reduce fuel consumption by more than 50%, from 1–1.5 L/h to 0.5 L/h, while maintaining the same level of discharge (Bom et al., 2001). Second, market availability of energy-efficient and lower-cost small pumpsets can be improved. However, spare-parts availability for small pumpsets is currently insufficient, so this is an important consideration in efforts to increase irrigation efficiency. Third, energy loss to friction in lateral distribution piping can be reduced by making strategic use of gravity flow through lay-flat pipes or elevated tanks with attached delivery pipes. Larger-diameter delivery pipes can also be used to reduce friction loss, though potential trade-offs with the cost of larger pipes requires further investigation. Fourth, local technicians can use better and cheaper shallow tubewell construction techniques to reduce fuel consumption by 30% per unit of water pumped (Gert Jan Bom & van Steenbergen, 1997) – e.g., using mosquito nets as screens to cover pipe suction holes, or cleaning the well borehole after construction to increase inflow.

In general, guidelines on pump choice, irrigation system design and well construction could be made broadly available through trainings, pump dealers, and state extension services, but gaps between existing information and local knowledge of technical systems must also be addressed. Social marketing campaigns could also be aimed at addressing perception biases towards larger pumps, to educate farmers as to the benefits of smaller and more efficient pumpsets. The development of a common database as a clearinghouse of independently measured technical specification for pumps could also help farmers and policy makers with irrigation investment decisions. Without accurate knowledge of the importance of proper pump choice, pump dealers currently have limited capacity to promote energy-efficient solutions. Shopkeepers are happy to sell more expensive and less efficient pumps, and farmers who lack information will continue to purchase them. Our surveys indicated that pump mechanics also recommend less efficient and more costly pumps to farmers because they receive commissions from machinery dealers. Thus, efforts are needed to align the supply of technically sound information on pump efficiency with incentives to promote the use of smaller but ultimately more suitable pumpsets for shallow tubewells and *kharif* rice irrigation in the Terai.

Support pathway 2: improving community-level water markets

Our data demonstrate that most farmers in the Terai have several small and scattered fields. This is a clear challenge to irrigation coordination, although it also represents an opportunity to share resources with neighbouring farmers through water markets, i.e., users renting out their pumping equipment, their borehole, or both to other users. In all districts, delay factors associated with borrowing pumpsets and queuing for shallow tubewell use are common. Informal water markets with monopolistic pricing schemes erode the profitability of groundwater use for many farmers (Shah et al., 2009; Sugden, 2014). In addition to the well-known price distortions that may contribute to late irrigation scheduling, the observed coordination problems suggest two more factors that limit the efficiency and equitability of groundwater markets where renters could theoretically profit. First, the high land fragmentation increases coordination difficulties, as farmers need to arrange irrigation access in advance with several shallow tubewell owners. Second, farmers with limited financial resources experience financial illiquidity, as they have already invested their available cash in raising and transplanting rice seedlings. The latter situation in particular challenges the development of efficient water markets to overcome moisture constraints in the face of precipitation uncertainty.

Another issue that requires attention is male out-migration from rural areas. In our survey, this is mainly noticeable through the time delays reported by farmers in finding and hiring agricultural labourers in Rupandehi, the district with the largest migration rate (Ministry of Labor and Employment, 2017). Since irrigation of field crops is predominantly a male task, households with migrated men rely on relatives or neighbours to assist, or they need to wait for a male household member to return home after work. More policy attention to the gender context is required as this trend intensifies. While there is support for expanding the role of women as decision makers in agriculture, service provision provides another approach to address this delay factor.

The first pathway requires improving the organizational components of informal water markets to address the prohibitive cost of irrigation. The provision of low-interest financial services to farmers, so they may overcome within-season cash constraints prior to harvest and the sale of rice grain, is likely to be important in efforts to move farmers to 'productivity-enhancing' shallow tubewell irrigation use (Bhandari & Pandey, 2006). Lastly, and perhaps most importantly, taking an anticipatory approach and agreeing on terms and conditions for pump rental and irrigation well before the season, instead of when large soil cracks appear, is also likely to be crucial, especially given the organizational difficulties posed by land fragmentation. To this end, seasonal precipitation forecasts deployed through meteorological and extension services could help 'trigger' irrigation arrangements and the provision of fuel for pumps well in advance of the months when rainfall deficits could occur. But such forecasts require much skill to be taken seriously by farming communities (Hewitt et al., 2012).

Support pathway 3: regional investment prioritization – selectively increasing infrastructure density

Turn-taking for shallow tubewells and pumpset rental are major delay factors and therewith inefficiency of the informal water markets in all three surveyed districts. This means that mere access to irrigation is not a sufficient condition for timely irrigation. Better organization at the community level can only reduce delays to a certain extent. For more substantial delay reductions, increasing the density of shallow tubewells becomes crucial in areas where infrastructure is present but insufficient. Targeting criteria can be developed from this perspective so that pumps are prioritized where they are most needed, increasing the return on public investment in irrigation infrastructure. This approach can be employed together with data on aquifer characteristics such that shallow tubewell infrastructure development is also prioritized in areas where higher levels of water abstraction can be sustained.

Initiatives to encourage better use of groundwater resources for irrigation are also likely to require policy action to address land ownership patterns. This is because there are fewer incentives for tenant farmers and landlords to invest in shallow tubewell infrastructure on rented land (Sugden, 2014). Where higher pumpset penetration is desired, awareness raising and opportunities for the private sector to encourage market growth are likely to be crucial. For example, supporting spare-parts markets, assuring sufficient mechanic services and increasing commercial availability of pumpsets were prerequisites for the rapid and transformative growth of tubewell use in nearby Bangladesh (Qureshi et al., 2015). Convincing farmers that bigger pumpsets are not always better, e.g., by highlighting the fuel efficiency of small pumpsets, as well as identifying and promoting smaller pumpsets that require less frequent repair and maintenance, could be critical steps to increase pumpset ownership among smaller and marginal farmers.

Electric pumps: rural electrification and solar-powered irrigation

Given the large reductions in costs and CO_2 emissions achievable with electric pumpsets, rural electrification or solar-powered irrigation may be the best long-term solution for the sustainable intensification of *kharif* rice. Against the backdrop of groundwater overexploitation due to de facto free electricity in north-west India, it is important to keep in mind that metering electricity consumption can incentivize farmers to irrigate effectively (Mukherji et al., 2009). But the current rural electricity network in the Terai has significant problems, including frequent power cuts. Even when power is available, voltage fluctuations commonly damage pumpsets. And areas with relatively good access to electricity are especially vulnerable to power cuts and voltage fluctuations because irrigation pumping is often the largest consumer of irrigation during short periods, and often overloads the system. Besides the physical availability of electricity, access is often difficult to secure because the permit system in Nepal is highly political. And while the price of solar irrigation is falling, it still remains prohibitive for many farmers.

The benefit of reliable agricultural electrification in the Terai is expected to be enormous for broader policy goals such as food security, poverty alleviation and climate change adaptation. But despite the potential benefits of expanding the grid to rural areas, officials in Nepal report that electricity for irrigation is not a development priority, as irrigation would only constitute 1–2% of potential new revenue sources, with very high installation costs required to reach widely scattered villages (Nepal Electricity Authority, 2017). Thus, improving the efficiency and reach of the diesel-based shallow tubewells continues to make sense as a near-term development priority.

Groundwater governance: reiterating the case for sustainable and evidence-based management

Groundwater depletion has become a global concern (Famiglietti, 2014), but the opposite is true in the Eastern Gangetic Plains, including Nepal's Terai. Nepal's government considers the Terai's groundwater resources underdeveloped and estimates that 88% of the groundwater that could be abstracted on a sustainable basis (based on annual recharge) is not utilized, providing ample space for increased groundwater use for productivity-enhancing irrigation. In other words, encouraging productive use of water resources within regionally and locally defined sustainable abstraction boundaries will be a key element in reaching sustainable intensification targets in Nepal's Terai (Steffen et al., 2015).

Establishing a regional evidence base for groundwater governance in Nepal is difficult because the low-resource and low-technology environment poses great challenges for gathering recurrent monitoring data on the highly complex aquifer systems. But implementing a system that identifies persistent groundwater decline is a good first step; since groundwater dynamics are localized, local countermeasures such as managed aquifer recharge can be implemented. Regionally, such planning could leverage the Ganges Water Machine initiative, which is being used as a conceptual tool for managing the Indo-Gangetic Plain's complex and erratic hydrological cycle (Bharati et al., 2016; Shah, Ray, & Lele, 2018).

Conclusions

This study has established and ranked the importance of the key factors influencing the use of shallow tubewells for supplementary irrigation in *kharif*-season rice cultivation in the Terai region of Nepal. In areas where diesel pumps predominate, the factors most limiting shallow tubewell use are poor coordination among water users, delays in pump and tubewell availability, and financial constraints coupled with risk aversion towards cash investment. The electric grid permits the use of lower-cost pumps, and solar-powered irrigation systems could reduce operation costs. Electrification may overcome some of these delay factors, but the grid reaches only a small fraction of fields at present, and solar-powered irrigation is likely to remain beyond the financial means of most farmers in the Terai.

Our work indicates that a multiscalar strategy to encourage diesel-pump-based shallow tubewell irrigation, to move Nepali rice farmers from 'life-saving' to 'productivity-enhancing' irrigation use, could contribute to climate resilience, higher yields and higher profits. At the farm level, raising awareness of the importance of timely irrigation can be coupled with efforts to increase operational efficiency (e.g., pump maintenance, pump sizing, forecast-based irrigation scheduling) to overcome aversion to cash investments. At the community level, better preparation for irrigation events through organization of water markets before the start of the season could reduce transaction costs and delays during the season itself. At the regional level, government support programmes can target areas where tubewell and pumpset density is not yet high enough to ensure that all farmers have timely access to irrigation through water markets.

Acknowledgments

This research was conducted under the Cereal Systems Initiative for South Asia, Phases II and III, funded by USAID and the Bill and Melinda Gates Foundation. Additional support to Timothy J. Krupnik was provided by the Climate Services for Resilient Development project in South Asia, supported by USAID, and to Anton Urfels through Stiftung fiat panis. The contents and opinions expressed herein are those of the author(s) and do not necessarily reflect the views of the Bill and Melinda Gates Foundation, USAID, the US government or Stiftung fiat panis and shall not be used for advertising or product endorsement purposes. The authors declare no conflict of interest. Data can be obtained as referenced. Data owned by government departments can be obtained through their consent only. The household survey data for the ethnographic decision tree models are being deposited in the International Maize and Wheat Improvement Center dataverse repository and will be made available before publication.

Disclosure statement

No potential conflict of interest was reported by the authors.

Funding

This work was supported by the Bill and Melinda Gates Foundation; Stiftung fiat panis; United States Agency for International Development.

ORCID

Anton Urfels http://orcid.org/0000-0003-2920-8721
Andrew J. McDonald http://orcid.org/0000-0002-2660-3470
Timothy J. Krupnik http://orcid.org/0000-0001-6973-0106
Pieter R. van Oel http://orcid.org/0000-0001-7740-0537

References

Banerjee, A. V., Chandrasekhar, A., Duflo, E., & Jackson, M. O. (2018). Using gossips to spread information: Theory and evidence from two randomized controlled trials. *The Review of Economic Studies, 86* (6), 2453–2490.

Bhandari, H., & Pandey, S. (2006). Economics of groundwater irrigation in Nepal: Some farm-level evidences. *Journal of Agricultural and Applied Economics, 38*(1), 185–199.

Bharati, L., Sharma, B., & Smakthin, V. (2016). *The Ganges River basin: Status and challenges in water, environment and livelihoods.* New York: Routledge.

Bom, G. J., van Raalten, D., Majundar, S., Duali, R. J., & Majumder, B. N. (2001). Improved fuel efficiency of diesel irrigation pumpsets in India. *Energy for Sustainable Development, 5*(3), 32–40.

Bom, G. J., & van Steenbergen, F. (1997). Fuel efficiency and inefficiency in private tubewell development. *Energy for Sustainable Development, 3*(5), 46–50.

Bonsor, H. C., MacDonald, A. M., Ahmed, K. M., Burgess, W. G., Basharat, M., Calow, R. C., ... Zahid, A. (2017). Hydrogeological typologies of the Indo-Gangetic basin alluvial aquifer, South Asia. *Hydrogeology Journal, 25*(5), 1377–1406.

Central Bureau of Statistics. (2012). *Nepal: National sample census of agriculture 2011-2012.* Kathmandu, Nepal: National Planning Commission Secretariat, Government of Nepal.

Ciais, P., Reichstein, M., Viovy, N., Granier, A., Ogee, J., Allard, V., ... Valentini, R. (2005). Europe-wide reduction in primary productivity caused by the heat and drought in 2003. *Nature, 437*(7058), 529–533.

Famiglietti, J. S. (2014). The global groundwater crisis. *Nature Climate Change, 4*(11), 945–948.

Fitzpatrick, R. G. J., Parker, D. J., & Willetts, P. D. (2016). Assessing the level of spatial homogeneity of the agronomic Indian monsoon onset. *Geophysical Research Letters, 43*(22), 11, 867–11, 874.

Hartung, H., & Pluschke, L. (2018). *The benefit and risks of solar powered irrigation: A global overview.* Rome: FAO & GIZ.

Hewitt, C., Mason, S., & Walland, D. (2012). The global framework for climate services. *Nature Climate Change, 2,* 831–832.

Jain, M., Singh, B., Srivastava, A. A. K., Malik, R. K., McDonald, A. J., & Lobell, D. B. (2017). Using satellite data to identify the causes of and potential solutions for yield gaps in India's Wheat Belt. *Environmental Research Letters, 12*(9), 094011.

Kishore, A., Sharma, B., & Joshi, P. K. (2014). *Putting agriculture on the takeoff trajectory: Nurturing the seeds of growth in Bihar, India.* New Delhi: International Food Policy Research Institute and International Water Management Institute.

Massuel, S., Riaux, J., Molle, F., Kuper, M., Ogilvie, A., Collard, A. L., ... Barreteau, O. (2018). Inspiring a broader socio-hydrological negotiation approach with interdisciplinary field-based experience. *Water Resources Research, 54,* 2510–2522.

Ministry of Labor and Employment. (2017). *Labor migration for employment: A status report for Nepal: 2015/2016 & 2016/2017.* Kathmandu, Nepal: Author.

Mukherji, A., Das, B., Majumdar, N., Nayak, N. C., Sethi, R. R., & Sharma, B. R. (2009). Metering of agricultural power supply in West Bengal, India: Who gains and who loses? *Energy Policy, 37* (12), 5530–5539.

Nepal Agriculture Survey. (2012). Kathmandu, Nepal: Central Bureau of Statistics.

Nepal Electricity Authority. (2017). *A year in review.* Kathmandu, Nepal: Author.

Paudel, G., Maharjan, S., Guerena, D., Rai, A., & McDonald, A. J. (2017). *Nepal rice crop cut & survey data 2016.* Texcoco, Mexico: CIMMYT Research Data & Software Repository Network.

Pretty, J., & Bharucha, Z. P. (2014). Sustainable intensification in agricultural systems. *Annals of Botany, 114*(8), 1571–1596.

Qureshi, A. S., Ahmad, Z. U., & Krupnik, T. J. (2015). Moving from resource development to resource management: Problems, prospects and policy recommendations for sustainable groundwater management in Bangladesh. *Water Resources Management, 29*(12), 4269–4283.

Rai, R. K., Bhatta, L. D., Acharya, U., & Bhatta, A. P. (2018). Assessing climate-resilient agriculture for smallholders. *Environmental Development, 27,* 26–33.

Roth, H., & Botha, N. (2009). *Using ethnographic decision tree modelling to explore farmers' decision-making processes: A case study* . 2009 Conference, August 27–28, 2009, Nelson, New Zealand 97157, New Zealand Agricultural and Resource Economics Society.
Shah, T., Ray, C., & Lele, U. (2018). How to clean up the Ganges? *Science, 362*(6414), 503.
Shah, T., Ul Hassan, M., Khattak, M. Z., Banerjee, P. S., Singh, O. P., & Rehman, S. U. (2009). Is irrigation water free? A reality check in the Indo-Gangetic basin. *World Development, 37*(2), 422–434.
Shrestha, S. R., Tripathi, G. N., & Laudari, D. (2018). Groundwater resources of Nepal: An overview. In A. Mukherjee (Ed.), *Groundwater of South Asia* (pp. 169–193). doi:10.1007/978-981-10-3889-1_11
Steffen, W., Richardson, K., Rockström, J., Cornell, S. E., Fetzer, I., Bennett, E. M., … Sörlin, S. (2015). Planetary boundaries: Guiding human development on a changing planet. *Science, 347* (6223), 1259855.
Sudhir-Yadav, Humphreys, E., Kukal, S. S., Gill, G., & Rangarajan, R. (2011). Effect of water management on dry seeded and puddled transplanted rice. *Field Crops Research, 120*(1), 123–132.
Sugden, F. (2014). *Landlordism, tenants and the groundwater sector: Lessons from Tarai-Madhesh, Nepal.* IWMI research report 62. doi:10.5337/2015.204
Turner, A. G., & Annamalai, H. (2012). Climate change and the South Asian summer monsoon. *Nature Climate Change, 2*(8), 587–595.
United Nations Development Programme. (2014). *Nepal human development report 2014: Beyond geography, unlocking human potential.* Kathmandu: Author.
Wilson, K. (2002). Small cultivators in Bihar and 'new' technology: Choice or compulsion? *Economic and Political Weekly, 37*(13), 1229–1238.

Groundwater policies and irrigation development: a study of West Bengal, India, 1980–2016

Tapas Singh Modak

ABSTRACT
After a prolonged agrarian impasse, agricultural production in West Bengal, India, grew at an unprecedented rate in the 1980s, a process in which the expansion of groundwater irrigation played an important role. The growth of groundwater irrigation and agricultural production, however, decelerated from the early 1990s. This paper argues that the deceleration in groundwater irrigation development in West Bengal since 1995 is primarily attributed to the adoption of different regulatory policies in two ways: directly, through controlling the overall installation of tubewells or decelerating tubewell electrification; and indirectly, through high electricity pricing for water use in agriculture.

Introduction

Irrigation in India underwent major transformation after Independence. The 1950s and 1960s were marked by significant investment in public irrigation schemes and an increase in the area irrigated under public surface irrigation. From the late 1970s, however, private investment in groundwater irrigation registered rapid increase throughout the country. After the 'Green Revolution' began in India, the policies adopted by different state governments focused on promoting groundwater use by means of electricity supply at subsidized rates or free of cost, subsidized institutional credit for tubewell installation, and expansion of the public tubewell programme. These policies contributed significantly to an increase in the area irrigated by groundwater throughout the country. Groundwater emerged as a major source of irrigation in India, accounting for 62.3% of the net irrigated area in 2013. It is of note that although the development of groundwater irrigation contributed significantly to Indian agriculture, private groundwater irrigation is also associated with greater inequality in ownership of irrigation sources and equipment (Bhatia, 1992; Modak, 2018; Prakash, 2005).[1]

This paper studies groundwater irrigation in West Bengal, a water-abundant state in India. West Bengal provides an interesting site to research this topic because the issues related to groundwater use for agriculture have certain specific features that are unique to itself and makes for a case quite distinct from other parts of India that are dependent on groundwater. The major groundwater-irrigated states other than eastern India are Punjab, Haryana, Rajasthan, Tamil Nadu and Karnataka. In general, studies on groundwater use in the country focus on studying the overexploitation of groundwater. Such overexploitation is often caused by free or highly subsidized electricity for agriculture and by other policies relating to the ownership of groundwater (Bhatia, 1992; Cullet, 2014; Dubash, 2002; Janakarajan, 1993; Moench, 1992; Nagraj & Chandrakanth, 1997; Shah, 1993).

The discussions on groundwater use, however, in West Bengal and in general for the rest of eastern India are similar, but also differ in important ways. First, barring a few districts, West Bengal is rich in groundwater resources, and the resource is underutilized. In 2013, only 45% of rechargeable groundwater resource was used in the state. In contrast, overexploitation of groundwater resources is a common feature in other groundwater dependent states. Second, diesel is still the predominant source of energy for groundwater irrigation pumps in West Bengal. As a result, the consumption of electricity for agriculture was incredibly low. In comparison, 80% of all tubewells are electrified in the other groundwater-irrigated states. Third, despite abundant groundwater resources in West Bengal, the operational cost of tubewell irrigation is higher than in other groundwater-irrigated states.

Until the late 1970s, the underdevelopment of water control because of the lack of public intervention was identified as a key constraint leading to the stagnation of agricultural growth and production in West Bengal (Rawal & Swaminathan, 1998). In the 1980s, however, public interventions by the state, in terms of two important institutional changes – the implementation of land reforms and empowerment of local governance, and investment in rural infrastructure played a crucial role in inducing the mobilization of private capital in irrigation, which led to high agricultural growth in West Bengal (Rawal, 1999).[2]

But subsequently, there has been a deceleration in the expansion of groundwater irrigation as well as in the growth of agricultural output in West Bengal since the mid-1990s. In this context, this paper will analyses the policies related to groundwater and electricity use in agriculture in West Bengal, specifically to understand how the deceleration in groundwater irrigation development from mid-1990s is linked with different policies adopted by state and central government.

The paper uses data from the various state annual plans, five-year plans and economic reviews published by the Government of West Bengal between 1985 and 2015. In addition, the paper also uses the documents related to different irrigation programmes undertaken by the state, along with projections made by the Agricultural Census and Minor Irrigation Census.

Irrigation development in West Bengal, 1975–2011

Though statistics on irrigation in West Bengal are largely inconsistent and limited, the Agricultural Census provides a profile of irrigation development by all the sources at five-year intervals (Modak, 2020). We can draw a trend in area irrigated by different sources over the time in the state. The proportion of net area irrigated in total operational holding in the state increased from 24.1% in 1980–81 to 53.3% in 1995–96 to 62.2% in 2010–11. The net irrigated area was 1.3 million ha in 1980–81, which increased to 3.0 million ha in 1995–96 to 3.4 million ha in 2010–11. Groundwater emerged as a major source of irrigation and contributed to major expansion of irrigated area in the state, like the

Table 1. Annual average growth (%) of net irrigated area by sources, between 1976–77 and 2010–11, in West Bengal, India.

Sources	1975–85	1985–95	1995–2005	2005–11
Canal	1.3	0.0	−1.0	−4.1
Tank	4.4	0.4	−0.5	−4.4
Well	−3.5	10.8	−1.2	−1.3
Tubewell	57.5	17.7	1.8	2.8
Others sources	20.1	0.2	4.0	4.1
All sources	7.4	5.7	1.0	0.9

Sources: For 1975–85, cited in Rawal and Swaminathan (1998); All-India Report on Agriculture Census, Government of India 1992, 1998, 2003, 2008, 2012 and 2015.

country as a whole (Modak, 2018). The share of groundwater irrigation in total net area irrigated in West Bengal was only 6% in 1970–71 and that increased to 66% in 2010–11. In contrast, surface water irrigation, including canal and tank accounted for 87% of total net irrigated area in 1970–71, but sharply declined to only 21% in 2010–11.

There are two distinct phases in the expansion of irrigation in West Bengal between 1975 and 2011 (Table 1). During the first phase, between 1975 and 1995, the state showed a high growth in annual net irrigated area, at 7.4% between 1975 and 1985 and at 5.7% between 1985 and 1995. It is noteworthy that the expansion in tubewell irrigation was highest among all sources, at 57.5% in annual net irrigated area between 1975 and 1985. But it slowed down at 17.7% per annum between 1985 and 1995.

The second phase, between 1995 and 2011, showed a significant decline in the growth of irrigation development in the state. The annual growth rate of net irrigated area came down to only 1% in the same period. Surface water (canal and tank) and well irrigation had experienced negative growth. Importantly, the growth rate of groundwater irrigation development stagnated in this period.

Thus, the contribution of groundwater irrigation was the major factor for high growth in the expansion of irrigation in the 1980s, while its slowdown significantly affected the aggregate declining growth of irrigation in West Bengal from the mid-1990s. There can be two possible reasons. First, the scope of further increase in irrigation was restricted due to natural limits of groundwater exploitation in the state. But, as mentioned, the stage of groundwater development in the state as a whole was only 45% in 2013 (Central Ground Water Board, 2017). This implies that overuse of groundwater in West Bengal is not a big concern as it is in Punjab, Haryana, Karnataka, Rajasthan and Tamil Nadu. Thus, there is scope for further groundwater use in various districts, particularly those which have moderate irrigation levels. However, arsenic contamination of groundwater is often a serious quality problem for some parts of the state, which must also be considered. Second, there can be economic factors and policy constrains for the slowdown of irrigation development after the mid-1990s.

In the proceeding discussion, this paper analyses the state policies undertaken between 1980 and 2011 related to groundwater and electricity use in agriculture in West Bengal. There were two types of policy discourses. First, since the 1980s there was an emphasis on promoting minor irrigation, particularly groundwater use. Second, though the promotional policies continued, policies for regulation of groundwater use have also found currency since the mid-1990s.

The first phase, 1980–95: policies to promote minor irrigation in state plans

In this phase, the policy instruments used to pursue promoting minor irrigation were public minor irrigation schemes, agricultural electrification and credit support for the installation of minor irrigation structures.

The Left Front government, after assuming office in West Bengal in 1977, had stated in the state plans that the development of irrigation was one of the important non-land inputs for agricultural growth (Government of West Bengal, 1987). It is to be noted that 1983–84 was a landmark year in the state marking high agricultural production of 91.7 *lakh* tonnes of foodgrains, which continued in the subsequent years. The average annual foodgrain production in the state was 73.74 *lakh* tonnes during 1970–75 and 80.24 *lakh* tonnes during 1975–80. However, it was observed that this high agricultural growth was not uniform across different regions of the state in the early 1980s.[3] The regions that faced climatic vagaries experienced lower agricultural productivity and production, and here the main reasons for this uneven growth were attributed to inadequate irrigation and drainage coverage along with low investments in other inputs. The State Plan of 1986–97 specifically suggested greater emphasis on irrigation and drainage programmes to address the problems of weather aberrations.

There was an emphasis on minor irrigation schemes in the state policies since the 1980s. While the justification for such an emphasis is often based on the assumption that minor irrigation suits the interests of small farmers who have been incentivized by land reforms to farm, there were other macroeconomic reasons that favoured these policies. There was a paucity of sufficient funds to establish new major irrigation schemes in the state.[4] In the 1980s, the state policies on minor irrigation focused direct investment for two broad aims: to create public minor irrigation structures and on infrastructure, most importantly in rural electrification for agriculture.

Public minor irrigation structures

The West Bengal Minor Irrigation Project, an important initiative of public investment in groundwater irrigation, was taken up in 1985–86 with strong budgetary support from the World Bank. The main objective of the project was large-scale investment in public groundwater development to support the marginal and small farmers in the state, and also to complement the growing private investment in groundwater irrigation. Under this project, about 1.09 *lakh* ha of gross minor irrigation potential was created through the installation of 9750 tubewells with varying capacities, modernization of 200 river-lift irrigation schemes and the installation of 7170 open dug wells.[5] It is to be noted that under the project, Rs6402 *lakh* were allocated to the West Bengal State Electricity Board for electrifying the schemes and establishing a reliable power supply mechanism in the agricultural sector.[6] While the project could not achieve the targeted number of installing various types of groundwater structures, it was successful in achieving higher irrigation intensity and yields than what was initially anticipated as well as establishing economically viable tubewell irrigation systems (World Bank, 1995). An important aspect of the implementation of these schemes was the progressive involvement of *panchayat*s, which were tasked with the operation and maintenance of the structures and primarily run by beneficiary committees. It is also noted in a World Bank review report that the strong *panchayat* system in West Bengal favoured large investment in public groundwater irrigation development, since the self-government system at the grassroots level facilitated the implementation and sustainability of the schemes (World Bank, 1995). The project ended in 1994, having used 95% of the available credit. The remaining incomplete works, in particular the electrification of minor irrigation schemes by the West Bengal State Electricity Board, were completed with the support from the National Bank for Agriculture and Rural Development.[7]

Apart from this, there were several attempts to create public minor irrigation assets under different programmes. In 1988–89, the Government of India launched a programme for the installation of shallow tubewells under the crop husbandry sector which was funded in equal shares by the central and state government to assist small and marginal farmers to increase their agricultural production.[8] Under the Command Area Development programme, different types of tubewells were installed in the regions of the command area where flow irrigation did not reach. In addition, the minor irrigation schemes were created under different rural employment generation programmes. The most important of these was the Million Wells Schemes under Jawahar Rozgar Yozana, which was introduced in 1988–98. It is noteworthy that the special area programmes, focusing on backward regions, such as North Bengal and Jhargram, also paid special attention to the development of minor irrigation schemes.

The state government had adopted a strategy for the operation and maintenance of the irrigation projects at the local level through the involvement of project beneficiaries under the supervision of *panchayats*. It is noted in the Annual Plan of 1997–98 that local public action established a foundation that most importantly improved the management of the projects and their efficient and cost-effective functioning.[9]

Rural electrification for agriculture

In India, one important feature of agriculture is the complementarity between public and private investment. Ramakumar (2012, p. 95) argues that, 'in a growing economy, public investment induces or "crowds in" private investment by farmers'. Rural electrification is one of the most important components of indirect public investment in irrigation development and has important consequences in mobilizing private finance for investment in irrigation (Rawal, 1999). The Government of West Bengal had invested substantially in the power sector in the 1980s. While there was no investment by the state government in building infrastructure for the power sector in the 1970s, the state annual expenditure on the power sector increased from Rs4 *crores* in 1980–81 to about Rs14.6 *crores* (in 1980 prices) in 1990–91 (Rawal, 1999).

It stated that besides the electrification of all state-owned groundwater irrigation schemes, an attempt would be made to electrify privately owned groundwater structures.[10] As a result, there was a significant increase in the number of electrified pumpsets between 1977 and 1995 in West Bengal. For instance, the number of electrified pumpsets increased from 10,701 in 1976–77 to 99,252 in 1995–96. In the same period, the growth rate of the number of tubewells in West Bengal was about 10.3%, the third highest among major states in the country, as compared with 7.3% in the country as a whole. Similarly, the share of electricity consumption in the agriculture sector out of the total consumption in the state had increased from only 1.2% in 1980–81 to 10.3% in 1998–99.

It points towards the conclusion that the special attention of policies on rural electrification in the state after 1980, in particular for the agriculture sector, increased electricity use in agriculture for irrigation.

Institutional credit for tubewell installation

In general, the growth in institutional credit for agriculture had stagnated in West Bengal in the 1980s and 1990s (Rawal & Swaminathan, 1998). After banks were nationalized, there was substantial expansion in banking infrastructure in rural West Bengal in terms of the number of bank branches and credit. However, the credit disbursement to agriculture remained low in comparison with other states and in relation to need (Rawal, 2005). The amount of short-term and term loans per ha of gross cropped area disbursed by scheduled commercial banks was only Rs33.7 for West Bengal in 1980–81, while in the same year it was Rs234.7 in Kerala, Rs226.5 in Punjab, Rs151.5 in Tamil Nadu and Rs58.9 in India as a whole. In 1997–98, the credit disbursement had marginally improved in West Bengal (Rs259.3), but was still far from Kerala (Rs2251.3), Tamil Nadu (Rs2281.4) and even India as whole (Rs477.1) (Rawal, 2005). In this context, the S. R. Sen Committee, on reviewing the constraints of agricultural growth in eastern India, also noted that 'financing of inputs and on-farm investment have been adversely affected by the constraints in the flow of institutional credit in West Bengal' (Reserve Bank of India, 1994).

Specifically, institutional credit for the installation of private tubewells was substantially low in West Bengal. The data from the Minor Irrigation Census suggest that only 12.2% of the total shallow tubewells in the state were installed with any kind of support from bank loans in 1993–94.

The second phase, 1995–2011: policy for regulation of groundwater use

In the second phase, as discussed, the growth of groundwater irrigation development in West Bengal decelerated. While the progress of two important policy measures of the first phase (1980–95), promoting public minor irrigation and rural electrification for agriculture, had also stagnated from the mid-1990s. After the West Bengal Minor Irrigation Project supported by the World Bank ended in 1994, there was no further attempt for large-scale development of public groundwater irrigation in West Bengal. As a result, growth in the number of public minor irrigation schemes in the state decelerated after the mid-1990s. Data from the Minor Irrigation Census indicate that the number of new additions to public minor irrigation schemes, particularly shallow and deep tubewells, were very low between 2000–01 and 2013–14 (Table 2). In the case of dug wells and surface flow irrigation, there was a decline in the total number of schemes between 1993–94 and 2013–14, which may be attributed to the lack of maintenance and modernization.

After 1995, similarly, West Bengal could not continue the pace of electrification of agriculture. The data clearly indicate that the growth rate of rural electrification in agriculture has stagnated since 1995 (Figure 1). The compound annual growth rate of electrification of pumpsets in West Bengal has come down to only 1.1% between 1995 and 2010, even lower than the country average of 2.8%. As a result, the compound annual growth rate of electricity consumption in agriculture has declined to only 1.7% between 1995 and 2011, while it was about 20% between 1980 and 1995. The share of electricity

Table 2. Distribution (*n*) of minor irrigation schemes owned by public institutions in West Bengal, 1993–94 to 2013–14

Period	Dug well	Shallow tubewell	Deep tubewell	Surface flow irrigation	Surface lift irrigation
1993–94	8718	16,181	4038	9949	7335
2000–01	5554	22,000	5107	5332	8753
2006–07	3757	23,783	5165	5636	7949
2013–14	1470	25,842	6828	4803	13,576

Source: Minor Irrigation Census, Government of India.

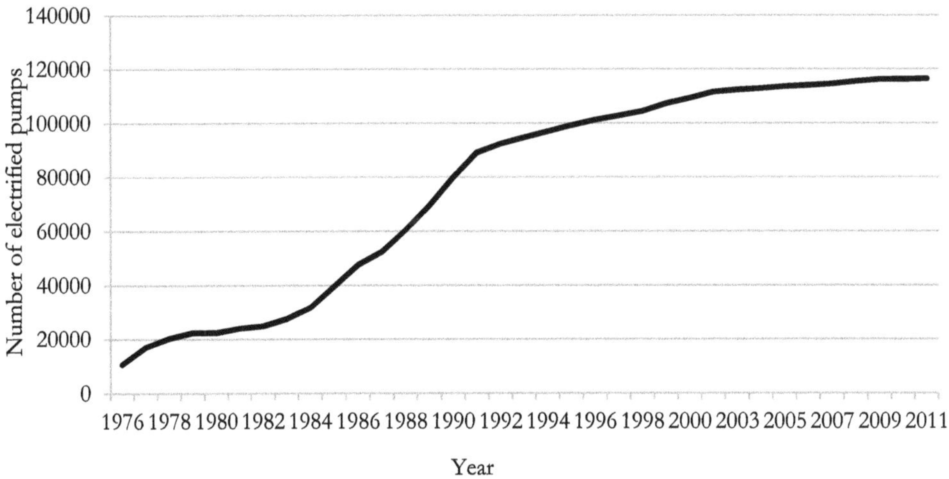

Figure 1. Number of electrified agricultural pumpsets in West Bengal, 1976–2011. Source: Public electricity supply.

consumption in the agriculture sector out of total consumption in the state has declined by half, from 10.3% in 1998–99 to 4.9% in 2010–11 (Figure 2).

The erstwhile West Bengal State Electricity Board was primarily responsible for rural electrification in West Bengal until 2007. In state plans of mid-1990s, it was noticed that the board was facing acute constraints in funding, which was one of the reasons why the targets of electrified pumpsets could not be achieved.[11] The Annual Plan of 1997–98 stated:

> The West Bengal State Electricity Board has a command area which does not include State's major industrial and commercial establishment. So, it is rather difficult for West Bengal State Electricity Board to set aside sufficient fund for rural electrification from its revenue collection after meeting its obligations, such as debt, servicing, purchase of power from other utilities, system augmentation etc. (p. 156)

Policy for regulation of groundwater use in West Bengal, 1995–2011

Apart from fund crisis, it is found that the state policies undertaken between 1993 and 2011 regulated the use of groundwater and restricted the electrification of tubewells had the effect of slowing down the growth of groundwater irrigation in West Bengal.

After Independence, policies in India have primarily focused on increasing groundwater use rather than regulating its use. Until the mid-1990s, groundwater use was

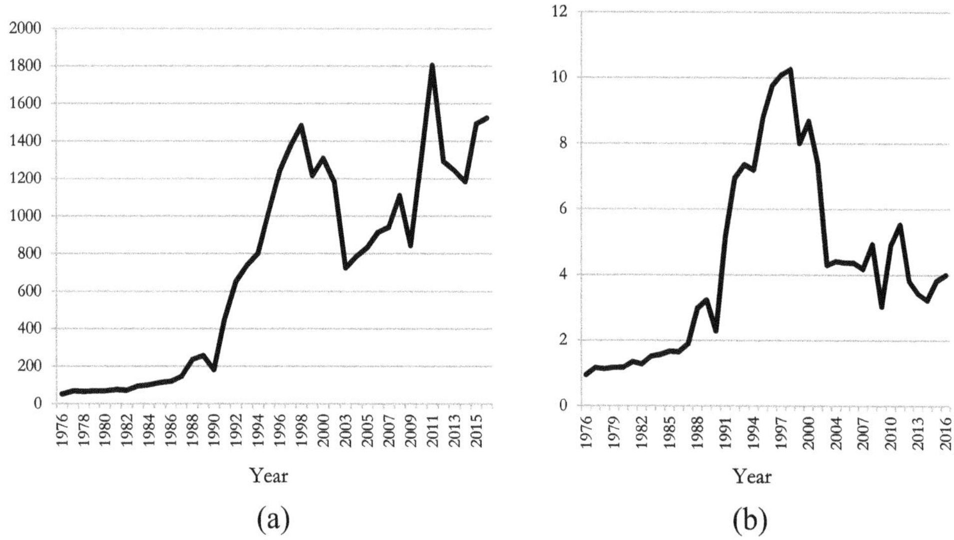

Figure 2. (a) Electricity consumption (GWh) in the agricultural sector in West Bengal, 1976–2016; and (b) electricity consumption (%) in the agriculture sector out of total electricity consumption in West Bengal, 1976–2016. Source: Public electricity supply.

recognized as a conventional approach of 'development', though some studies pointed out the threat of overuse, particularly in semi-arid and arid areas of the country (Dhawan, 1993; Moench, 1992). After the mid-1990s, there has been slow change in the policy of groundwater, from a development approach to a management approach (Shankar et al., 2011). It has been also argued that India lacks adequate regulation for groundwater use, which could also be a major factor for overexploitation in many parts of the country (Cullet, 2014; Koonan, 2016). The Government of India prepared the Model Bill to Regulate and Control the Development and Management of Groundwater in 1970. Cullet (2014) noted that this model bill offers states a framework of regulation that can be adapted according to the requirements of individual states. In general, the main objective of the model bill is to establish a groundwater authority under direct control of the government. The role of this authority is to identify areas where monitoring is required to control the development groundwater by means of regulatory policies. However, the state is the final authority for taking the decision. The model bill was revised several times in 1974, 1992, 1996 and 2005. Many state legislatures have adopted a bill on groundwater uses, some of which have adopted a special focus to drinking water (see Table A1 in Appendix A).

Before 1993, there was no restriction on groundwater use in West Bengal. The report of the first Minor Irrigation Census in 1986–87, which was published in 1993, indicated a rising trend of groundwater use in the state. In this context, the Government of West Bengal directed the West Bengal State Electricity Board to impose a restriction through an executive order on electricity connections of privately owned low duty tubewells fitted with submersible pumps starting from April 1993. According to this order, tubewell owners were supposed to obtain an approval from the State Water Investigate Directorate before applying for an electric connection to West Bengal State Electricity Board.[12] This restriction was applicable in only seven districts – Barddhaman, Birbhum, Hugli, Murshidabad, Nadia, Purba (East) Medinipur, Paschim (West) Medinipur and North 24 Parganas.

Following the Model Bill to regulate and control the development and management of groundwater as initiated by the central government, the Government of West Bengal passed a bill in the legislation, titled the West Bengal Groundwater Resources (Management, Control and Regulation) Act, 2005. Before that, a standing committee was formed to discuss the Act in 2004–05, and finally it was adopted in July 2005. The rules under the Act were in effect from 1 August 2006. The State Water Investigate Directorate, Government of West Bengal, was the principal authority for implementing the act which outlined three levels of authorities: (1) District Level Groundwater Resources Development Authorities; (2) One Corporation Level Groundwater Resources Development Authority; and (3) State Level Groundwater Resources Development Authority.

Permissions for the installation and electrification of groundwater structures

Under the West Bengal Groundwater Resources (Management, Control and Regulation) Act, 2005, it was compulsory for any users of groundwater across the state to obtain a registration certificate from the State Water Investigate Directorate for sinking a new tubewell or for existing tubewells fitted with any mechanized or electric device for extracting water (installed before the Act). The District Level Groundwater Resources Development Authorities was the responsible authority to issue the permit for irrigation, domestic purposes for sinking a new tubewell or issuing a certification for existing tubewells which extracted water not exceeding 50 m^3/h. In case the rate of extraction exceeded 50 m^3/h, the District Level Groundwater Resources Development Authorities forwarded the application to the State Level Groundwater Resources Development Authority with its comments. The Corporation Level Groundwater Resources Development Authority was given authority to issue a new tubewell that extracted water not exceeding the rate of 100 m^3/h. The State Level Groundwater Resources Development Authority adopted some rules for implementing the Act effectively, the most important of which was the 'safe distance criteria', the minimum permissible distance between two groundwater structures (Table 3).

Any users in the safe administrative blocks, which are recognized by the Central Groundwater Board, could apply for permission to extract groundwater for any purpose. For irrigation purposes, however, use was restricted strictly in semi-critical, critical and overexploited blocks in the state until 2011.

Table 3. Minimum distance for obtaining permission for a tubewell, by capacity of tubewells, in West Bengal.

Capacity of well (m^3/h)	Permissible distance (m)
≤ 50	200
> 50 to 100	400
> 100	600

Source: State Water Investigate Directorate, West Bengal.

It is often argued by some researchers that getting permission from the State Water Investigate Directorate for installing a new tubewell or permission for existing tubewells according to the Groundwater Act 2005 created bureaucratic hurdles for farmers (Jayaraman, 2015; Modak & Bakshi, 2017; Mukherji et al., 2009, 2018). On the ground, however, it was observed that the number of rejections from the State Water Investigate Directorate was significant across the districts in the state. Table 4 shows that about 64.1% new applications for installing tubewells were rejected by the State Water Investigate Directorate on different grounds between 2007 and 2010. The rate of rejection was different across the districts, ranging from 25% in Howrah to 98.4% in Darjeeling. Importantly, the number of existing tubewells applying for registration certificate was incredibly low between 2007 and 2010, comprising only about 10% of total tubewells in the state in 2007. It also implies that the awareness of registration for existing tubewells was uncommon among the farmers. The rejection rate was about 54.3% for existing tubewells for the same period.

In addition, the electricity distribution authority has apparently made the clearance certificate from the State Water Investigate Directorate mandatory for any electrification of tubewells in the state. It is to be noted that the number of electrified tubewells was much smaller than the number of permits given by the State Water Investigate Directorate between 2000 and 2010. For example, the numbers of newly electrified tubewells were less than 1000 between 2002 and 2010. Thus, it can be explained from the demand side also that the farmers were least interested in electrified tubewells. This was because to get their tubewells electrified, farmers had to pay the full capital cost; this included the cost of wire, cemented poles and a 10-kVa transformer. After submitting an

Table 4. Permission status as per Groundwater Act 2005 by SWID, 2007–10.

	Registration for new tubewells		Registration for tubewells constructed before 2005	
Districts	Applications received (*n*)	Application rejected (%)	Applications received (*n*)	Application rejected (%)
Bankura	2038	48.6	4215	26.9
Barddhaman	890	73.6	5911	73.5
Birbhum	2406	70.6	6448	47.4
Koch Behar	–	–	–	–
Dakshin Dinajpur	1856	81.5	153	90.8
Darjiling	63	98.4	0	
Hughli	1361	43.6	2812	47.1
Haora	136	25.0	15	46.7
Jalpaiguri	264	96.6	39	71.8
Maldah	1038	92.0	2541	88.6
Murshidabad	1953	79.7	9657	76.9
Nadia	263	27.0	1943	2.0
North 24 Parganas	439	41.5	367	40.3
Paschim Medinipur	6036	76.7	6708	40.9
Purba Medinipur	1116	53.7	2487	34.0
Puruliya	35	62.9	7	42.9
South 24 Parganas	125	56.0	110	80.9
Uttar Dinajpur	2878	27.1	4	50.0
West Bengal	22,897	64.1	43,417	54.3

Source: Mukherji et al. (2018).

application to the electricity board, applicants were provided a quotation for the required amount for money for deposit, which was not generally less than Rs1 *lakh.*

Therefore, these two interlinked regulation policies and high capital costs were the main reasons that decelerated the progress of electrification of tubewells in the state after the mid-1990s.

Power sector reform in West Bengal and regulation of groundwater use

By the beginning of the 1990s, the state electricity boards of the country broadly were plagued by technical and financial underperformance (World Bank, 2010, cited in Badiani & Jessoe, 2015; Chatterjee, 2018). In response, as part of reform policy adopted by the Government of India in 1990, an initiative to bring reforms to the power sector began through the opening up of private investment or independent power producers into the electricity sector (Dubash & Rajan, 2001). The reform in the power sector was comprehensively enacted through the implementation of the Electricity Act, 2003, which replaced all electricity-related legislations and allowed for a fundamental restructuring of the electricity sector.

In West Bengal, the West Bengal State Electricity Board, formed in May 1955, was the vertically integrated state body that included all state-owned power generation, distribution and transmission. Consequently, with the implementation of the Electricity Act, 2003 by the centre, the state government constituted a state-level committee chaired by Subimal Sen, a member of the State Planning Board, to examine the issues related to restructuring the power sector and also to review the reform process adopted in other states of India.[13] In 2007, West Bengal became the fourth state in India, after Maharashtra, Assam and Gujarat, to enact the recommendations of restructuring the integrated state electricity board. The integrated West Bengal State Electricity Board was spilt into two utilities: a transmission utility that is the West Bengal State Electricity Transmission Company Ltd, and a single-distribution company that is the West Bengal State Electricity Distribution Company Ltd. The West Bengal State Electricity Distribution Company Ltd was the responsible authority for distributing power in the state at the 33-kV level and below.

With respect to electricity use in agriculture, the important consequences of reform in the power sector were the implementation of universal metering in the state. As per a memorandum signed on 5 May 2001 between the Ministry of Power of the Government of India and the Government of West Bengal, the state government had agreed for universal metering for all consumers. The objective was to identify the transmission and distribution losses in different sectors.[14] In addition, there was pressure from donor agencies, in particular the World Bank, to revert to metering in agricultural sectors as a precondition for financial support to the power sector (Dubash & Rajan, 2001; Meenakshi et al., 2011). Metering as a precondition for electricity supply was also stated in the Electricity Act, 2003.[15] West Bengal was one of the states that adopted universal metering for the agriculture sector. The West Bengal State Electricity Distribution Company Ltd initiated metering for all electric tubewells in 2007, and by 2011 about 90% of all tubewells were metered in the state (Mukherji & Das, 2012).

In 2007, importantly, dynamic pricing was introduced in meter-based electricity tariffs for agriculture. It is a demand-side management technique, while the primary objective of this system is to reduce the use of electricity during peak hours and encourage its use

Table 5. Power tariff (*paisa*/kWh) in agriculture in West Bengal for temporary and permanent connection.

Period	Normal hours (6 am–5 pm)	Peak hours (5–11 pm)	Off-peak hours (11 pm–6am)
2007–08	137	475	75
2008–09	130	490	74
2009–10	140	510	79
2010–11	218	588	152

Source: West Bengal Electricity Regulatory Commission.

during off-peak hours. Accordingly, the three shifts for tariffs, the time of the day, in agriculture was introduced in 2007–08. They include the normal hours from 6 am to 5 pm, peak hours from 5 pm to 11 pm, and off-peak hours from 11 pm to 6 am. By these norms, the tariff was highest for peak hours and lowest for off-peak hours (Table 5). Further, the time-of-the-day system was extended to other consumers in the state, such as for domestic and commercial uses.

The West Bengal State Electricity Board also provided electricity connections to tubewell owners for agricultural use on a temporary basis for a period of 105 days between February and May. The objective was to ensure power supply for irrigation during *boro* paddy cultivation, a water-intensive crop in the state. The diesel-powered shallow tubewells are usually inoperable in the summer season with respect to the seasonal fluctuation of the water table and also not cost-effective because of increasing diesel prices. These temporary connections were generally provided from the domestic electric transformer and bamboo poles were used to connect the wire to the tubewells. The tubewell owners had to pay a security deposit as well as the connection fees. At the end of the season, the actual charges of electricity uses were adjusted with the deposit money. The bills were generated on the basis of actual consumption of electricity by installing a prepaid time-of-the-day meter. However, the charges were higher for temporary connections than the metered connections. The average tariff for temporary connections was 452.3 *paisa*/kWh, higher than the permanent connection (319.3 *paisa*/kWh) in 2010–11. However, the tubewell owners primarily found the temporary connection helpful for two reasons. First, they could excuse themselves from investing a large capital cost for extending the distribution infrastructure (including transformer, wire and poles) needed for a permanent connection; and second, no permissions were required from the State Water Investigator Directorate for such temporary connections (Shah et al., 2012).

Changes in policy of groundwater use for agriculture after 2011

To address the low rates of tubewell electrification, the Government of West Bengal removed the restrictions on issuing permits for tubewell installation in 2011. The Water Resources Investigation and Development Department issued a notification on 3 November 2011 that directed that the installation of tubewells with a discharge limit of up to 30 cubic metres/h and using 5-HP (horse power) pumps did not require permits from the district authority of the State Water Investigation Directorate. However, this was not applicable for 38 blocks of the state that were categorized as semi-critical as per Central Ground Water Board in 2008–09.[16]

At the same time, the West Bengal State Electricity Distribution Company Ltd reduced the electricity connection fee ranging from Rs1000 to Rs30,000, that is, it depends on the connection load, while earlier, farmers had to bear the full capital cost; this included the cost of wire, cemented poles and a 10-kVa transformer (Mukherji & Das, 2012). In addition, the Department of Agriculture, Government of West Bengal, launched a scheme called the One Time Assistance for Electrification of Agricultural Pump-sets on 30 November 2012. Under this scheme, small and marginal farmers received a one-time assistance of Rs8000 to meet the electricity connection charges for five HP agricultural pumps.

These changes in policy significantly increased the number of electrified tubewells in the state. Between 2011 and 2016, data from the public electricity supply show that the number of electrified tubewells more than doubled (Table 6). It is noteworthy that the share of electric shallow tubewells in the state increased from 18.2% in 2006–07 to 32.1% in 2013–14 (Minor Irrigation Census).

Though the number of electrified tubewells in the state increased after the implementation of the recent policies, other problems have since emerged. There has been a rapid hike in electricity tariffs for irrigation in the recent period in West Bengal (Figure 3)

Table 6. Number of electrified pumpsets in West Bengal, 2011–16.

Year	Electrified pumpsets (*n*)	Increase (%)
2011	1,16,343	–
2012	1,16,348	0.004
2013	1,16,458	0.1
2014	2,12,745	82.7
2015	2,43,027	14.2
2016	2,69,395	10.8

Source: Public electricity supply.

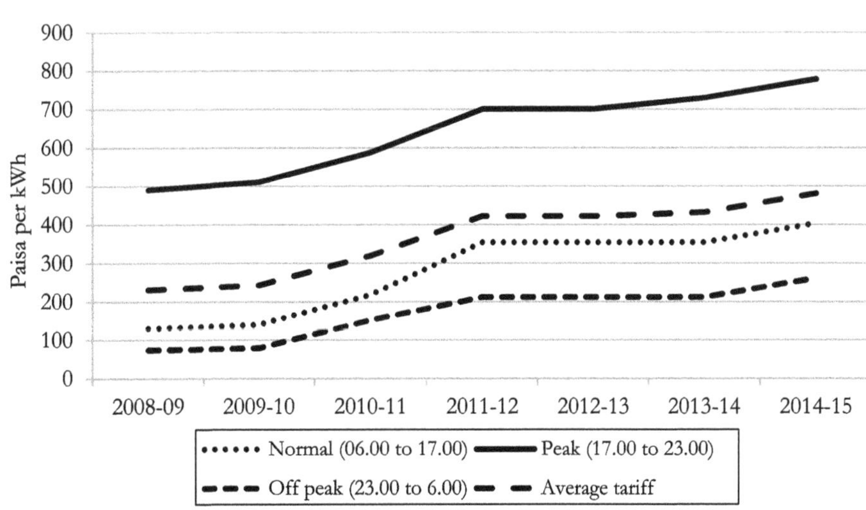

Figure 3. Average tariff (*paisa*/kWh) in agriculture in West Bengal, from 2008–09 to 2014–15 Source: West Bengal Electricity Regulatory Commission.

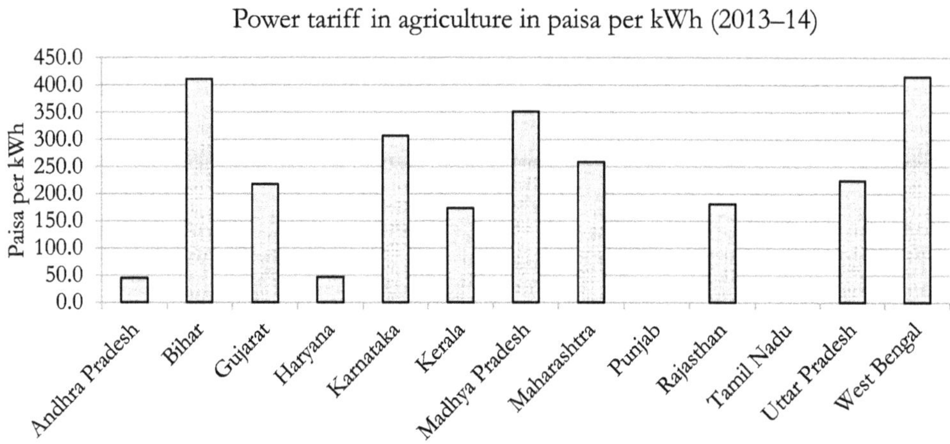

Figure 4. Average power tariff (*paisa*/kWh) in agriculture in major groundwater-irrigated states of India, 2013–14. Source: Ministry of Power, Government of India.

(Shah & Chowdhury, 2017). By assigning equal weights to the three tariff segments, it is observed that the average tariff increased about 50%, from Rs3.19/kWh in 2010–11 to Rs4.81/kWh in 2014–15. In addition, the yearly flat-rate electricity cost for unmetered tubewells with submersible pumps also increased from Rs13,176 in 2010–11 to Rs21,600 in 2015–16. The power tariff for agriculture in West Bengal is now the highest among all major groundwater-irrigated states in the country (Figure 4).

Thus, the abolition of state control over the allocation of electric pumps has not benefited farmers in West Bengal. This is because while the fixed cost of irrigation, that is, the pumpset installation charges and bureaucratic hurdles, has reduced; there has been no decline in variable costs, that is, electricity prices. Hence, there has been no marginal cost reduction. It is noteworthy that with the increasing number of electric tubewells in the state as a source of irrigation, the rapid hike in electricity prices for agriculture would have a greater implication on agriculture. Such high rates of tariffs do not augur well for agricultural production because they increase the costs of production.

Conclusions

This paper discusses different state policies, specifically those related to installation and electrification of tubewells, to understand the reasons for the deceleration of groundwater irrigation development in West Bengal after the mid-1990s.

It is clearly noted that there was a close relationship between the policies on groundwater use adopted in the state and two distinct phases of groundwater irrigation development in West Bengal. Between 1975 and 1995, groundwater irrigation for agriculture in West Bengal was promoted. The state government aimed to develop minor irrigation schemes by means of creating public minor irrigation structures and generating public investment in rural infrastructure, particularly in electrification for agriculture. It resulted a remarkable increase in the area irrigated by tubewells in the state. During the second phase (1995–2011), however, West Bengal could not continue the pace of growth in groundwater irrigation development. After the completion of the West Bengal Minor Irrigation Project in 1994, no further large-scale public groundwater irrigation scheme was implemented in West Bengal. At the same time, the growth of electrification in agriculture had also slowed down after 1995.

Apart from the fund crisis, the Government of West Bengal implemented two policies to regulate groundwater use, particularly in agriculture, from the mid-1990s. These policies involved direct control over installing groundwater structures and indirect control through regulating electricity connections and pricing. In addition, the electrification of a tubewell was not economically viable for a large number of small and marginal farmers in the state because a tubewell owner had to pay full capital costs for an electricity connection of a tubewell. It is argued here that these regulation policies adopted since 1995 and the high capital cost for electrification were the major reasons for the deceleration in the growth of groundwater structures in the state, and in particular the progress of electrification of tubewells between 1995 and 2010.

After 2011, the Government of West Bengal relaxed the restrictions on groundwater use and electrification of tubewells, resulting in a substantial increase in the number of electrified tubewells between 2011 and 2016. But the rapid hike in the electricity tariff for irrigation would have substantially increased the irrigation cost and adversely affected the agricultural production, particularly for those small and marginal farm households in West Bengal buying water from private water markets.[17] The official statistics published in the Statistical Abstracts of West Bengal also show that between 2010 and 2014, as a result, there was steady decline in area under *boro* paddy cultivation, the main irrigated crop, in West Bengal, at a rate of −2.5% annual compound growth (Modak, 2017).

In conclusion, based on the foregoing analysis of state policies and secondary data in light of West Bengal's experience, this paper highlights that by the early 1990s, the major policy prescription with regard to groundwater development in India was the restriction of the unregulated growth of groundwater structures and groundwater exploitation. This directive was based on the situation in the northern, western and southern regions of India. The main reasons for the introduction of such regulatory measures in these regions (and particularly in the states of Punjab, Haryana, Karnataka, Tamil Nadu, Gujarat and Rajasthan) were the falling water table and overexploitation of aquifers, and consequent water shortages for agriculture and other purposes. This restriction of groundwater exploitation became national policy and was enforced by means of different measures in different states.

In West Bengal, the medicine was administered, although the disease did not exist. The stage of groundwater development (i.e., the ratio of annual ground water draft to net annual ground water availability) was less than 45% (and therefore nowhere near overexploitation). Nevertheless, in seven districts of the state, policies were implemented to slow down the electrification of wells (and therefore the extension of groundwater irrigation) in 1993. Restrictions were extended to all districts in 2005. The effect of the policy was to put brakes on the development of groundwater irrigation and, consequently, on the growth of agricultural production.

This paper argues that West Bengal's experience with respect to groundwater irrigation highlights the need for region-specific, watershed-specific and sometimes even aquifer-specific policies. The implementation in West Bengal of a policy directive based on the experience and needs of other regions cost West Bengal agriculture dear. In practice, one size did not fit all.

Notes

1. Groundwater ownership rights in India are tied to land ownership. As the distribution of land across households of different classes and castes is extremely unequal, there is an inherent inequality in the ownership of and access to groundwater.
2. It is argued by several scholars that public investment in Indian agriculture played an important role in inducing private investment (Rao, 1993; Dhawan, 1997; Shetty, 1990).
3. State Plan Proposal 1986–87, Government of West Bengal, 1985, p. 14.
4. The Seventh and Eighth Five-Year Plans (1985–90 and 1992–97) acknowledge the scope for a significant increase in irrigated area through major and medium irrigation projects was limited.
5. Annual Plan 1997–98, p. 140.
6. Annual Plan 1990–91 and Eighth Five-Year Plan 1992–97, Vol I, p. 222.
7. Annual Plan 1997–98, p. 140.
8. Eighth Five-Year Plans 1992–97, Annual Plan 1992–93 and Vol I, p.189.
9. It has been argued that poor operation and maintenance is the main reason for the shrinking of public irrigation systems in India (Dhawan, 1997; Shah, 2008; Namboodiri & Gandhi, 2009; Mukherji, 2016).
10. Eighth Five-Year Plan, 1990–95, Annual Plan 1991–92 and Vol I, p. 142.
11. Annual Plan 1995–96 and Vol I, p. 120.
12. State Plan 1992–97, p. 190.
13. Annual Plan 2006–07, p. 69.
14. Annual Plan, West Bengal 2002–03, p. 130.
15. 'No licensee shall supply electricity, after the expiry of two years from the appointed date, except through installation of a correct metre in accordance with the regulations to be made in this behalf of a correct metre in accordance with the regulations to be made in this behalf by the authority' (Electricity Act 2003, Article 55, p. 1).
16. Seehttp://wbwridd.gov.in/swid/appellate.html.
17. The statistics published by the Commission for Agricultural Costs and Prices (CACP) indicate that there was an increase in irrigation cost for paddy cultivation in West Bengal during this period.

Acknowledgments

This research paper is a part of authors' submitted PhD dissertation on 'Groundwater Irrigation in West Bengal after 1990'. The author is grateful to Aparajita Bakshi, V. K. Ramachandran, Madhura Swaminathan and an anonymous referee of this journal for useful comments on the paper.

Disclosure statement

No potential conflict of interest was reported by the author.

References

Badiani, R., & Jessoe, K. (2015). Electricity prices, groundwater and agriculture: The environmental and agricultural impacts of electricity subsidies in India. World Bank and University of California – Davis.

Bhatia, B. (1992, December 19). Lush fields and parched throats: Political economy of groundwater in Gujarat. *Economic and Political Weekly*, *27*(51–52). https://www.epw.in/journal/1992/51-52/review-agriculture-review-issues-specials/lush-fields-and-parched-throats

Central Ground Water Board. (2017). *Dynamic groundwater resources in India (as on March 31, 2013).* Ministry of Water Resources, River Development, and Ganga Rejuvenation, Government of India, Faridabad. http://cgwb.gov.in/Documents/Dynamic%20GWRE-2013.pdf

Chatterjee, E. (2018). The politics of electricity reform: Evidence from West Bengal, India. *World Development, 104*(2018), 128–139. https://doi.org/10.1016/j.worlddev.2017.11.003

Cullet, P. (2014). Groundwater law in India: Towards a framework ensuring equitable access and aquifer protection. *Journal of Environmental Law, 26*(1), 55–81. Retrieved May 9, 2018, from http://jel.oxfordjoirnals.org/content/26/1/55

Dhawan, B. D. (1993). *Trends and new tendencies in Indian irrigated agriculture.* Commonwealth Publishers.

Dhawan, B. D. (1997, June 28). Large-scale canal irrigation: How cost effective?. *Economic and Political Weekly, 32*(26). https://www.epw.in/journal/1997/26/review-agriculture-review-issues-specials/large-scale-canal-irrigation-how-cost

Dubash, N. K. (2002). *Tubewell capitalism: Groundwater development and Agrarian change.* Oxford University Press.

Dubash, N. K., & Rajan, S. C. (2001). Power politics: Process of India's power sector reform. *Economic and Political Weekly, 36*(35), 3367–3387+3389–3390. https://www.epw.in/journal/2001/35/special-articles/power-politics.html

Government of West Bengal. (1987). *State plan proposals 1988-89, Volume I.* Kolkata: Department and Planning Department.

Janakarajan, S. (1993). Triadic exchange relations: An illustration from South India. *IDS Bulletin, 24*(3), 75–82. Institute of Development Studies, Brighton. https://doi.org/10.1111/j.1759-5436.1993.mp24003011.x

Jayaraman, T. (2015, September 11–13). Climate change and agriculture: Some considerations for West Bengal. *Presented at the Symposium on Results from Village Surveys, organised by the Foundation for Agrarian Studies.*

Koonan, S. (2016). Revamping the groundwater legal regime in India: Towards ensuring equity and sustainability. *Socio-Legal Review, 12*(2), 45. http://www.ielrc.org/content/a1602.pdf

Meenakshi, J. V., Mukharji, A., & Gupta, A. (2011). Does marginal cost pricing of electricity affect groundwater pumping behavior of farmer?: Evidence from natural experiment of West Bengal. IWMI.

Modak, T. S. (2017). *Electricity Tariffs and Boro Rice Cultivation in West Bengal.* blog at Foundation for Agrarian Studies. https://fas.org.in/blog/?s=tapas

Modak, T. S. (2018). From public to private irrigation: Implications for equity in access to water. *Review of Agrarian Studies, 8*(1), 28–64. http://ras.org.in/f071c902dea772c447c66730b0cb911c

Modak, T. S. (2020). *Groundwater irrigation in West Bengal after 1990: Policies, institutional arrangements and its implications for farm households.* PhD thesis submitted to School of Social Sciences, Tata Institute of Social Sciences.

Modak, T. S., & Bakshi, A. (2017). Changes in groundwater markets: A case study of Amarsinghi Village, 2005 to 2015, *Review of Agrarian Studies, 7*(2), 91–110. Retrieved May 9, 2018, from http://ras.org.in/changes_in_groundwater_markets

Moench, M. (1992). Chasing the water table: Equity and sustainability in groundwater management. *Economic and Political Weekly, 27*(s), 51–52. https://www.epw.in/journal/1992/51-52/review-agriculture-review-issues-specials/chasing-watertable-equity-and

Mukherji, A. (2016, December 28). Evolution of irrigation sector. *Economic and Political Weekly, 51*(52). https://www.epw.in/journal/2016/52/water-governance/evolution-irrigation-sector.html

Mukherji, A., Banerjee, P. S., & Biswas, D. (2018). Private investments in groundwater irrigation and smallholder agriculture in West Bengal: Opportunities and constraints. In A. Mukherjee (Ed.), *Groundwater of South Asia*(pp. 657–674). Springer.

Mukherji, A., & Das, A. (2012). How did West Bengal bell the proverbial cat of agricultural metering?: The economics and politics of groundwater. *Water Policy Research.* IWMI-Tata. https://www.iwmi.cgiar.org/iwmi-tata/PDFs/2012_Highlight-02.pdf

Mukherji, A., Das, B., Majumdar, N., Nayak, N. C., Sethi, R. R., & Sharma, B. R. (2009). Metering of agricultural power supply in West Bengal, India: Who gains and who loses? *Energy Policy, 37* (12), 5530–5539. https://doi.org/10.1016/j.enpol.2009.08.051

Nagraj, N., & Chandrakanth, M. G. (1997). Intra and inter-generational equity: Effects of irrigation well failures. *Economic and Political Weekly, 32*(13), A41–44. https://www.epw.in/journal/1997/13/review-agriculture-review-issues-specials/intra-and-inter-generational-equity

Namboodiri, N. V., & Gandhi, V. (2009). Institutional analysis of the performance of surface water institutions in India. In L. Crase & V. Gandhi (Eds.), *Reforming institutions in water resource management: Policy and performance for sustainable development*, Earthscan.

Prakash, A. (2005). *The dark zone: Groundwater irrigation, politics and social power in North Gujarat.* Orient Longman.

Ramakumar, R. (2012, July). Large-scale investments in agriculture in India. *IDS Bulletin, 43*, 92–103. https://doi.org/10.1111/j.1759-5436.2012.00351.x

Rao, P. S. (1993). *Review of selected literature on indicators of irrigation performance.* IIMI.

Rawal, V. (1999). *Irrigation development in West Bengal* [PhD thesis]. Indira Gandhi Institute of Development Research.

Rawal, V. (2005). Banking and credit relations in rural West Bengal. In V. K. Ramachandran & M. Swaminathan (Eds.), *Financial liberalization and rural credit in India.* Tulika Books.

Rawal, V., & Swaminathan, M. (1998). Changing trajectories: Agricultural growth in West Bengal, 1950 to 1996. *Economic and Political Weekly, 33*(40), 2595–2602. https://www.epw.in/journal/1998/40/special-articles/changing-trajectories-agricultural-growth-west-bengal-1950-1996

Reserve Bank of India. (1994). *Agricultural productivity in Eastern India.* Report of the Committee on Agricultural Productivity in Eastern India, Vol. I and II, Reserve Bank of India

Shah, T. (1993). *Groundwater markets and irrigation development: Political economy and practical policy.* Oxford University Press.

Shah, T. (2008). Groundwater management and ownership: Rejoinder. *Economic and Political Weekly, 43*(17). https://www.epw.in/journal/2008/17/discussion/groundwater-management-and-ownership-rejoinder.html

Shah, T., & Chowdhury, S. D. (2017). Farm power policies and groundwater markets: Contrasting Gujarat with West Bengal (1990–2015). *Economic and Political Weekly, LII*(25), 39–47. https://www.epw.in/journal/2017/25-26/review-rural-affairs/farm-power-policies-and-groundwater-markets.html

Shah, T., Giordano, M., & Mukherji, A. (2012). Political economy of the energy–groundwater nexus in India: Exploring issues and assessing policy options. *Hydrogeology Journal, 20*(5), 995–1006. https://doi.org/10.1007/s10040-011-0816-0

Shankar, P. S. V., Kulkarni, H., & Krishnan, S. (2011, January 8). India's groundwater challenge and the way forward. *Economic & Political Weekly, xlvi*(2), 37–45. https://www.epw.in/system/files/pdf/2011_46/02/Indias_Groundwater_Challenge_and_the_Way_Forward.pdf

Shetty, S. L. (1990, February 17). Investment in agriculture: A brief review of recent trends. *Economic and Political Weekly.* https://www.epw.in/journal/1990/7-8/special-articles/invest ment-agriculture-brief-review-recent-trends.html

World Bank. (1995). West Bengal minor irrigation project (Credit-1619-IN). Agriculture Operations Division, South Asia Regional Office.

Appendix A

Table A1. Acts and regulations related to groundwater in some major states.

State	Acts and regulations
Andhra Pradesh	Andhra Pradesh Ground Water (Regulation for Drinking Water Purposes) Act, 1996 Andhra Pradesh Water, Land and Trees Act and Rules, 2002 Andhra Pradesh Water Resources Regulatory Commission Act, 2009
Assam	Assam Ground Water Control and Regulation Act, 2012
Bihar	The Bihar Groundwater (Regulation and Control of Development and Management) Act, 2006
Chhattisgarh	Chhattisgarh Ground Water (Regulation and Control of Development and Management) Bill, 2012
Goa	The Goa Ground Water Regulation Act, 2002
Gujarat	Gujarat Ground Water Authority (GGWA) has been constituted by the state government in 2001 Gujarat Irrigation and Drainage Act, 2013
Haryana	Haryana Preservation of Subsoil Water Act, 2009 Haryana State Groundwater Management & Regulation Bill, 2013
Himachal Pradesh	Himachal Pradesh Ground Water (Regulation and Control of Development and Management) Rules, 2007
Karnataka	Karnataka Ground Water (Regulation and Control of Development and Management) Act, 2011
Kerala	The Kerala Ground Water (Control and Regulation) Act, 2002
Madhya Pradesh	Madhya Pradesh Pariyojana Ke Karan Visthapit Vyakti (Punhsthapan) Adhiniyam, 1985 was enacted with special focus to protection of drinking water. The model bill for groundwater is under consideration.
Maharashtra	Maharashtra Groundwater (Development and Management) Act, 2009 Maharashtra Water Resources Regulatory Authority Act, 2005/2011 Maharashtra Management of Irrigation Systems by the Farmers Act, 2005
Odisha	Odisha Groundwater (Regulation, Development and Management) Bill, 2011
Punjab	The Punjab Ground Water (Control and Regulation) Act, 1998 was prepared, but the State Committee observed that the draft is too harsh on users and also is not in the larger interest of farmers. However, the Government of Punjab adopted some regulations to regulate uses of irrigation water, such as the Preservation of Subsoil Water Act, 2009 which restricts new electricity connections in 'dark' areas and introduces water-saving technologies
Rajasthan	The Rajasthan Ground Water Management Bill, 2006 is under consideration of a select committee.
Tamil Nadu	Chennai Metropolitan Area Groundwater (Regulation) Act, 1987 Tamil Nadu Ground Water (Development and Management) Act, 2003 on 4 March 2003, which includes provision of Tamil Nadu Ground Water Authority to regulate and control water development in the State of Tamil Nadu
Uttar Pradesh	The Uttar Pradesh Groundwater Conservation, Protection and Development (Management, Control and Regulation) Bill, 2010
West Bengal	The West Bengal Ground Water Resources (Management, Control and Regulation) Act, 2005

Source: Central Ground Water Board, International Environmental Law Research Centre.

Part V

Transboundary aquifers

Transboundary groundwater governance in the Guarani Aquifer System: reflections from a survey of global and regional experts

Zachary P. Sugg, Robert G. Varady, Andrea K. Gerlak and Rafael de Grenade

To understand transboundary groundwater governance in the South American Guarani Aquifer System, we surveyed global and regional experts about the region's groundwater quantity and quality, ownership and rights, and regulation and administration. Respondents (1) perceived groundwater quality and withdrawal as under-regulated, and relevant information and data as inadequate; (2) suggested that contamination and overdrafting remain mostly incipient and localized along international borders; and (3) viewed groundwater as a shared resource administered by the state for the public, rather than as private property. Respondents suggested that while there is progress towards implementing a formal transboundary aquifer agreement, local-to-national-scale governance is important.

Introduction

As much as 98% of the earth's total freshwater that is not in ice caps is in aquifers (Foster & Chilton, 2003). Aquifers around the globe, though invaluable sources of water for humans and ecosystems, have largely escaped scrutiny, integration into legal frameworks, and management. Thousands of aquifers (water-bearing geologic formations) and groundwater basins (systems of one or more aquifers) have been identified, though mapping is recent and incomplete. Some 450 aquifers and basins merit scrutiny because they underlie two or more nations (IGRAC, 2012).

Visibility, transnationality and governance of aquifers

In many regions full allocation (and over-allocation) of surface water has increased pressure on groundwater as an alternative supply. In spite of rising extraction costs – especially for energy used in pumping and conveyance – groundwater may be the sole option available. In some places, growing competition for diminishing and degraded water has turned attention to governance, especially in transboundary and transjurisdictional settings. Where robust institutions are lacking, absence of rules around transboundary waters – along with prevailing uncertainty, indecision and differing agendas – has engendered competition, crisis, and occasionally conflict (Dietz, Ostrom, & Stern, 2003).

As with international rivers, transboundary groundwater governance is challenged by national sovereignty, upstream–downstream hegemony, quality control and, most significantly, water rights. Uniquely for aquifers, however, the water remains unseen and thus difficult to chart and quantify. Issues of sovereignty, property rights and ownership – already thorny in the case of surface water – are further complicated by this 'invisibility' of the resources and their boundaries. Despite the number of transboundary groundwater systems and their importance to human development, international water policy makers, conventions and agreements have paid scarce attention to these resources (Puri & Aureli, 2005).

The invisibility of and lack of attention to transboundary groundwater may be about to change. Propelled by increasing demand for groundwater, there is growing interest in its management and an emerging need for cross-border cooperation (Foster & Chilton, 2003; GEF/TWAP, n.d.; UNESCO, 2009). A notable example is the ambitious multi-year project, Groundwater Governance: A Global Framework for Country Action (2011–2015), funded by the Global Environment Facility (GEF), which seeks to gain a better understanding of groundwater governance and to protect the resource's sustainability. Accordingly, GEF has commissioned explorations of policies, laws, practices, and other aspects of governance (GEF/FAO/UNESCO/IAH/WB, 2013). This enterprise recognizes explicitly the potential benefits of 'soft-path' approaches to water management. For several decades after World War II, development approaches featured infrastructure-intensive, 'hard-path' solutions such as dams, reservoirs, water transfers and physical plants (Brandes & Brooks, 2007).

Touting non-infrastructural methods, Eckstein (2011) highlights a number of recently signed transnational agreements and activities relating to groundwater cooperation. International initiatives to map groundwater and to formulate legal frameworks, such as the UN International Law Commission's Draft Articles on the Law of Transboundary Aquifers (Yamada, 2011), also have emerged. Policy scholars call on the global water community to improve water governance – especially in transboundary settings (Dinar, Dinar, & McCaffrey, 2007; Jarvis, 2014; Linton & Brooks, 2011; Wolf, 2007; Wouters, 2013). In response, the GEF project and other programmes such as the International Groundwater Resources Assessment Centre, UNESCO's Internationally Shared Aquifer Management initiative, and the United Nations Economic Commission for Europe are creating and enhancing knowledge of the physical, societal, legal and policy aspects of transboundary groundwater governance (IGRAC, 2012; ISARM, http://www.isarm.org/; Margat & Van Der Gun, 2013; UNECE, 2014). These initiatives have included informal and formal, or non-diplomatic and diplomatic, arrangements at the state, national and transnational levels (de los Cobos, 2012).

The case of the Guarani Aquifer System

The 2010 Agreement on the Guarani Aquifer System is a notable example of formal regional cooperation between aquifer-sharing nations. One of the world's largest groundwater bodies, the Guarani Aquifer System (SAG)[1] underlies four countries (Argentina, Brazil, Paraguay and Uruguay) and has benefited from heightened attention, sustained research studies and important development projects. In the Guarani, as in other large transnational aquifers, there exist multiple challenges to effective groundwater governance. As a framework for our analysis – and building on previous research – we highlight three key elements of governance: (1) groundwater quantity and quality issues; (2) ownership and rights; and (3) regulation and administration. Further, we contend that groundwater governance requires examination at two scales: regional, given the specific geographical settings and geologic heterogeneity of aquifers; and global, to provide context and perspective.

To better understand transboundary groundwater governance in the SAG, we developed a survey questionnaire to elicit the perspectives and experiences of two sets of informants: (1) individuals with knowledge of transboundary issues in the SAG region, i.e. officials, managers and other stakeholders actively involved in SAG groundwater governance or familiar with on-the-ground conditions and practices; and (2) 'global experts', i.e. representatives of international and nongovernmental organizations, law and policy scholars, hydrologic professionals and other observers who are broadly conversant with groundwater management and governance, especially in transboundary settings. We supplemented our questionnaire with individual interviews with 10 experts on the 2010 agreement and related political and governance issues. These interviews clarified perceptions of SAG science and policy, and apprised us of recent national and international developments.

Potential application of this study

Ultimately, we hope our empirical data will improve transboundary groundwater governance in three ways.

1. By addressing the knowledge gap that inevitably lags legal and mapping efforts, this study elicits perceptions of stakeholders as a step towards developing a holistic understanding of transboundary groundwater governance, policies and institutions.
2. Consistent with soft-path inquiries and the recent focus on informal shared agreements based on norms and rules (e.g. Blatter & Ingram, 2001), this knowledge can help clarify whether governance improves or worsens in a particular setting.
3. An assessment of formal agreement efficacy in addressing specific transboundary governance challenges will enable better cross-border resources management.

Our study may help other researchers and professionals interpret and assess transnational groundwater agreements. Prospective drafters of such instruments may benefit from more in-depth knowledge of a prominent transboundary aquifer. Our case, built on the perspectives of regional and international experts on shared governance challenges, can inform management of other transboundary aquifers.

Transboundary groundwater governance

Groundwater governance has gradually pervaded the literature on resources management policy over the last decade (see e.g. Conca, 2005; Kemper, 2004; Moench, Dixit, Janakarajan, Rathore, & Mudrakartha, 2003; Mukherji & Shah, 2005; Rogers & Hall, 2003; Shah, 2008; UNDP, 2004; Varady, Meehan, & McGovern, 2009). Its application to international transboundary situations also has emerged. Efforts such as UNESCO's Internationally Shared Aquifer Resources Management project (ISARM, http://www.isarm.org/) – to identify, situate and assess aquifers – have helped draw attention to transboundary groundwater governance as an issue of international consequence (Aureli

& Eckstein, 2011; Aureli et al., in press; Linton & Brooks, 2011; Margat & Van Der Gun, 2013; Puri & Aureli, 2009).

Notwithstanding rising attention, functional definitions of groundwater governance have remained elusive. Megdal, Gerlak, Varady, and Huang (2014, p. 2) define groundwater governance as "the overarching framework of groundwater use laws, regulations, and customs, as well as the processes of engaging the public sector, the private sector, and civil society" that "shapes how groundwater resources are managed and how aquifers are used". This framework operates among multiple jurisdictional levels. *Transboundary* groundwater governance may be seen as the subset of groundwater governance that applies to aquifers that underlie and cross political boundaries (Figure 1).

Transboundary groundwater governance often faces legal and cultural constraints, and practical limitations such as ineffective organizations and insufficient human and financial resources (Varady et al., 2013). Effective water governance is characterized by institutions that operate accountably, efficiently, responsively and sustainably (Rogers & Hall, 2003). Overall, effective governance of transboundary aquifers can be pursued by establishing appropriate scales-of-interest articulation and decision making, and by involving assorted non-state actors, as well as state agencies (Linton & Brooks, 2011). Yet, stakeholder participation demands good, symmetric, transparent and reliable information to facilitate cooperation (Mukherji & Shah, 2005; Puri, 2001). Groundwater governance is intimately linked to the development and implementation of norms and principles that promote change in actor behaviour across management scales (Pahl-Wostl, Gupta, & Petry, 2008). In addition to more formal processes and institutions, informal rules, institutions and practices may be key elements of effective governance (e.g. Blatter & Ingram, 2001; Sehring, 2009). Further, interplay between formal and informal institutions – across and within scales – is an important institutional and political element of governance (Lebel, Garden, & Imamura, 2005; Young, 2002).

Although innovative insights into groundwater governance are emerging (e.g. de Gooijer et al., 2009; Kemper, 2007; Theesfeld, 2010), gaps remain in understanding of the workings of more contextual and cross-scalar arrangements. Further, the existing literature generally lacks instructive examples of transboundary groundwater situations in which such arrangements are pursued (but see e.g. Eckstein & Hardberger, 2007; Eckstein, 2013). Especially lacking are views and opinions concerning some important elements of transboundary groundwater governance, such as availability of sufficient quantities of acceptable-quality water, types of ownership

Figure 1. Transboundary groundwater governance as a set of nested concepts.

regimes, and questions of regulation and administration type. For example, results from the first multinational survey of socio-economic dimensions of groundwater use were reported only as recently as 2006 in a study that solicited information from tube-well owners in various South Asian countries (Shah, Singh, & Mukherji, 2006), though this study did not identify transnational issues. And while the international and comparative study by Sagala and Smith (2008) addressed questions of legal ownership and control of groundwater, it similarly did not address transboundary groundwater governance.

These governance questions merit consideration because of the relative infancy and unknown effectiveness of formal and informal transboundary groundwater agreements and other multilateral instruments. As indicated earlier, we examined three basic elements of transboundary groundwater governance in the SAG and globally: (1) groundwater quantity and quality; (2) ownership and water rights; and (3) regulation and administration.

Groundwater quantity and quality

Aquifer depletion has remained geographically uneven; that is, in some places around the globe, only a small fraction of the estimated total fresh groundwater resources has been tapped for human uses, while in other cases there has been a 'revolution' in groundwater use, especially for agricultural purposes (Giordano & Villholth, 2007; Llamas & Martínez-Santos, 2005). The severity of groundwater depletion in a region may reflect access to aquifers, rising water demand driven by population growth and urbanization, availability of financing and infrastructure, and full allocation of existing surface-water supplies. The rapid and severe exploitation of groundwater in basins around the world creates management challenges, including controlling overdraft and water-quality problems (Giordano, 2009).

Governance and management of groundwater present special challenges. Few would contest the need for responsiveness to local and regional conditions, problems, constituencies, perspectives and interests. When such considerations are disregarded, tensions are exacerbated and solutions become more elusive (Ingram, Milich, & Varady, 1994). When resources cross (above or beneath) national boundaries, national interests necessarily come into play and often dominate resource governance.

Identifying salient cross-border groundwater issues, however, is far more daunting than doing so for surface water, partly because of the immense cost and inherent difficulty of quantifying the resource's physical parameters. As importantly, data and information essential to effective decision making and implementation may be inadequate. Furthermore, crafting rules specifying who can and should collect such information, to whom to distribute it and how is not trivial – especially when entangled with transnational politics.

Ownership and rights

In most nations, groundwater has been considered as either government or state property; or a communally owned resource; or private property that may be owned by individuals or private-sector entities. In both civil law and common law countries, groundwater may be considered the private property of individual landowners (Burchi & Nanni, 2003). In such cases, active or collective governance, management and regulation may be minimal. Users may perceive groundwater as property that can be owned or exploited without deference to neighbouring users. In contrast, groundwater can be viewed as a shared resource, where individuals see it not as ownable but as legally trust-held by the state. Then, in some places, groundwater is owned collectively by civil society and subject to regulation, as per South Africa's 1998 Water Act, Argentina's Civil Code, and by Islamic law and tradition in Islamic countries (Burchi, 1999; Burchi & Nanni, 2003).

The need for governance mechanisms to establish transboundary management processes underlines the importance of who 'owns' or *ought to* own groundwater. State and regional jurisdictions, as well as nations, may apply varying legal doctrines of usufruct and resource-ownership rights. Yet researchers and policy makers may not know whether stakeholders in transboundary settings know which of these regimes might be in effect or most appropriate. Questions of sovereignty, differences in views about ownership, and laws and policies determining ownership rights are part of the groundwater governance process.

Regulation and administration

Groundwater quantity and quality and ownership-regime heterogeneity of groundwater basins are often connected to our third significant element of transboundary groundwater governance: cross-border administration and regulation. Boundaries – whether international or interjurisdictional – complicate natural resources management. Administrative procedures, legal statutes and regulatory mechanisms all are artefacts of the societies and political units within which they are embedded and function. Transnational settings can render issues of ownership and sovereignty even more byzantine than intra-national settings by introducing an additional web of institutions, obligations, rights, neighbours' rivalries, and other complications. Some sharing countries may have undeveloped or unenforced regulatory regimes for groundwater within their borders, complicating transnational coordination and agreements.

In regard to water resources, concerns over sovereignty, national security, budgetary resources and other central-government prerogatives strongly favour the predominance of national agencies. But water management typically responds to local issues and concerns that may transcend or contradict national interests. This disjuncture is also reflected in data collection and control – data are typically hard to come by, considered valuable, and commonly sought by national agencies.

As Moench (2004) argues, provision of the information that matters most to users should include local-scale measurements rather than regional-scale assessments. Schlager (2007) further emphasizes that groundwater information promotes management solutions more effectively when it involves active (i.e. local) user participation. When users are on opposite sides of an international border, however, sovereignty and security concerns can hamper coordination.

Groundwater – like surface water – is a flowing resource that ignores political boundaries, and this complicates transboundary groundwater governance, administration, allocation and regulation. With transboundary aquifers, a common question is: Who does or should perform these critical functions? We seek to provide new empirical evidence to examine stakeholders' views on administration in a transboundary context.

Figure 2. The Guarani Aquifer region. Map credit: Lily House-Peters.

The Guarani Aquifer System: description and context

Our research focuses on the Guarani Aquifer System (SAG, Figure 2). The SAG is one of the most spatially extensive groundwater reserves in the world, covering 1.1 million km^2 and underlying parts of four nations in the mid-eastern region of South America.[2] The SAG's freshwater storage volume is approximately 30,000 km^3 (Foster, Hirata, Vidal, Schmidt, & Garduño, 2009), qualifying it as one of the world's largest aquifers by volume.

The SAG comprises Mesozoic sandstones ranging in thickness from less than 50 m to more than 800 m (Kemper, Mestre, & Amore, 2003). The aquifer is mostly confined by basalt layers (Tujchneider, Perez, Paris, & D'Elia, 2007), although there is approximately 125,000 km^2 of outcrop zones around the periphery that contain areas of both recharge and discharge (Amore & Tröger, 2010). The system includes two sub-basins: the Central Paraná and the south-western Chaco–Lower Paraná.

Natural groundwater quality is generally considered very good; approximately 90% is estimated to be potable (Kemper et al., 2003). The system is 'storage dominated', with the estimated 45–55 km^3 of yearly natural recharge being less than 0.2% of the SAG's total freshwater storage (Foster et al., 2009). Current withdrawal rates are approximately 1.04 km^3 per year, with around 94% used in Brazil, 3% in Uruguay, 2% in Paraguay and 1% in Argentina (Foster et al., 2009). SAG groundwater use is primarily urban, with the major uses being 66% for public water supplies, 16% for industrial, and 13% for recreation and tourism (Amore & Tröger, 2010).

The SAG has been the target of well-funded efforts to formulate and adopt a multinational protocol for information sharing, management, and conflict resolution. Gerring (2007) refers to the SAG as an "extreme" case in terms of attention and resources devoted and extent of transboundary cooperation.[3] In turn, this prominence has facilitated the implementation of transnational agreements for sustainable management and dispute resolution. As a result, the SAG has been studied hydrologically, legally and socioeconomically – work that has been funded by the GEF, World Bank, Organization of American States, and Mercosur (Foster et al., 2009).

The SAG's significance owes much to its location in a densely populated, rapidly developing area of south-central South America. Some 15–24 million inhabitants live within the area delimited by the Guarani Basin, and 70–90 million people live in areas influenced by the aquifer (Foster et al., 2009; OAS, 2005). Because of the enormous stores of surface water and groundwater, the area not only supplies irrigation water but also holds major potential for municipal and industrial use and for generation of additional hydro-geothermal energy (the Itaipu hydroelectric facility, near SAG's centre, is the world's second-largest hydropower plant). The area is rapidly urbanizing and faces intensive land-use-change pressures, principally the conversion of naturally forested lands to soybean cultivation. These land-use changes contaminate the runoff that recharges the aquifer, and affect groundwater quality (Hirata & Vidal, 2012).

Since at least the late 1980s to early 1990s, observers of world water consumption have noted a rising demand for finite supplies of freshwater (e.g. Falkenmark, 1990). Not far from the SAG, São Paulo's 2015 drought and water-shortage crisis demonstrates the intensity of population-related need for ever-growing supplies of water (Romero, 2015). In the SAG, localized areas of pollution and overexploitation evoked concern from communities, researchers and institutions in the four countries. This concern – leveraged with the support and leadership of scientists and stakeholders – prompted transnational cooperation towards an aquifer-management strategy. That effort led to serious multinational water-management negotiations amongst the water officials of the four nations. The framework for this collaborative process built on the 1969 La Plata Basin Treaty to establish principles for governance, conflict resolution and information exchange for the basin's surface waters (del Castillo Laborde, 1996; Walter, 2013a).

In 2003, with GEF and World Bank support, the four SAG countries began discussions and negotiations on joint management and protection of the aquifer's groundwater resources. The initiative aimed towards sustainable management and use of the SAG via a management framework based on "appropriate technical, scientific, institutional, legal, economical and environmental guidance" (OAS, 2005, p. 2). These negotiations took place in an atmosphere of water-management decentralization and democratization that featured scientific research and involvement of stakeholders at various levels in the four countries (Villar & Ribeiro, 2011; Walter, 2013a).

The 2003–2009 Project for the Protection and Sustainable Development of the Guarani Aquifer System aimed to: (1) increase scientific knowledge about the SAG; (2) establish a permanent monitoring system for the groundwater resources; (3) create a transboundary diagnostic analysis and a strategic action plan; and (4) identify basic guidelines for a framework for integrated management of the resources. The scientific outcomes included regional hydrogeological assessments and maps, hydrodynamic models of the aquifer system, surveys and geo-referencing of water-wells, and a piezometric monitoring system to identify 'hotspots' (OAS, 2005; Walter, 2013a). The countries also developed science-based, local-level operational arrangements. The cities of Salto, Uruguay, and Concordia, Argentina, for example, adopted common drilling practices and regulation of waste disposal and minimum distances between wells, and the Riberão Preto municipality implemented groundwater regulations to protect recharge areas. The project increased awareness of the SAG and stimulated discussion and cooperative groundwater management within the four countries at national, provincial and community levels (Villar & Ribeiro, 2011).

The project culminated in the development of the transnational Guarani Aquifer Agreement, signed in August 2010. The agreement is a cooperative-governance model that functions at diverse scales. It is considered the world's first such agreement covering a major transboundary aquifer (Sindico, 2011). Furthermore, it is the first such accord encouraged by the UN Convention on the Laws of Transboundary Aquifers – a document presented to the UN General Assembly in 2008 and submitted for ratification in 2011 (Sindico, 2011; Villar & Ribeiro, 2011) – and the first that directly incorporates some of the principles contained in the Draft Articles (Eckstein & Sindico, 2014). The agreement is characterized as a "more general cooperative regime" (Eckstein & Sindico, 2014, p. 32); it does not prescribe specific regulatory duties but leaves them open to future development (Cassuto & Sampaio, 2011). Our survey results and detailed expert interviews include recent updates on the agreement's implementation within the four nations.

Methods: a survey and interviews of experts

Our case-study survey methodology distinguished two kinds of experts: global and regional. Global experts, drawn from around the world, are persons with broad professional experience studying and working on groundwater. The focus of their responses was not the SAG but rather transboundary groundwater governance in general. Our second set of informants, regional experts, drew primarily from the SAG region. Although our general objective was to survey SAG experts, the survey instruments allowed us to see whether and how regional experts' understanding of SAG governance diverged from the global experts' views.

To identify respondents, the sampling methodology combined respondent-driven (snowballing) recruitment with literature searches for authors of relevant publications. The survey was administered online between June and December 2011. Tailored (yet broadly similar) versions of the survey were created, one for each respondent group. Questions for the global experts were worded so as to extract academic, experiential and comparative knowledge from a broader context; targeted and case-specific questions were presented to the regional experts. The survey questions – structured questions (dichotomous and Likert-scale, including filter/contingency questions) and unstructured or open-ended, short-answer-text-response questions – elicited respondents' perceptions on groundwater quantity and quality issues, ownership and rights, and regulation and administration. Text responses were coded using qualitative-data-analysis software and frequencies were analyzed with statistical analysis software. As a follow-up procedure, short, detailed interviews were conducted via email or videoconference with 10 SAG experts in August 2014 to clarify survey responses and update the status of the multinational 2010 agreement.

Results

Although the sample sizes of the two groups (global and regional experts) were comparable, and the overall response rate of 30% was favourable (Table 1), the methodology encountered constraints. Sample sizes were limited by the volume of

Table 1. Overall respondent totals and response rates.

Respondent group	Received survey	Returned survey
Global experts	143	54 (38%)
Guarani regional respondents	186	43 (23%)
Total	329	97 (30%)

published literature, reliance on authors' personal networks, and availability of print- and Internet-based contact information. We were thus able to query only a partial, not fully representative sample of stakeholders and experts – typically men, with post-graduate degrees, working in research or government. Researchers and physical scientists together made up the majority of respondents in both groups (58% of global experts and 60% of regional experts). Other professions included engineers, lawyers, managers and planners. Vocational-sector proportions were similar for both groups, with the two most common categories being academic/research organization (50% of global experts, 38% of regional experts) and government (20% of global experts, 41% of regional experts). The third-largest category included NGO and intergovernmental-organization workers (18% of global experts, 12% of regional experts). The majority (58%) of the 43 SAG regional experts were from Brazil, with 12% in Argentina, 5% in Paraguay and 7% in Uruguay. This distribution roughly corresponds with the geographical distribution of the aquifer, of which, as noted above, approximately 70% lies within Brazil. Sixteen per cent of the regional experts were located outside the four sharing countries.

The following sections report results from questions related to the three elements of transboundary groundwater governance identified earlier: groundwater quantity and quality; ownership and rights; and regulation and administration.

Groundwater quantity and quality

Global experts were asked to identify whether contamination, well interference and drawdown, and several other issues (Table 2) were prevalent in the transboundary groundwater systems in their areas of experience and expertise. Regional experts were asked to indicate which of the same set of issues had occurred within the last decade in the SAG. Groundwater contamination and well interference/drawdown were the two issues most frequently cited by both groups of respondents.

In text responses, several SAG respondents characterized the problems of contamination and depletion as localized, emergent phenomena. An especially illustrative statement was, "The contentious issues have been circumscribed to urban centres located on the borders between countries. There has been no conflict at the level of the entire aquifer." Another regional expert indicated that drawdown resulting from intensive pumping was limited to municipal boundaries and did not extend to the level of transboundary conflict. One respondent described competition between wells, and local pollution in border areas, as "emerging" rather than "existing" problems.

Table 2. Frequencies and percentages of group respondents' responses indicating the occurrence of several transboundary groundwater issues.

Transboundary groundwater issue	Occurrence reported by global experts (n = 51)	Occurrence reported by Guarani regional respondents (n = 34)
Contamination	37 (73%)	21 (62%)
Well interference/ drawdown	38 (75%)	19 (56%)
Quantification of water rights	**31 (61%)**	11 (32%)
Access	21 (41%)	9 (27%)
Inter-agency jurisdictional conflict	22 (43%)	16 (47%)
Regulatory disputes	23 (45%)	12 (35%)
No contentious issues	5 (10%)	**10 (29%)**

Note. Values in bold are significantly greater at $p < .05$ in a two-sided test of equality for column proportions.

Ownership and rights

To better understand the extent to which groundwater is perceived as private property or as a shared resource, both global experts and SAG respondents were provided the statement, "Groundwater can be viewed as private property or a shared resource." The global experts were asked to indicate how strongly they agree or disagree with the statement, "In my experience, groundwater is generally viewed as a shared resource" (which could be based on any of a wide variety of countries and locations, depending on the respondent's experience). SAG respondents were asked how strongly they personally agree with the statement, "I view groundwater as a shared resource."

The agreement and disagreement of the two groups with these statements are reported in the net stacked distribution shown in Figure 3. Views of regional respondents diverge in two ways from those of the global experts. First, a higher proportion of global experts disagreed that groundwater has been generally viewed as a shared resource (a score of −21) compared with SAG respondents (a score of −3). Second, a strong majority of SAG respondents (82%) agreed or strongly agreed that they view groundwater as a shared resource rather than private property (data not shown). However, this view is not expressed as frequently by global experts, as shown in Figure 3, where SAG respondents were twice as likely to select 'strongly agree' than the global experts (66 compared to 30). This proportion of SAG respondents who selected 'strongly agree' was significantly larger ($p < .05$) than in the global experts (data not shown).

To better understand the existing systems of groundwater ownership, we asked respondents to identify what kinds of users possess legal rights to groundwater as property. Global experts were asked, "If you have observed cases where groundwater is individually owned, what types of users tend to possess it?" Respondents mostly cited individual users and landowners, and the farming/agriculture sector (Figure 4). Industrial and non-farm commercial users made up the third-largest category of 'owners.'

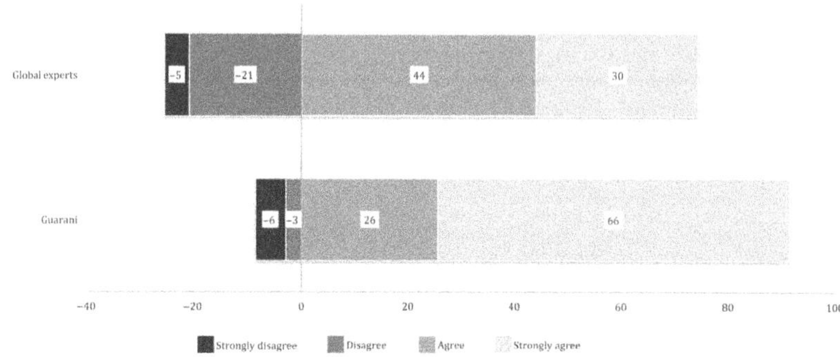

Figure 3. Net stacked distribution showing agreement or disagreement with the view of groundwater as a shared resource rather than private property. Global experts' responses reflect individual impressions of how groundwater is generally viewed based upon their personal experience. Guarani respondents' responses express their personal views. Positive (agree or strongly agree) and negative (disagree or strongly disagree) responses are shown as a percentage of total non-neutral responses (neutral responses are omitted).

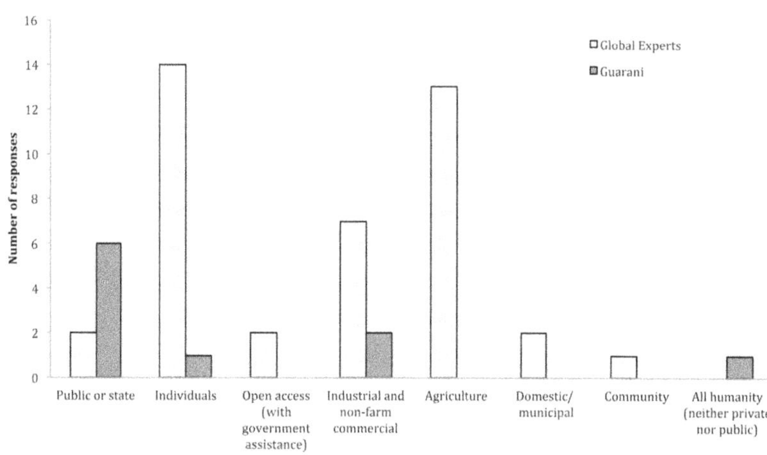

Figure 4. Varieties of groundwater rights owners reported by respondents. Guarani experts' responses pertain only to the Guarani aquifer system, while global experts reflect property rights regimes outside the Guarani. Respondents were allowed to cite more than one kind of user.

Drawing on the informants' open-ended responses, these owners are most commonly associated with Roman law–based countries, where groundwater is a property right attached to land ownership, especially in the case of irrigation farmers and commercial users. Some respondents observed that sometimes the state officially owns the resource but exercises little regulation or oversight – especially where the resource is plentiful. One global expert remarked that, in general, the notion of groundwater as a shared resource is "not yet widely perceived."

SAG respondents were asked, "If you view groundwater as property, please identify who possesses it." Two respondents mentioned that companies doing explorative work to develop water for human consumption may have ownership rights to groundwater, but the rest of the comments (6 of 10 total) were mostly to the effect that the aquifer "is a public good of four countries, according to the national constitutions [of] the states".

Regulation and administration

We asked respondents to report their perceptions of transboundary groundwater regulation (of production and quality) and administration (including data and information collection and availability). Overwhelmingly, both global experts and regional respondents indicated that there is not enough regulation of either quality or quantity (Table 3).

Additionally, strong majorities of respondents in both groups indicated that not enough information exists on groundwater quality or quantity. In fact, the strongest point of agreement overall for this question was that technical information and data sharing on transboundary aquifers are inadequate, an opinion shared by over 90% of global experts and 81% of SAG respondents.

Of the 27 global experts who offered 'additional' issues, 9 mentioned ones related to transboundary groundwater information, citing information gaps, lack of expertise, and/or a lack of shared understanding of physical aspects of transboundary aquifers. SAG respondents also expressed concerns about an overall lack or inadequacy of information. In their open-ended responses, they noted the need for better information and coordinated management; lack of laws and regulations; and existence of physical problems being contingent on local settings.

To better understand the relevance of institutions at different jurisdictional levels, we asked respondents to rate the relative importance of those institutions in regulating groundwater quality and extraction. Global experts were asked to appraise the importance of several levels of government jurisdictions in terms of each one's role in regulating groundwater quality and use. SAG respondents were asked to identify the institutions

Table 3. Respondents' perspectives on transboundary groundwater regulation, information and access.

Transboundary groundwater issue		Respondent group: Global experts (n = 53)	Respondent group: Guarani regional respondents (n = 36)
Quality	Not enough regulation	38 (76%)	27 (77%)
	Too much regulation	1 (2%)	2 (6%)
	Groundwater quality not a major issue	11 (22%)	6 (17%)
Quantity	Not enough regulation	43 (90%)	24 (77%)
	Too much regulation	2 (4%)	0 (0%)
	Groundwater quantity not a major issue	3 (6%)	**7 (23%)**
Information	Inadequate data sharing and technical info.	48 (91%)	25 (81%)
	Availability of information not an issue	5 (9%)	6 (19%)
Access	Access to groundwater is an issue	31 (66%)	14 (50%)
	Access to groundwater not a major issue	16 (34%)	14 (50%)

Note: Value in bold is significantly greater at $p < .05$ in a two-sided test of equality for column proportions.

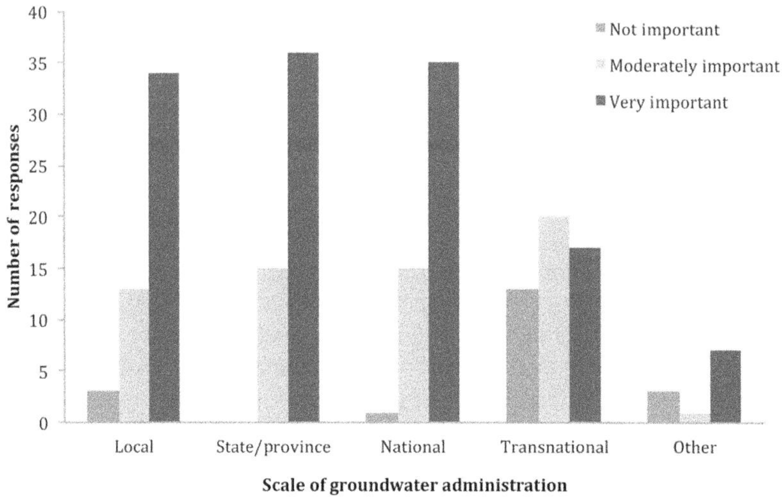

Figure 5. Views of global experts on the importance of different jurisdictional and administrative levels for groundwater regulation. Responses are not mutually exclusive; respondents could rate the importance of any or all administrative levels.

involved, and to then name the one with the most influential role in regulating quality and use of groundwater in the system.

Global experts indicated that local, state, and national agencies are of about equal importance, with a majority rating each as being 'very important' and very few rating any of them as 'not important' (Figure 5). In contrast, only 33% rated transnational institutions as 'very important'. 'Other' institutions rated as 'very important' by global experts included user groups and associations, NGOs, market parties, individual users and civil society, and aboriginal authorities, where relevant.

SAG respondents most commonly identified 'municipalities and local governments' as the most influential in regulating groundwater quality and use at the local level, accounting for 52% of the responses for local agencies. The second-most frequent response was 'state or provincial agencies or authorities', at 22%. At the state/provincial level, 90% of Guarani respondents cited state-level agencies as having the largest regulatory role for groundwater. At the national level, most respondents named the respective national water agencies (e.g. the Brazilian Agência Nacional de Águas), though other agencies also were mentioned (e.g. Corposana, Paraguay's state water company, and Brazil's mining agency).

At the transnational jurisdictional scale, 6 out of 14 respondents (43%) stated that no transnational body has a role in groundwater regulation. Three informants mentioned interstate commissions (especially the Intergovernmental Coordinating Committee of the Countries of the Plata Watershed), and two named the multilateral commission authorized by the agreement (though not yet formed).

Discussion

Groundwater quantity and quality issues

One aim of our research was to identify and characterize common problems associated with groundwater development in transboundary areas. The open-ended responses from survey respondents, coupled with our expert interviews, allow us to more effectively contextualize quality and quantity issues in transboundary areas.

Survey responses and interviews generally agreed that while localized quantity and quantity issues do exist in transboundary areas of the system, these problems are not yet perceived as urgent or severe. Almost one-third of regional respondents felt that there were no *contentious* problems (Table 2). Legal scholar Gabriel Eckstein (personal communication, 11 August 2014) points out that only recently have experts appreciated the SAG's massive spatial extent and volume, and that exploitation of groundwater remains at a relatively low level in most of the aquifer. Interviewees agreed that relatively little information exists on the aquifer's physical characteristics, and, consonant with the regional respondents, concerns over quantity and quality (contamination and well interference/drawdown, for instance) are not widespread but limited to local, especially urban areas.

The absence of serious or high-profile transboundary groundwater quantity and quality issues in the SAG prompt the question: What led to the Guarani Aquifer Agreement? The agreement is notable because it came about not because of intra-national or international conflict over water contamination or scarcity but out of localized concern over *future* conflicts among the countries. Villar and Ribeiro (2013) describe the role of scientists as critical in the cooperation process. Researchers in the region, mindful of the aquifer system's extensive nature and of the potential of future conflicts, promoted awareness and organized the first international meetings concerning the SAG's transboundary nature. These efforts attracted the attention of national governments and international institutions, including the World Bank and the Organization of American States, two of the agreement's central facilitators. Grass-roots movements, building on experiences of prior international cooperative agreements, and the buy-in of respected and powerful international institutions, facilitated the treaty process.

While the SAG lacks major high-profile problems, groundwater resource challenges have the potential to worsen in the future. Some SAG respondents suggested that these dormant challenges are mainly localized and urban, and not yet transboundary. This response reflects the nearly universal tendency to respond to problems only after they become crises. Yet identifying incipient and emerging *transboundary* groundwater-resource challenges within the SAG, as well as globally, is critical because the impacts of drawdown on human and natural systems may happen far in advance of major threats to the resource (Giordano, 2009). Thus, advance understanding of conditions and accurate diagnostic information can help avert problems. Because of groundwater's inherently unseen nature, problems may not be detected until the aquifers are in advanced stages of depletion.

Seen in this light, the Guarani Aquifer Agreement is a notable culmination of a movement that began with local concern over local water problems. These local concerns were leveraged through a network of informed scientists, mobilization of stakeholders, support and funding by international institutions, prior cooperative-agreement experience by national governments, successful realization of the Project for the Protection and Sustainable Development of the Guarani Aquifer System, and approval of the United Nations Law of Transboundary Aquifers.

These effective cooperative processes created the momentum needed to sign the agreement (Villar & Ribeiro, 2013; M. Walter, personal communication, 28 August 2014). The participating nations have had the opportunity to discuss the challenges, agree on a basis for decision making, negotiate use rights and conflict-resolution strategies, and craft guidelines before major problems arise. The agreement is also evidence of rare recognition of the need to develop accords *before* transboundary groundwater challenges become severe – even if the effectiveness of this agreement in the event of future conflict has been called into question (F. Sindico, personal communication, 14 August 2014).

Ownership and rights

Our results indicate that the prevailing view amongst regional and global informants is that groundwater is considered a shared resource rather than private property (at least within countries). Perhaps because global experts reflected a highly diverse array of experiences in different parts of the world, they reported much more frequently than did the SAG respondents that groundwater is not viewed as a shared resource. SAG respondents emphatically viewed groundwater as a shared resource rather than private property (Figure 3), and most of those who saw groundwater as property indicated that the state or public is the owner, rather than individuals (Figure 4). This contrasts with the experiences of many global experts, who primarily identified individuals, industrial users and agriculture as by far the most common individual groundwater owners. While regional respondents clearly affirmed that groundwater in SAG countries is owned by the public or the state, there is still some disagreement about transboundary-scale ownership.

Even if all parties involved have a common knowledge of the aquifer's physical attributes, variations in each country's prevailing legal regimes may not be readily amenable to international arrangements. For example, Cassuto and Sampaio (2011, p. 668) argue that "Brazil's constitutional predicament with respect to transboundary groundwater, as well as other domestic legal hurdles, further complicates an already difficult multilateral management process." Domestic laws and rules are therefore not immaterial at the multinational level of discussion, and in fact, according to one Brazilian government official (personal communication, 15 August 2014), can make or break international agreements.

Further, even if two sharing countries apply identical ownership laws – a rare situation – differences in hydrogeology suggest there is no guarantee that an existing doctrine will facilitate more effective governance. In the western US states, for instance, a particular ownership doctrine (e.g. prior appropriation) can have very different effects depending on whether the groundwater in question is tributary or non-tributary (Schlager, 2006). Sharing agreements at the international level must be able to overcome this sort of conundrum.

Another issue raised by Eckstein (personal communication, 11 August 2014) is that even when a nation's constitution deems groundwater a public good and not real property, in a legal sense, countries might not apply that general principle of ownership at the international level. Instead, they may assert and defend their sovereign ownership over the groundwater within their national boundaries. McCaffrey (2011) has strongly criticized the principle of national sovereignty over shared aquifers as codified in Article 3 of the United Nations International Law Commission Draft Articles on the Law of Transboundary Aquifers and Article 2 of the Guarani Agreement, arguing that it can lead to adoption of what he calls a "drill first and ask questions later" strategy by individual sharing countries. In his view, "A state simply cannot have the exclusive ownership that sovereignty implies in something that is shared with another state" (p. 570). The Guarani Agreement addresses sovereignty directly, as a (purported) safeguard for the aquifer (Villar & Ribeiro, 2013). Sindico (personal communication, 14 August 2014) believes that "protection" of the resource from interests outside the aquifer's physical borders (e.g. transnational corporations, other countries) is as important to the Guarani Aquifer countries as sovereignty among them; in other words, these countries are as concerned about pressure from without as they are about trouble from within.

Regulation and administration

Overall, both regional and global respondents report that the quality and quantity of transboundary groundwater resources are *under*-regulated. There was also strong agreement in the surveys and interviews that availability of technical information on transboundary groundwater and data sharing are inadequate, complicating determination of how much and what kind of regulation is appropriate to a given setting. This raises questions about who can and should regulate and collect information. These issues become especially problematic at the transboundary level because of differing ownership regimes and administrative procedures, and the complications of delineating the roles of agencies at various jurisdictional levels. Further, cross-border collaborative management can be stymied if water managers on the two sides cannot agree on characterization of an aquifer, have different priorities, or pursue unilateral national policies (Milman & Ray, 2011).

Like data collection, most groundwater regulation is the responsibility of national, not subnational, agencies. With some exceptions, e.g. in Brazil, regional experts most frequently cited federal institutions as having primacy over regulation, relative to local and provincial ones. With the SAG, although there are numerous national regulatory organizations and actors, capacity is quite low. Thus, there are few institutions tasked with regulating groundwater extraction and quality across national borders. Such institutions may be unnecessary if the level of exploitation is low and there are no contentious transboundary groundwater problems (personal communications from G. Eckstein, 11 August 2014; F. Sindico, 14 August 2014). To address transboundary groundwater regulation, some global experts highlight the need for regulation at multiple levels – local, state or provincial, national, and even transnational.

This multi-scalar approach to regulation conforms to recent thinking about multilevel governance that involves cross-jurisdictional coordination of policies to address gaps in funding, accountability, and policy, and to reduce institutional fragmentation (Akhmouch, 2012; Charbit, 2011). It also fits well with research on international rivers that calls for collective solutions to public problems. For instance, joint efforts by institutions, states, civil society and business – from the local to the global level – hold promise for achieving cross-border and cross-jurisdictional solutions to resource management issues (Finger, Tamiotti, & Allouche, 2006).

Because respondents identified a need for greater information (data) availability and sharing, they were asked for information about any such SAG transboundary groundwater-related activities since the agreement's adoption. Although the agreement promotes cooperation around the aquifer – e.g. information and data sharing – no such initiative has been implemented.[4] Further, a Web-based SAG database network has not been developed, and the original agreement-process files remain unavailable to universities and government agencies (R. Hirata, personal communication, 18 August 2014). Despite these perceived limitations in implementation, there are indications that information sharing among country experts occurs informally (M. Walter, personal communication, 28 August 2014). In addition, because the agreement frames SAG management within the 1969 La Plata River Basin Treaty framework, it could provide a mechanism for data sharing and conflict resolution (personal communications from G. Eckstein, 11 August 2014; F. Sindico, 14 August 2014; M. Walter, 28 August 2014; A. Pastori, 3 September 2014).

Since survey respondents indicated a need for more regulation of groundwater quantity and quality, we also asked about the (post-agreement) development of new regulations that might affect the Guarani Aquifer at a transboundary level. We found little indication that the four countries have worked to formally 'harmonize' their legislation. Ricardo

Hirata, director of the Groundwater Research Centre at the University of São Paulo, Brazil, maintains that since the signing, no change in any legislation or specific regulation has been implemented in any Brazilian state within the SAG (personal communication, 18 August 2014). Yet, in São Paulo State, a draft proposal is under consideration for protecting the active recharge (outcrop) zone, a vulnerable, unconfined region of the aquifer. And Brazil's National Water Agency is finalizing a pollution-vulnerability study in the outcrop zone, which might prompt other states to take analogous action (R. Hirata, personal communication, 18 August 2014).

Likewise, regulation has remained a local, regional and state matter. According to Eckstein (personal communication, 11 August 2014), "In South America, local and national governance rather than international agreements are preferred." For one Brazilian official (anonymous, personal communication, 15 August 2014), the local scale also seems most appropriate: "The Guarani shows that the most appropriate scale to address most groundwater issues is at the local level." A consultant for the Infrastructure and Environment Department at the Inter-American Development Bank, Martin Walter (personal communication, 28 August 2014), articulates the relation of local to international governance:

> Concrete problems need to be solved at the local level, and they give rise to formal and informal arrangements. However, local problems, especially when they occur near national borders, can require international support. These are often places removed from [financial and administrative] resources and decision-making centres. The internationalization of issues is a way of mobilizing support and resources functional to local-level resolution, though it can also be a problem. When you escalate the issue, [financial and human] resources can be diluted. The international agreement can serve to motivate state- and local-level policy changes, but it can also have a life of its own.

Finally, while Article 15 of the agreement called for establishment of a multilateral commission, such a commission has yet to be created (Villar & Ribeiro, 2013; Walter, 2013b). Lilian del Castillo Laborde (personal communication, 17 April 2015), professor of law at the University of Buenos Aires, explained that although Argentina and Uruguay both passed statutes approving the agreement in 2012, it is still not in force due to lack of internal approval by Brazil and Paraguay; consequently, the proposed intergovernmental commission has not been established. In the absence of any real controversies or threats, the countries have no motive to act further (personal communications from anonymous Brazilian government official, 15 August 2014; G. Eckstein, 11 August 2014; F. Sindico, 14 August 2014). International legal scholar Francesco Sindico suggests that the current deadlock in transboundary cooperative efforts may be due to at least three factors: the lack of urgency because of the aquifer's plentiful reserves; tensions between Paraguay and Brazil that have influenced Paraguay's decision not to ratify the agreement; and other internal priorities within Brazil, a country with many pressing natural resource challenges (personal communication, 14 August 2014).

In synthesis, Sindico (personal communication, 14 August 2014) explained,

> As a trajectory, the Guarani tells a story of local concern, spurred by the GEF international project, with four-country interest. This is important because it brought diverse interest groups together, including the four nations' legal and political communities, which eventually led to the Agreement's signing. Provisions 13–14 contain legal grounds for the commission to adopt specific measures in so-called critical areas. This implies an acknowledgment that something as huge as the Guarani Aquifer would require differential treatment across its extent. Any 'real' issues will likely happen on the borders. I think that this has led to the

realization that there will need to be legal grounds for these critical areas. The agreement provided an umbrella structure, but left many things open; it has a loose and general framework. If and when challenges are faced, the countries will have a document in place.

Status of the Guarani Aquifer Agreement

The 2010 agreement may be seen as a progressive negotiation amongst four member countries to collaborate on information sharing, regulation and administration of groundwater resources. Since its signing, Argentina and Uruguay have internally ratified the agreement. However, Brazil has taken little action, and Paraguay has refused to ratify, citing perceived limits to national sovereignty (personal communications from L. del Castillo Laborde, 8 August 2014; G. Eckstein, 11 August 2014; M. L. Lins Brzezinski, 16 August 2014) – though several of the informants agreed that political tension between Brazil and Mercosur also influenced the decision. In Brazil, the agreement is under consideration at the Ministry of the Environment, where, according to an informant, parliamentary discussion typically "proceeds slowly". As an international treaty, this accord is subject to the rules of the Vienna Convention on the Law of Treaties, which states that if a country signs a treaty, it is bound to not defeat the treaty's objectives. Paraguay's refusal to ratify could negate its signature and its position on the agreement.

A Brazilian official with foreign affairs expertise argues that Paraguay's refusal has created challenges in the process (personal communication, 15 August 2014). Some Paraguayan stakeholders are calling for the terms of the agreement to be re-evaluated by all four countries, but Argentina and Uruguay are unwilling to revisit the matter because they have already ratified. Brazil has yet to make an internal ratification decision. In sum, conflict between the countries and questions over resource sovereignty may have inhibited the internal ratification of the agreement in two of the four SAG countries.

Conclusions

Groundwater governance needs to address a number of issues, including lack of physical data and technical understanding of aquifer mechanics, varying administrative jurisdictions, insufficient regulation of quantity and quality, and disagreements over access or allocation. However, in the case of transboundary groundwater governance, few agreements exist, and the literature has relatively little to say about the perceptions of stakeholders on important institutional aspects of governance. This article has reported on responses by global experts and regional Guarani Aquifer System specialists, focusing on issues of quantity and quality, ownership and rights, and regulation and administration. We found that:

1. Contamination and overdrafting in the SAG remain mostly incipient and limited to local areas and municipalities situated on international borders.
2. SAG respondents strongly view groundwater as a shared resource administered by the state on behalf of the public, rather than as private property. Although this principle is reflected in the constitutions of the four basin countries, it does not extend in the same way to the international level in the context of the Guarani Aquifer Agreement; that is, the agreement affirms and protects individual states' sovereignty over the portions of the SAG within their boundaries.
3. There is a general perception that groundwater quality and extraction are in need of greater regulation, and that availability and sharing of information and data are

insufficient. Local, provincial and national governing entities such as government agencies are perceived as important regulators of groundwater, but despite the agreement, transnational entities and agreements are not.

The agreement illustrates an attempt at a governance architecture intended to harmonize multinational interests, while allowing for locally specific institutional arrangements. Based on evidence collected in this study, we find that one of the agreement's most important features is that it provides foundational principles and a means for resolving conflicts, should the need arise. Importantly, the existence of such an agreement does not preclude the adoption of other local and regional arrangements between countries.

Indeed, Article 13 of the Guarani Aquifer Agreement provides clear legal grounds for the adoption of specific, local-level management measures. By including such a provision, future multinational groundwater agreements afford a legal basis for addressing problems at the appropriate geographic scales. As the SAG case demonstrates, informal local and regional governance mechanisms are most likely to be appropriate in transboundary settings where groundwater resource challenges are few and remain geographically discrete.

But the gathered momentum leading up to the signing of an international treaty could diminish rapidly, according to the regional and international experts we interviewed. The activities conducted under the auspices of the agreement, should it succeed in overcoming current obstacles, will be illustrative. To be functional, the agreement will need to develop rules and regulations to avoid and resolve transboundary groundwater conflict, while respecting national sovereignty. So long as the agreement remains inactive, however, transboundary groundwater within the aquifer's multinational extent will be regulated according to individual national practices.

Significantly, the SAG case can help clarify the role of governance by highlighting specific challenges in local/regional contexts. Are such issues addressed effectively through multilateral agreements? This sort of approach can facilitate comparative investigations and provide examples from which to draw. Information derived from such studies should prove helpful, given the early stage of development of transnational groundwater agreements in general.

Further explorations of groundwater governance in the SAG and in other important basins should poll water users from all relevant sectors, at multiple levels (Puri & Aureli, 2005). One template is the type of survey deployed by Shah et al. (2006) for South Asia. International spearhead organizations like UNESCO-IHP may be able to lend support and coordination for such exercises. Additionally, given the proliferation of conferences, workshops and forums, a high degree of communication exists in the world's groundwater epistemic community. As these professionals and stakeholders discuss mutual interests and incorporate salient ideas from the literature, we may anticipate promising strategies for multinational governance of transboundary groundwater resources.

Acknowledgements

This work could not have been accomplished without the strong cooperation of UNESCO's International Hydrological Programme, especially by Dr Alice Aureli, the chief of UNESCO's Groundwater Systems and Human Settlements Section. We are very much indebted to the 100 or so experts, officials, stakeholders and other informants who generously consented to personal interviews or participated in our survey. Finally, we would like to acknowledge Robert Merideth and Christopher Scott of the Udall Center, and Lily House-Peters, a PhD candidate at the School of Geography and Development, University of Arizona, for their insightful critiques of and suggestions for improvements to the article; Emily McGovern, who assisted in editing an early version of the manuscript; and Lynn Massey, a University of Arizona undergraduate intern, who early in the process helped assemble baseline information.

Funding

This work was partially supported by the Inter-American Institute for Global Change Research [SGP-CRA #005], supported by the National Science Foundation [GEO-1138881]; the National Oceanic and Atmospheric Administration Climate-Society Interactions Program [NA11OAR4310143]; a grant from Lloyd's Register Foundation for research as part of the International Water Security Network; and annual workplan support from the Morris K. Udall and Stewart L. Udall Foundation. The authors confirm that they have no conflict of interest with any individuals or institutions, nor any financial interests in or potential benefits from any of the findings or outcomes of this article or from any direct application of the research therein.

Notes

1. This article uses the Spanish and Portuguese acronym for Guarani Aquifer System.
2. In Brazil 850,000 km^2, in Paraguay 70,000 km^2, in Argentina 225,000 km^2, and in Uruguay 45,000 km^2 (Foster et al., 2009; OAS, 2005).
3. Gerring's (2007) typology of cases is most appropriate when quantitative indicators of the characteristics of the full population of interest are available. Research on transboundary-governance arrangements does not yet offer such indicators.
4. Transboundary-aquifer data-sharing agreements have been successfully initiated elsewhere, including in the Nubian Sandstone Aquifer System and the North-Western Sahara Aquifer System (G. Eckstein, personal communication, 11 August 2014; FAO, n.d.)

References

Akhmouch, A. (2012). Water governance in Latin America and the Caribbean. Retrieved from http://rioc.org/IMG/pdf/GOV_LAC_WP_4-2.pdf

Amore, L., & Tröger, U. (2010). Transboundary Guarani Aquifer System and groundwater management mechanisms. In *Proceedings of the ISARM 2010 International Conference "Transboundary aquifers: Challenges and new directions,"* December 6–8, 2010. Paris: UNESCO.

Aureli, A., & Eckstein, G. (2011). Strengthening cooperation on transboundary groundwater resources. *Water International*, *36*(5), 549–556. doi:10.1080/02508060.2011.615137

Aureli, A., Ganoulis, J., Varady, R. G., Megdal, S. B., Van Der Gun, J., Nijsten, G.-J., & Merla, A. (in press). Addressing the groundwater governance challenge. In I. Buchan (Ed.), *Water governance for the 21st century.* London, UK: IWA Publications.

Blatter, J., & Ingram, H. (Eds.). (2001). *Reflections on water: New approaches to transboundary conflicts and cooperation.* Cambridge, MA: MIT Press.

Brandes, O. M., & Brooks, D. B. (2007). The soft path for water in a nutshell. *Friends of the Earth Canada.* Retrieved from http://poliswaterproject.org/publication/23

Burchi, S. (1999). National regulations for groundwater: Options, issues and best practices. In S. M. Salman (Ed.), *Groundwater: Legal and policy perspectives: proceedings of a World Bank seminar* 456. Washington, DC: World Bank.

Burchi, S., & Nanni, M. (2003). How groundwater ownership and rights influence groundwater intensive use management. In R. Llamas & E. Custodio (Eds.), *Intensive use of groundwater: Challenges and opportunities* (pp. 227–240). Leiden, Netherlands: Balkema.

Cassuto, D. N., & Sampaio, R. S. R. (2011). Keeping it legal: Transboundary management challenges facing Brazil and the Guarani. *Water International*, *36*(5), 661–670. doi:10.1080/02508060.2011.599779

Charbit, C. (2011). *Governance of public policies in decentralised contexts: The multi-level approach.* OECD Publishing. Retrieved from http://ideas.repec.org/p/oec/govaab/2011-4-en.html

Conca, K. (2005). *Governing water: Contentious transnational politics and global institution building.* Cambridge, MA: MIT Press.

de Gooijer, G., Lofgren, R., Granit, J., Jagerskog, A., Aureli, A., & Renck, A. (2009). *Innovations in groundwater governance in the MENA region*. Stockholm: Stockholm International Water Institute.

de los Cobos, G. (2012). *L'Eau sans frontière: Quarante ans d'une gestion partagée de la nappe d'Eau souterraine du Genevois* (p. 195). Genève: Editions Slatkine.

del Castillo Laborde, L. (1996). Legal regime of the Rio De La Plata. *Natural Resources Journal*, *36* (2), 251–295.

Dietz, T., Ostrom, E., & Stern, P. C. (2003). The struggle to govern the commons. *Science*, *302* (5652), 1907–1912. doi:10.1126/science.1091015

Dinar, A., Dinar, S., & McCaffrey, S. (2007). *Bridges over water: Understanding transboundary water conflict, negotiation and cooperation*. Singapore: World Scientific.

Eckstein, G., & Hardberger, A. (2007). State practice in the management and allocation of transboundary ground water resources in North America. *Yearbook of International Environmental Law*, *18*, 96–125. doi:10.1093/yiel/18.1.96

Eckstein, G. E. (2011). Managing buried treasure across frontiers: The international Law of Transboundary Aquifers. *Water International*, *36*(5), 573–583. doi:10.1080/02508060.2011.598642.

Eckstein, G. E. (2013). Rethinking transboundary ground water resources management: A local approach along the Mexico-US border. *Georgetown International Environmental Law Review*, *25* (1), 95–128.

Eckstein, G. E., & Sindico, F. (2014). The law of transboundary aquifers: Many ways of going forward, but only one way of standing still. *Review of European, Comparative & International Environmental Law*, *23*(1), 32–42. doi:10.1111/reel.2014.23.issue-1

Falkenmark, M. (1990). Population growth and water supplies: An emerging crisis. *People*, *17*(1), 18–20.

FAO (Food and Agricultural Organization). n.d. II. Groundwater agreements. Retrieved September 3, 2014, from http://www.fao.org/docrep/008/y5739e/y5739e05.htm

Finger, M., Tamiotti, L., & Allouche, J. (2006). *The multi-governance of water: Four case studies*. Albany, NY: SUNY Press.

Foster, S., & Chilton, P. (2003). Groundwater: The processes and global significance of aquifer degradation. *Philosophical Transactions of the Royal Society of London. Series B: Biological Sciences*, *358*(1440), 1957–1972. doi:10.1098/rstb.2003.1380

Foster, S., Hirata, R., Vidal, A., Schmidt, G., & Garduño, H. (2009). *The Guarani Aquifer initiative —towards realistic groundwater management in a transboundary context*. Sustainable groundwater management: Lessons from practice. Case profile collection no. 9. GW-MATE/The World Bank. Retrieved from http://siteresources.worldbank.org/INTWAT/Resources/GWMATE_English_CP_09.pdf

GEF/FAO/UNESCO/IAH/WB (Global Environment Facility/Food and Agriculture Organisation/UNESCO/International Association of Hydrologists/World Bank). (2013). *Groundwater governance—A global framework for action (2011–2014)*. Retrieved from http://www.groundwatergovernance.org/

GEF/TWAP (Global Environment Facility/Transboundary Waters Assessment Programme). n.d. *Global assessment of transboundary aquifers and SIDS groundwater systems: An overview for partners*. Retrieved from http://www.isarm.org/dynamics/modules/SFIL0100/view.php?fil_Id=352At

Gerring, J. (2007). *Case study research: Principles and practices*. Cambridge, UK: Cambridge University Press.

Giordano, M. (2009). Global groundwater? Issues and solutions. *Annual Review of Environment and Resources*, *34*(1), 153–178. doi:10.1146/annurev.environ.030308.100251

Giordano, M., & Villholth, K. G. (2007). *The agricultural groundwater revolution: Opportunities and threats to development*. Wallingford, UK: CABI.

Hirata, R., & Vidal, A. (2012). *Transboundary: Groundwater management issues for Guarani aquifer*. Global Water Partnership Toolbox Case Study #38. Retrieved August 1, 2014, from http://www.gwp.org/en/ToolBox/CASE-STUDIES/Americas–Caribbean/CES-Solidaridad-and-GWP-Peru-and-GWP-South-America-infogwpsudamericaorg/

IGRAC (International Groundwater Resources Assessment Centre). (2012). *Transboundary Aquifers of the World 2012*. Retrieved August 18, 2013, from http://www.un-igrac.org/publications/456/

Ingram, H., Milich, L., & Varady, R. G. (1994). Managing transboundary resources: Lessons from Ambos Nogales. *Environment: Science and Policy for Sustainable Development*, *36*(4), 6–38. doi:10.1080/00139157.1994.9929996
Jarvis, W. T. (2014). *Contesting hidden waters: Conflict resolution for groundwater and aquifers*. Abingdon, UK: Routledge.
Kemper, K. (2004). Groundwater—From development to management. *Hydrogeology Journal*, *12* (1), 3–5. doi:10.1007/s10040-003-0305-1
Kemper, K. (2007). Instruments and institutions for groundwater management. In M. Giordano & K. G. Villholth (Eds.), *The agricultural groundwater revolution: Opportunities and threats for development* (pp. 153–172). Cambridge, MA: CABI.
Kemper, K., Mestre, E., & Amore, L. (2003). Management of the Guarani Aquifer System. *Water International*, *28*(2), 185–200. doi:10.1080/02508060308691684
Lebel, L., Garden, P., & Imamura, M. (2005). The politics of scale, position, and place in the governance of water resources in the Mekong region. *Ecology and Society*, *10*(2), 18. Retrieved from http://www.ecologyandsociety.org/vol10/iss2/art18/
Linton, J., & Brooks, D. B. (2011). Governance of transboundary aquifers: New challenges and new opportunities. *Water International*, *36*(5), 606–618. doi:10.1080/02508060.2011.599312
Llamas, M. R., & Martínez-Santos, P. (2005). Intensive groundwater use: Silent revolution and potential source of social conflicts. *Journal of Water Resources Planning and Management*, *131* (5), 337–341. doi:10.1061/(ASCE)0733-9496(2005)131:5(337)
Margat, J. F., & Van Der Gun, J. (2013). *Groundwater around the world: A geographic synopsis*. Leiden, Netherlands: CRC Press.
McCaffrey, S. C. (2011). The international law commission's flawed draft articles on the law of transboundary aquifers: The way forward. *Water International*, *36*(5), 566–572. doi:10.1080/02508060.2011.597094
Megdal, S. B., Gerlak, A. K., Varady, R. G., & Huang, L.-Y. (2014). Groundwater governance in the United States: Common priorities and challenges. *Groundwater*. doi:10.1111/gwat.12294
Milman, A., & Ray, I. (2011). Interpreting the unknown: Uncertainty and the management of transboundary groundwater. *Water International*, *36*(5), 631–645. doi:10.1080/02508060.2011.599782
Moench, M. (2004). Groundwater: The challenge of monitoring and management. In P. H. Gleick (Ed.), *The world's water 2004–2005: The biennial report on freshwater resources*. Washington, DC: Island Press.
Moench, M., Dixit, A., Janakarajan, S., Rathore, M. S., & Mudrakartha, S. (2003). *The fluid mosaic: Water governance in the context of variability, uncertainty and change (a synthesis paper)*. Kathmandu: Nepal Water Conservation Foundation.
Mukherji, A., & Shah, T. (2005). Groundwater socio-ecology and governance: A review of institutions and policies in selected countries. *Hydrogeology Journal*, *13*(1), 328–345. doi:10.1007/s10040-005-0434-9
OAS (Organization of American States). (2005). *Guarani aquifer system: Environmental protection and sustainable development of the Guarani Aquifer System*. Washington, DC: Organization of American States Office for Sustainable Development & Environment. Retrieved from http://www.oas.org/dsd/Events/english/Documents/OSDE_7Guarani.pdf
Pahl-Wostl, C., Gupta, J., & Petry, D. (2008). Governance and the global water system: A theoretical exploration. *Global Governance*, *14*, 419–435.
Puri, S. Ed. (2001). *Internationally shared (Transboundary) Aquifer Resources Management—A framework document*. (IHP-VI Non Serial Documents in Hydrology). Paris, France: UNESCO.
Puri, S., & Aureli, A. (2005). Transboundary aquifers: A global program to assess, evaluate, and develop policy. *Ground Water*, *43*(5), 661–668. doi:10.1111/j.1745–6584.2005.00100.x
Puri, S., & Aureli, A. (2009). Atlas of transboundary aquifers: Global maps, regional cooperation and local inventories. Paris: UNESCO.
Rogers, P., & Hall, A. W. (2003). *Effective water governance* (Vol. 7). Global Water Partnership Stockholm. Retrieved from http://www.tnmckc.org/upload/document/bdp/2/2.7/GWP/TEC-7.pdf
Romero, S. (2015, February 16). Taps start to run dry in Brazil's largest city: São Paulo water crisis linked to growth, pollution and deforestation. *New York Times*. Retrieved from http://www.nytimes.com/2015/02/17/world/americas/drought-pushes-sao-paulo-brazil-toward-water-crisis.html?_r=0
Sagala, J. K., & Smith, Z. A. (2008). Comparative groundwater management: Findings from an exploratory global survey. *Water International*, *33*(2), 258–267. doi:10.1080/02508060802027182

Schlager, E. (2006). Challenges of governing groundwater in U.S. western states. *Hydrogeology Journal*, *14*(3), 350–360. doi:10.1007/s10040-005-0012-1

Schlager, E. (2007). Community management of groundwater. In M. Giordano & K. G. Villholth (Eds.), *The agricultural groundwater revolution: Opportunities and threats to development* (Vol. 3, pp. 131–152). Wallingford, UK: CABI.

Sehring, J. (2009). Path dependencies and institutional bricolage in post-Soviet water governance. *Water Alternatives*, *2*(1), 61–81.

Shah, T. (2008). *Taming the anarchy: Groundwater governance in South Asia*. Washington, DC: Resources for the Future.

Shah, T., Singh, O. P., & Mukherji, A. (2006). Some aspects of South Asia's groundwater irrigation economy: Analyses from a survey in India, Pakistan, Nepal Terai and Bangladesh. *Hydrogeology Journal*, *14*(3), 286–309. doi:10.1007/s10040-005-0004-1

Sindico, F. (2011). The Guarani Aquifer System and the international law of transboundary aquifers. *International Community Law Review*, *13*, 255–272. doi:10.1163/187197311X585338

Theesfeld, I. (2010). Institutional challenges for national groundwater governance: Policies and issues. *Ground Water*, *48*(1), 131–142. doi:10.1111/gwat.2010.48.issue-1

Tujchneider, O., Perez, M., Paris, M., & D'Elia, M. (2007). The Guarani Aquifer System: State-of-the-art in Argentina (Chapter 18). In L. Chery & G. de Marsily (Eds.), *Aquifer systems management: Darcy's legacy in a world of impending water shortage* (pp. 239–252). London: Taylor and Francis.

UNDP (United Nations Development Programme. (2004). *Water governance for poverty reduction —Key issues and the UNDP response to MDG*. New York, NY: UNDP.

UNECE (United Nations Economic Commission for Europe). (2014). *Model provisions on transboundary groundwaters* (pp. 1–28). New York, NY: United Nations. Retrieved from http://www.unece.org/fileadmin/DAM/env/water/publications/WAT_model_provisions/ece_mp.wat_40_eng.pdf

UNESCO (United Nations Educational, Scientific and Cultural Organization. (2009). *Atlas of transboundary aquifers*. Paris: UNESCO.

Varady, R. G., Meehan, K., & McGovern, E. (2009). Charting the emergence of 'global water initiatives' in world water governance. *Physics and Chemistry of the Earth, Parts A/B/C*, *34*(3), 150–155. doi:10.1016/j.pce.2008.06.004

Varady, R. G., van Weert, F., Megdal, S. B., Gerlak, A., Iskandar, C. A., & House-Peters, L. (2013). *Groundwater governance: A global framework for country action* (No. 5). Rome: FAO/Global Environment Facility.

Villar, P. C., & Ribeiro, W. C. (2011). The Agreement on the Guarani Aquifer: A new paradigm for transboundary groundwater management? *Water International*, *36*(5), 646–660. doi:10.1080/02508060.2011.603671

Villar, P. C., & Ribeiro, W. C. (2013). The Agreement on the Guarani Aquifer: Cooperation without conflict. *Global Water Forum* (Discussion Paper No. 1344). UNESCO/Australian National University. 9 pp. Retrieved November 25, 2013, from http://www.globalwaterforum.org/2013/09/02/the-agreement-on-the-guarani-aquifer-cooperation-without-conflict/

Walter, M. (2013a). *Explaining the emergence of transboundary groundwater management: The cases of Guarani Aquifer System, the Hueco and Mesilla Bolsón Aquifers, and the Génévoise Aquifer* (Unpublished doctoral dissertation). Northwestern University, Evanston, IL.

Walter, M. (2013b). Towards the joint-management of transboundary groundwaters: Hydrogeology and the Guarani Aquifer System. *Global Water Forum* (Discussion Paper No. 1310). UNESCO/Australian National University. 5 pp. Retrieved November 25, 2013, from http://www.globalwaterforum.org/2013/03/11/towards-the-joint-management-of-transboundary-groundwaters-hydrogeology-and-the-guarani-aquifer-system/

Wolf, A. T. (2007). Shared waters: Conflict and cooperation. *Annual Review of Environment and Resources*, *32*, 241–269. doi:10.1146/annurev.energy.32.041006.101434

Wouters, P. (2013). International law–facilitating transboundary water cooperation. *TEC Background Papers*, 17. Retrieved from http://www.indiaenvironmentportal.org.in/files/file/Facilitating%20Transboundary%20Water%20Cooperation.pdf

Yamada, C. (2011). Codification of the law of transboundary aquifers (groundwaters) by the United Nations. *Water International*, *36*(5), 557–565. doi:10.1080/02508060.2011.597345

Young, O. R. (2002). *The institutional dimensions of environmental change: Fit, interplay, and scale*. Cambridge, MA: MIT Press.

OPEN ACCESS

A methodology to identify vulnerable transboundary aquifer hotspots for multi-scale groundwater management

Christina M. Fraser, Robert M. Kalin, Modesta Kanjaye and Zione Uka

ABSTRACT

Thirty-eight aquifer units are shared between Malawi and its neighbouring countries. It is essential to prioritize those transboundary aquifers that require immediate attention. A methodology of identifying hotspots in the transboundary aquifers of Malawi that may be at risk of depletion or contamination has been developed. There are 11 local-scale and three national-scale hotspots of transboundary concern in Malawi. Fiscal and planning measures can now be taken to assess these areas in more detail, fostering transboundary cooperation between stakeholders at both local and national scales.

Introduction

Groundwater is a vital natural resource accounting for 97% of the non-frozen freshwater resources available across the globe (IGRAC & UNESCO-IHP, 2015). It provides almost half of the drinking water worldwide, alongside 40% of the water used for irrigated agriculture and 33% used in industry (Smith et al., 2016; UN-Water, 2018). It is essential for sustaining ecosystems and providing baseflow to rivers (Kelly et al., 2019a). Climate change and increasing human impacts on the environment are putting pressure on groundwater resources, causing quality and quantity issues (UN-Water, 2018). Over the last decade, groundwater abstraction has tripled and continues to increase at around 2% per year (Van der Gun, 2012). This issue is exacerbated in many parts of Africa, where groundwater is often the only reliable source of water. Up to 75% of the population on the continent relies on groundwater as their main source of drinking water (UNECA et al., 2000). Groundwater is also essential in these areas for rural livelihoods such as livestock rearing and agricultural crop cultivation (Villholth, 2013 and Foster et al., 2008 in Nijsten et al., 2018). The importance of water was recognized in the United Nations Sustainable Development Goals adopted in 2015. Goal 6 is dedicated to clean water and sanitation, including groundwater. Target 6.5 calls for the implementation of Integrated Water resources Management, at a transboundary level where appropriate (Mccracken, 2017).

Both surface waters and groundwater can be transboundary, meaning that rivers, lakes and aquifers have the ability to store and transmit water across a state or border (Fraser et al., 2020). Whereas transboundary surface waters have been extensively studied worldwide, transboundary aquifers have received much less attention (Rivera & Candela, 2018). Transboundary aquifer assessments have become more common across the globe since 2000 due to substantial efforts from the Internationally Shared Aquifer Resources Management initiative of the International Association of Hydrogeology and UNESCO (Rivera & Candela, 2018). There are currently 592 transboundary aquifers identified worldwide, 226 of which are groundwater bodies defined under the EU Water Framework Directive. There are 80 transboundary aquifers identified in mainland Africa (IGRAC & UNESCO-IHP, 2015). The identification and basic hydrogeological understanding of these aquifers has been driven by the need for transboundary aquifer management and agreements for these vulnerable shared resources (Fraser et al., 2018; Nijsten et al., 2018; Rivera & Candela, 2018).

A recent study that took a detailed and national-border-based approach to transboundary aquifer assessments in Malawi identified 38 transboundary aquifer units shared between Malawi and its neighbours, in contrast to a previous estimate of three (Fraser et al., 2020, 2018). It is likely that other countries, like Malawi, have many more transboundary aquifers than previously thought. As the number of such aquifers rises around the world, prioritizing which ones receive resources and funds for in-depth assessment and specialized management will become essential. Inherently, not all transboundary aquifers are at risk for reduced quality and over-abstraction. Identifying which aquifers require attention and which do not is essential under Integrated Water Resources Management and the Sustainable Development Goal agenda. Furthermore, assessing whether a transboundary aquifer requires local, national or international cooperation and management will allow the best policy instruments, legal agreements and management mechanisms to be selected.

This article presents a methodology for identifying hotspots in transboundary aquifers that may be vulnerable to groundwater quality and quantity issues. Asset management data collected as part of the Climate Justice Fund's Water Futures Programme were utilized in this study through a spatial analysis site selection technique to identify key areas in transboundary aquifers in Malawi that may be at risk. These areas of risk were then classified into local and national scale transboundary hotspots. Aided with this information, policy makers and governmental officials have the ability to select those transboundary aquifers (or areas of transboundary aquifers) that require further investigation, directed cross border management and potentially, transboundary agreements governing them.

Methodology

Study area

Malawi is a small country in south-eastern Africa and is part of the Southern African Development Community (SADC). It is landlocked by Tanzania, Mozambique and Zambia. Groundwater is a vital resource in Malawi, providing 82% of drinking water in rural areas as well as agriculture and industry needs (Chavula, 2012). Malawi has many challenges and limitations for its groundwater supply, particularly in rural areas (Kalin et al., 2019). Threats to aquifers include decreasing groundwater levels, high salinity levels in alluvial deposits, contamination from pit latrine location and -agricultural runoff, an increasing population and the effects of land degradation and climate change

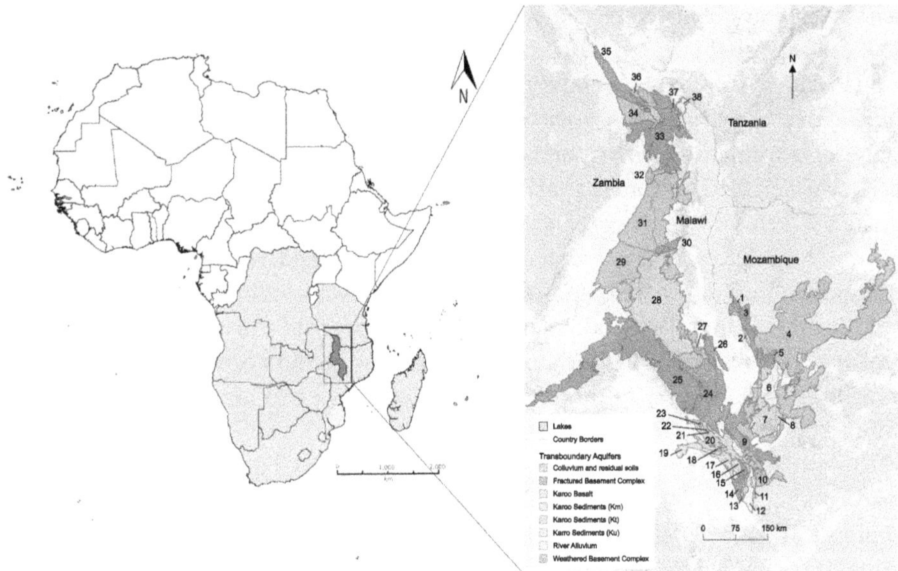

Figure 1. Transboundary aquifers shared between Malawi and its neighbours in the wider SADC (adapted from Fraser et al., 2020, 2018).

on groundwater recharge (Addison et al., 2020; Back et al., 2018; Bath, 1980; Kelly et al., 2019b; Monjerezi et al., 2011; Rivett et al., 2019, 2018a, 2018b; Smedley, 2004; Smith-Carrington & Chilton, 1983).

Three transboundary aquifers in Malawi were first identified by the Global Environment Facility's Transboundary Waters Assessment Programme, which aimed to provide the first global-scale assessment of all transboundary waters (Fraser et al., 2018; ILEC et al., 2016). More recently, Fraser et al. (2020) increased this number to 38 by including smaller, more locally significant aquifers in their assessment. Figure 1 illustrates the current understanding of transboundary aquifers shared between Malawi and its neighbours.[1] Aquifer lithologies include widespread weathered and fractured basement complex, Karoo sediments and basalts, and quaternary alluvium and colluvium. Not all of these transboundary aquifers will be at risk to over-abstraction or reduced water quality. The method presented in this article seeks to identify which of these 38 aquifers, and indeed which areas in these aquifers, pose a potential risk.

Methodological approach

The method presented here is based on the concept of fuzzy logic first developed by L. A. Zadeh in 1965. It is based on the assumption that people do not think in exact variables ('yes' or 'no') but distinguish a range of 'blurry' values (for example 'very', 'somewhat', 'perhaps', or 'yes and no'). Fuzzy logic can be used as an overlay analysis technique in geographical information systems (GIS) to solve traditional overlay analysis applications such as site selection and suitability models. The fuzzy logic method assigns membership values from 0 to 1. Fuzzy logic site selection is different from other site selection methods because it represents the possibility of an ideal site. Fuzzy overlay allows the user to overlay the various reclassified layers to analyze the possibility of a specific occurrence. This can then be used to identify a site based on a series of criteria.

Spatial analysis of transboundary aquifers in Malawi was undertaken using fuzzy logic and GIS overlays to generate 'hotspot maps' of areas with the greatest risk of transboundary mismanagement. The hotspot map is generated using multiple input raster layers that each have weighted attributes attached to them based on the parameter being modelled, which are then summed to produce a combined raster layer. The higher the cell weight, the worse the case. From this, a heat map is produced where the areas of greatest accumulation of highly weighted categories are shown as 'hotspots'. Then shapefiles of the known transboundary aquifers in Malawi were overlaid and eliminated based on their local or national hotspot significance. Although this method is relatively common, it has never been applied to a transboundary aquifer context to assess areas of risk due to diminished water quality and quantity.

The spatial co-location of areas of higher risk for reduced quality and quantity of water corresponding to transboundary aquifers provides a first indication that the aquifer requires further evaluation. A hotspot that spans across the majority of a transboundary aquifer is deemed to be of a national scale. If it is only focused along the border, then it is considered to be of local scale. Hotspots not within a transboundary aquifer, or within a transboundary aquifer but not directly linked to the border, were not considered.

Three important characteristics of the method should be noted. First, any data or criteria could be used. This means that each user (or country, or community, etc.) can define what they believe the biggest transboundary risks to be and how they should be weighted. This empowers countries to make their own decisions about their transboundary resources. If arsenic contamination, for example, is deemed to be an issue, it can be added as a layer. Second, the map is data-driven, but given its flexibility, as described above, a user can have as few as two layers, or as many as needed to define the risks. The more data, the more accurate the result, but even with little available data, results can still be achieved. This makes the method inherently flexible and accessible to data-limited countries. Third, although this is a transboundary context, the method was only applied to one country, rather than two or more. This was chosen to show that previous transboundary cooperation is not necessary for countries to start to look at their own transboundary circumstances; on completion of the hotspot analysis, a country could then approach its neighbours with a mandate for future cooperation.

Data collection

The case study and the chosen area should dictate what data are used for the creation of the hotspot maps. The data layers for this study were selected in consultation with the government of Malawi's Ministry of Agriculture, Irrigation and Water Development, based on groundwater quality and quantity concerns. Malawi, like most of Southeast Africa, had limited data for certain parameters on which to base groundwater assessments (Adelana & MacDonald, 2008). Instead, we used proxies for these parameters which we believe to be suitable.

The primary data source for this assessment is from the Futures Programme asset management data set of the Climate Justice Fund (CJF), hosted on the mWater platform (www.mwater.co). Data were collected through a series of water point functionality and household-level surveys using a method often called 'water point mapping'. The information gathered included water point functionality, geographical location, accessibility, reliability and communities served. In total, 120,935 water points, 278,045 sanitation points and 10,297 waste sites were assessed, with five pieces of site data collected per water point, 81 questions per water point functionality survey and up to 23 questions per sanitation survey. A total of 93,000 water and sanitation points (and data from four of the specific questions) were used in this study, supplemented by two open source shapefiles for hydrology and land use (MASDAP, 2019; Persits et al., 2002; Upton et al., 2018)

Data selection and weighting

The hotspot spatial analysis used six combined raster files. Each data point in these six files was given a weigh between 0 and 1 based on a selected criterion. The higher the weight, the worse the case and thus the higher the risk rating. Table 1 summarizes the parameters selected for this study and the weight selected for each attribute.

Layer 1 is water point type. These data were collected through observation during the CJF water point functionality survey. Water point type is used here as a proxy for water supply reliability. The attributes are (1) hand-dug well, which are considered to have the lowest water supply reliability and thus are given a weight of 1, indicating high risk; (2) mechanically drilled borehole, which are considered more reliable and given a weight of 0.5 (MacDonald et al., 2019); and (3) piped supply, which is considered to have the highest reliability in this scenario and is given a weight of 0, indicating low risk.

Layer 2 is the hydrogeology of the aquifer. This is open source shapefile data from the British Geological Survey (Upton et al., 2018). Hydrogeology is used as a proxy of water supply availability, as some aquifer types provide better yields of water than others (Upton et al., 2018). Attributes in this layer are (1) basement lithology, which has low yields and thus is given a weight of 1, indicating high risk; (2) Karoo sediments and basalts, which have higher yields and so are given a weighting of 0.5; and (3) alluvial and colluvial lithologies, which have the highest yields and therefore a weight of 0, indicating low risk (Smith-Carrington & Chilton, 1983).

Table 1. Input layers for hotspot map.

Layer	Attribute	Weight	Data type / location
Water point type	(1) Hand-dug well	1	mWater point data
	(2) Hand pump/borehole	0.5	
	(3) Piped supply	0	
Hydrogeology type	(1) Basement	1	GIS shapefile
	(2) Karoo	0.5	
	(3) Alluvial/colluvium	0	
Users per water point	(1) Above Malawi guidelines (> 250 per borehole, 120 per tap)	1	mWater point data
	(2) At or below Malawi guidelines	0	
Proximity to pit latrine	(1) Outside Malawi guidelines (<30 metres)	1	mWater point data
	(2) Within Malawi guidelines (>30 metres)	0	
Land Use	(1) Settlement/cropland/industry	1	GIS shapefile
	(2) Forest/grassland/wetland	0	
Seasonal fluctuations	(1) Yes	1	mWater point data
	(2) No	0	
No data		0.5	

Layer 3 is number of users per water point. The data were collected during the CJF water point functionality survey through questioning of a local community representative. This layer is used as a proxy for stress on the aquifer's water supply. When a water point exceeds the recommended number of users, this can cause stress on the aquifer. In layer 3, when the number of users is above the government guidelines of 250 users per borehole or 120 per tap (Stoupy & Sugden, 2003), a weight of 1 is given; otherwise a weigh of 0 is given.

Layer 4 is proximity to a pit latrine. The pit latrine and water point GPS data were collected through observation during the CJF water point functionality survey. GIS was then used to represent the data spatially and determine the distance between water point and pit latrine. Drinking water sources close to a pit latrine are more susceptible to faecal contamination, so distance from a pit latrine is a proxy for water quality (Back et al., 2018). The government guideline is that pit latrines should not be within 30 metres of a water point (Ministry of Irrigation and Water Development, 2008). In layer 4, water points within 30 metres of a pit latrine are given a weight of 1, indicating the highest risk; otherwise they are given a weight of 0.

Layer 5 is land use. This is open source shapefile data obtained from Malawi's Spatial Data Platform (MASDAP, 2019). Different uses of land can cause a variety of pressures on groundwater. Here, we use land use as a proxy for potential nitrate contamination, as heavy industry and agricultural land use can raise nitrate concentrations in the environment (Lapworth et al., 2017; Wick et al., 2012; Wongsanit et al., 2015). In layer 5, areas of settlement, cropland or industry and are given a weight of 1, indicating a high risk of nitrate contamination; forest, grassland and wetland are given a weight of 0.

Finally, layer 6 is seasonal fluctuation. These data were collected during the CJF water point functionality survey through questioning of a local community representative. Groundwater levels in Malawi fluctuate due to seasonal recharge and baseflow impacts (Kelly et al., 2019b). At times, the levels can drop below borehole and shallow dug well depths. Seasonal fluctuation of water availability from a water point is used as a proxy for groundwater reliability. A source that is seasonally fluctuating is considered less reliable than a source that supplies water year round. In layer 6, water points that fluctuate seasonally are given a weight of 1; otherwise, they are given a weight of 0.

These weights were summed, so each data point had a range from 0 to 6. This overall weighting was then presented spatially in GIS.

Existing methodologies

Methods to prioritize transboundary aquifers for directed management already exist. Davies et al. (2013) sought to identify the transboundary aquifers in most need within the Southern African Development Community. The method aimed to classify 14 transboundary aquifers in the region based on how 'troublesome' they are. The criteria included groundwater flow and vulnerability/susceptibility, groundwater knowledge and understanding, governance capability, socio-economic/water-demand capability and environmental issues. Only two of the 14 aquifers assessed were classified as troublesome. Curiously, neither has been subject to any further study. Instead, three of the other 12 aquifers have had projects dedicated to their assessment and management: the Shire Valley Alluvial Aquifer through the Shire CONWAT project, the South West Kalahari/Karoo Basin Aquifer through the GGRETA project, and the Zeerust-Ramotswa-Lobatse Dolomite Basin Aquifer through the Ramotswa Project (International Water Management Institute, 2019; Southern African Development Community-Groundwater Management Institute, 2018; UNESCO-IHP, 2016). This method is limited by its applicability to only entire transboundary aquifers. Often, large parts of transboundary aquifers are not utilized, especially those that expand across large areas or are not at risk due to their small population (Fraser et al., 2018). This method also risks misrepresenting parts of transboundary aquifers based on data from potentially only one area.

The term 'transboundariness' was coined by Sanchez et al. (2018a) to describe a method that identifies and prioritizes transboundary aquifers based on a political and socio-economic criteria. The data used include population metrics, water quality, data/research availability, political recognition as transboundary, cooperation efforts and other issues governing the agenda. The approach was applied to the Texas–Mexico case, where transboundary aquifers are fairly well understood and often recognized constitutionally (Sanchez & Eckstein, 2017; Sanchez et al., 2016, 2018b). The results split the aquifers into priority groups based on their 'transboundariness'. The more 'transboundary' an aquifer is, the higher the priority it is given. 'Priority' refers to the level of attention or importance given to an aquifer. Thus, the approach gives the highest priority to those aquifers with the most attention and the best data availability. The approach does not prioritize aquifers based on need but rather highlights those which have been given priority until now (Sanchez et al., 2018b). Although effective in identifying which aquifers are already receiving the most attention, this method does not identify which may be at risk of contamination or reduction in water quantity in the future. Furthermore, the method is not applicable in countries where transboundary aquifers are not as well studied, understood or recognized.

Finally, Sanchez et al. (2020) developed an approach to identify the priority areas within a single aquifer using pumping well locations and density as indicators. Areas with a high density of wells are termed 'effective transboundary aquifer areas'. These areas can be considered priority zones by both local and national governments for more specific and specialized transboundary management between riparian states. In focusing only on abstraction, this method does not consider the water quality in transboundary aquifers and thus makes it unsuitable for areas where transboundary aquifers are at risk of reduced water quality, as in the Malawi case.

Our method seeks to highlight the importance of both water quantity and quality in the SADC transboundary aquifer context, where financial resources and groundwater data are often limited. Water point mapping is common practice in the region, and through an alternative use of these data we can fill the gap left by the lack of comprehensive aquifer studies in the region. Furthermore, we can single out individual problematic water points for rehabilitation.

Results

Hotspot analysis

Figure 2 shows the results of the hotspot analysis. Darker spots indicate a higher risk to groundwater. The inset gives more detail. Each of the over 93,000 water points has an individual rating. The spread of the higher-risk areas can be said to be directly related to population. The population is much higher in the central and southern regions of the country than in the north. These areas inherently require a greater number of water points for access to groundwater. This however does not mean the northern areas may not be at risk. Many areas in the north have smaller but equally at-risk water points. Here, pressure from contamination may be more of a driving factor than over-abstraction risk.

It is important to highlight that this risk analysis is confined to within Malawi. Where hotspots are indicated along the border in Malawi, it cannot be assumed that the situation is the same across the border. However, it provides Malawi's government reasons to responsibly notify a neighbour that they share a transboundary aquifer that has been highlighted as at risk for groundwater contamination or over-abstraction. In areas where the groundwater flow direction is from Malawi to a neighbouring country, the risk for that neighbouring country can be deemed just as high as in Malawi. This is because (1) any contamination of the aquifer in Malawi has the potential to travel across the border

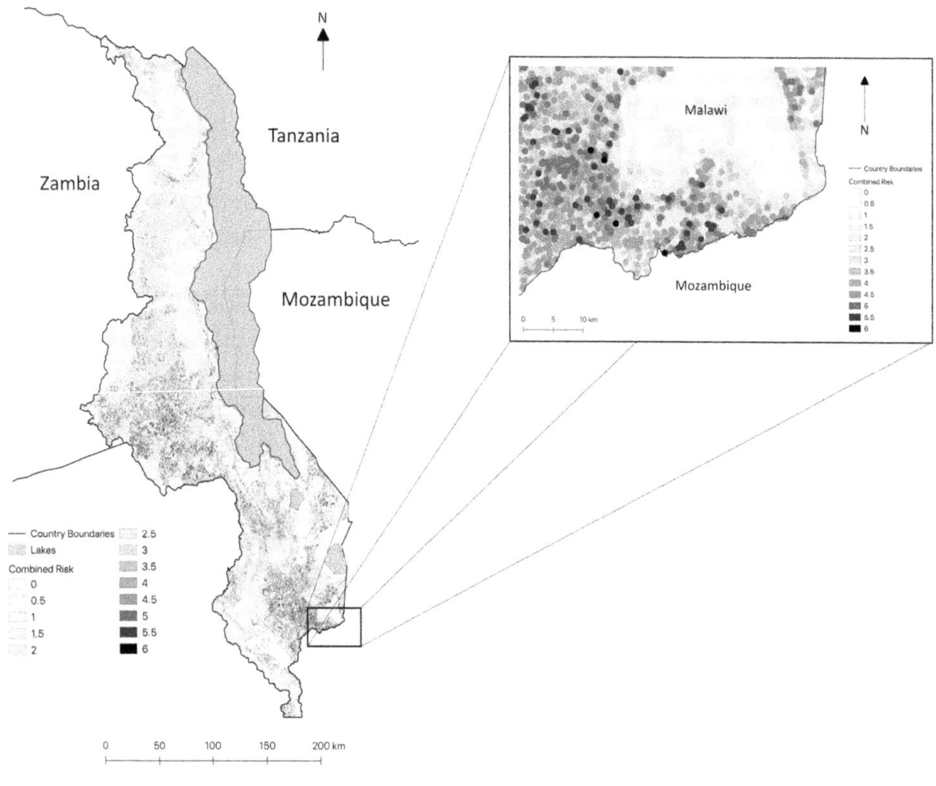

Figure 2. Risk hotspot analysis results.

into the neighbouring country, and (2) over-abstraction from an aquifer in Malawi could change the groundwater flow pattern or even reverse it. These would cause groundwater quality and quantity issues in the neighbouring country.

The hotspot analysis also highlights areas in Malawi where groundwater is not currently being used. Areas lacking water point data indicate that either surface water or a piped supply system, likely from a surface reservoir, is in place. This information could be useful to the government, as these areas could be exploited in the future for greater water availability and thus water security, especially in areas with high-yielding aquifers.

The insert in Figure 2 covers the Mulanje district of Malawi, which borders Mozambique. At this scale, individual water points can be seen. This can help identify the exact location of the water points at the most risk. Isolating these water points and ensuring they are managed effectively could greatly reduce the risk to the overall aquifer.

The next two sections will illustrate which of the 38 previously identified transboundary aquifers (Figure 1) have potential national and local-scale hotspots of transboundary risk. Transboundary aquifers that do not exhibit any sign of national or local-scale hotspots are dropped from further discussion. Although these transboundary aquifers should continue to be monitored, we deem them not to pose a current transboundary risk.

National-scale hotspots

Figure 3 shows the hotspot map overlaid with selected transboundary aquifers from Figure 1 that have national-scale hotspots, which span the majority of a transboundary aquifer. Just three transboundary aquifers exhibit a national scale of risk. The largest is 37,000 km^2, and the smallest is 3800 km^2. Due to the large spread and high density of the water points in these hotspots, they must be considered and dealt with at the national, not the local level.

The national-scale hotspots are all in the central or southern region of Malawi, where the population density is greatest. This emphasizes that as the population rises, the risk to aquifers' water quality and quantity also rises. Also, the transboundary aquifers aligned with these hotspots are all of basement complex lithology, so large populations are relying on low-yielding aquifer types in these areas. Descriptions of the national-scale aquifers are given in Section 1 of the online supplemental material.

Local-scale hotspots

Figure 4 shows the hotspot map overlaid with transboundary aquifers that have local-scale hotspots, which are focused along the border and do not spread further inland into the aquifer. There are 11 of these. Due to their isolated and smaller nature, they can be considered a more local risk and managed as such.

The local-scale hotspots are spread fairly evenly throughout Malawi. They range in size from 230 km^2 to 8600 km^2. Their lithology varies from basement complex to colluvial and alluvial superficial deposits. The hotspots in these aquifers are all much more isolated than the nationals-scale hotspots. There could be multiple reasons for this:

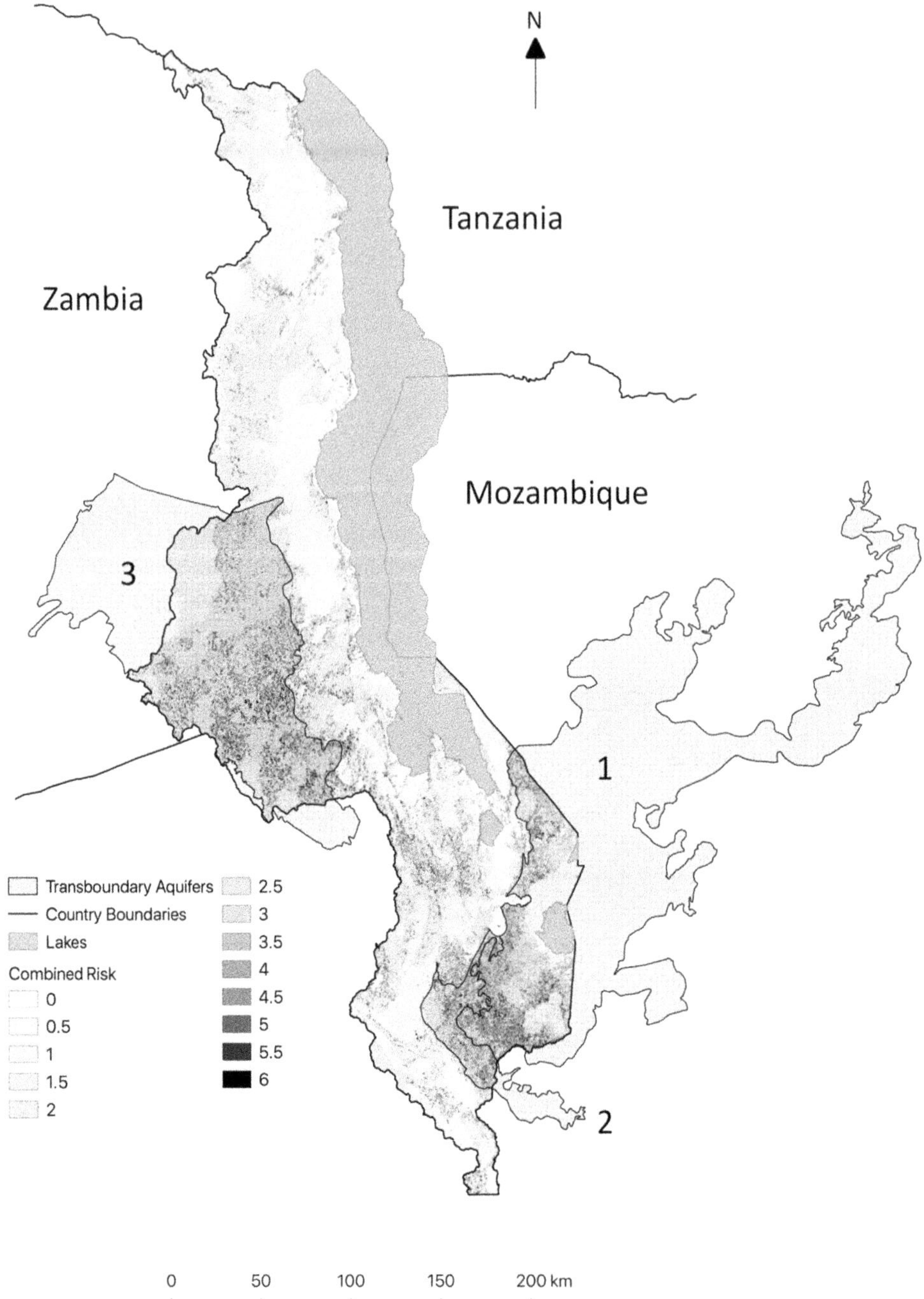

Figure 3. National-scale hotspots and corresponding transboundary aquifers.

these areas tend to be more rural and secluded, with no large populations close to the hotspots; and specific issues relating to particular communities could influence them, such as poor pit latrine location planning, location on a low-yielding aquifer, or a specific land use type, such as agriculture.

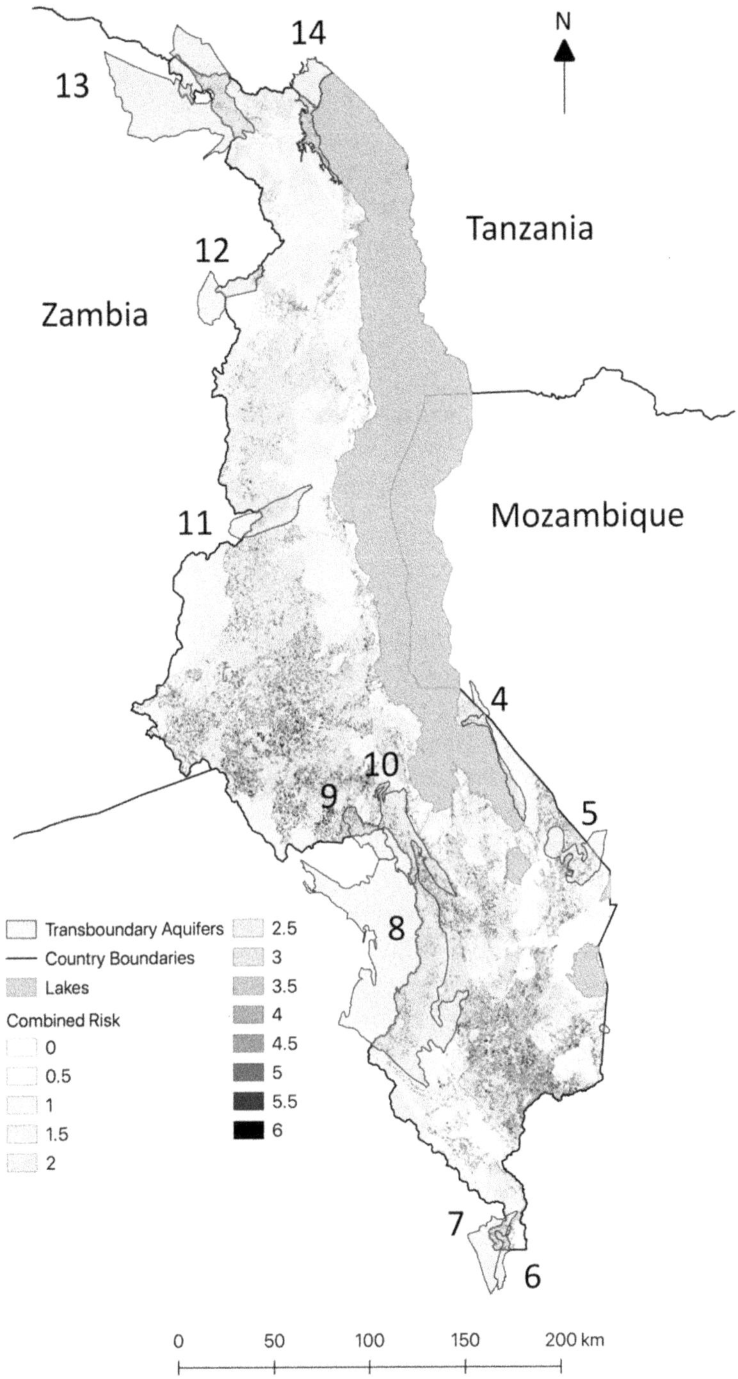

Figure 4. Local-scale hotspots and corresponding transboundary aquifers.

Transboundary responsibility

Of the 14 aquifers identified as exhibiting risk hotspots, only four have groundwater flowing from Malawi into a neighbouring country. The other 10 have groundwater flowing into Malawi from a neighbouring country. Groundwater flow direction can influence transboundary impacts and thus management. Aquifers 2, 6, 7 and 14 are considered high-risk transboundary aquifers with transboundary flow from Malawi to a neighbouring country (see Section 2 of the online supplemental material). The Malawi government must recognize that impacts on these aquifers will affect their downstream neighbours. Ensuring good water quality and sustainable abstraction rates will help prevent negative impacts on bordering countries. Conversely, it is also in Malawi's interest to ensure that the other 10 transboundary aquifers, which flow into Malawi from neighbouring countries, are managed sustainably to prevent any contaminated groundwater from crossing into Malawi's portion of the transboundary aquifer, alongside ensuring that groundwater levels do not fall over time.

Multi-scale management

No two aquifers, including transboundary aquifers, are the same (Eckstein, 2013). Lithology, transmissivity, storage capability, abstraction rates, population dependency and seasonal fluctuations are just some of the factors that may affect how a transboundary aquifer is used. The use of an aquifer should then inform how it is managed (Eckstein, 2013). For example, a small aquifer in a highly productive lithology used by a limited number of people will require a different management strategy from a large aquifer with a low recharge rate used by a large population. Classifying what type of management is required of an aquifer is particularly important in countries like Malawi, where resources and financial reserves are limited. In a transboundary context, it is even more essential to ensure that the selected management type is established in cooperation with neighbouring countries. Available resources should be directed to the aquifers in the most need of management on a case-by-case basis, but there are limited methods currently to identify which aquifers in a country like Malawi should get the highest priority. It is difficult to identify which aquifers may require a specific level of management (e.g., local vs. national) or which currently are not at risk for over-abstraction or contamination and can be left as is.

National-border-based assessments (Fraser et al., 2020) generate data that allow governments to evaluate their transboundary aquifers in detail. Hotspot analysis of these identified transboundary aquifers, as conducted in this study, highlight which aquifers require immediate attention and at what scale. But what management is appropriate for these transboundary aquifers? Figure 5 highlights the relationship between the information generated from different transboundary aquifer assessment types and transboundary management they could inform.

Regional assessments are useful in identifying large transboundary aquifers that may be impacted on a regional scale. For example, the Global Environment Facility's Transboundary Waters Assessment Programme identified 80 transboundary aquifers that are of high importance across south-eastern Africa (ILEC et al., 2016). These aquifers are best managed under intergovernmental organizations, like the river basin

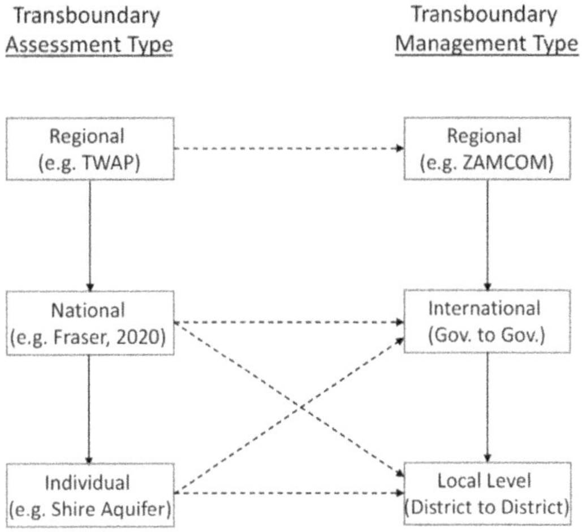

Figure 5. Flow diagram illustrating the relationship between different transboundary aquifer assessments and management types.

organization ZAMCOM, that bring together riparian states to facilitate and promote the sustainable management of water resources in the region, in view of their large scale and regional importance (ZAMCOM, 2019).

On the other hand, national-border-based assessments can inform international (government-to-government) and local (district-to-district) management (Figure 5). The transboundary aquifers in Malawi that have been identified as having national-scale hotspots are good candidates for international management. These aquifers present a significant enough risk at the national level to warrant the government of Malawi to notify and cooperate with its neighbouring-country governments to ensure management of these shared aquifers. This is best done through an international transboundary agreement or an international institutional governance mechanism. International agreements over shared groundwater are however still in their infancy; globally, there are only six transboundary aquifer agreements governing the use and management of shared groundwater (Rivera & Candela, 2018), for only 1% of identified transboundary aquifers worldwide. Each is unique to the aquifer in question. They include the Genevese Aquifer agreement, which governs the management and allocation of water from the aquifer on the French–Swiss border (De Los Cobos, 2018); the Nubian Sandstone and North-Western Sahara Aquifer agreements, which facilitate data sharing between parties (Nubian Sandstone Aquifer System, 2002; SASS, 2002, in Eckstein, 2011) and the Al-Sag/Al-Disi agreement, which created a protected area of 10 km on either side of the border for the fossil (non-recharging) aquifer (Burchi, 2018). Most recently, the UN Draft Articles on the Law of Transboundary Aquifers were developed to provide guidelines for countries wanting to foster agreements over the management of their transboundary aquifers. They cover all aspects of potential transboundary scenarios, including transboundary recharge and fossil aquifers (Sanchez et al., 2016; United Nations, 2008). The draft articles assisted in development of the Guarani Aquifer Agreement, which draws from the draft articles covering equitable and reasonable use, the obligation to cause no harm, and the exchange of technical data (Sindico et al., 2018; Villar & Ribeiro, 2011). A transboundary agreement over any of the identified nationally at-risk aquifers in Malawi could build on the lessons learnt from these agreements. There are currently very few institutional mechanisms that act to govern the use and management of transboundary aquifers in place of a formal agreement at the international and binational level, particularly in the African context (Fraser et al., 2020). Moving forward, Malawi could look to set up a binational commission to govern the use and management of the three national-scale transboundary hotspots we identified, but this would require a strong mandate and willingness to cooperate from all involved stakeholders, which can take time to develop.

The transboundary aquifers in Malawi that have been identified as having only local-scale hotspots are probably too small to justify a full international agreement or binational commissions. These issues could be addressed locally particularly because Malawi's water regulations are decentralized to the district level (Truslove et al., 2020). This approach has been successful in other countries. Local-scale agreements include the Hueco Bolson Aquifer underlying the cities of El Paso and Juárez on the Mexico–US border (Juárez-El Paso MoU, 1999 in Eckstein, 2011), the Abbotsford–Sumas Aquifer between the US State of Washington and the Canadian province of British Columbia (Abbotsford–Sumas MoA, 1996), and the 1999 Memorandum of Understanding between the Municipal Water and Sanitation Board of the City of Juarez, in Chihuahua, Mexico (Eckstein, 2013). Local-scale management has also been suggested as an alternative to fostering formal agreements along the border between the US and Mexico (Eckstein, 2013). A local agreement could be informal or in the form of a memorandum of understanding, but there is nothing to stop parties creating an official agreement on a smaller scale.

Local-scale management can be advantageous for multiple reasons. Local communities often have the biggest stake in good transboundary management; small changes to a local aquifer can have large impacts on the community that depends on it. Communities are often best informed about local issues. Local communities along the border region are also more likely to have links to individuals, communities or businesses on the other side of the border (Eckstein, 2013). In Malawi, for example, many communities close to the border speak the same language and practice the same religion as their cross-border counterparts, due to precolonial tribal connections (Kaspin, 1997).

There is already evidence of local-scale transboundary cooperation in Malawi. The district of Mulanje, on the south-east border of Malawi, is party to semi-regular meetings with the neighbouring district, Milange, in Mozambique. Representatives from the two districts meet and discuss shared issues such as trade, border control and management of their shared river, the Ruo, which forms the border between the two districts. One could make a case for the addition of transboundary groundwater to this working group's mandate. A local-scale agreement over the use and management of transboundary aquifers in Malawi might include clauses for regular exchange of data, joint monitoring of boreholes, and the inclusion of local stakeholders such as communities and local businesses.

Conclusions and recommendations

This study has identified transboundary aquifers in Malawi that are at risk of over-abstraction or reduced water quality (hotspots). Using fuzzy logic and GIS overlays, we pinpointed specific areas in the country that may be at transboundary risk. The method provides the Malawi government with a tool to prioritize which transboundary aquifers require management and international agreements. In Malawi, where water management is decentralized, it also gives an opportunity to identify areas where local-scale transboundary management can play an important role in managing cross-border water resources.

The results of this study can inform the Malawi government where to direct resources for transboundary groundwater management. We highlight three transboundary aquifers thought to be at high risk of over-abstraction and contamination and which require consideration for transboundary agreements. Another 11 hotspots on smaller-scale transboundary aquifers could be managed at the local level. A transboundary diagnostic analysis and coherent conceptualization of these transboundary aquifers could improve our understanding of what management practices are needed. This should ideally be done in cooperation with neighbouring local stakeholders and governments.

Notes

1. Readers of the print article can view the figures in colour online at https://doi.org/10.1080/02508060.2020.1832747.

Acknowledgments

The authors would like to acknowledge financial support from the Scottish government, through the Climate Justice Fund's Water Futures Programme (research grant HN-CJF-03), and the University of Strathclyde. They would also like to acknowledge their partners in the government of Malawi and BASEflow for their support of this research.

Disclosure statement

No potential conflict of interest was reported by the authors.

ORCID

Christina Fraser https://orcid.org/0000-0003-0282-5770

Robert M Kalin https://orcid.org/0000-0003-3768-3848

References

Abbotsford–Sumas MoA. (1996, October 10). Memorandum of agreement related to referral of water right applications related to the transboundary Abbotsford–Sumas Aquifer between the State of Washington as represented by the Department of Ecology and the Province of British Columbia as represented by the Minister of Environment, Lands and Parks.

Addison, M. J., Rivett, M. O., Robinson, H., Fraser, A., Miller, A. M., Phiri, P., Mleta, P., & Kalin, R. M. (2020). Fluoride occurrence in the lower East African Rift System, Southern Malawi. *Science of the Total Environment*, *712*, 136260. https://doi.org/10.1016/j.scitotenv.2019.136260

Adelana, S. M. A., & MacDonald, A. M. (2008). *Applied groundwater studies in Africa (IAH special publications in hydrogeology* (Vol. 13). CRC Press/Balkema.

Back, J. O., Rivett, M. O., Hinz, L. B., Mackay, N., Wanangwa, G. J., Phiri, O. L., Songolo, C. E., Thomas, M. A. S., Kumwenda, S., Nhlema, M., Miller, A. V. M., & Kalin, R. M. (2018). Risk assessment to groundwater of pit latrine rural sanitation policy in developing country settings. *Science of the Total Environment*, *613-614C*, 592–610. https://doi.org/10.1016/j.scitotenv.2017.09.071

Bath, A. H. (1980). *Hydrochemistry in groundwater development: Report on an advisory visit to Malawi* (Report WD/OS/80/20). British Geological Survey.

Burchi, S. (2018). Legal frameworks for the governance of international transboundary aquifers: Pre- and post-ISARM experience. *Journal of Hydrology: Regional Studies*, *20*, 15–20. https://doi.org/10.1016/j.ejrh.2018.04.007

Chavula, G. M. S. (2012). Malawi. In P. Pavelic, M. Giordano, B. Keraita, V. Ramesh, & T. Rao (Eds.), *Groundwater availability and use in Sub-Saharan Africa: A review of fifteen countries* (pp. 78–90). International Water Management Institute.

Davies, J., Robins, N. S., Farr, J., Soresnsen, J., Beetlestone, P., & Cobbing, J. E. (2013). Identifying transboundary aquifers in need of international resource management in the Southern African Development Community region. *Hydrogeology Journal*, *21*(2), 321–330. https://doi.org/10.1007/s10040-012-0903-x

de los Cobos, G. (2018). The Genevese transboundary aquifer (Switzerland-France): The secret of 40 years of successful management. *Journal of Hydrology: Regional Studies*, 1–12. https://doi.org/10.1016/j.ejrh.2018.02.003

Eckstein, G. (2011, September). Managing buried treasure across frontiers: The international law of transboundary aquifers. *Water International*, *36*(5), 573–583. https://doi.org/10.1080/02508060.2011.598642

Eckstein, G. (2013). Rethinking transboundary ground water resources management: A local approach along the Mexico-U.S. border. *Georgetown International Environmental Law*, *25*(1), 95. https://ssrn.com/abstract=2254081

Economic Commission for Africa, Organization of African Unity, & African Development Bank. (2000). Safeguarding life and development in Africa: A vision for water resources management in the 21st century. In *Africa Caucus Presentations, Second World Water Forum* (pp. 45).

Foster, S., Tuinhof, A., & Garduno, H. (2008). Groundwater in sub-Saharan Africa: A strategic overview of developmental issues. In S. Adelana & A. MacDonald (Eds.), *Applied groundwater studies in Africa: IAH selected papers on hydrogeology* (pp. 9–24). Taylor & Francis. https://doi.org/10.1201/9780203889497.ch

Fraser, C. M., Kalin, R. M., Kanjaye, M., & Uka, Z. (2020). A national border-based assessment of Malawi's transboundary aquifer units: Towards achieving sustainable development goal 6.5.2. *Journal of Hydrology: Regional Studies*, *31*. https://doi.org/10.1016/j.ejrh.2020.100726

Fraser, C. M., Kalin, R. M., Rivett, M. O., Nkhata, M., & Kanjaye, M. (2018). A national approach to systematic transboundary aquifer assessment and conceptualisation at relevant scales: A Malawi case study. *Journal of Hydrology: Regional Studies*, special issue on International Shared Aquifer Resources Assessment and Management, *20*, 35–48. https://doi.org/10.1016/j.ejrh.2018.04.001

IGRAC & UNESCO-IHP. (2015). *Transboundary aquifers of the world* [map]. Scale 1:50 000 000, edition 2015. IGRAC.

ILEC, UNEP-DHI, UNESCO-IHP, UNESCO-IOC, & UNEP. (2016). Water system information sheets: Eastern & Southern Africa. In L. Talaue-McManus (Ed.), *Transboundary waters: A global compendium, Volume 6, Annex G*. United Nations Environment Programme.

International Water Management Institute. (2019). Joint strategic action plan for the Ramotswa transboundary aquifer area. Project Report. Retrieved March 28, 2020, from http://conjunctivecooperation.iwmi.org/systems/ramotswa-ngotwane-system/reports-and-publications/

Juárez-El Paso MoU. (1999, December 6). Memorandum of understanding between City of Juárez, Mexico Utilities and the El Paso Water Utilities Public Services Board of the City of El Paso, Texas.

Kalin, R. M., Mwanamveka, J., Coulson, A. B., Robertson, D. J. C., Clark, H., Rathjen, J., & Rivett, M. O. (2019). Stranded assets as a key concept to guide investment strategies for sustainable development goal 6. *Water*, *11*(4), 702. https://doi.org/10.3390/w11040702

Kaspin, D. (1997). Tribes, regions and nationalism in democratic Malawi. In *Nomos: American Society for Political and Legal Philosophy: Vol. 39. Ethnicity and group rights*, pp. 464–503.

Kelly, L., Bertram, D., Kalin, R., & Ngongondo, C. (2019b). Characterization of groundwater discharge to rivers in the Shire River Basin, Malawi. *American Journal of Water Science and Engineering, 5*(4), 127–138. https://doi.org/10.11648/j.ajwse.20190504.11

Kelly, L., Kalin, R. M., Bertram, D., Kanjaye, M., Nkhata, M., & Sibande, H. (2019a). Quantification of temporal variations in base flow index using sporadic river data: Application to the Bua Catchment, Malawi. *Water, 11*(5), 901. https://doi.org/10.3390/w11050901

Lapworth, D. J., Nkhuwa, D. C. W., Okotto-Okotto, J., Pedley, S., Stuart, M. E., Tijani, M. N., & Wright, J. (2017). Urban groundwater quality in sub-Saharan Africa: Current status and implications for water security and public health. *Hydrogeology Journal, 25*(4), 1093–1116. https://doi.org/10.1007/s10040-016-1516-6

MacDonald, A. M., Bell, R. A., Kebede, S., Azagegn, T., Yehualaeshet, T., Pichon, F., Young, M., McKenzie, A. A., Lapworth, D. J., & Black, E. (2019). Groundwater and resilience to drought in the Ethiopian Highlands. *Environ. Res. Lett., 14*(9), 095003. https://doi.org/10.1088/1748-9326/ab282f

MASDAP. (2019). *Land use, land cover.* Raster Data. http://www.masdap.mw/layers/geonode:Malawi_Sentinel2_LULC2016

Mccracken, M. (2017). Measuring transboundary water cooperation sustainable development goal target 6.5. *Background Papers*, Vol. 88, Issue 23. Global Water Partnership.

Ministry of Irrigation and Water Development, Government of Malawi. (2008). *The National Sanitation Policy.* MoIWD, Lilongwe. Retrieved August 16, 2020, from https://www.ircwash.org/sites/default/files/824-MW06-19182.pdf

Monjerezi, M., Vogt, R. D., Aagaard, P., & Saka, J. D. K. (2011). Hydro-geochemical processes in an area with saline groundwater in lower Shire River valley, Malawi: An integrated application of hierarchical cluster and principal component analyses. *Applied Geochemistry, 26*(8), 1399–1413. https://doi.org/10.1016/j.apgeochem.2011.05.013

Nijsten, G., Christelis, G., Villholth, K. G., Brauned, E., & Gayee, C. B. (2018). Transboundary aquifers of Africa: Review of the current state of knowledge and progress towards sustainable development and management. *Journal of Hydrology: Regional Studies, 20*, 21–34. https://doi.org/10.1016/j.ejrh.2018.03.004

Nubian Sandstone Aquifer System. (2002). *Programme for the development of a regional strategy for the utilisation of the Nubian Sandstone Aquifer System (NSAS).* Terms of reference for the monitoring and exchange of groundwater information of the Nubian Sandstone Aquifer System, 2002. Tripoli, 5 October 2000 [online]. Retrieved February 18, 2020, from http://www.fao.org/docrep/008/y5739e/y5739e05.htm

Persits, F., Ahlbrandt, T., Tuttle, M., Charpentier, R., Brownfield, M., & Takahashi, K. (2002). *Map showing geology, oil and gas fields and geologic provinces of Africa* (Ver 2.0) (USGS Open File report 97-470 A). U.S. Geological Survey.

Rivera, A., & Candela, L. (2018). Fifteen-year experiences of the Internationally Shared Aquifer Resources Management Initiative (ISARM) of UNESCO at the global scale. *Journal of Hydrology: Regional Studies, 20*(2018), 5–14. https://doi.org/10.1016/j.ejrh.2017.12.003

Rivett, M. O., Budimir, L., Mannix, N., Miller, A. V. M., Addison, M. J., Moyo, P., Wanangwa, G. J., Phiri, O. L., Songola, C. E., Nhlema, M., Thomas, M. A. S., Polmanteer, R. T., Borge, A., & Kalin, R. M. (2019). Responding to salinity in a rural African alluvial valley aquifer system: To boldly go beyond the world of hand-pumped groundwater supply? *Science of the Total Environment, 653*, 1005–1024. https://doi.org/10.1016/j.scitotenv.2018.10.337

Rivett, M. O., Halcrow, A. H., Schmalfuss, J., Stark, J. A., Truslove, J. P., Kumwenda, S., Harawa, K. A., Nhlema, M., Songola, C., Wanangwa, G. J., Miller, A. V. M., & Kalin, R. M. (2018a). Local scale water-food nexus: Use of borehole-garden permaculture to realise the full potential of rural water supplies in Malawi. *Journal of Environmental Management, 209*, 354–370. https://doi.org/10.1016/j.jenvman.2017.12.029

Rivett, M. O., Robinson, H. L., Wild, L. M., Melville, J., McGrath, L., Phiri, P., Flink, J., Wanangwa, G. J., Mleta, P., MacLeod, S. S. P., Miller, A. V. M., & Kalin, R. M. (2018b). Arsenic occurrence in Malawi groundwater. *Journal of Applied Sciences & Environmental Management, 22*(11), 1807–1816. https://doi.org/10.4314/jasem.v22i11.16

Sanchez, R., & Eckstein, G. (2017). Aquifers shared between Mexico and the United States: Management perspectives and their transboundary nature. *Groundwater*, *55*(4), 495–505. https://doi.org/10.1111/gwat.12533

Sanchez, R., Lopez, V., & Eckstein, G. (2016). Identifying and characterizing transboundary aquifers along the Mexico-US border: An initial assessment. *Journal of Hydrology*, *535*(711), 101–119. https://doi.org/10.1016/j.jhydrol.2016.01.070

Sanchez, R., Rodriguez, L., & Tortajada, C. (2018a). The transboundariness approach and prioritisation of transboundary aquifers between Mexico and Texas. *Journal of the Human Environment*, *47*(7), 760–770. https://doi.org/10.1007/s13280-018-1015-1

Sanchez, R., Rodriguez, L., & Tortajada, C. (2018b). Transboundary aquifers between Chihuahua, Coahuila, Nuevo Leon and Tamaulipas, Mexico and Texas, USA: Identification and categorization. *Journal of Hydrology, Regional Studies*, special issue on International Shared Aquifer Resources Assessment and Management, *20*, 74–102. https://doi.org/10.1016/j.ejrh.2018.04.004.

Sanchez, R., Rodriguez, L., & Tortajada, C. (2020). Effective transboundary aquifer areas: An approach for transboundary groundwater management. *Journal of the American Water Resources Association*, *56*(3), 360–378. https://doi.org/10.1111/1752-1688.12836

SASS. (2002). *Establishment of a consultation mechanism for the Northwestern Sahara Aquifer System (SASS)* [online]. Rome, 19–20 December; endorsed 6 January 2003 (Algeria), 15 February 2003 (Tunisia), 23 February 2003 (Libya). Retrieved February 17, 2020, from http://www.fao.org/docrep/008/y5739e/y5739e05.htm#bm05.2.1

Sindico, F., Hirata, R., & Manganelli, A. (2018). The Guarani Aquifer System: From a Beacon of hope to a question mark in the governance of transboundary aquifers. *Journal of Hydrology: Regional Studies, 20*, 49–59. https://doi.org/10.1016/j.ejrh.2018.04.008

Smedley, P. 2004. *Groundwater quality: Malawi*. British Geological Survey.

Smith, M., Cross, K., Paden, M., & Laban, P. (Eds.). (2016). *Spring: Managing groundwater sustainably*. IUCN. 978-2-8317-1789-0.

Smith-Carrington, A. K., & Chilton, P. J. (1983). *Groundwater resources of Malawi*. Overseas Development Administration Institute of Geological Sciences. Department of Lands Valuation and Water, Republic of Malawi.

Southern African Development Community – Groundwater Management Institute. (2018). *Conjunctive transboundary water resource management*. Project Homepage. Retrieved March 28, 2020, from http://sadc-gmi.org/shire-river/

Stoupy, O., & Sugden, S. (2003): *Halving the number of people without access to safe water by 2015: A Malawian perspective. Part 2: New indicators for the Millennium Development Goal*. WaterAid.

Truslove, J. P., Coulson, A. B., Nhlema, M., Mbalame, E., & Kalin, R. M. (2020). Reflecting SDG 6.1 in rural water supply tariffs: Considering 'affordability' versus 'operations and maintenance costs' in Malawi. *Sustainability*, *12*(2), 744. https://doi.org/10.3390/su12020744

UNESCO-IHP. (2016). *Stampriet transboundary aquifer system assessment: Governance of groundwater resources in transboundary aquifers (GGRETA) - Phase 1*. Technical Report.

United Nations. (2008). *Draft articles on the law of transboundary aquifers*. Retrieved July 9, 2018, from http://legal.un.org/ilc/texts/instruments/english/draft_articles/8_5_2008.pdf

UN-Water. (2018). Groundwater overview: Making the invisible visible. Produced by International Groundwater Resources Assessment Centre, in cooperation with UNESCO-IHP, IAH, IWMI and with contributions of many UN-Water members and partners.

Upton, K., Ó Dochartaigh, B. É., Chunga, B., & Bellwood-Howard, I. (2018). *Africa groundwater atlas: Hydrogeology of Malawi. Geology of Malawi at 1:5 million scale*. British Geological Survey. Retrieved February 6, 2020, from http://earthwise.bgs.ac.uk/index.php/Hydrogeology_of_Malawi

van der Gun, J. (2012). *Groundwater and global change: Trends, opportunities and challenges*. United Nations World Water Assessment Programme. https://unesdoc.unesco.org/ark:/48223/pf0000215496

Villar, P. C., & Ribeiro, W. C. (2011). The agreement on the Guarani Aquifer: A new paradigm for transboundary groundwater management? *Water International, 36*(5), 646–660. https://doi.org/10.1080/02508060.2011.603671

Villholth, K. G. (2013). Groundwater irrigation for smallholders in Sub-Saharan Africa: A synthesis of current knowledge to guide sustainable outcomes. *Water International, 38*(4), 369–391. https://doi.org/10.1080/02508060.2013.821644

Wick, K., Heumesser, C., & Schmidb, E. (2012, November 30). Groundwater nitrate contamination: Factors and indicators. *Journal of Environmental Management, 111*(3), 178–186. https://doi.org/10.1016/j.jenvman.2012.06.030

Wongsanit, J., Teartisup, P., Kerdsueb, P., Tharnpoophasiam, P., & Workakhunpist, S. (2015). Contamination of nitrate in groundwater and its potential human health: A case study of lower Mae Klong river basin, Thailand. *Environmental Science and Pollution Research, 22*(15), 11504–11512. https://doi.org/10.1007/s11356-015-4347-4

ZAMCOM. (2019). The Zambezi Watercourse Commission. Retrieved February 18, 2020, from http://www.zambezicommission.org/about-zamcom/about-zamcom

OPEN ACCESS

Binational reflections on pathways to groundwater security in the Mexico–United States borderlands

Rosario Sanchez, José Agustin Breña-Naranjo, Alfonso Rivera, Randall T. Hanson, Antonio Hernández-Espriú, Rick J. Hogeboom, Anita Milman, Jude A. Benavides, Adrian Pedrozo-Acuña, Julio Cesar Soriano-Monzalvo, Sharon B. Megdal, Gabriel Eckstein and Laura Rodriguez

ABSTRACT

Shared groundwater resources between Mexico and the United States are facing unprecedented stressors. We reflect on how to improve water security for groundwater systems in the border region. Our reflection begins with the state of groundwater knowledge, and the challenges groundwater resources face from a physical, societal and institutional perspective. We conclude that the extent of ongoing cooperation frameworks, joint and remaining research efforts, from which alternative strategies can emerge, still need to be developed. The way forward offers a variety of cooperation models as the future offers rather complex, shared and multidisciplinary water challenges to the Mexico–US borderlands.

Introduction

Groundwater is estimated to account for 97% of available global freshwater resources (World Resources Institute, 1990). Worldwide, the increase in population coupled with water scarcity and the degradation of water quality have highlighted the need for transboundary water management agreements (Draper & Kundell, 2007). The efforts supporting the development of joint assessments of equitable allocation of surface water and groundwater, with particular attention to sustainability, vulnerability to contamination, and management of wildlife and ecosystem preservation, are slowly but diligently elevating the priorities of the global water agenda on transboundary water resources (UNECE & UNESCO, 2021).

While some international laws governing transboundary surface water do exist, shared groundwater resources have long been neglected by both the international legal regime and most transboundary water management agreements. The 'invisibility' of the resource, combined with its natural physical complexities and the fact that sustainable groundwater use has not drawn much attention until recent decades (Gleeson et al., 2019), significantly contributed to the lack of policy development at the cross-border level. Additionally, the lack of transboundary coordination at a worldwide level has been highlighted in the literature as the main cause of poor multinational groundwater management approaches (Eckstein, 2017). Recently, due to the impacts of climate change, population growth and widespread pollution of surface water, policymakers and water managers are moving towards a paradigm of water being 'more protected underground' (UNECE, 2014). As a result, conservation measures, as well as the conjunctive use of surface and groundwater resources, have become more important and strategic at the transboundary level.

In the case of the waters shared between Mexico and the United States, two complementary binational surface water agreements are in place: the 1906 Convention between the United States and Mexico, Equitable Distribution of the Waters of the Rio Grande, and the 1944 United States and Mexico Treaty of the Utilization of Waters of the Colorado and Tijuana Rivers and of the Rio Grande (together referred as 'the Water Treaties'). The Water Treaties provide the legal framework governing the three major binational rivers and their tributaries shared between Mexico and the United States, but do not cover other transboundary rivers (such as the Santa Cruz and San Pedro rivers, just to mention a couple) or any shared groundwater resources. The Water Treaties also do not recognize the physical linkages between surface water and groundwater in transboundary watersheds and aquifers. In addition, climatic variability and climate change, as well as expectations of population growth, were not considerations when the Water Treaties were developed. Furthermore, the total volumes of water in the river basins managed under these treaties were quantified according to measured water values recorded in the early 1900s when the region was less arid.

With demand increasing for new sources of water, greater pressure is now being placed on the region's groundwater resources, thereby intensifying the need for cooperation over transboundary aquifers (TBAs) (Rivera et al., 2015). However, in the border region of Mexico and the United States, as in many other borders around the world, cooperation related to groundwater resources faces challenges that need to be recognized and addressed before any new policies and approaches are suggested at a binational level, or even proposed for incorporation into the existing Water Treaties. Several elements essential to sustainable and long-term water management are required. These include: bilateral effective communication that incentivizes the exchange of data and information on current groundwater conditions (e.g., hydraulic head records, abstraction rates, land uses, surface–groundwater interactions and water quality); second, joint criteria for aquifer and watershed delineation and monitoring; third, a comprehensive plan for surface–groundwater conjunctive use; fourth, conservation and protection of shared endangered habitats; and fifth, governing legal principles for management and sustainability of these shared resources.

This paper describes the challenges in the border region and makes recommendations as to how to build and improve water security for surface and groundwater systems in the border region between Mexico and the United States. We begin by describing the state of groundwater knowledge in the region and its implications for water security. Main topics include: (1) the current state of knowledge about groundwater; (2) groundwater use and competition; (3) water security challenges related to groundwater and surface water; (4) hydrological and water resources models, from numerical and governance-management approaches to support holistic decision-making; and (5) a proposed research agenda and path forward. We conclude that increased funding availability, research development and data generation, and prioritization within a coordinated binational agenda are needed to advance in terms of water security for groundwater systems in the border region.

While there are many transboundary connections that must be bridged, this paper unearths several relevant lessons to enhance the stewardship of transboundary water resources between Mexico and the United States.

Transboundary groundwater resources in the Mexico–US border region

Geographical setting

There are numerous TBAs along the 3145 km-long Mexico–US border supporting ecosystems, socioeconomic development and the well-being of more than 10 million people in 30 cities and communities (Figure 1a). However, there is no consensus on the exact number of TBAs in the border region nor on their geographical delineations. Some

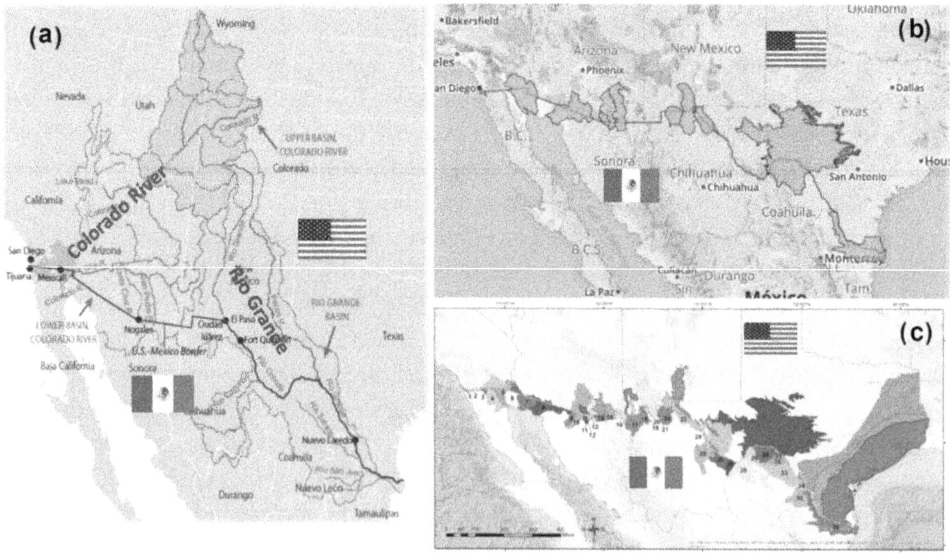

Figure 1. Map showing (a) the 3145 km international border with the main rivers between Mexico and the United States (CRS, 2018); (b) transboundary aquifers along the Mexico–US border according to UNESCO (2010); and (c) transboundary aquifers along the Mexico–US border according to Sanchez et al. (2016).

authors list 11 TBAs (UNESCO, 2010) (Figure 1b), while others suggest that there might be up to 36 TBAs (Sanchez et al., 2016) (Figure 1c). Differences in definitions, criteria and methodologies account for much of these numerical disparities.

Assessment of these TBAs reveals marked differences between the two countries in terms of subsurface characterization, available information, monitoring schemes, management models and outreach policies.

While the two Water Treaties address surface waters in the border region, Minute 242 (an amendment to the 1944 Treaty that focuses on an aquifer in the Sonora–Arizona border) is the only legally binding agreement between Mexico and the United States to address binational groundwater resources (Sanchez & Eckstein, 2020a). There are no agreements between the two countries that address conjunctive use of surface and groundwater. Further, different water policies, water ownership regimes and governance approaches at the national and subnational levels within each country pose challenges for potentially managing transboundary waters (Megdal & Scott, 2011).

Gaps in knowledge exacerbate the challenge of managing TBAs along the borderlands. The three fundamental unknowns include knowledge of groundwater extraction and use, natural and artificial groundwater recharge, and land use, all of which are key elements of the groundwater budget (Kim & Jackson, 2012; Rivera et al., 2015). While the general perception is that groundwater is being extracted at rates greater than natural and artificial recharge, there are no datasets to assess or confirm this claim outside of inferences from groundwater-level declines at a border-wide scale. Existing data are limited to the political boundaries of each country or subnational state.

On the US side, states such as New Mexico have required monthly pumping reports since 2009 for every well (Hanson et al., 2020). However, land use is not regularly mapped to track agricultural, municipal or ecosystem water uses. Aquifer levels are monitored in established networks, and in California, groundwater use is estimated within specific groundwater basins under the California Statewide Groundwater Elevation Monitoring Program (CASGEM) (2021). These data are generated either by the federal government (US Geological Survey – USGS), or by the states along the border (California, Arizona, New Mexico and Texas). However, there is no integration of withdrawal rates and other data into a single continuous database for the entire border that transcends political jurisdictions. Arizona, for example, has established an active management area in the Santa Cruz TBA that has withdrawal reporting requirements, but only for the Arizona part of the aquifer. Additionally, there is no withdrawal reporting for the San Pedro TBA on either side of the border (Megdal & Scott, 2011).

Within Mexico, groundwater uses by private sectors are implemented as groundwater concessions (groundwater rights permitted from the federal government), however these concessions are not systematically monitored or measured for all users and uses. Once a concession is officially granted, users do not always report their water use as required by CONAGUA, the centralized federal water agency in Mexico, and, in many cases, extract more groundwater than the amount officially allowed. Thus, estimates of groundwater extraction are inexact and can be subject to large biases and errors (Sanchez et al., 2010). CONAGUA applies a water balance for 'officially delineated basins', which includes both surface water and groundwater boundaries as 'administrative units' (Sanchez et al., 2016). However, in most cases, the water balance is incomplete and biased as statistics are based on old and limited data instead of annually updated estimates. CONAGUA has applied management measures such as the Aquifer Protection Act, which limits pumping rates and restricts pumping areas (*zonas de veda*) where abstraction has been reported to exceed annual recharge. However, these efforts have not been enough to constrain consistent declines in groundwater tables, and groundwater quality degradation has increasingly been reported in the border region (CONAGUA, 2015). Additionally, the amount of illegal groundwater pumping on the Mexico side is believed to account for a substantial portion of water pumping in the region (Loch et al., 2020).

In terms of research, from the centre to the south-east of the border to the Lower Rio Grande/Rio Bravo basin, important binational studies have been conducted, such as for the Santa Cruz (ongoing) and San Pedro watersheds and the Lower Rio Grande basin (Callegary et al., 2018, 2016; Hanson et al., 2013), all of which were funded by the Transboundary Aquifer Assessment Program (TAAP). Additional regional studies were developed for the Mimbres–Las Palmas aquifer (Hanson et al., 1994), the transboundary region of the Lower Rio Grande basin in the Rincon Valley, the Mesilla Bolsón, the Conejos–Médanos aquifer (Hanson et al., 2020; Ritchie et al., 2018; Sweetkind, 2017; Sweetkind et al., 2017), the Hueco Bolsón (Heywood & Yager, 2003), and the Allende–Piedras Negras (APN) aquifer (Rodriguez et al., 2020). The Tijuana River Watershed is part of an ongoing USGS projects and, in the case of the Lower Colorado River Basin, the integration of surface water and groundwater at the transboundary level with a conceptual model is currently being pursued as a first step in a bilateral scientific assessment (Cital et al., 2021). However, participation in these efforts from the Mexico side has been very limited (Danskin, 2020). Current assessments can only be considered as initial steps towards a more refined integration of binational data for potential transboundary groundwater management. Current modelling results tend to be limited to a portion of the aquifer, and other studies are often limited to general assessments of physical conditions of the aquifers. While some more in-depth binational management approaches are under development (Callegary et al., 2018; Sanchez & Eckstein, 2020a), data-sharing, joint studies and monitoring are largely absent.

While there have been a few isolated efforts to develop the state of knowledge on Mexico–US TBAs, in general there is a significant lack of binational cooperation at the technical, institutional and managerial levels. The TAAP, which was funded by the US government and facilitated by the International Boundary and Water Commission (IBWC), has promoted the development of protocols and principles for transboundary groundwater data-sharing for priority TBAs in Texas, New Mexico and Arizona (La Mesilla and Hueco Bolsón, Santa Cruz, and San Pedro) on the US side, and Sonora and Chihuahua on the Mexican side (Senate Reports, 2006). However, the funding and framework for coordinated monitoring networks and maintaining data-sharing and developing data protocols that are facilitated by TAAP are still limited.

While there have been some case studies in which groundwater budgets were estimated for particular TBAs using conceptual and/or numerical models, only four model studies that straddle the border have been published: for the Hueco Bolsón (Heywood & Yager, 2003), the Lower Rio Grande basin (Hanson et al., 2020), the Mimbres Basins (Hanson et al., 1994) and the APN aquifer (Rodriguez et al., 2020). Groundwater budgets for other TBAs on the border are based on inferred or basin-wide mass-balance estimates (Rivera et al., 2015).

Groundwater-use estimates for the 11 TBAs shown in Figure 1a are recognized by the two countries (Table 1), however most of these assessments were compiled during the International Shared Aquifer Resources Management (ISARM) Americas Network meetings held in 2010 (Rivera et al., 2015), and some are from the above-mentioned models. These estimates vary in accuracy, timescales and level of detail, and are incomplete. As shown in Table 1, information on groundwater extraction rates and potential overexploitation threats is one of the most important knowledge gaps in the region. The estimates collectively indicate that most TBAs along the border are overexploited with groundwater extraction rates slightly exceeding recharge rates by 14%. Additionally, many smaller TBAs are not catalogued, monitored, assessed or managed (Sanchez et al., 2016).

Understanding groundwater use and competition

Agriculture is the most important and largest anthropogenic user of groundwater along the border; other uses include municipal, domestic, mining and industrial. For example, the Mexicali Valley in Baja California, located in the southernmost part of the Lower Colorado River Basin Transboundary Aquifer (LCRB), has been one the most important agricultural regions on the Mexican side of the border since the 1950s. The LCRB is primarily recharged from the Colorado River, although induced recharge from agriculture has also played an important role for the aquifer. Prior to 2010, Mexican agriculture in the Mexicali Valley suffered from water logging conditions as a result of leakage from the All-American Canal. However, that leakage also provided about 74×10^6 m^3/year of artificial recharge to the TBA (Hanson et al., 2015). With the lining of the canal completed in 2010, less water leaked into the Mexicali Valley, which put farmers in Mexico under pressure to extract more groundwater more quickly, thereby overexploiting the aquifer, deteriorating water quality and inducing saltwater intrusion from the Mar de Cortés (Gulf of California). This case provides an example of the magnitude of the interlinkages of a shared surface–groundwater system and how they can be impacted at different scales by a unilateral measure on one side of the border.

As with the growth of water used in agriculture, municipal and industrial uses of groundwater have steadily increased over the last few decades due to growing metropolitan areas and expansion of *Maquiladoras* (manufacturers assembling products from raw materials primarily for export) in northern Mexico along the border. International companies have been attracted by a good-quality workforce and inexpensive labour and, in most cases, the *Maquiladoras* rely on local groundwater from TBAs. In addition, there are other less evident but important growing uses for groundwater, such as mining along the Arizona/Sonora border, beer production in Coahuila and Sonora, and hydraulic fracturing activities in Texas (Rodriguez et al., 2020). Similar stresses and competition for surface water in the northern states of Mexico are expected to induce collateral damage and conflicts for Mexican agriculture in the Conchos River basin in Chihuahua, and for farmers in the Mexicali Valley of Baja California where declining surface water availability will need to be supplemented with groundwater from TBAs.

In addition, recent advances in horizontal drilling and hydraulic fracturing techniques that require the injection of a large volume of water-based pressurized fluid to increase shale permeability and enhance oil recovery have allowed profitable large-scale

Table 1. Officially recognized transboundary aquifers along the Mexico–US border.

Official UNESCO designation	Estimated recharge (m^3/year)	Estimated extraction (m^3/year)	Area (km^2) and population (number of inhabitants)	Groundwater users
8 N San Diego–Tijuana	16×10^6 (both sides)	17×10^6 (Mexican side) $< 6 \times 10^6$ (US side)	Mexican side: 300 km^2; 2.8 million US side: 250 km^2; 400 000	Mexican side: supply to Tijuana; minor agriculture use US side: public supply and minor agricultural use
9 N Lower Colorado River Basin	620×10^6 (both sides)	826×10^6 (Mexican side)	Mexican side: 14,602 km^2; 900 000 (2005) US side: 15,198 km^2	Mexican side: main use agriculture US side: agriculture and public supply
10 N Sonoyta–Pápagos	41.4×10^6 (both sides)	46.3×10^6 (Mexican side)	Mexican side: 43 000	Mexican side: main use agriculture US side: mainly domestic
11 N Nogales	5.2×10^6 (both sides)	0.8×10^6 (Mexican side)	Mexican side: 120 km^2; 157 000 (2005) US side: 6400 km^2; 21 000 (2005)	Domestic, public supply and industrial use on both sides
12 N Santa Cruz	33×10^6 (Mexican side)	28×10^6 (Mexican side) 4.4×10^6 (US side)	Mexican side: 950 km^2; 25 000 US side: 980 000	Mexican side: mainly domestic use, minor agricultural use US side: mainly agricultural and municipal use, minor industrial use
13 N San Pedro	41.2×10^6 (Mexican side)	22.3×10^6 (Mexican side)	Mexican side: 62 000 (2005) US side: 130 000 (2005)	Mexican side: Mainly industrial use US side: mainly agriculture
14 N Conejos Médanos–Bolsón de la Mesilla and Rincon Valley (Hanson et al., 2020)	1.6×10^6 (Mexican side) 2.39×10^8 (US side)	1.6×10^6 (Mexican side) 2.37×10^7 (US side)	Total area: 10,000 km^2	Manly agriculture use, public supply and minor domestic use
15 N Bolsón de Hueco–Valle de Juárez (Heywood & Yager, 2003)	5.03×10^7 (US side)	310×10^6 (Mexican side)	Mexican side: 1600 km^2 US side: 6400 km^2	Mexican side: domestic and agriculture US side: minor use of groundwater
16 N Edwards–Trinity–El Burro	Unknown	10×10^6 (Mexican side) 72×10^6 (US side)	Mexican side: 22,000 km^2, 111 000 (2005) US side: no data, 49 000 (2005)	Mexican side: agriculture US side: agroindustry
17 N Cuenca Baja del Río Bravo–Grande	198×10^6 (Mexican side)	26×10^6 (Mexican side)	Mexican side: 17,500 km^2, 1.4 million (2005) US side: no data, 1.2 million (2005)	Mexican side: main use agriculture, minor use industrial US side: main use agriculture, minor use domestic
18 N Los Mimbres–Las Palmas (Hanson et al., 1994)	90×10^6 (US side)	29.5×10^6 (Mexican side) 180×10^6 (US side)	Mexican side: 11,400 km^2, 10 000 (2005) US side: 3800 km^2, 27 000 (2005)	Both sides: main use agriculture, minor uses public supply and mining industry

Sources: Data sources are from Rivera et al. (2015) and UNESCO (2010), unless noted otherwise.

production of unconventional oil and gas, which in some plays reaches approximately 1 million barrels of oil/day. That is the case of the Eagle Ford Shale, a transboundary hydrocarbon-rich formation crossing from south Texas into north-east Mexico. Because of the large volumes of water needed in the hydraulic fracturing process, domestic and international concerns have been raised regarding water scarcity and depletion, as well as the environmental, geological and hydrogeological impacts of hydraulic fracturing development and produced water disposal. These latter concerns include induced seismicity, changes in land use, landscape fragmentation and aquifer pollution, to name a few.

Water extracted to satisfy hydraulic fracturing production has progressively increased over the last few years. In Texas, water use in the Eagle Ford Shale between 2009 and 2011, in 2016, and between 2015 and 2017 was in the order of about 16,000, about 33,000, and about 38,000 m^3/frac well, respectively (Hernández-Espriú et al., 2019; Ikonnikova et al., 2017). The accumulated volume of water used in the Eagle Ford play to fracture more than 5400 wells drilled in 2015–17 totalled about 190 Mm^3 (Groundwater Protection Council et al., 2011). See Hernández-Espriú et al. (2019, tab. 1, p. 147) for a summary of relevant research on the water footprint associated with hydraulic fracturing production. A total of 80% came from fresh surface and groundwater sources, while about 20% originated from brackish aquifers (Ikonnikova et al., 2017). These figures reveal that unconventional oil and gas development has become a high water-demanding user in the region. On the Mexican side of the Eagle Ford play (correlative units within the Sabinas and Burgos basins), while unconventional resources remain unexploited, their potential development will rely almost entirely on local aquifers. Although groundwater use for fracturing might not be a significant user at the regional level, it does represent an important consumptive use at local levels with considerable impacts on local groundwater systems (Hernández-Espriú et al., 2019).

Groundwater and water security

Water supply from the basins of the Colorado, Tijuana and the Rio Grande/Bravo rivers is increasingly becoming scarce, thereby threatening the region's water security (Cohen, 2005). According to the United Nations Council on Water Security in the context of transboundary waters, water security can be defined as:

> the capacity of a population to safeguard sustainable access to adequate quantities of acceptable quality water for sustaining livelihoods, human well-being, and socio-economic development, for ensuring protection against water-borne pollution and water-related disasters, and for preserving ecosystems in a climate of peace and political stability. (UN Water, 2013)

The most important water security challenges identified in the transboundary basins in the Mexico–US borderlands are the following:

- Water over-allocation: the Colorado River Basin is considered one of the most over-allocated river systems in the world (Castle et al., 2014; Christensen et al., 2004). It has a total area of 637,137 km^2, of which almost 98% is in the United States. It has an average runoff of 19,735 hm^3/year and an estimated water allocation among the different consumptive uses of about 21,586 hm^3/year. Reservoir storage capacity across the basin represents about five times the annual naturalized flow, hence water allocation exceeds mean annual availability at the expense of storage from surface reservoirs and groundwater aquifers. Over-allocation is due in part to allocation levels having been set in 1922 during a period of above-average streamflow relative to more recent long-term streamflow records (Scanlon et al., 2015). Castle et al. (2014) suggest that future water allocated volumes might be jeopardized as water availability continue to decline. On the Rio Grande/Bravo, the basin has a total drainage area of 471,928 km^2, of which just over half is located in the United States. It has an estimated average runoff of 10,467 hm^3/year, of which almost 5762 hm^3 runs off in Mexico, and an estimated annual water over-allocation of 7076 hm^3 (CONAGUA, 2020). The implications of having international river basins with over-allocated water rights and growing water scarcity conditions puts additional pressure on the binational agenda as water rights preferences and priorities at transboundary levels become more complex and require a shared binational management vision that has yet to be developed for the region.

- Global climate change: the most prominent hydroclimatic change in the Colorado River Basin over the past 40 years has been a substantial warming trend, which has altered land surface hydrology and induced a decline in water availability. The latest studies report that runoff from the Colorado River has decreased by 9.3% for each degree increase in temperature since the 1910s (Milly & Dunne, 2020), whereas climate models based on the RCP8.5 scenario (the worst-case climate change scenario) suggest that streamflow is expected to fall between 19% and 31% by 2065 (Lukas & Payton, 2020), mostly due to dwindling snowpack. In the Rio Grande/Bravo basin, the scenario is not very different. In recent decades, the snowpack has decreased by about 25% and the region is expected to experience a water supply deficit of almost 740 hm^3 by 2060 (USBR, 2016, 2017). The current climate change scenarios as a main binational challenge, compared with the over-allocation challenge, represent an even more complicated issue given the inherent uncertainty of the phenomena, lack of control, and the corresponding preparedness and adaptation strategies that are required at the transboundary level to mitigate climatic impacts in the region.
- Anthropogenic alteration of the water cycle: human activities exert both significant direct and indirect pressure on terrestrial water availability with negative consequences for the region, not only on the quality of life of the population, but also on natural ecosystems and biodiversity (Wu et al., 2020). This pressure is caused, in part, by rapid population growth, urbanization and economic development in the region, which are expanding demand and, thereby, exacerbating water scarcity. The pressure on terrestrial water availability is also the product of an increase in the frequency and intensity of climate variability and extreme hydrometeorological events, which led to the development of hydraulic infrastructure for water storage and hydropower generation that altered the natural water cycle regime and adversely affected local ecosystems (Pitt et al., 2017). This latter situation is especially concerning where the traditional engineering approach of 'grey infrastructure' development, such as dams, dikes and reservoirs, has been utilized.

- Water quality degradation: urbanization, land-use changes, agricultural runoff, lack of monitoring and regulation of illegal discharges into surface water bodies, and unregulated extraction of groundwater in the region have degraded water quality of both surface and groundwater bodies at different scopes and scales. Examples of degraded water quality can be seen in the increasing salinity levels in the Hueco Bolsón (Heywood & Yager, 2003), the Lower Rio Grande regions (Hanson et al., 2020) and the Mimbres–Las Palmas region (Hanson et al., 1994), as well as surface water and groundwater quality degradation from sewage effluent discharge in the Santa Cruz and Tijuana River watersheds (IBWC Minute 320). Public health concerns as well as irrigated agriculture that depend on international basins as their main water source have dominated the binational agenda in terms of cooperation efforts and investment on water treatment (IBWC, 2021; Sanchez & Eckstein, 2020b).

Surface water and groundwater interactions

Addressing the interactions between surface water and groundwater is critical for achieving water security along the border. Groundwater pumping can affect base flows, while surface water diversions can influence groundwater recharge (Alley et al., 1999). Along the Mexico–US border, concerns related to surface–groundwater interactions arise on several fronts, including the impacts of groundwater abstractions on surface water rights, the timing and quantity of flows needed to support ecosystems, increasing water extraction levels due to climate change/variability and impacts on surface water availability, as well as population growth and other shifts in demands for water (Cuthbert et al., 2019; Famiglietti, 2014).

Understanding these interactions is often difficult, especially in regions that span various system boundaries, whether those systems are natural, political or socioeconomic. These complications stem not only from the international political boundary between Mexico and the United States, but also from other political boundaries at the state, county and city levels. For example, current and past locations of the deltas of the Rio Colorado and Rio Grande rivers, encompass terrains with a variety of fragile, yet highly productive ecosystems and microecosystems whose health depend highly on sustained, periodic freshwater and instream flows. Rapid development and resulting socioeconomic changes throughout both basins, coupled with increasing water withdrawals for municipal and irrigation uses, have significantly reduced instream and freshwater inflows in the most downstream portions of these rivers.

Both rivers are flood-pulse flow-dominant systems that are highly impacted by natural and anthropogenic factors that reduce the average annual flows, impacting groundwater recharge and discharge into sensitive ecosystems highly dependent on groundwater for their sustainability (e.g., Big Bend National Park). Flows in these rivers are currently low with only periodic, and often short-lived, higher flow periods that occur during occasional heavy rainfalls over their contributing watersheds. The predominantly arid to semi-arid climate conditions found throughout these subtropical and middle-latitude basins result in high evaporation losses and frequent drought periods, further reducing runoff and the baseflows of both river systems. Additionally, dams, reservoirs and diversions have altered the natural hydrological processes in both river systems by reducing mean and peak streamflows throughout their reaches. These reduced flows, coupled with ever-increasing urban and agricultural withdrawals and consumption, have led to both basins, especially the Colorado River basin, being listed among the most overallocated in the world (Christensen et al., 2004). The natural consequence has been an increasing demand on groundwater in the region and an increasing trend towards the utilization of brackish groundwater desalination to meet growing municipal water demands, such as in the case of the Lower Rio Grande Valley (Meyer et al., 2014). Therefore, the need for a better understanding of surface water and groundwater interactions for economic and environmental sustainability on both sides of the Mexico–US border is a necessary priority (USBR, 2016, 2017).

While there is evidence that reduced streamflows in both the Rio Grande and Colorado River have impacted deltaic ecosystems, water availability and policy issues, it is less clear what role they have had on local groundwater resources (e.g., Kennedy et al., 2017). Data collection has been limited by funding, and lack of collaboration efforts and institutional trust has impeded access to information related to infiltration to and connectivity between surface water and groundwater. However, it is increasingly suspected that connections between the two played vital roles historically in maintaining ecosystem health as well as surface water hydrography, quantity and quality. As demand for water grows in both basins, gathering data to confirm this historical connection, what linkages persist today, and the effects of decoupling them from one another is becoming paramount for water security, ecosystem sustainability, and exploring options for cross-border water policy moving forward.

In terms of management and policy development, the interactions between surface water and groundwater are complex and substantial data and analysis are needed to determine the relationships between groundwater uses, groundwater flow, surface water uses and surface water flows (Kalbus et al., 2006). When change needs to occur within an institutional management system, often (though, not always) the starting point for negotiations is demonstration of impacts, such that the evidence establishes who is responsible for what actions and what the impacts of those actions may be. Understanding the institutional system limitations, gaps and needs is a crucial element for assessing these impacts, yet, throughout the borderlands, the knowledge base needed to understand the system is incomplete.

This knowledge uncertainty hampers the development of a legal and regulatory framework for addressing surface water and groundwater interactions that is also fraught with knowledge gaps. The Water Treaties do not address groundwater or local runoff (e.g., irrigation return flows and small arroyos in sub-watersheds) within the watersheds (Mumme, 2000; Sanchez & Eckstein, 2020b), and the polycentric and multilevel structure of authority over groundwater and surface waters in both countries (Milman & Scott, 2010) means that there is no consolidated framework for addressing surface water and groundwater interactions. In Mexico, the federal government oversees groundwater throughout the country (Hernandez, 2003; Ramírez, 1967), albeit with some limited decentralization of authority to watershed councils and groundwater technical committees (COTAS) (Wester et al., 2009). In the United States, groundwater quantity and allocation are primarily managed by the states and enforced by the Supreme Court (Hesser, 2011), whereas the federal government has authority on issues connected to interstate commerce, including groundwater quality and endangered species (Mumme, 2005; Thompson et al., 2000). As a result, each state has its own system for addressing surface water and groundwater interactions.

While there are many barriers to addressing surface water and groundwater interactions along the borderlands, progress has been made over the past two decades. At the binational level, some of these policy actions include Minutes 242, 316, 319 and 323, which allude to the need to consider groundwater pumping and instream flows at the binational level. Additionally, within the United States, the listing and protection of species under the Endangered Species Act is leading to actions that address surface water and groundwater interactions (e.g., see the Upper Colorado Endangered Fish Recovery Program, the San Juan River Basin Recovery Implementation Program and the Lower Colorado Multispecies Conservation Program). State-of-the-art numerical modelling in the Mexico–US border region has been conducted for several TBAs. In the Lower Rio Grande River segment located at the intersection of New Mexico, Texas and Chihuahua (Figure 2), a joint USGS and US Bureau of Reclamation (USBR) effort analyses conjunctive use with an integrated hydrological transboundary model (Hanson et al., 2020) that provides a physically based flow model embedded in a supply-and-demand framework (Boyce et al., 2020; Hanson et al., 2014).

This modelling effort started with the TAAP (Hanson et al., 2013) and was then used by the USBR (Ferguson & Llewellyn, 2015; USBR, 2016, 2017) to assess climate change and confirm the efficacy of the 2008 Operating Agreement for the Rio Grande Project (USBR, 2008) and the 1906 Treaty (IBWC, 1906). The holistic framework combines science (data and modelling), policy and governance to produce a more dynamic operating management capacity for conjunctive use. The result is both short- and long-term sustainability of water uses that enhances water security and increases resilient capacity in the face of climate change/variability. This model, however, does not include Mexican groundwater data for the Conejos–Médanos aquifer, therefore its application is valid only for the US side of the aquifer.

Another effort is the numerical model developed for the APN TBA. This is one of the few TBA systems where cross-formational groundwater flow along the Mexico–US border has been scientifically studied by means of numerical modelling approaches (Rodriguez et al., 2020). The APN management unit is a strategic aquifer for Mexico because it supplies 85% of the total water needs in the APN aquifer region (Rodriguez et al., 2020). Moreover, this aquifer will likely experience additional water demand (about 56–128 Mm^3/year) to satisfy future shale gas/oil development in the area (Hernández-Espriú et al., 2019). Despite its importance, very little formal research has been conducted regarding local, regional and transboundary aspects of the APN aquifer. This model, however, does include data from the Mexican side that have been compiled under the memorandum of understanding between Texas AgriLife Research and the Instituto Mexicano de Tecnología del Agua (IMTA), which developed a data-sharing portal in 2019 (TWRI, 2019).

Several basic uncertainties still exist that clearly limit binational numerical and management approaches in transboundary systems, including: (1) inconsistency in Mexican and US aquifer delineation methodologies, as the former is based on arbitrary limits that ignore geological features/structures/contacts, while the latter is based on geological

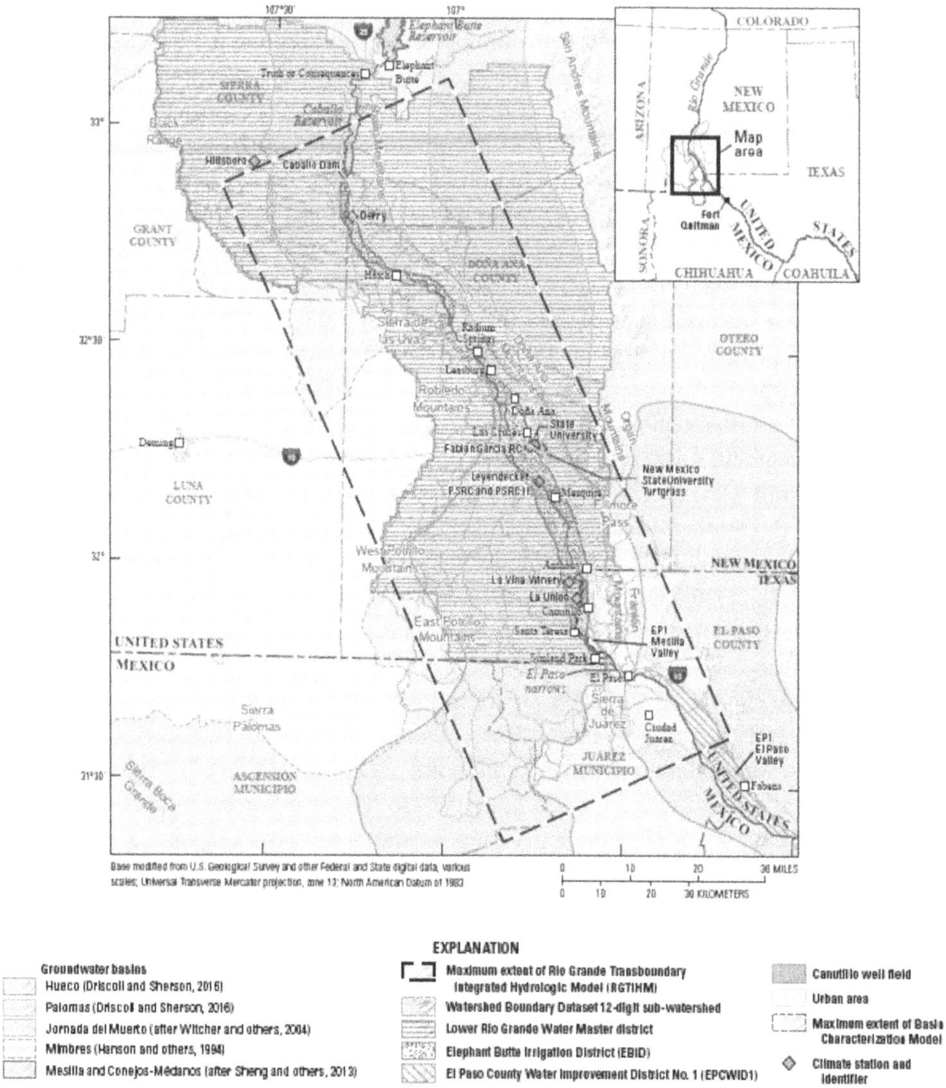

Figure 2. Areas modelled by the Rio Grande Transboundary Integrated Hydrologic model (RGTIHM) in the Transboundary Rio Grande River area of New Mexico, Texas and Mexico, along with an active hydrologic model grid, selected climate stations and rivers for specific areas showing the total extent of the model area, including watersheds and groundwater basins. Source: Hanson et al. (2020).

boundaries; (2) poorly understood outcropping and sub-cropping correlative units along the border region's hydrostratigraphy; and (3) different unification or separation criteria used to characterize groundwater flow and/or aquifer systems between the two countries.

Proposed research agenda and the way forward

Our review of TBA-related topics laid bare some pressing knowledge gaps that need to be addressed to improve water security in the border region. Here, we identify and provide recommendations on a path that could guide researchers and scholars on the way forward.

Improving transboundary coordination and collaboration

According to the perspectives presented in this study and the conclusions of previous sections, the path forward will need to incorporate research and data on surface and groundwater interactions and conjunctive use, baseflow and ecosystem needs, and transboundary governance and policy development to overcome water security challenges and account for the need to adapt to climate change/variability. Some modifications through the minute system of the 1944 Treaty have included elements related to groundwater (Minutes 242, 316, 319 and 323; IBWC, 1973, 2010, 2012, 2017b), drought contingency for surface water deliveries and salinity limits on the Colorado River (Minute 323; IBWC, 2017b), emergency deliveries of Colorado River water to Tijuana (Minute 322; IBWC, 2017a), and discharge of pulse flow for environmental restoration in the Colorado River delta (Minute 319; IBWC, 2012). However, the use and joint management of shared groundwater resources, as well as the need to adapt to or cope with climate variability at the transboundary level, have yet to be addressed. In addition, while it is true that transboundary coordination on information development and sharing has occurred in some cases, such as for the San Pedro and Santa Cruz aquifers (Callegary et al., 2018), these efforts need additional funding and governmental support to further develop joint monitoring and transboundary modelling along the entire Mexico–US border and to facilitate more comprehensive cooperative management.

Another major task at hand is to find common ground in water management that bridges scientific, technical, political and governance frameworks. This requires facilitating the development of mutual respect, diplomacy and dialogue, and acknowledging the rights of neighbouring countries and jurisdictions. It also requires a binational discussion to define the meaning of sustainable use of groundwater within the context of TBAs, as well as in a conjunctive use framework of sound land management and water use within transboundary watersheds.

Thus, the path forward requires the development of economic, diplomatic and scientific frameworks for TBAs to facilitate joint monitoring and modelling, sharing of resources, and development of protocols that promote sustainability. Some progress has been made in the science of TBAs (mapping and assessment), but insufficient progress has been made in terms of governance, policy and management of TBAs. Issues that must be addressed in the coming years include defining transboundary groundwater units along the Mexico–US border that need to be managed; designing and operating frameworks for appropriate governance and shared TBA management; hosting shared and mutually accessible servers with information databases; and increasing cooperation by means of institutional agreements, arrangements, memoranda of understandings and other types of mutually agreed-upon mechanisms for cooperation. This latter point could even include separate arrangements between individual Mexican states, along with coordination with relevant institutions, such as the IBWC, CONAGUA, North America Development Bank (NADBank) and US Environmental Protection Agency (EPA), among others.

Disclosure statement

No potential conflict of interest was reported by the authors.

ORCID

Rosario Sanchez http://orcid.org/0000-0002-3012-5408
Alfonso Rivera http://orcid.org/0000-0002-2973-0889
Randall T. Hanson http://orcid.org/0000-0002-9819-7141
Rick J. Hogeboom http://orcid.org/0000-0002-5077-4368
Anita Milman http://orcid.org/0000-0002-5712-9388
Jude A. Benavides http://orcid.org/0000-0001-8524-8707
Adrian Pedrozo-Acuña http://orcid.org/0000-0001-6921-4363
Sharon B. Megdal http://orcid.org/0000-0001-7781-297X
Gabriel Eckstein http://orcid.org/0000-0002-9393-2362
Laura Rodriguez http://orcid.org/0000-0003-4224-9763

References

Alley, W. M., Reilly, T. E., & Franke, O. L. (1999). *Sustainability of ground-water resources*. U.S. Geological Survey Circular 1186. U.S. Geological Survey .

Boyce, S. E., Hanson, R. T., Ferguson, I., Schmid, W., Henson, W., Reimann, T., Mehl, S. M., & Earll, M. M. (2020). *One-water hydrologic flow model: A MODFLOW based conjunctive-use simulation software*. (Techniques and Methods 6–A60). U.S. Geological Survey. https://doi.org/10.3133/tm6a60.

Callegary, J. B., Megdal, S. B., Villaseñor, E. T., Petersen-Perlman, J. D., Sosa, I. M., Monreal, R., Gray, F., & Noriega, F. G. (2018). Findings and lessons learned from the assessment of the Mexico–United States transboundary San Pedro and Santa Cruz aquifers: The utility of social science in applied hydrologic research. *Journal of Hydrology: Regional Studies, 20*, 60–73. https://doi.org/10.1016/j.ejrh.2018.08.002

Callegary, J. B., Sosa, I. M., Villaseñor, E. T., Dos Santos, P., Monreal, R., Grijalva Noriega, F., Huth, A., Gray, F., Scott, C., & Megdal, S. B. (2016). *Binational study of the transboundary San Pedro aquifer.* Binational report. International Boundary and Water Commission. http://www.ibwc.gov/Files/Binational_Study_Transboundary_San_Pedro_Aquifer.pdf

CASGEM. (2021). *Groundwater monitoring (CASGEM)* https://water.ca.gov/Programs/Groundwater-Management/Groundwater-Elevation-Monitoring-CASGEM

Castle, S. L., Thomas, B. F., Reager, J. T., Rodell, M., Swenson, S. C., & Famiglietti, J. S. (2014). Groundwater depletion during drought threatens future water security of the Colorado River Basin. *Geophysical Research Letters, 41*(16), 5904–5911. https://doi.org/10.1002/2014GL061055

Christensen, N. S., Wood, A. W., Voisin, N., Lettenmaier, D. P., & Palmer, R. N. (2004). The effects of climate change on the hydrology and water resources of the Colorado River basin. *Climatic Change, 62*(1), 337–363. https://doi.org/10.1023/B:CLIM.0000013684.13621.1f

Cital, F., Rivera, A., Rodríguez-Burgueño, E., & Ramírez-Hernández, J. (2021). A governance panorama of an aquifer in a semiarid region, Mexico. In S. H. Choi, E. Shin, A. K. Makarigakis, O. Sohn, C. Clench, & M. Trudeau (Eds.), *The role of sound groundwater resources management and governance to achieve water security* (pp. 216–238). UNESCO. https://unesdoc.unesco.org/ark:/48223/pf0000379093

Cohen, M. A. (2005). El agua en la frontera México-Estados Unidos: Reto político-ambiental. *Espacio Abierto, 14*(2). https://www.redalyc.org/pdf/122/12214202.pdf

CONAGUA. (2015). *Statistics on water in Mexico. 2015.* Ministry of the Environment and Natural Resources.

CONAGUA. (2020). *Public registry of water rights (REPDA).* CONAGUA https://app.conagua.gob.mx/consultarepda.aspx

CRS. (2018). *Sharing the Colorado River and the Rio Grande: Cooperation and conflict with Mexico.* (No. R45430). Congressional Research Service. https://www.everycrsreport.com/reports/R45430.html#_Ref521665121

Cuthbert, M., Gleeson, T., Moosdorf, N., Befus, K. M., Schneider, A., Hartmann, J., & Lehner, B. (2019). Global patterns and dynamics of climate–groundwater interactions. *Nature Climate Change, 9*(2), 137–141. https://doi.org/10.1038/s41558-018-0386-4

Danskin, W. (2020). *San Diego hydrogeology. California Water Science Center.* U.S. Geological Survey. https://ca.water.usgs.gov/projects/sandiego/contact/scientists_current.html

Draper, S. E., & Kundell, J. E. (2007). Impact of climate change on transboundary water sharing. *Journal of Water Resources Planning and Management, 133*(5), 405–415. https://doi.org/10.1061/(ASCE)0733-9496(2007)133:5(405)

Eckstein, G. (2017). *The international law of transboundary groundwater resources.* Routledge.

Famiglietti, J. S. (2014). The global groundwater crisis. *Nature Climate Change, 4*(11), 945–948. https://doi.org/10.1038/nclimate2425

Ferguson, I. A., & Llewellyn, D. (2015). *Simulation of Rio Grande project operations in the Rincon and Mesilla Basins: Summary of model configuration and results.* (Technical Memorandum 86-68210–2015–05). 56. U.S. Department of the Interior, Bureau of Reclamation.

Gleeson, T., Villholth, K., Taylor, R., Perrone, D., & Hyndman, D. (2019). Groundwater: A call to action. *Nature, 576*(7786), 213. https://doi.org/10.1038/d41586-019-03711-0

Groundwater Protection Council, & Interstate Oil and Gas Compact Commission. (2011). *FracFocus.* https://fracfocus.org

Hanson, R. T., Boyce, S. E., Schmid, W., Hughes, J. D., Mehl, S. M., Leake, S. A., Maddock, T., III, & Niswonger, R. G. (2014). *MODFLOW-One-Water Hydrologic Flow Model (OWHM).* (Techniques and Methods 6-A51). 122. U.S. Geological Survey.

Hanson, R. T., McLean, J. S., & Miller, R. S. (1994). *Hydrogeologic framework and preliminary simulation of ground-water flow in the Mimbres Basin, southwestern New Mexico.* (Water-Resources Investigations Report 94-4011). 118. U.S. Geological Survey. https://doi.org/10.3133/wri944011

Hanson, R. T., Ritchie, A. B., Boyce, S. E., Galanter, A. E., Ferguson, I. A., Flint, L. E., Flint, A., & Henson, W. R. (2020). *Rio Grande transboundary integrated hydrologic model and water-availability analysis, New Mexico and Texas, United States, and northern Chihuahua, Mexico.* (Scientific Investigations Report 2019–5120). 186. U.S. Geological Survey. https://doi.org/10.3133/sir20195120

Hanson, R. T., Schmid, W., Knight, J., & Maddock, T., III. (2013). Integrated hydrologic modeling of a transboundary aquifer system – Lower Rio Grande. Paper presented at *MODFLOW and more 2013 – Translating science into practice conference*, Golden, CO, June 2–5, 2013. Colorado School of Mines, Integrated Groundwater Modeling Center. 5. https://www.researchgate.net/publication/258698983

Hanson, R., Chávez-Guillen, R., Tujchneider, O., Alley, W. M., Rivera, A., Dausman, A., Batista, L., & Espinoza, M. (2015). Enfoques para Implementar la Estrategia Regional. UNESCO.

Hernández-Espriú, A., Wolaver, B., Arciniega-Esparza, S., Scanlon, B. R., Young, M. H., Nicot, J.-P., Macías-Medrano, S., & Breña-Naranjo, J. A. (2019). A screening approach to improve water management practices in undeveloped shale plays, with application to the transboundary Eagle Ford Formation in northeast Mexico. *Journal of Environmental Management*, *236*, 146–162. https://doi.org/10.1016/j.jenvman.2018.11.123

Hernandez, A. (2003). Water law in the Republic of México. *US–Mex. Law Journal*, *11*(1), 15. https://doi.org/10.25922/0zrx-xr72.

Hesser, J. N. (2011). The nature of interstate groundwater resources and the need for states to effectively manage the resource through interstate compacts. *Wyoming Law Review*, *11*(1), 25. https://scholarship.law.uwyo.edu/cgi/viewcontent.cgi?article=1245&context=wlr.

Heywood, C. E., & Yager, R. M. (2003). *Simulated ground-water flow in the Hueco Bolsón, an alluvial-basin aquifer system near El Paso, Texas.* (Water-Resources Investigation Report 2002–4108). 73. U.S. Geological Survey. https://doi.org/10.3133/wri024108

Hogeboom, R. J., De Bruin, D., Schyns, J. F., Krol, M. S., & Hoekstra, A. Y. (2020). Capping human water footprints in the world's river basins. *Earth's Future*, *8*(2), e2019EF001363. https://doi.org/10.1029/2019EF001363

IBWC. (1906). Convention between the United States and Mexico – Equitable distribution of the waters of the Rio Grande. *International Boundary Water Commission.* Gov. Printing Off. Washington, DC.

IBWC. (1973). Minute 242, Permanent and definitive Solution to the International Problem of the Salinity of the Colorado River https://www.ibwc.gov/Files/Minutes/Min242.pdf

IBWC. (2010). *Minute 316, utilization of the Wellton-Mohawk bypass drain and necessary infrastructure in the United States for the conveyance of water by Mexico and non-governmental organizations of both countries to the Santa Clara wetland during the Yuma desalting plant pilot run.* (https://www.ibwc.gov/Files/Minutes/Min316.pdf

IBWC. (2012). *Minute 319, Interim International Cooperative Measures in the Colorado River Basin through 2017 and Extension of Minute 318 Cooperative Measures to Address the Continued Effects of the April 2010 Earthquake in the Mexicali Valley, Baja California.* (https://www.ibwc.gov/Files/Minutes/Min319.pdf

IBWC. (2017a). *Minute 322, Extension of the Temporary Emergency Delivery of Colorado River Water for Use in Tijuana, Baja California* (https://www.ibwc.gov/Files/Minutes/Min322.pdf

IBWC. (2017b). *Minute 323, Officials with the International Boundary and Water Commission, United States and Mexico, today announced the conclusion of a new Colorado River Agreement, Minute 323, "Extension of Cooperative Measures and Adoption of a Binational Water Scarcity Contingency Plan in the Colorado River Basin"* https://www.ibwc.gov/Files/Minutes/Min323.pdf

IBWC. (2021). *Minutes between the United States and Mexican Sections of the IBWC.* https://www.ibwc.gov/Treaties_Minutes/Minutes.html

Ikonnikova, S. A., Male, F., Scanlon, B. R., Reedy, R. C., & McDaid, G. (2017). Projecting the water footprint associated with shale resource production: Eagle Ford shale case study. *Environmental Science & Technology, 51*(24), 14453–14461. https://doi.org/10.1021/acs.est.7b03150

Kalbus, E., Reinstorf, F., & Schirmer, M. (2006). Measuring methods for groundwater–surface water interactions: A review. *Hydrology and Earth System Sciences, 10*(6), 873–887. https://doi.org/10.5194/hess-10-873-2006

Karandish, F., Hoekstra, A. Y., & Hogeboom, R. J. (2018). Groundwater saving and quality improvement by reducing water footprints of crops to benchmarks levels. *Advances in Water Resources, 121*, 480–491. https://doi.org/10.1016/j.advwatres.2018.09.011

Kennedy, J., Rodríguez-Burgueño, J. E., & Ramírez-Hernández, J. (2017). Groundwater response to the 2014 pulse flow in the Colorado River Delta. *Ecological Engineering, 106*, 715–724. https://doi.org/10.1016/j.ecoleng.2016.10.072

Kim, J. H., & Jackson, R. B. (2012). A global analysis of groundwater recharge for vegetation, climate, and soils. *Vadose Zone Journal, 11*(1). https://doi.org/10.2136/vzj2011.0021RA

Loch, A., Perez-Blanco, C., Carmody, E., Felbab-Brown, V., Adamson, D., & Seidl, C. (2020). Grand theft water and the calculus of compliance. *Nature Sustainability, 3*(12), 1012–1018. https://doi.org/10.1038/s41893-020-0589-3

Lukas, J., & Payton, E. (2020). *Colorado River Basin climate and hydrology: State of the science.* Technical report. Western Water Assessment Cooperative Institute for Research in Environmental Sciences. https://doi.org/10.25810/3hcv-w477

Megdal, S. B., & Scott, C. A. (2011). The importance of institutional asymmetries to the development of binational aquifer assessment programs: The Arizona–Sonora experience. *Water, 3*(3), 949–963. http://dx.doi.org/10.3390/w3030949

Meyer, J., Croskrey, A., Wise, M., & Kalaswad, S. (2014). *Brackish groundwater in the Gulf Coast Aquifer, Lower Rio Grande Valley, Texas.* Texas Water Development Board. http://www.twdb.Texas.gov/publications/reports/numbered_reports/doc/R383_BrackishGW.pdf

Milly, P. C., & Dunne, K. A. (2020). Colorado River flow dwindles as warming-driven loss of reflective snow energizes evaporation. *Science, 367*(6483), 1252–1255. https://doi.org/10.1126/science.aay9187

Milman, A., & Scott, C. A. (2010). Beneath the surface: Intranational institutions and management of the United States – Mexico transboundary Santa Cruz aquifer. *Environment and Planning. C, Government & Policy, 28*(3), 528–551. https://doi.org/10.1068/c0991

Mumme, S. P. (2000). Minute 242 and beyond: Challenges and opportunities for managing transboundary groundwater on the Mexico–US border. *Natural Resources Journal 40* , 341–378 https://digitalrepository.unm.edu/nrj/vol40/iss2/8.

Mumme, S. P. (2005). Advancing binational cooperation in transboundary aquifer management on the US–Mexico border. *Colorado Journal of International Environmental Law and Policy, 16*, 77.

Pitt, J., Kendy, E., Schlatter, K., Hinojosa-Huertaf, O., Flessa, K. W., Shafroth, P. B., Ramirez-Hernandez, J., Nagler, P. L., & Glenn, E. P. (2017). It takes more than water: Restoring the Colorado River Delta. *Ecological Engineering, 106*(B), 629–632. https://doi.org/10.1016/j.ecoleng.2017.05.028

Ramírez, R. C. (1967). Aspectos legales del agua subterránea en México. *Ingeniería Hidráulica en México, 23*, 253–255.

Ritchie, A. B., Hanson, R. T., Galanter, A. E., Boyce, S. E., Damar, N. A., & Shephard, Z. M. (2018). *Digital hydrologic and geospatial data for the Rio Grande transboundary integrated hydrologic model and water-availability analysis, New Mexico and Texas, United States, and Northern Chihuahua, Mexico.* (U.S. Geological Survey data release). https://doi.org/10.5066/P9J9NYND

Rivera, A., Amore, L., Kettelhut, J., Chávez-Guillén, R., Vences Mejía, J. A., Arias, M., & Rivera, A. (Ed.). (2015). *Book on Regional Strategy for the Assessment and Management of Transboundary Aquifer Systems (TAS) in the Americas. UNESCO/OEA-ISARM AMERICAS, Book IV (Estrategia Regional para la Gestión de los* Sistemas de Acuíferos Transfronterizos *(SAT) en las Americas.* UNESCO. https://unesdoc.unesco.org/ark:/48223/pf0000235394?posInSet=7&queryId=32a47d34-c6c2-47d0-bc1c-6a9756cb7847

Rodriguez, L., Sanchez, R., Zhan, H., & Knappett, P. S. (2020). The transboundary nature of the Allende–Piedras Negras aquifer using a numerical model approach. *JAWRA Journal of the American Water Resources Association*, *56*(3), 387–408. https://doi.org/10.1111/1752-1688.12843

Sanchez, R., & Eckstein, G. (2020a). Groundwater management in the borderlands of Mexico and Texas: The beauty of the unknown, the negligence of the present, and the way forward. *Water Resources Research*, *56*(3), e2019WR026068. https://doi.org/10.1029/2019WR026068

Sanchez, R., & Eckstein, G. (2020b). The path towards groundwater management in the borderlands of Mexico and Texas. *Wiley Interdisciplinary Reviews: Water*, *7*(1), e1399. https://doi.org/10.1002/wat2.1399

Sanchez, R., Kaiser, R., & Giardino, J. R. (2010). *Water planning under an unsustainable condition: Challenges for the Laguna Region, Mexico. World Environmental and Water Resources Congress 2010: Challenges of Change.*

Sanchez, R., Lopez, V., & Eckstein, G. (2016). Identifying and characterizing transboundary aquifers along the Mexico–US border: An initial assessment. *Journal of Hydrology*, *535*, 101–119. https://doi.org/10.1016/j.jhydrol.2016.01.070

Scanlon, B. R., Zhang, Z., Reedy, R. C., Pool, D. R., Save, H., Long, D., Chen, J., Wolock, D. M., Conway, B. D., & Winester, D. (2015). Hydrologic implications of GRACE satellite data in the Colorado River Basin. *Water Resources Research*, *51*(12), 9891–9903. https://doi.org/10.1002/2015WR018090

Senate Reports. (2006). *Public Law 109-448-Dec.22, 2006. United States–Mexico Transboundary Aquifer Assessment Act.* https://www.congress.gov/109/plaws/publ448/PLAW-109publ448.pdf

Sweetkind, D. S., Hanson, R. T., Ritchie, A. B., & Hawley, J. W. (2017). *Data release of three-dimensional hydrogeologic framework model of the Rio Grande transboundary region of New Mexico and Texas, USA and northern Chihuahua, Mexico.* (U.S. Geological Survey data release). https://doi.org/10.5066/F7JM27T6.

Sweetkind, D. S. (2017). *Three-dimensional hydrogeologic framework model of the Rio Grande transboundary region of New Mexico and Texas, USA, and northern Chihuahua, Mexico.* (Scientific Investigations Report 2017–5060). 49. U.S. Geological Survey. https://doi.org/10.3133/sir20175060.

Thompson, B., Sax, J., Leshy, J., & Abrams, R. H. (2000). *Legal control of water resources.* St. Paul: West Publishing.

TWRI. (2019). *Transboundary water portal.* Texas Water Resources Institute https://transboundary.tamu.edu/

UN Water. (2013). *What is water security? Infographic.* United Nations. https://www.unwater.org/publications/water-security-infographic/

UNECE & UNESCO. (2021). *Progress on transboundary water cooperation – 2021 update.* Global status of SDG indicator 6.5.2 and acceleration needs. *United Nations.* https://www.unwater.org/publications/progress-on-transboundary-water-cooperation-652-2021-update/

UNECE. (2014). *Model provisions on transboundary groundwaters.*

UNESCO. (2010). *Aspectos socioeconomicos, ambientales y Climaticos de los sistemas acuiferos transfronterizos de las Americas.*

USBR. (2008). *Operating agreement for the Rio Grande project. U.S. Department of the Interior. Bureau of Reclamation.* http://www.usbr.gov/uc/albuq/rm/RGP/pdfs/Operating-Agreement2008.pdf

USBR. (2016). *Continued implementation of the 2008 operating agreement for the Rio Grande Project, New Mexico and Texas – Final environmental impact statement. United States Bureau of Reclamation.* Retrieved March, from https://www.usbr.gov/uc/envdocs/eis/pdf/2008OperatingAgreementRioGrandeEIS_Final.pdf

USBR. (2017). *Rio Grande operating agreement environmental impact statement. United States Bureau of Reclamation.* Retrieved February, from https://www.usbr.gov/uc/albuq/rm/RGP/EIS/RGOA-EIS-BkgOvervInfo.pdf

Wester, P., Hoogesteger, J., & Vincent, L. (2009). *Local IWRM organizations for groundwater regulation: The experiences of the Aquifer Management Councils (COTAS) in Guanajuato, Mexico. (Ed.), (Eds.).* Natural Resources Forum.

World Resources Institute. (1990). *World Resources 1990–1991*. Oxford University Press.

Wu, W.-Y., Lo, M.-H., Wada, Y., Famiglietti, J. S., Reager, J. T., Yeh, P. J.-F., Ducharne, A., & Yang, Z.-L. (2020). Divergent effects of climate change on future groundwater availability in key mid-latitude aquifers. *Nature Communications*, *11*(1), 1–9. https://doi.org/10.1038/s41467-020-17581-y

A critical review of the transboundary aquifers in South-Eastern Europe and new insights from the EU's water framework directive implementation process

Charalampos Skoulikaris, Jacques Ganoulis and Alice Aureli

ABSTRACT

The object of the research is an advanced review and comparative analysis of the last two decades' main international programmes and initiatives on the management of Transboundary Aquifers in South-Eastern Europe. The aquifers' delineation and occurrence, identified pressures, and water quality and use are compared with those derived by the EU's Water Framework Directive implementation process. The final outputs reveal a solid conceptual follow-up of the programmes and initiatives with a gradual enrichment of information related to the status of transboundary aquifers over the years. Groundwater bodies' data analysis validates the initial outputs and provides enhanced information at finer scales.

Introduction

In the field of water resources, it is only since the beginning of the 21st century that the undoubtful importance of transboundary groundwater resources for meeting human demands, for development purposes and for supporting environmental services has been rightfully recognized on a global scale. A pioneering role in this recognition process is attributed to the United Nations Educational, Scientific and Cultural Organization (UNESCO) through its Intergovernmental Hydrological Programme (IHP) and the launch in 2000 of the ongoing International Shared Aquifer Resources Management (ISARM) Programme (UNESCO, 2001). The principal aim of the UNESCO IHP/ISARM programme was the compilation of a world inventory of transboundary aquifers (TBAs), where scientific, legal, environmental, socioeconomic, and institutional guidelines and recommendations for the shared management of TBAs would be available to the co-riparian states (Puri & Aureli, 2005). The programme continues to provide new perspectives on identifying the challenges and setting new directions on TBAs management (UNESCO, 2010), strengthening cooperation (Aureli & Eckstein, 2011) and guiding the development of regional strategies for managing TBAs (Burchi, 2018; Rivera & Candela, 2018). Currently, various programmes and initiatives focusing on different aspects of transboundary waters (e.g., Talaue Mcmanus et al., 2016; UNECE, 2007, 2011) and numerous publications have been articulated on the thematic subject of TBAs management (Altchenko & Villholth, 2013; Fraser et al., 2020; Gerlak et al., 2011; Rivera & Candela, 2018; Sanchez et al., 2016; Wada & Heinrich, 2013).

The Draft Articles on the Law of Transboundary Aquifers of the United Nations International Law Commission (UNILC), conceptualized under the ISARM and UNESCO-IHP umbrella (Stephan, 2011), define a transboundary aquifer or a transboundary aquifer system as 'an aquifer or aquifer system, parts of which are situated in different States' respectively (UNESCO-IHP, 2009). Among other issues the Articles focus on the importance of the TBAs' spatial delimitation, their equitable and reasonable utilization in terms of water uses and abstractions, the identification of recharge and discharge mechanisms and zones, the long-term benefits and useful lifespan of aquifers, as well as underlining the importance of protecting and preserving the ecosystems that are dependent on them (Altchenko & Villholth, 2013; Burchi, 2018).

The existence of accurate data as well as data exchange is pivotal to equitable and sustainable management of transboundary waters (Article 6 of the Convention on the Protection and Use of Transboundary Watercourses and International Lakes (UNECE, 1992), Article 9 of the UN Convention on the Law of Non-Navigational Uses of International Watercourses (UN, 1997), Article 8 of the Draft Articles on the Law of Transboundary Aquifers (UNESCO-IHP, 2009)). The European Union's Water Framework Directive (EU's WFD) (EC, 2000) not only considers the sharing of information as a pillar of integrated water resources management, but also has created legislative and technical infrastructures for the way that the exchange of data should be accomplished, e.g., the INSPIRE (INfrastructure for SPatial InfoRmation in Europe) Directive. In this perspective, UNESCO-IHP funds programmes and workshops where objectives include the sharing of available data on transboundary watercourses (Gerlak et al., 2011) and the creation of global and regional inventories (Ganoulis et al., 2014; IHP-ISARM /INWEB, 2008; UNESCO-IHP Water Information Network System (IHP-WINS)., 2017).

The paper has multiple research objectives aiming at i) reviewing the main international programmes and initiatives, which are perceived as data sources and are related to TBAs of South-Eastern Europe (SEE), for setting the conceptual and logical linkages among these data sources, ii) analysing these data sources and depicting potential continuities and/or duplications and/or gaps among them in terms of TBA's delineation, identified problems and groundwater usage, iii) identifying and quantifying groundwater use both at TBA and national level for each data source, and iv) demonstrating the new insights on these international water as derived from the EU's WFD implementation process. For the latter, new features regarding the extent, pressures and quality and quantitative status of the groundwaters bodies (GWBs) of the EU countries located in SEE are explicitly illustrated and are compared with the relevant data of the TBAs under investigation. The output is conceived as a comprehensive and up-to-date guide to the status of transboundary aquifers and groundwaters in SEE that can facilitate and foster shared groundwater management.

Materials and methods

In this section synoptic information about the SEE, and the TBAs' data sources that were retrieved, exploited, analysed and further processed are given. The sources are i) the ISARM and UNESCO Chair's International Network of Water-Environment Centres for the

Balkans (INWEB) TBAs' inventory in the Balkans of 2008, ii) the First and Second Assessment of Transboundary Rivers, Lakes and Groundwaters of 2007 and 2011, respectively, iii) the Dinaric Karst Aquifer System (DIKTAS) and the Transboundary Waters Assessment Programme (TWAP) programmes of 2010–2014 and 2013–2015 respectively, iv) the Maps of the Transboundary Aquifers of the World Eds. 2015 and Eds. 2020, and v) the EU's WFD implementation process outputs of 2018.

Case study area

The SEE, also known as the Balkans Peninsula, geographically comprises Albania, Bulgaria, Greece, Romania, the European part of Turkey, and the countries of ex-Yugoslavia which collapsed in 1992, i.e., Bosnia and Herzegovina, Croatia, Kosovo, Montenegro, North Macedonia, Serbia and Slovenia; hence it is less than 30 years that SEE has comprised so many states. The total population of the Balkan region in 2019 was 73,070,777 inhabitants,[1] a figure that includes the 11,961,338 Turkish[2] inhabitants located in the EU part of Turkey. The total area of the Balkan region is 789,703 km^2. The largest country both in terms of area and population is Romania with 238,369 km^2 and 19,414,458 inhabitants. The smallest country is Kosovo, with an area of 10,908 km^2 while Montenegro with 622,182 inhabitants is the least populated country. Currently and in chronological order of access Greece, Slovenia, Romania, Bulgaria and Croatia are member-states of the EU. This means that these countries have ratified the EU environmental legislation as their national legislation and the management of their water resources is implemented and based on the EU's WFD.

ISARM/INWEB's inventory

During the UNESCO-IHP/ISARM's initial Phase 2000–2008, UNESCO Chair INWEB, with the support of UNESCO-IHP and of United Nations Economic Commission for Europe (UNECE), produced in 2008 the TBAs inventory for the Balkans (hereinafter ISARM/INWEB's inventory) giving descriptive and analytical information for 65 TBAs, together with their approximate boundaries (IHP-ISARM/INWEB, 2008). The procedure for collecting information and data on the TBAs was based on formulated questionnaires that were distributed to national experts representing the countries of the Balkan region, including Turkey, as well as to representatives of Hungary and Moldova. The two latter countries, although not part of the Balkan area, were included since several TBAs are shared among them and Balkan countries. The questionnaire had a specific predefined template and was organized in 4 major thematic categories (IHP-ISARM /INWEB, 2008): 1) Aquifers' properties, 2) Monitoring data, 3) Significant issues affecting the aquifer system, and 4) Policy and legislative properties. In total, the questionnaires contained 40 fields to be filled in by the national experts.

UNECE's first and second assessment

The First Assessment of Transboundary Rivers, Lakes and Groundwaters (UNECE, 2007), hereinafter First Assessment, is a derivative of the Convention on the Protection and Use of Transboundary Watercourses and International Lakes (UNECE, 1992) commonly known as the Water Convention. The First Assessment highlights the importance of transboundary waters and provides a comprehensive analysis of the pressure factors, the quantity and quality status, as well as responses and future trends of transboundary waters. Among other transboundary waters, the First Assessment integrates 70 transboundary aquifers located in the UNECE region, with 51 of them being in SEE.

The Second Assessment of Transboundary Rivers, Lakes and Groundwaters (UNECE, 2011), hereinafter Second Assessment, is the logical follow up of the First Assessment. Apart from various amendments in the descriptive information of the TBAs included in the First Assessment, the Second Assessment contained an increased number of transboundary waters, e.g., about 200 transboundary groundwaters. Focusing on SEE TBAs, the assessment incorporates 103 transboundary aquifers, however descriptive information is provided only for 74 TBAs located in SEE, hence only the latter TBAs are elaborated in this research.

DIKTAS and TWAP programmes

The DIKTAS programme was initiated in 2010 as a Global Environment Facility (GEF) regional programme, implemented by the United Nations Development Programme (UNDP) and executed by UNESCO-IHP, aiming at a first ever coordinated action to introduce sustainable integrated management principles in transboundary karst freshwater aquifer systems. To do so, an innovative Transboundary Diagnostic Analysis (TDA) approach coupled with a Strategic Action Programme (SAP) (DIKTAS, 2013) was implemented to facilitate the equitable and sustainable utilization of the transboundary karst aquifer system (Bonacci et al., 2013). The outputs of the specific data source are included in the Maps of the Transboundary Aquifers of the World (see below) and are used in the research.

The TWAP Programme is an another GEF funded initiative, coordinated by UNEP, aiming at providing a baseline assessment of identifying and evaluating the changes in transboundary water systems caused by human activities and natural processes, as well as the consequences these have on dependent human populations (Talaue Mcmanus et al., 2016). TWAP's area of groundwater research, which was implemented and coordinated by UNESCO-IHP (UNESCO-IHP & UNEP, 2016) included, amongst other, 199 transboundary aquifers. In the SEE region, the Upper Pannonian Thermal Aquifer shared among Slovakia, Austria, Hungary and Slovenia was thoroughly investigated (ILEC, UNEP-DHI, UNESCO-IHP, UNESCO-IOC & UNEP, 2016), with the outputs forming part of the present research.

Map of the transboundary aquifers of the world – editions 2015 and 2020

The Map of Transboundary Aquifers of the World of 2015 (IGRAC & UNESCO-IHP, 2015), which was created by UNESCO's Global Groundwater Centre IGRAC (International Groundwater Resources Assessment Centre), integrates the outputs of various projects and programmes related to groundwaters and conducted around the world till the year 2015, as well as the designated GWBs of the WFD implementation process. For this research, the 90 transboundary aquifers and groundwater bodies located in SEE are taken into consideration.

The Map of Transboundary Aquifers of the World of 2020 is an updated version of the one published in 2015. It contains minor updates to the 2015 edition TBA map, while it does not integrate the transboundary GWBs of EU's WFD implementation process. Additionally, it doesn't contain approximately 20 relatively small TBAs, that are mainly shared between Greece and Bulgaria, Slovenia and Austria and Slovenia and Italy. The 57 TBAs identified within the Balkan Peninsula are used in the present research. Both the TBAs Map of 2015 and 2020 are available digitally on the IGRAC's Global Groundwater Information System.[3]

Water framework directive

The WFD is state-of-the art environmental legislation of the EU aiming at the protection of its water resources. The main instruments for its implementation are the River Basin Management Plans (RBMPs) that not only contain extensive information for all water types (rivers, lakes, groundwater, coastal and transitional waters), but also propose measures for achieving the WFD's aims of good water quality (Skoulikaris & Krestenitis, 2020). In WFD, a GWB means a distinct volume of groundwater within an aquifer or aquifers, with the directions for their delineation to be given in the literature (EC, 2003). For groundwaters, Member States are obliged to define and designate GWBs within their national area. However, the Directive specifies that for international GWBs, the establishment of threshold values is subject to coordination between the involved Member States (Lipponen & Chilton, 2018).

In this research, the national GWBs that are located at the borders of the SEE countries and derived from the EU's Water Information System for Europe (WISE) database are processed. Specifically, after retrieving the GWBs in digitized format for the whole of the EU, the GWBs that intersect the boundaries of SEE countries were extracted and thereafter linked with attributes tables that integrate information about the GWBs' productivity, chemical status and quantitative status. The spatial information is derived from the 2nd cycle of the development of the national RBMPs, which was concluded in 2016.[4] The descriptive information of the GWBs, i.e., pressures, chemical and quantitative status, is based on data from 2018 (EEA, 2018).

Results and derived information on TBAs

Outputs from the ISARM/INWEB's inventory

In the ISARM/INWEB's inventory 65 TBAs(Figure 1) were identified in the SEE (Table A1 in the Appendix). In terms of absolute numbers of TBAs, Croatia and Serbia share the largest number of aquifers with 21 and 17 TBAs respectively. Bulgaria, Greece and North Macedonia also share a double figures number of TBAs, as depicted in Table 1. Of the 65 TBAs, 62 aquifers are shared between two riparian countries, while 3 aquifers are shared by three countries (see Table A1 in the Appendix). Moreover, 42 TBAs are karstic aquifers, and 23 aquifers are alluvial formations.

Regarding the extent of the aquifers, it should be mentioned that in the inventory the aquifers' area is given at a national scale, thus the total area is the sum of the given answers (see Table A1 in the Appendix). This is the common assessment practice also used in all the other programmes investigated in the research. There are cases where

Table 1. Identified TBAs per country in the initiatives/programmes/frameworks related to transboundary aquifers in the Balkans' area.

Shared TBA per country	TBAs in ISARM/INWEB's Inventory	TBAs in UNECE's First Assessment	TBAs in UNECE's Second Assessment	Map of World's TBAs (2015)	Map of World's TBAs (2020)	GWBs in WFD implementation process[d]
Albania	10	7	9	8	8	
Bosnia & Herzegovina	10	10	10	12	12	
Bulgaria	15	7	16	9	5	59
Croatia	21	23	35	23	24	23
North Macedonia	11	7	11	9	9	
Greece	14	9	18	15	7	33
Hungary	7	6	15	13	7	53
Moldova	1	1	3	3	3	
Montenegro	5	6	6	8	8	
Romania	9	6	16	22	7	40
Serbia	17	11	16	17	19	
Slovenia	9	10	38	10	11	18
Turkey	4	2	5	3	3	
Austria[a]			4	1	2	13
Italy[a]			12			8
Kosovo[b]			2			
Ukraine[a]			2	2	2	
Slovakia[c]			2	1	1	

[a]In the Second Assessment, the Balkan's TBAs shared with Austria, Italy and Ukraine are included in the assessment. [b]In the Second Assessment, Kosovo is denoted as a new state, namely a UN administered territory under UN Security Council Resolution 1244. However, Kosovo is not mentioned in the Maps of World's TBAs (2015 and 2020). [c]Slovakia is referred in the TBAs World Map (2015 and 2020) as one of the sharing parties of the "Upper Pannonian Thermal Aquifer". [d]The column "GWBs in WFD implementation process" numbers only the GWBs that intersect the national border lines of the Balkan countries.

information on an aquifer's area were given only by one riparian party, as well as cases where the aquifer's extent was not attributed by either of the countries sharing the aquifer. More specifically there is no information on the aquifers' boundaries for 3 TBAs, while in 32 aquifers the extent is given by only one riparian party.

As far the usage of the TBAs' waters is concerned, the primary use is for domestic water supply, i.e., drinking water. The coverage of the other water demands, including agricultural demand on water, is rather limited. For example, Croatia shares 9 TBAs with Slovenia, and in both the Croatian and Slovenian parts of the TBAs not only is the water used for water supply in 8 and 9 aquifers respectively, but also in most cases more than 75% of the available water is used for that purpose (Figure 2a). Similarly, Croatia shares 3 TBAs with Hungary and in all cases water supply is the primary use (Figure 2b). In almost in all cases of the 7 TBAs shared between Croatia and Bosnia and Herzegovina, the aquifers' water is consumed for drinking water supply (Figure 2c). Romania and Serbia share 3 TBAs (Figure 2d), with the aquifers' water being used for drinking water in all the aquifers located within Serbia. Similar water usages percentages may be found for the 4 TBAs shared by Bulgaria and Serbia (Figure 2e), the 3 TBAs shared by North Macedonia and Serbia (Figure 2f), the 3 TBAs shared between Albania and Greece (Figure 2g), while the only case where the drinking water is not the main use is in the 2 TBAs shared by Greece and North Macedonia (Figure 2h).

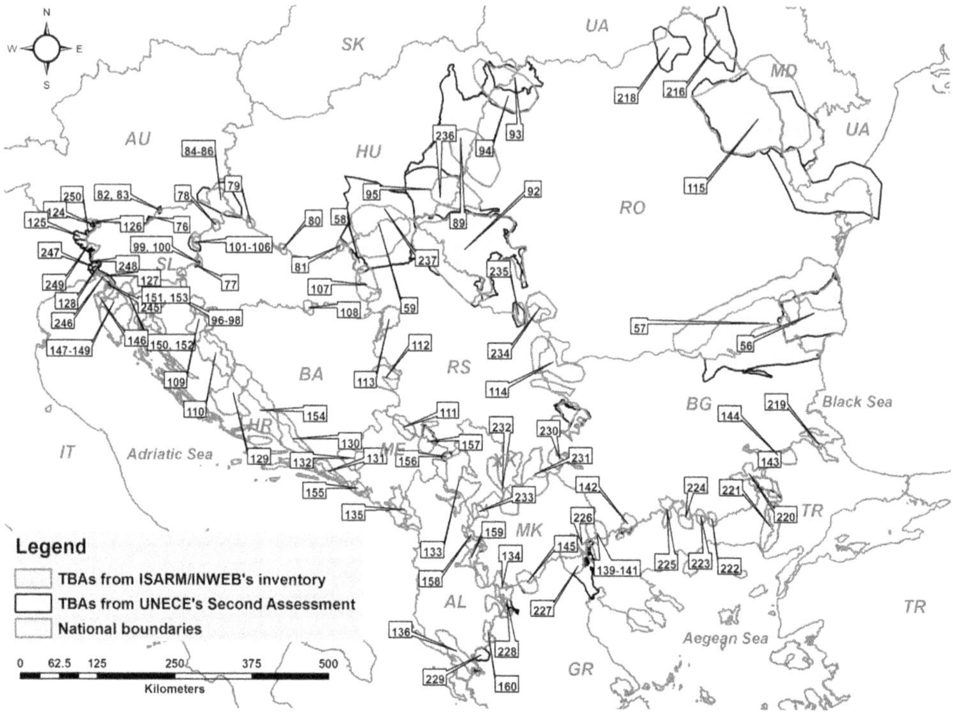

Figure 1. TBAs included in the Second Assessment of UNECE (yellow polygons) and their spatial correlation with the TBAs identified in the ISARM/INWEB's inventory (red polylines) in the Balkans Peninsula. The numbering of the TBAs corresponds to the numbering utilized in the Second Assessment.

Regarding the problems identified on the water quantity of the TBAs, increased pumping is the most frequently reported problem in the TBAs of Hungary, Romania and Serbia with six, four and four identified cases respectively. The decrease of borehole yield is mostly denoted in the national parts of the TBAs of North Macedonia and of Hungary with seven and six cases respectively. Changes in groundwater levels are reported in various Croatian and North Macedonian TBAs, while pressures due to sea water intrusion is only mentioned in two of Montenegro's TBAs. The most common problem, which was acknowledged by eight countries, is the degradation of ecosystems. In any case, most of the countries declared that they are not aware of any problems related to water quantity within their national parts of the TBAs.

In terms of water quality, nitrogen species in the groundwaters is reported by all countries, i.e., in 29 national parts of TBAS, apart from Montenegro. The second most common problem is the presence of pathogens in the groundwaters, i.e., described in 24 national parts of TBAS. Heavy metals were detected in 19 national parts of TBAs, with Serbia leading the identified number of cases (7 cases). Problems attributed to pesticides were recognized in 14 national parts of TBAs, while salinization and industrial organic compounds were identified in 10 aquifers. However, the quality of most of the national parts of the aquifers (63 cases) was not known during the period that the inventory was compiled.

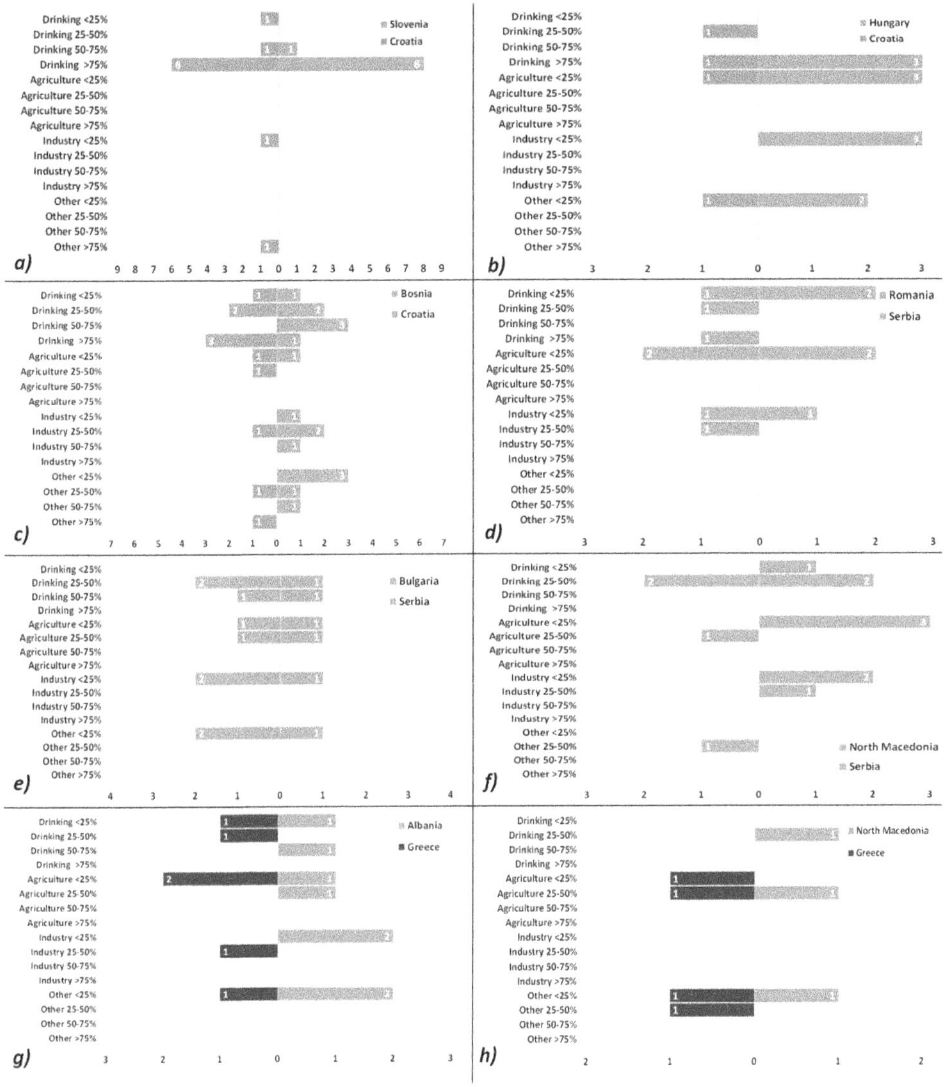

Figure 2. Percentages of aquifers' water usages in the case of the a) 9 TBAs shared between Croatia and Slovenia, b) 3 TBAs shared between Croatia and Hungary, c) 7 TBAs shared between Croatia and Bosnia & Herzegovina, d) 3 TBAs shared between Serbia and Romania, e) 4 TBAs shared between Serbia and Bulgaria, f) 3 TBAs shared between Serbia and North Macedonia, g) 3 TBAs shared between Greece and Albania, and h) 2 TBAs shared between Greece and North Macedonia.

First and second assessment of transboundary rivers, lakes and groundwaters

The First Assessment includes 51 TBAs located in the Balkans area, Table 1. All the TBAs, as well as their descriptive information, apart from three TBAs are coming from the ISARM/INWEB's inventory. The correlation of the First Assessment TBAs with those of the ISARM/INWEB's inventory, together with the limited identified differences are depicted in Table A1 in the Appendix. The differences are mainly related to the TBAs' area (four cases) and the merging of specific neighbouring TBAs as unique TBAs (two cases). On the other hand, the characteristics, such as the water uses percentages as well as the pressures on the quality and quantity of the TBAs' water are identical between these two data sources, since they come from the ISARM/INWEB's inventory.

In the Second Assessment, 103 TBAs, where at least one of the sharing countries is a Balkan country, are included, Table 1. This listing comprises the First Assessment TBAs, as well as those shared between Slovenia with Italy, and Slovenia with Austria. Moreover, Kosovo is considered a new state, thus its aquifers shared with neighbouring countries are also included in the listing (Figure 1). However, the assessment provides descriptive information only for 74 TBAs located in the SSE region, hence only these TBAs are utilized in the development of this research. Among these 74 TBAs, 51 are derived from the First Assessment, thus a thorough investigation was only made for the 23 new TBAs. The alterations in comparison to the 51 TBAs of the First Assessment as well as facts about the 23 newly identified TBAs are provided in the following analysis (See also Table A2 in the Appendix).

Starting with the 51 common TBAs between First and Second Assessment, the characteristics and description of 27 TBAs are identical in the two assessments. In the Second Assessment, 23 TBAs of the First Assessment have amendments or changes made to their description. Specifically, 2 TBAs have different names, 2 TBAs have different names together with different extents and 19 TBAs have differentiations in their extent, (Table A2 in the Appendix). In terms of the water utilization of the TBAs, i) in 42 out of 74 TBAs, which are described in the Second Assessment, the water uses are identical with those presented in the First Assessment, ii) new information is given for 29 TBAs which either differs from that of the First Assessment or is introduced for the first time in the Second Assessment, and iii) the water uses are not defined in 3 TBAs. Further analysis of the water uses in these 29 TBAs, (Table 2), demonstrates that in 25 cases the water is used for water supply in at least one national part of the aquifer. The maintenance of the ecosystem services is the second most frequent use, followed by coverage of irrigation demands.

Map of transboundary aquifers of the world edition 2015 and edition 2020

In the IGRAC's Map of Transboundary Aquifers of the World (Edition 2015) 90 transboundary aquifers or transboundary aquifer systems or transboundary groundwater bodies (Table 1), have been designated (Table A3 in the Appendix). The term 'aquifers systems' is used since some of the aquifers identified are essentially a system of neighbouring but separate aquifers with different names, however, the extent of these aquifers is expressed using a single figure. In this research, 7 of the TBA systems identified are considered as 7 TBAs. The aquifers identified are shared among the same countries as those stated in the Second Assessment, i.e., all the Balkan countries as well as Austria, Hungary, Moldova, Slovakia, Turkey and Ukraine (Table A3 in the Appendix). The most recent information about the TBAs in the SEE is provided by the Map of Transboundary Aquifers of the World (Edition 2020) (Figure 3), which includes updated information about the TBAs occurrence.

Table 2. Use of TBAs' water in the national parts of the transboundary aquifers.

No/C[a]	Sharing countries	Water uses in % (Country 1)[b]				Water uses in % (Country 2)[c]			
		Drinking	Irrigation	Industry[d]	Ecosystem[e]	Drinking	Irrigation	Industry[d]	Ecosystem[d]
56	Bulgaria, Romania	>75			<25				
57	Bulgaria, Romania	>75	<25	<25	<25	>75	<25	<25	
76	Austria, Slovenia	<25	<25		<25	<25	<25	<25	<25
82	Austria, Slovenia								<50
83	Austria, Slovenia					<25	<25		<25
84	Slovenia, Hungary	>50	<25		<25				
85	Slovenia, Austria, Hungary[f]			<75					
86	Slovenia, Hungary, Croatia[g]	<25	<25		<25				
87	Romania, Hungary	<25	<25						
88	Romania, Hungary	<25		<25					
89	Romania, Hungary	<25		<25					
92	Serbia, Romania	>75				>50	<25	>25	
93	Romania, Hungary	>25	<25	>50		>75	<25	<25	<25
97	Croatia, Slovenia	>50							
100	Slovenia, Croatia	>50							
102	Slovenia, Croatia	<50							
103	Croatia, Slovenia					<50			
104	Croatia, Slovenia					<50			
105	Croatia, Slovenia					<50			
106	Croatia, Slovenia					<50			
112	Serbia, Bosnia & Herzegovina	<25				>75	<25		<25
140	Bulgaria, Greece				>75				
141	Bulgaria, Greece	<25	<25	<25	<25				
142	Bulgaria, Greece	<25	<25		<25	<25	<25	<25	<25
148	Croatia, Slovenia					>50			
149	Croatia, Slovenia					>50			
151	Croatia, Slovenia					>50			
152	Croatia, Slovenia					>50			
153	Croatia, Slovenia					>50			

[a]No/C is the numbering of the TBAs in the Second Assessment; [b]This refers to the 1st country mentioned in the column "Sharing countries"; [c]This refers to the 2nd country mentioned in the column "Sharing countries"; [d]The industrial use of water might include small scale hydropower, fish breeding, thermal spa and heating purposes; [e]This refers to the coverage of ecosystem services; [f]The Mura – Zala basin/Radgona – Vaš aquifer(No.85) is shared among three countries, nevertheless no information is provided for either Austria or Hungary; [g]The Kot aquifer (No.86) is shared among three countries, nevertheless no information is provided for Hungary and Slovenia.

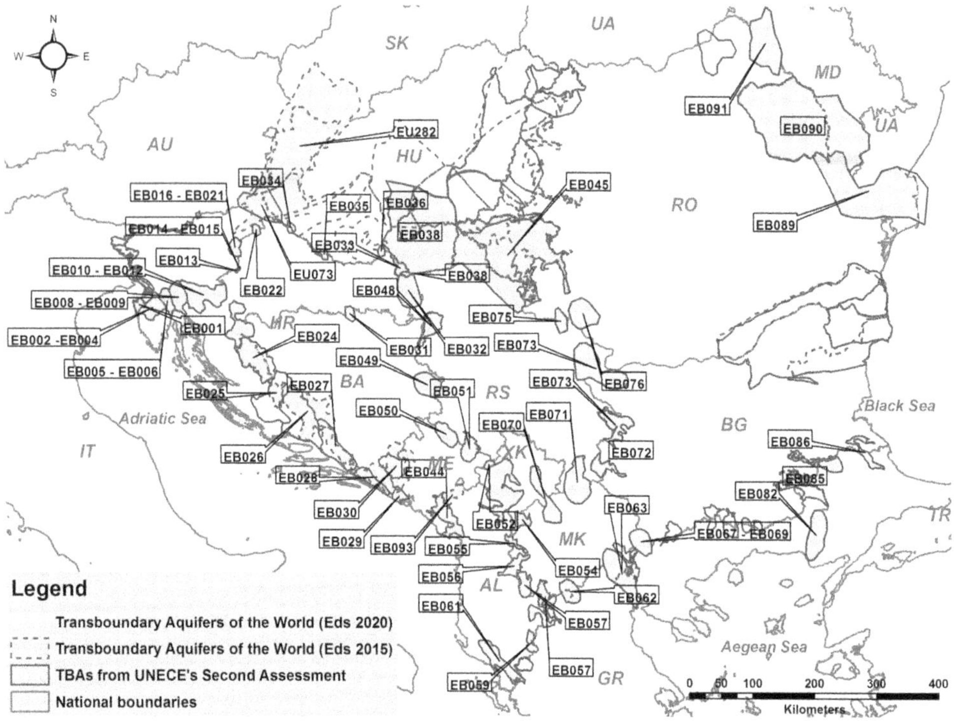

Figure 3. Illustration of the TBAs of the world transboundary map edition 2020 (beige coloured polygons) overlayed by the TBAs of the world transboundary map edition 2015 (dotted red lines) and by the TBAs of UNECE's Second Assessment (solid blue lines). The numbering of the TBAs corresponds to the numbering in the latest world transboundary map edition 2020.

GWBs in the WFD implementation process

The analysis demonstrated that 247 GWBs (of horizons 1, 2, 3 and 4) are located at the borders of the EU countries in SEE, with Figure 4 depicting the spatial distribution of these GWBs per country. The 247 GWBs are those GWBs that intersect i) the boundaries of Greece, Bulgaria, Romania, Slovenia and Croatia, ii) the boundaries of Hungary with Croatia, Romania, Serbia and Slovenia respectively, iii) the boundaries of Austria with Slovenia, and iv) the boundaries of Italy with Slovenia. On the other hand, and based on the WFD implementation process outputs of 2016, i.e., the 2nd cycle of the RBMPs, only 103 (out of the identified 247) GWBs are characterized as transboundary ones by the EU Member-States, while 144 GWBs, although are located on the border line, are not denoted as those (Figure 4).

Bulgaria has the largest number of GWBs, i.e., 59 GWBs, at its border line followed by Hungary, which has 53 GWBs at its borders with the other SEE countries (Figure 5). These figures are much higher from the GWBs denoted as transboundary ones within the 2nd cycle of RBMPs, especially for Bulgaria where only 5 GWBs are officially mentioned as transboundary bodies (Figure 5). In the research we identified 23 Croatian GWBs lay on the country's boundaries, a figure that is very close to the 21 GWBs that are designated as transboundary ones. Similar outputs were derived for Italy, where the identified 8 GWBs are identical with the ones are designated as transboundary ones. In contrary, none of 40

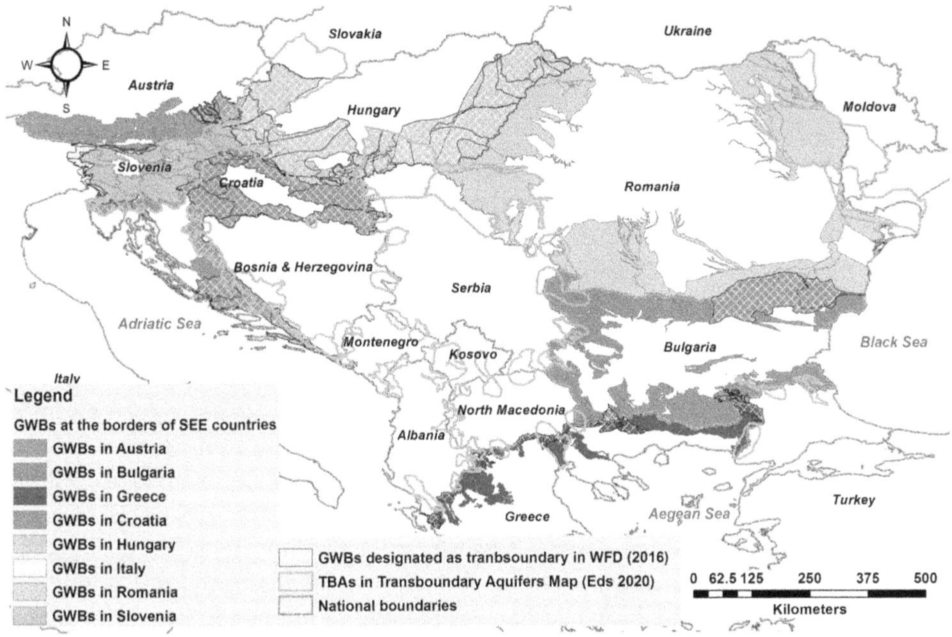

Figure 4. Thematic map illustrating the occurrence of GWBs at the borders of the EU countries located in SEE and the GWBs designated as transboundary in the WFD implementation process outputs of 2016, overlayed by the TBAs of the transboundary aquifers map (edition 2020).

identified Romanian GWBs is designated as transboundary in the WFD implementation process. To avoid topology biases, it should be mentioned that (Figure 5) shows only the GWBs whose boundaries intersect the border lines of the other Balkans' countries. For example, the 59 Hungarian GWBs, are solely those GWBs located at the borders with Slovenia, Croatia, Serbia and Romania.

As far as the identified pressures on GWBs is concerned, and based on the WFD classification of pressures on groundwaters (WFD Reporting Guidance, 2016), the analysis revealed that no anthropogenic pressures were observed in 123 GWBs, i.e.,

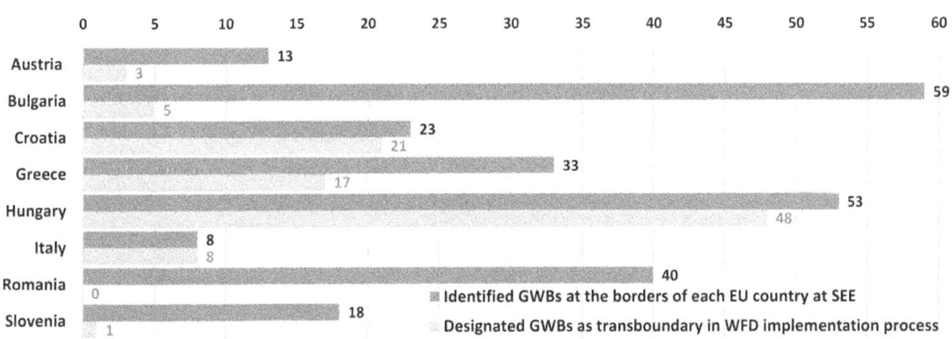

Figure 5. Distribution per country of the i) 247 GWBs located at the border line of the EU countries located in the South-Eastern Europe (SEE), and ii) the GWBs that are characterized as transboundary in the 2nd cycle of RBMPs of WFD. The figures for Italy, Hungary and Austria show the GWBs bordering only the SEE countries.

they belong to the category 'P0-No significant anthropogenic pressure'. Point source pressures (category 'P1-Point sources') exist in 49 GWBs, pressures from diffuse sources (category 'P2-Diffuse sources') are apparent in 81 GWBs, while abstraction pressures, (category 'P3-Abstractions') are found in 62 GWBs. Pressures related to hydromorphology (category"P4-Hydromorphology") are found only in one groundwater body. The analysis does not include information on the Austrian GWBs, since no relevant descriptive information was available; thus the analysis performed only included 234 out of 247 GWBs, i.e., excluded the 13 Austrian GWBs.

Regarding the chemical status of the GWBs' waters, i.e., the qualtiy index of a GWB which indicates the lack or not of hazardous substances, as well as the limitation or not of all other pollutants entering groundwaters (EEA, 2018), it is revealed that the chemical status of 196 GWBs is characterized as good, 37 GWBs fail to achieve a good chemical status, while the chemical status of 1 GWB is unknown (for the 13 Austrian GWBs no information is available) (Figure 6). Most of the problematic GWBs, in terms of chemical status, are to be found in Bulgaria (15 cases). On the other hand, Croatia, Greece and Slovenia have identified two or fewer GWBs whose chemical quality is less than good.

Finally, the quantitative status of the GWBs is the corresponding index that indicates whether the available groundwater resource is reduced or not by the long-term annual average rate of abstraction. The analysis outputs (Figure 7), show that the quantitative status of 212 GWBs is classified as good and 22 GWBs fails to achieve a good quantitative status, while this type of information is not available for the 13 Austrian aquifers.

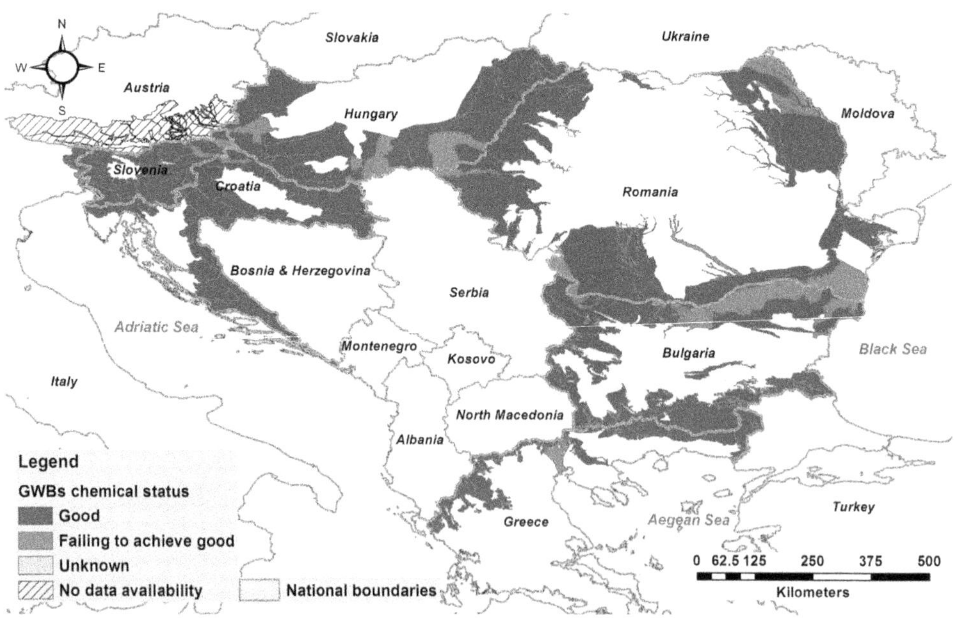

Figure 6. Chemical status of the GWBs in the EU countries located in SEE. The GWBs with good chemical status are shown in blue, those that fail to achieve a good chemical status in red, and those whose chemical status is unknown in pink, those for which no information is available with hatched lines.

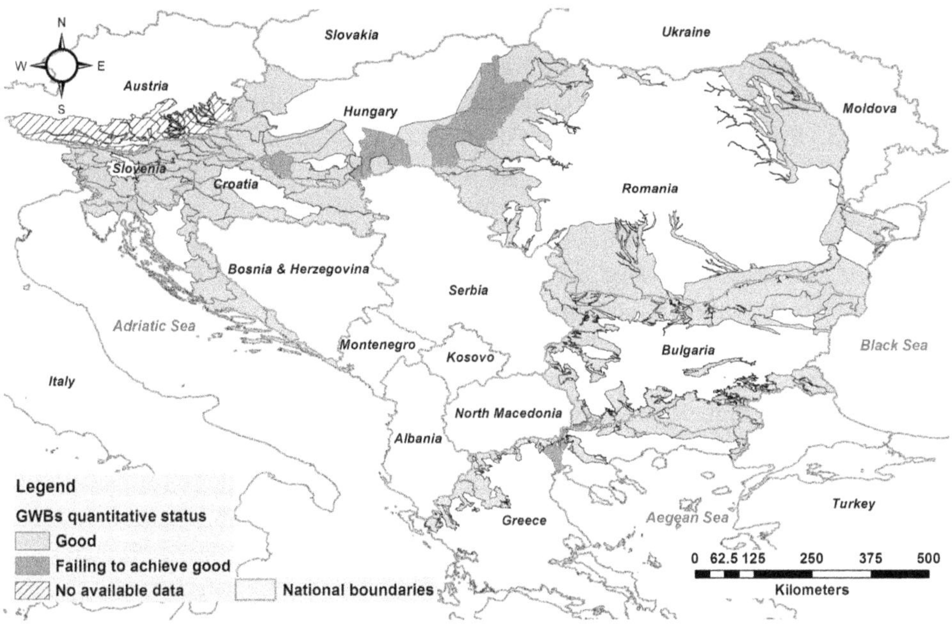

Figure 7. Quantitative status of the GWBs in the EU countries located in SEE. The GWBs with good status are shown in light green, those which fail to achieve a good quantitative status in red, and those for which no information are available with hatched lines.

Discussion

The research presents a comparative analysis of the status of TBAs derived from the main international programmes and initiatives focused on shared groundwater management in SEE. It provides the conceptual links between the ISARM/INWEB's inventory with the First and Second Assessment and with the Maps of Transboundary Aquifers of the World of 2015 and 2020. Furthermore, the research focuses on the TBAs' pressures and on the use of the groundwaters, concluding on the importance of the TBAs for drinking water supply on a regional level. The new insights coming from the WFD implementation process, i.e., delineated groundwater bodies, attributed pressures and water quality status, are for the first time demonstrated by the specific research and are correlated with the outputs of the TBAs' programmes and initiatives.

Interpretation of findings

The research reveals the continuity of the programmes and initiatives related to the management of TBAs in the Balkan Peninsula. Although the difficulty on TBAs' delineation, since hydraulic and geometric properties of aquifers are primarily based on classical geological and hydrogeological survey data, borehole tests, tracing experiments, geophysical observations and discharge measurements which often results in mere approximations, particularly in karst systems (Kovács et al., 2005), the overlaying of the TBAs area retrieved by the various data sources (Figures 1 and 3) demonstrate a quite precise spatial delineation of the TBAs through the various programmes. Taking as a reference the

TBAs of the Second Assessment, which include the work conducted in the ISARM/INWEB's inventory and adopted by the First Assessment (Figure 1), the dedicated DIKTAS and TWAP programmes contributed to improving the level of spatial knowledge on the TBAs occurrence in SEE, as depicted in the Map of TBAs of the World Eds. 2015 and Eds. 2020. The delineation of the aquifers in GWBS through the WFD implementation process in the EU countries of SEE also demonstrates the relatively good spatial correlation of the designated TBAs and the GWBs, notably in the western part of the case study area (Figure 4). The increased accuracy of information about the GWBs could contribute to a more precise mapping in cases with approximations, such as the TBAs of Greece shared with its northern riparian parties.

The research also highlights the continuity of the geological formations in the study region and demonstrates that the national administrative boundaries are not consequent boundaries of the GWBs. Spatial analysis techniques on the GWBs revealed that 247 GWBs intersect the border lines of the EU countries of SEE, with only 103 of these GWBs to be officially designated as transboundary GWBs by the Member-States. However, almost all GWBs at the border lines of specific countries, such as Bulgaria, Romania and Slovenia, are not designated as transboundary ones, even if their neighbouring riparian states identify transboundary groundwater formations in their borders. The case of Slovenia and Croatia is an additional example, where although the two countries have officially identified TBAs in the various international programmes, the designation of the relevant GWBs by both countries as transboundary ones is not apparent within the WFD.

The principal and by far the greatest use of the TBAs' waters in the Balkans Peninsula is for domestic water supply, with all the investigated data sources reaching the same conclusion. Although the primary use of groundwater varies enormously worldwide (WWAP, 2015), similar outputs and figures are detailed in the literature at global scale (Grönwall & Danert, 2020; Zektser & Everett, 2004), at continent scales (Foster et al., 2006; Lee et al., 2018) as well as at national scales (De Stefano et al., 2015). In the ISARM/INWEB's inventory, the use of water for drinking purpose is apparent in all the TBAs with relevant information on water uses (58 out of 65 TBAs). Water is similarly used for domestic water supply in each of the 29 new TBAs of the Second Assessment, as depicted in Table 2.

Focusing on the pressures that the TBAs are subjected to, initial facts and figures on the identified pressures were provided in the ISARM/INWEB's inventory, with the degradation of the ecosystems being the most prevalent stressing factor. However, most of the inventory's TBAs did not identify problems and pressures due to the national experts' lack of knowledge at the time the inventory was compiled. The amount of available information was significantly improved in the Second Assessment together with the outputs of DIKTAS programme.

Based on the analysis of the GWBs' pressures, it is concluded that more than half (~53%) of the GWBs (123 out of 233 GWBs) are not subject to anthopogenic pressures (category P0 of pressures). This figure is slightly better than the one (~50%) that corresponds to the GWBs that are not subject to anthopogenic pressures at an EU scale (11,203 out of the 22,260 identified GWBs in the EU) (EEA, 2018). In the remaining 110 GWBs in SEE, the pressures attributed to point (category P1) and diffuse (category P2) sources correspond to ~45% and ~74% respectively, while abstraction pressures are present in ~56% of the GWBs. The latter figures exceed 100% since there are several

GWBs where more than one type of pressure has been identified. At an EU scale, point sources, diffuse sources, and abstractions correspond to 25.5%, 46.7% and 20.6% of the identified GWBs' pressures. By comparing the abstraction figures of SEE (~56%) with those of the EU (20.6%) groundwater is an important water resource for covering water demands in SEE, since its abstraction (or over-abstraction) is more than double that of the EU.

Cooperation perspectives on the management of the TBAs in the SEE

In the Balkans, all countries have ratified the Water Convention, and even though this framework agreement does not replace existing bilateral agreements at a transboundary scale, it fosters transboundary cooperation and supports further developments. On the other hand, the fact that 25%,i.e. 63 out of 247, of the EU's GWBs are tangential to the border lines of non-EU countries in SEE demonstrates the potential difficulties of the common management of transboundary water bodies under the common legislative framework of the EU's WFD.

However, all the non-EU countries in SEE have started or are about to start processes for their accession to the EU. North Macedonia is taking advantage of the EU's Instruments for Pre-Accession Assistance (IPA) and receives financial and technical support (Skoulikaris & Zafirakou, 2019). Efforts by Turkey and Albania to reform their national water related strategies and policies have started to be referenced (Delipınar & Karpuzcu, 2017). The latest EC Progress Reports on Serbia (EC, 2016), on Bosnia and Herzegovina (EC, 2019a), and on Montenegro (EC, 2019b) demonstrate the progress that has been conducted on the water management sector. Hence, the harmonization of the legislation of non-EU countries with the EU acquis is bound to improve the perspectives for the joint management of shared groundwaters under the same legislative and technical framework (Skoulikaris, 2021).

A core element in transboundary cooperation is the effective exchange mechanisms of data and information (Ganoulis et al., 2011; UNESCO-WWAP, 2009). In our case study, all the data sources are open-access sources, in other words freely available databases for knowledge and data exchange. The IGRAC's Global Groundwater Information System (GGIS), includes much of the information provided in this research, while at an EU level and relevant to groundwater bodies, the EU has developed its own system for water related data sharing, namely WISE. Hence, the use of public official data could shed light on transboundary conflict areas, foster cooperation as well as set the foundations for joint management of the shared waters.

To sum up, the management of transboundary aquifers may be further jeopardized and the pressures on groundwaters may be worsened due to climate change. Although water is the transversal element at the heart of climate change (Stephan, 2017), IPCC has not given extensive emphasis to the impacts of climate change on groundwater because it is still difficult to evaluate and distinguish these impacts from human pressure and because of large areas of uncertainty regarding the impact of climate change on groundwater (Jiménez Cisneros et al., 2014). Whatever the case, the buffer capacity of aquifers can be conceived as a natural adaptation to climate change, thus intensified efforts to properly share, manage and maintain TBAs' water at high quality and quantity status is of great importance.

Conclusions

The conducted research demonstrates the continuity of collecting and sharing information related to transboundary aquifers in SEE, which has been achieved through a series of back-to-back international programmes and initiatives. In all data sources, the groundwater use for drinking water supply is dominant, while the pressures on groundwaters are mainly due to diffuse pollution sources.

The accuracy of the information integrated in the international programmes and examined in this research is validated by the EU's WFD implementation process. In terms of TBAs' occurrence, the GWBs identified in the Balkans demonstrate the relevant precise delimitation of the TBAs' boundaries. The associated pressures on the GWBs show that point and diffuse source pollution are the main pressures on groundwater systems, while many GWBs are characterized as natural, i.e., with no significant anthropogenic pressures. The quality and quantitative status of GWBs together with information from the international programmes and initiatives implemented by UNESCO-IHP, render the findings of this research an important guide for the common management of the transboundary groundwater resources in the case study region.

Notes

1. Population of 1 January. ec.europa.eu/eurostat. Eurostat. Retrieved 14 January 2021.
2. Turkey: Administrative Division. Retrieved 14 January 2021
3. Map of the Transboundary Aquifers of the World. Retrieved 15 January 2021
4. WISE WFD Reference Spatial Datasets reported under Water Framework Directive 2016 – PUBLIC VERSION – version 1.4

Disclosure statement

No potential conflict of interest was reported by the author(s).

ORCID

Charalampos Skoulikaris http://orcid.org/0000-0002-1319-2449

References

Altchenko, Y., & Villholth, K. G. (2013). Transboundary aquifer mapping and management in Africa: A harmonised approach. *Hydrogeology Journal*, *21*(7), 1497–1517. https://doi.org/10.1007/s10040-013-1002-3

Aureli, A., & Eckstein, G. (2011). Strengthening cooperation on transboundary groundwater resources. *Water International*, *36*(5), 549–556. https://doi.org/10.1080/02508060.2011.615137

Bonacci, O., Željković, I., & Galić, A. (2013). Karst rivers' particularity: An example from dinaric karst (croatia/bosnia and herzegovina). *Environmental Earth Sciences*, *70*(2), 963–974. https://doi.org/10.1007/s12665-012-2187-9

Burchi, S. (2018). Legal frameworks for the governance of international transboundary aquifers: Pre-and post-ISARM experience. *Journal of Hydrology: Regional Studies*, *20*, 15–20. https://doi.org/10.1016/j.ejrh.2018.04.007

De Stefano, L., Fornés, J. M., López-Geta, J. A., & Villarroya, F. (2015). Groundwater use in Spain: An overview in light of the EU water framework directive. *International Journal of Water Resources Development, 31*(4), 640–656. https://doi.org/10.1080/07900627.2014.938260

Delipınar, Ş., & Karpuzcu, M. (2017). Policy, legislative and institutional assessments for integrated river basin management in Turkey. *Environmental Science & Policy, 72*, 20–29. https://doi.org/10.1016/j.envsci.2017.02.011

DIKTAS. 2013. *Protection and sustainable use of the dinaric karst transboundary aquifer system - Transboundary diagnostic analysis.* Programme Technical Report. pp 154. Retrieved 23 November 2020 from http://diktas.iwlearn.org/resources/projectdocuments/transboundart-diagniostics-analysis/transboundary-diagnostic-analysis-tda/view.

EC. (2000, December 22). Directive 2000/60/EC of the European parliament and of the council of 23 October 2000 establishing a framework for community action in the field of water policy (water framework directive). In *Off J Eur communities L327* (pp. 1–72).

EC. (2003). Common implementation strategy for the water framework directive (2000/60/EC). *Guidance document n°2. Identification of water bodies.* European Communities. 28.

EC. (2016). *European commission staff working document: Serbia 2016 report.* Communication from the Commission to the European Parliament, the Council, the European Economic and Social Committee and the Committee of the Regions, 2016 Communication on EU Enlargement Policy. 9.11.2016SWD(2016)361 final.

EC (2019a). *European commission staff working document: Analytical report.* Communication from the Commission to the European Parliament and the Council, Commission Opinion on Bosnia and Herzegovina's application for membership of the European Union. 29.5.2019SWD (2019) 222 final.

EC (2019b). *European commission staff working document: Montenegro 2019 report.* Communication from the Commission to the European Parliament, the Council, the European Economic and Social Committee and the Committee of the Regions. 2019 Communication on EU Enlargement Policy. 29.5.2019SWD(2019) 217 final.

EEA. (2018). *European waters-Assessment of status and pressures 2018.* European Environmental Agency Report No 7/2018. https://doi.org/10.2800/303664

Foster, S., Tuinhof, A., & Garduño, H. (2006). *Groundwater development in sub-Saharan Africa: A strategic overview of key issues and major needs.* Sustainable groundwater management, lessons from practice, Case profile collection number 15. World Bank.

Fraser, C. M., Kalin, R. M., Kanjaye, M., & Uka, Z. (2020). A methodology to identify vulnerable transboundary aquifer hotspots for multi-scale groundwater management. *Water International, 45*(7–8), 865–883. https://doi.org/10.1080/02508060.2020.1832747

Ganoulis, J., Aureli, A., & Fried, J. (2011). *Transboundary water resources management: A multidisciplinary approach.*

Ganoulis, J., Quartano, K., & Skoulikaris, C. (2014). Promoting cooperation for transboundary water security: The experience of the UNESCO Chair/INWEB. In Grafton, R. Q., Wyrwoll, P., White, C., and Allendes, D. (Eds.), *Global water: Issues and insights* (pp. 1–378). Australian National University Press.

Gerlak, A. K., Lautze, J., & Giordano, M. (2011). Water resources data and information exchange in transboundary water treaties. *International Environmental Agreements: Politics, Law and Economics, 11*(2), 179–199. https://doi.org/10.1007/s10784-010-9144-4

Grönwall, J., & Danert, K. (2020). Regarding groundwater and drinking water access through a human rights lens: Self-Supply as A norm. *Water, 12*(2), 419. https://doi.org/10.3390/w12020419

IGRAC & UNESCO-IHP. (2015). Transboundary aquifers of the world [map]. Edition 2015. In *Scale 1: 50 000 000* (pp. 2015). IGRAC.

IHP-ISARM/INWEB. (2008). *Phase I of ISARM/Balkans Internationally shared aquifers in the Balkan region: Preliminary assessment.* Programme and meeting document, UNESCO-BRESCE /UNECE/INWEB, SC.2009/WS/31, pp. 150. Retrieved 23 January 2021 from https://unesdoc.unesco.org/ark:/48223/pf0000188566

ILEC, UNEP-DHI, UNESCO-IHP, UNESCO-IOC & UNEP. (2016). Water system information sheets: Eastern Europe. In L. Talaue-mcmanus (Ed.), *Transboundary waters: A global compendium, volume 6-Annex E.* United Nations Environment Programme (UNEP).

Jiménez Cisneros, B. E., Oki, T., Arnell, N. W., Benito, G., Cogley, J. G., Döll, P., & Mwakalila, S. S. 2014. Freshwater resources. In: Field, C.B., Barros, V.R., Dokken, D.J., Mach, K.J., Mastrandrea, M.D., Bilir, T.E., Chatterjee, M., Ebi, K.L., Estrada, Y.O., Genova, R.C., Girma, B., Kissel, E.S., Levy, A.N., MacCracken, S., Mastrandrea, P.R., and White, L.L. (Eds.), *Climate change 2014: Impacts, adaptation, and Vulnerability. Part A: Global and sectoral aspects. Contribution of working group II to the fifth assessment report of the intergovernmental panel on climate change* (Vol. (1 229–269)).Cambridge University Press.

Kovács, A., Perrochet, P., Király, L., & Jeannin, P. Y. (2005). A quantitative method for the characterisation of karst aquifers based on spring hydrograph analysis. *Journal of Hydrology, 303*(1–4), 152–164. https://doi.org/10.1016/j.jhydrol.2004.08.023

Lee, E., Jayakumar, R., Shrestha, S., & Han, Z. (2018). Assessment of transboundary aquifer resources in Asia: Status and progress towards sustainable groundwater management. *Journal of Hydrology:. Regional Studies, 20*, 103–115. https://doi.org/10.1016/j.ejrh.2018.01.004

Lipponen, A., & Chilton, J. (2018). Development of cooperation on managing transboundary groundwaters in the pan-European region: The role of international frameworks and joint assessments. *Journal of Hydrology: Regional Studies, 20*, 145–157. https://doi.org/10.1016/j.ejrh.2018.05.001

Puri, S., & Aureli, A. (2005). Transboundary aquifers: A global program to assess, evaluate, and develop policy. *Groundwater, 43*(5), 661–668. https://doi.org/10.1111/j.1745-6584.2005.00100.x

Rivera, A., & Candela, L. (2018). Fifteen-year experiences of the internationally shared aquifer resources management initiative (ISARM) of UNESCO at the global scale. *Journal of Hydrology: Regional Studies, 20*, 5–14. https://doi.org/10.1016/j.ejrh.2017.12.003

Sanchez, R., Lopez, V., & Eckstein, G. (2016). Identifying and characterizing transboundary aquifers along the Mexico–US border: An initial assessment. *Journal of Hydrology, 535*, 101–119. https://doi.org/10.1016/j.jhydrol.2016.01.070

Skoulikaris, C., & Krestenitis, Y. (2020). Cloud data scraping for the assessment of outflows from dammed rivers in the EU. A case study in South Eastern Europe. *Sustainability, 12*(19), 7926. https://doi.org/10.3390/su12197926

Skoulikaris, C., & Zafirakou, A. (2019). River basin management plans as a tool for sustainable transboundary river basins' management. *Environ Sci Pollut Res, 26*(15), 14835–14848. https://doi.org/10.1007/s11356-019-04122-4

Skoulikaris, C. (2021). Transboundary cooperation through water related EU Directives' implementation process. The case of shared waters between Bulgaria and Greece. *Water Resour Manage, 35*(14), 4977–4993. https://doi.org/10.1007/s11269-021-02983-4

Stephan, R. M. (2011). The draft articles on the law of transboundary aquifers: The process at the UN ILC. *International Community Law Review, 13*(3), 223–235. https://doi.org/10.1163/187197311X582287

Stephan, R. M. (2017). Climate change considerations under international groundwater law. *Water International, 42*(6), 757–772. https://doi.org/10.1080/02508060.2017.1351911

Talaue Mcmanus, L., Mahon, R., Aureli, A., Barbiére, J., Bertule, M., & Sherbinin, A. (2016). *Transboundary water systems - status and trends: Crosscutting analysis* (Vol. 6). United Nations Environment Programme (UNEP.

UN. (1997). *United Nations' Convention on the law of the non-navigational uses of international watercourses.*

UNECE. (1992). *Convention on the protection and use of transboundary watercourses and international lakes.* Retrieved 8 December 2020 from http://www.unece.org/fileadmin/DAM/env/water/pdf/watercon.pdf

UNECE. (2007). *First assessment of transboundary rivers, lakes and groundwaters.*

UNECE. (2011). *Second assessment of transboundary rivers, lakes and groundwaters.*

UNESCO-IHP & UNEP. (2016). *Transboundary aquifers and groundwater systems of small Island developing states: Status and trends, summary for policy makers.* United Nations Environment Programme (UNEP).

UNESCO-IHP Water Information Network System (IHP-WINS). 2017. Retrieved 12 February 2021 from: http://en.unesco.org/ihp-wins

UNESCO-IHP. (2009). *Transboundary aquifers managing a vital resource –The UNILC draft articles on the law of transboundary aquifers* (R. Stephan, Ed.). UNESCO.

UNESCO-WWAP. (2009). *The united nations world water development report 3: Water in a changing world.* UNESCO, and London: Earthscan; 2009.

UNESCO. (2001). Internationally shared (transboundary) aquifer resources management. *Their significance and sustainable management. A framework document. on serial document SC-2001/WS/40.* 74.

UNESCO. (2010). Transboundary aquifers: Challenges and new directions. *Abstracts, ISARM 2010 International Conference, Paris,* 6–8 December, UNESCO, Paris, pp. 188

Wada, Y., & Heinrich, L. (2013). Assessment of transboundary aquifers of the world—vulnerability arising from human water use. *Environmental Research Letters, 8*(2), 024003. https://doi.org/10.1088/1748-9326/8/2/024003

WFD Reporting Guidance. (2016). *Common implementation strategy for the water framework directive and the floods directive.* Technical report Final –Version 6.0.6. pp. 505. Retrieved 9 January 2021 from https://circabc.europa.eu/sd/a/5b969dc0-6863-4f75-b5d8-8561cec91693/Guidance%20No%2035%20-%20WFD%20Reporting%20Guidance.pdf.

WWAP. (2015). *The United Nations world water development report 5, water for a sustainable world facts and figures.* UNESCO.

Zektser, I. S., & Everett, L. G. (2004). *Groundwater resources of the world and their use, UNESCO-IHP-VI Series on groundwater No. 6.* UNESCO Publications.

APPENDIX

Table A1. Transboundary aquifers identified in ISARM/INWEB's inventory and in the First Assessment of Transboundary Rivers, Lakes and Groundwaters

No./A[a]	No./B[b]	Sharing countries	Aquifer	Area (km^2)
1	1	Slovenia, Croatia	Secovlje-Dragonja/Istra	*119*
2	2	Croatia, Slovenia	Mirna/Istra	214
3	3	Croatia, Slovenia	Opatija/Istra	302
4	4	Croatia, Slovenia	Mura	460
5	5	Croatia, Slovenia	Cerknica/Kupa	*375*
6	6	Slovenia, Croatia	Radovic-Metlika/Zumberak	*185*
7	7	Croatia, Slovenia	Bregana-Obrezje/Sava-Samobor	*58*
8	8	Slovenia, Croatia	Bizeljsko/Sutla	*192*
9		Slovenia, Croatia	Ormoz-Sredisce ob Drava/Drava-Varazdin[c]	*795*
10	9	Croatia, Slovenia	Dolinsko-Ravensko/Mura	449
11	10	Croatia, Hungary		300
12	11	Hungary, Croatia	Drava/Drava West	*359*
13	12	Croatia, Hungary	Baranja/Drava East	*1562*
14	13	Serbia, Croatia	[AQ28] SW Backa/Dunav[d]	2672
15		Serbia, Croatia	Srem-West Srem/Sava[c]	627
16	14	B&H, Croatia	Sava/Posavina I	*646*
17	15	Croatia, B&H	Kupa	452
18	16	B&H, Croatia	Una/Plesevica	*1700*
19	17	Croatia, B&H	Krka	*499*
20	18	B&H, Croatia	Cetina	*3237*
21	19	B&H, Croatia	Neretva right	*2982*
22	20	Croatia, B&H	Trebisnjica/Neretva left	*2242*
23	21	Montenegro, B&H,	Bileko Lake	1000
24	22	Montenegro, Croatia	Dinaric littoral west coast	200
25	23	Albania, Montenegro	Skadar/Shkodra Lake Dinaric East Coast	*650*
26	24	Serbia, Albania	Beli Drim/Drini Bardhe	*1170*
27	25	Serbia, Montenegro	Metohija	1350
28		Montenegro, Serbia	Pester[c]	557
29	26	Montenegro, Serbia	Lim	700
30	27	B&H, Serbia	Tara Massif	311
31	28	B&H, Serbia	Macva-Semberija	*1104*
32	29	Serbia, Hungary	Northeast Backa/Danube-Tisza Interfluve	*13,565*
33	30	Romania, Serbia	North and South Banat	*19,964*[e]
34	34	Bulgaria, Serbia	Stara Planina/Salasha Montana[f]	567
35	40	Albania, N. Macedonia	Korab/Bistra-Stogovo	140
36	41	Albania, N. Macedonia	Jablanica/Golobordo	250
37	44	Greece, Albania	Mourgana Mountain/Mali Gjere	*640*[g]
38	43	Albania, Greece	Nemechka/Vjosa-Pogoni	*900*
39	45	Albania, Greece, N. Macedonia	Prespes and Ohrid Lakes[h]	*1163*
40	47	N. Macedonia, Greece	Pelagonia/Florina[i]	607
41	48	N. Macedonia, Greece	Gevgelija/Axios-Vardar	–
42	49	Greece, N. Macedonia	Dojran Lake	*282*
43	50	Bulgaria, Greece, N. Macedonia	Sandansky-Petrich	768
44	51	Bulgaria, Greece	Goltze Delchev/Orvilos-Agistros	*298*[j]
45	56	Bulgaria, Greece, Turkey	Orestiada/Svilengrad-Stambolo Edirne	*1265*

(Continued)

(Continued).

No./A[a]	No./B[b]	Sharing countries	Aquifer	Area (km^2)
46	58	Bulgaria, Turkey	Topolovgrad Massif	249[k]
47	61	Romania, Hungary	Maros/Mures alluvial fan	*6339*
48	60	Romania, Hungary	Pleistocene Some/Szamos alluvial fan	*2356*
49	63	Romania, Moldova	Middle Sarmatian-Pontian	11,964
50	64	Bulgaria, Romania	Neogene-Sarmatian	*6628*
51	65	Romania, Bulgaria	Upper Jurassic-Lower Cretaceous	*26,903*
	31	Romania, Serbia	Miroc & Golubac	300
	32	Romania, Serbia	Dacian Basin	1500
	33	Bulgaria, Serbia	Timok	*337*
	35	Serbia, Bulgaria	Nishava and Tran/Vidlic	*465*
	36	Bulgaria, Serbia	Zemen	200
	37	N. Macedonia, Serbia	N. Macedonia-South-west Serbia	300
	38	N. Macedonia, Serbia	N. Macedonia-Central Serbia	100
	39	N. Macedonia, Serbia	Tetovo-Gostivar	*100*
	42	N. Macedonia, Albania	Ohrid Lake	–
	46	Albania, Greece	Galicia	–
	52	Bulgaria, Greece	Nastan-Trigrad	203
	53	Bulgaria, Greece	Smolyan	94
	54	Bulgaria, Greece	Rudozem	80
	55	Bulgaria, Greece	Erma Reka	40
	57	Greece, Turkey	Evros/Meric	2100
	59	Turkey, Bulgaria	Malko Tarnovo Massif	85
	62	Romania, Hungary	Lower Pleistocene alluvial fan	1390

Note: Areas shown in italics is the sum of the given national areas of the TBAs. B&H, Bosnia and Herzegovina; N. Macedonia, North Macedonia.

[a]No./A is the numbering of the TBAs in the First Assessment.

[b]No./A is the numbering of the TBAs in the ISARM/INWEB's inventory.

[c]There is no reference to this TBA in ISARM/INWEB's inventory.

[d]In ISARM/INWEB's inventory this is referred to as the West Serbia/Backa/Sava.

[e]In ISARM/INWEB's inventory the area is given as 15,639 km^2.

[f]This also includes the TBA 'Nishava and Tran/Vidlic', that is, no. 35 in ISARM/INWEB's inventory.

[g]In ISARM/INWEB's inventory the area is given as 700 km^2.

[h]This also includes the TBA 'Galicica', that is, no. 46 in ISARM/INWEB's inventory.

[i]In ISARM/INWEB's inventory the TBA is referred to as 'Pelagonia/Florina/Bitolsko'.

[j]In ISARM/INWEB's inventory the area is given as 356 km^2.

[k]In ISARM/INWEB's inventory the area is given as 311 km^2.

Table A2. Transboundary aquifers identified in the Second Assessment of Transboundary Rivers, Lakes and Groundwaters

No./C[a]	No./A[b]	Sharing countries	Aquifer	Area (km^2)
56	*50*	*Bulgaria, Romania*	*Dobrudja/Dobrogea Neogene-Sarmatian*	*5486*[c]
57	*51*	*Bulgaria, Romania*	*Dobrudja/Dobrogea Upper Jurassic-Lower Cretaceous*	*24,461*[c]
58	*14*	*Croatia, Serbia*	*South Western Backa/Dunav*	*441*[d]
59	*32*	*Hungary, Serbia*	*Northeast Backa/Danube-Tisza Interfluve or Backa/Danube-Tisza Interfluve aquifer*[e]	*9713*
76		Austria, Slovenia	Karstw asser-Vorkommen Karawanken/Karavanke aquifer	624
77	**9**	**Croatia, Slovenia**	**Ormoz-Sredisce ob Drava/Drava-Varazdin**	**795**
78	**10**	**Croatia, Slovenia**	**Dolinsko-Ravensko/Mura aquifer**	**449**
79	**11**	**Croatia, Hungary**	**Mura aquifer**	**398**
80	**12**	**Croatia, Hungary**	**Drava/Drava West aquifer**	**359**
81	**13**	**Croatia, Hungary**	**Baranja/Drava East Aquifer**	**1562**
82		Austria, Slovenia	Černeško-Libeliško aquifer	11
83		Austria, Slovenia	Kučnica aquifer	449
84		Slovenia, Hungary	Goričko aquifer	494
85		Slovenia, Austria, Hungary	Mura-Zala basin/Radgona-Vaš aquifer	494
86		Slovenia, Hungary, Croatia	Kot aquifer	449
87		Romania, Hungary	Körös-Crisuri Holocene, Pleistocene transboundary aquifer (Hort obágy-Nagykunság Bihar Nort hern Part)	15,700
88		Romania, Hungary	Hortobágy, Nagykunság, Bihar northern part	3148
89		Romania, Hungary	Körös valley, Sárrét, shallow Crişuri aquifer	8450
92	*33*	*Serbia, Romania*	*North and South Banat or North and Mid Banat aquifer*	*13,953*[d]
93	*48*	*Romania, Hungary*	*Somes/Szamos alluvial fan aquifer*	*24,254*[d]
94		Romania, Hungary	Nyírség, keleti rész/Nyírség, east margin	1240
95	*47*	*Romania, Hungary*	*Pleistocene-Holocene Mures/Maros alluvial fan aquifer*	*8894*[d]
96	**5**	**Croatia, Slovenia**	**Cerknica/Kupa aquifer**	**375**
97		Croatia, Slovenia	Kočevje Goteniška gora aquifer	595
98	**6**	**Croatia, Slovenia**	**Radovica-Metlika/Zumberak aquifer**	**185**
99	**7**	**Croatia, Slovenia**	**Bregana-Obrezje/Sava-Samobor aquifer**	**58**
100		Slovenia, Croatia	Bregana aquifer	16
101	**8**	**Slovenia, Croatia**	**Bizeljsko/Sutla aquifer**	**192**
102		Slovenia, Croatia	Boč aquifer	48
103		Croatia, Slovenia	Rogaška aquifer	178
104		Croatia, Slovenia	Atomske toplice aquifer	51
105		Croatia, Slovenia	Bohor aquifer	153
106		Croatia, Slovenia	Orlica aquifer	180
107	**15**	**Croatia, Serbia**	**Srem-West Srem/Sava aquifer**	**627**
108	*16*	*Croatia, B&H*	*Posavina I/Sava aquifer*	*396*[d]
109	*17*	*Croatia, B&H*	*Kupa aquifer*	*100*[d]
110	*18*	*Croatia, B&H*	*Pleševica/Una aquifer*	*1564*[d]
111	*29*	*Montenegro, Serbia*	*Lim aquifer*	*700*
112	*30*	*Serbia, B&H*	*Tara Massif Aquifer*	*311*
113	**31**	**B&H, Serbia**	**Macva-Semberija aquifer**	**1217**
114	*34*	*Bulgaria, Serbia*	*Stara Planina/Salasha Montana*	*3375*[d]
115	*49*	*Romania, Moldova*	*Middle Sarmantian Pontian*	*22,194*[d]

(Continued)

(Continued).

No./C[a]	No./A[b]	Sharing countries	Aquifer	Area (km^2)
129	*19*	*B&H, Croatia*	*Krka aquifer*	
130	**21**	**B&H, Croatia**	**Neretva Right coast aquifer**	**2462**
131	**22**	**B&H, Croatia**	**Trebišnjica/Neretva Left coast**	**2242**
132	**23**	**B&H, Montenegro**	**Bileko Lake aquifer**	**1000**
133	**26**	**Albania, Kosovo**	**Beli Drim/Drini Bardhe aquifer**	**1170**
134	*39*	*Albania, Greece, N. Macedonia*	*Prespa and Ohrid Lake aquifer*	*1525*[d]
135	*25*	*Albania, Montenegro*	*Skadar/Shkoder Lake, Dinaric East coast aquifer*	*1110*[d]
136	**38**	**Albania, Greece**	**Nemechka/Vjosa-Pogoni**	**920**
137	**41**	**N. Macedonia, Greece**	**Gevgelija/Axios-Vardar aquifer**	**8**
138	*42*	*N. Macedonia, Greece*	*Dojran/Doirani Lake aquifer*[e]	*212*
139	*43*	*Bulgaria, Greece, N. Macedonia*	*Sandansky-Petrich aquifer*	
140		Bulgaria, Greece	Sandansky valley aquifer	630
141		Bulgaria, Greece	Petrich valley aquifer	124
142	*44*	*Bulgaria, Greece*	*Orvilos-Agistros/Gotze Delchev*	*420*[d]
143	*45*	*Bulgaria, Greece, Turkey*	*Orestiada/Svilengrad-Stambolo/Edirne aquifer*	*1162*[d]
144	*46*	*Bulgaria, Turkey*	*Topolovgrad Massif aquifer*	*315*[d]
145	**40**	**Greece, N. Macedonia**	**Pelagonia-Florina/Bitolsko aquifer**	**180**
146	**1**	**Croatia, Slovenia**	**Secovlje-Dragonja/Istra aquifer**	**108**
147	*2*	*Croatia, Slovenia*	*Mirna/Istra aquifer*	*198*[d]
148		Croatia, Slovenia	Mirna aquifer	
149		Croatia, Slovenia	Območje izvira Rižane aquifer	277
150	*3*	*Croatia, Slovenia*	*Opatija/Istra aquifer*	*67*[d]
151		Croatia, Slovenia	Riječina-Zvir aquifer	70
152		Croatia, Slovenia	Notranjska Reka aquifer	315
153		Croatia, Slovenia	Novokračine aquifer	21
154	**20**	**B&H, Croatia**	**Cetina aquifer**	**3237**
155	**24**	**Croatia, Montenegro**	**Dinaric Littoral (West Coast) aquifer**	**200**
156	**27**	**Kosovo, Montenegro**	**Metohija aquifer**	**1350**
157	*28*	*Montenegro, Serbia*	*Pester aquifer*	*467*[d]
158	**35**	**Albania, N. Macedonia**	**Korab/Bistra-Stogovo aquifer**	**140**
159	**36**	**Albania, N. Macedonia**	**Jablanica/Golobordo aquifer**	**250**
160	**37**	**Albania, Greece**	**Mourgana Mountain/Mali Gjere**	**530**

Note: TBAs in bold come form the First Assessment (UNECE, 2007); TBAs in italic denotes that the data source is the First Assessment plus amendments included in the Second Assessment; TBAs in regular wording denotes that the data source is the Second Assessment (UNECE, 2011). B&H, Bosnia & Herzegovina; N. Macedonia, North Macedonia.

[a]No./C is the numbering of the TBAs in the Second Assessment.

[b]No./B is the numbering of the TBAs in the First Assessment (the same as in Appendix A).

[c]The name of the TBA and its extent is different to that in the First Assessment.

[d]The extent of the TBA is different to that in the First Assessment.

[e]The name of the TBA is different to that in the First Assessment.

Table A3. Transboundary aquifers identified in IGRAC's Maps of Transboundary Aquifers of the World Editions 2015 and 2020

No./D[a]	No./E[b]	No./C[c]	Sharing countries	Aquifer
EB1	EB001	146	Croatia, Slovenia	Secovlje-Dragonja/Istra Aquifer
EB2–EB4	EB002–EB004	147–149	Croatia, Slovenia	Mirna/Istra Aquifer, Mirna Aquifer, Obmocje izvira Rižane Aquifer
EB5, EB6	EB005–EB006	150, 152	Croatia, Slovenia	Opatija/Istra Aquifer, Notranjska Reka Aquifer
EB8, EB9	EB008–EB009	151, 153	Croatia, Slovenia	Rijecina-Zvir Aquifer, Novokracine Aquifer
EB10–EB12	EB010–EB012	96–98	Croatia, Slovenia	Cerknica/Kupa, Kocevje Goteniška, Radovica-Metlika/ Zumberak Aquifers
EB13	EB013	77	Croatia, Slovenia	Ormoz-Sredisce ob Drava/Drava-Varazdin Aquifer
EB14, EB15	EB014,EB015	100, 99	Croatia, Slovenia	Bregana Aquifer, Bregana-Obrezje/Sava-Samobor
EB16–EB21	EB016–EB021	101–106	Croatia, Slovenia	Bizeljsko/Sutla, Boc, Rogaška, Atomske toplice, Bohor, Orlica Aquifers
EB22	EB022	78	Croatia, Slovenia	Dolinsko-Ravensko/Mura Aquifer
EB23	EB023	109	B&H, Croatia	Kupa Aquifer
EB24	EB024	110	B&H, Croatia	Una[d]
EB25	EB025	129	B&H, Croatia	Krka
EB26	EB026	158	B&H, Croatia	Korab/Bistra-Stogovo Aquifer
EB27	EB027	130	B&H, Croatia	Neretva[e]
EB28	EB028	131	B&H, Croatia	Trebišnjica[f]
EB29	EB029	155	B&H, Croatia, Montenegro	Dinaric Littoral (West Coast aquifer)
EB30	EB030	132	B&H, Montenegro	Bilecko Lake
EB31	EB031	108	B&H, Croatia	Posavina I/Sava Aquifer
EB32	EB032	107	Croatia, Servia	Srem-West Srem/Sava Aquifer
EB33	EB033	58	Croatia, Servia	South Western Backa/Dunav Aquifer
EB34	EB034	79	Croatia, Hungary	Mura Aquifer
EB35	EB035	80	Croatia, Hungary	Drava/Drava West Aquifer
EB36	EB036	59	Croatia, Hungary[g]	Northeast Backa/Danube-Tisza Interfluve or Backa/ Danube-Tisza Interfluve Aquifer
EB38	EB038	237	Croatia, Hungary, Romania, Serbia	Northeast Backa/Danube-Tisza Interfluve Aquifer
EB44	EB044	135	Albania, Montenegro	Skadar/Shkoder Lake, Dinaric East Coast Aquifer
EB45	EB045	92	Hungary, Romania, Serbia	North and South Banat or North and Mid Banat Aquifer
EB48	EB048	113	B&H, Serbia	Macva-Semberija Aquifer
EB49	EB049	112	B&H, Serbia	Tara Massif
EB50	EB050	111	Montenegro, Serbia	Lim Aquifer
EB51	EB051	157	Montenegro, Serbia	Pester Aquifer
EB52	EB052	156	Montenegro, Serbia	Metohija Aquifer
EB53	EB053	133	Albania, Serbia	Beli Drim/Drini Bardhe Aquifer
EB54	EB054	233	N. Macedonia, Serbia	Tetovo-Gostivar
EB55	EB055	158	Albania, N. Macedonia	Korab/Bistra-Stogovo Aquifer
EB56	EB056	159	Albania, N. Macedonia	Jablanica/Golobordo Aquifer
EB57	EB057	134	Albania, Greece, N. Macedonia	Prespa and Ohrid Lake Aquifer
EB58			Greece	Triklariou Kastorias System
EB59	EB059	160	Albania, Greece	Mourgana Mountain/Mali Gjere

(Continued)

(Continued).

No./D[a]	No./E[b]	No./C[c]	Sharing countries	Aquifer
EB60			Greece	Systima Pogonianis
EB61	EB061	136	Albania, Greece	Nemechka/Vjosa-Pogoni Aquifer
EB62	EB062	145	Greece, N. Macedonia	Pelagonia-Florina/Bitolsko Aquifer
EB63	EB063	137	Greece, N. Macedonia	Axiou System[h]
EB67–EB69	EB067–EB069	139–141	Bulgaria, Greece, N. Macedonia	Sandansky-Petrich Aquifer, Sandansky Valley Aquifer, Petrich Valley Aquifer
EB70	EB070	232	N. Macedonia, Serbia	FYROM-Central Serbia
EB71	EB071	231	N. Macedonia, Serbia	FYROM-South-west Serbia
EB72	EB072	230	Bulgaria, Serbia	Zemen
EB73	EB073	114	Bulgaria, Serbia	Stara Planina/Salasha Montana
EB75	EB075	235	Romania, Serbia	Miroc & Golubac
EB76	EB076	234	Romania, Serbia	Dacian Basin
EB77	–	142	Greece	Agkistro System
EB82	EB082	221	Greece, Turkey	Evros/Meric
EB85	EB085	144	Bulgaria, Turkey	Topolovgrad Massif Aquifer
EB86	EB086	219	Bulgaria, Turkey	Malko Tarnovo Kasrt Waterbearing Massif
EB89	EB089	218	Romania, Moldova, Ukraine	Danube-Prut
EB90	EB090	115	Romania, Moldova, Ukraine[i]	Middle Sarmantian Pontian Aquifer
EB91	EB091	216	Romania, Moldova	Prut
EB92	EB092		Montenegro, B&H	Piva
EB93	EB093		Montenegro, Albania	Cijevna
EU282	EU282		Slovakia, Austria, Hungary, Slovenia	Upper Pannonian Thermal Aquifer
	EU073	84–86	Austria, Croatia, Hungary, Slovenia	Goričko aquifer; Mura-Zala basin/Radgona-Vaš aquifer; Kot aquifer
EB94			Romania	Valea lui Mihai (Campia de Vest)
EB95			Romania	Campia Carei
EB96			Romania	Conul Somesului
EB97			Romania	
EB98			Romania	
EB99			Romania	Crisuri (Campia de Vest)
EB100			Romania	Oradea (Campia de Vest)
EB101			Romania	Lovrin-Vinga
EB102			Romania	Timisoara
EB103			Romania	Banat
EB104			Romania	Gataia
EB105			Romania	Jam
EB106			Romania	Platforma Valaha
EB107			Bulgaria	
EB108			Bulgaria	
EB109			Romania	Cobadin-Mangalia
EB110			Bulgaria	BG3G0000T12034
EB111			Greece	Systima Doiranis
EB112			Bulgaria	BG3G000000N011
EB113			Greece	Orestiadas System
EB114			Greece	Systima Axiou
EB115			Romania	Podisul Central Moldovenesc
EB116			Greece	Systima Mourgianas

(*Continued*)

(Continued).

No./D[a]	No./E[b]	No./C[c]	Sharing countries	Aquifer
EB117			Greece	Systima Florinas
EU235			Hungary	Északnyugat-Dunántúl
EU252			Hungary	Északkelet-Alföld
EU262			Hungary	Délnyugat-Dunántúl
EU263			Hungary	Közép-dunántúli Termálkarszt
EU269			Hungary	Nyugat-Alföld
EU273			Hungary	Dél-Alföld
EU276			Hungary	Délkelet-Alföld

Note: B&H. Bosnia and Herzegovina; N. Macedonia, North Macedonia; FYROM, *former Yugoslav Republic of Macedonia.*
[a]No./D is the numbering of the Transboundary Aquifers of the World 2015 Edition.
[b]No./E is the numbering of the Transboundary Aquifers of the World 2020 Edition.
[c]No./C is the numbering of the TBAs in the Second Assessment (the same as that used in Appendix B).
[d]In the Second Assessment the TBA is called the Pleševica/Una aquifer.
[e]In the Second Assessment the TBA is called the Neretva Right coast aquifer.
[f]In the Second Assessment the TBA is called the Trebišnjica /Neretva Left coast.
[g]In the Second Assessment the TBA is shared between Hungary and Serbia.
[h]In the Second Assessment the TBA is called the Gevgelija/Axios-Vardar aquifer.
[i]In the Second Assessment the TBA is shared between Romania and Moldova.

Index

Note: Page numbers in **bold** refer to tables and those in *italics* refer to figures.

For Product Safety Concerns and Information please contact our EU
representative GPSR@taylorandfrancis.com
Taylor & Francis Verlag GmbH, Kaufingerstraße 24, 80331 München, Germany

www.ingramcontent.com/pod-product-compliance
Ingram Content Group UK Ltd.
Pitfield, Milton Keynes, MK11 3LW, UK
UKHW050053200926
475702UK00019B/2818

* 9 7 8 1 0 3 2 3 8 6 0 9 6 *